# Handbook of
# REGRESSION
# METHODS

# Handbook of
# REGRESSION METHODS

### Derek S. Young

University of Kentucky, Lexington

CRC Press
Taylor & Francis Group
Boca Raton  London  New York

CRC Press is an imprint of the
Taylor & Francis Group, an **informa** business

A CHAPMAN & HALL BOOK

CRC Press
Taylor & Francis Group
6000 Broken Sound Parkway NW, Suite 300
Boca Raton, FL 33487-2742

© 2017 by Taylor & Francis Group, LLC
CRC Press is an imprint of Taylor & Francis Group, an Informa business

No claim to original U.S. Government works

Printed on acid-free paper
Version Date: 20170613

International Standard Book Number-13: 978-1-4987-7529-8 (Hardback)

**Library of Congress Cataloging-in-Publication Data**

Names: Young, Derek S.
Title: Handbook of regression methods / Derek S. Young.
Description: Boca Raton : CRC Press, 2017. | Includes bibliographical references and index.
Identifiers: LCCN 2017011248 | ISBN 9781498775298 (hardback)
Subjects: LCSH: Regression analysis. | Multivariate analysis.
Classification: LCC QA278.2 .Y66 2017 | DDC 519.5/36--dc23
LC record available at https://lccn.loc.gov/2017011248

Visit the Taylor & Francis Web site at
http://www.taylorandfrancis.com

and the CRC Press Web site at
http://www.crcpress.com

Printed and bound in Great Britain by
TJ International Ltd, Padstow, Cornwall

*To Andri.*

# Contents

List of Examples      xiii

Preface      xv

**I   Simple Linear Regression**      **1**

**1   Introduction**      **3**

**2   Basics of Regression Models**      **7**
- 2.1   Regression Notation . . . . . . . . . . . . . . . . . . . . . . .   8
- 2.2   Population Model for Simple Linear Regression . . . . . . .   8
- 2.3   Ordinary Least Squares . . . . . . . . . . . . . . . . . . . .   10
- 2.4   Measuring Overall Variation from the Sample Line . . . . .   12
  - 2.4.1   $R^2$ . . . . . . . . . . . . . . . . . . . . . . . . . .   12
- 2.5   Regression Through the Origin . . . . . . . . . . . . . . . .   13
- 2.6   Distinguishing Regression from Correlation . . . . . . . .   14
- 2.7   Regression Effect . . . . . . . . . . . . . . . . . . . . . . . .   16
  - 2.7.1   Regression Fallacy . . . . . . . . . . . . . . . . . . . .   17
- 2.8   Examples . . . . . . . . . . . . . . . . . . . . . . . . . . . .   18

**3   Statistical Inference**      **25**
- 3.1   Hypothesis Testing and Confidence Intervals . . . . . . . .   25
- 3.2   Power . . . . . . . . . . . . . . . . . . . . . . . . . . . . . .   30
- 3.3   Inference on the Correlation Model . . . . . . . . . . . . .   33
- 3.4   Intervals for a Mean Response . . . . . . . . . . . . . . . .   35
- 3.5   Intervals for a New Observation . . . . . . . . . . . . . . .   36
- 3.6   Examples . . . . . . . . . . . . . . . . . . . . . . . . . . . .   41

**4   Regression Assumptions and Residual Diagnostics**      **49**
- 4.1   Consequences of Invalid Assumptions . . . . . . . . . . . .   50
- 4.2   Diagnosing Validity of Assumptions . . . . . . . . . . . .   51
- 4.3   Plots of Residuals Versus Fitted Values . . . . . . . . . . .   53
  - 4.3.1   Ideal Appearance of Plots . . . . . . . . . . . . . .   54
  - 4.3.2   Difficulties Possibly Seen in the Plots . . . . . . . .   56
- 4.4   Data Transformations . . . . . . . . . . . . . . . . . . . . .   57
- 4.5   Tests for Normality . . . . . . . . . . . . . . . . . . . . . .   59
  - 4.5.1   Skewness and Kurtosis . . . . . . . . . . . . . . . .   61

| | | |
|---|---|---:|
| 4.6 | Tests for Constant Error Variance | 63 |
| 4.7 | Examples | 66 |

**5 ANOVA for Simple Linear Regression**   **73**
| | | |
|---|---|---:|
| 5.1 | Constructing the ANOVA Table | 73 |
| 5.2 | Formal Lack of Fit | 77 |
| 5.3 | Examples | 78 |

## II  Multiple Linear Regression    83

**6 Multiple Linear Regression Models and Inference**   **85**
| | | |
|---|---|---:|
| 6.1 | About the Model | 85 |
| 6.2 | Matrix Notation in Regression | 87 |
| 6.3 | Variance–Covariance Matrix and Correlation Matrix of $\hat{\beta}$ | 92 |
| 6.4 | Testing the Contribution of Individual Predictor Variables | 94 |
| 6.5 | Statistical Intervals | 95 |
| 6.6 | Polynomial Regression | 96 |
| 6.7 | Examples | 99 |

**7 Multicollinearity**   **109**
| | | |
|---|---|---:|
| 7.1 | Sources and Effects of Multicollinearity | 109 |
| 7.2 | Detecting and Remedying Multicollinearity | 110 |
| 7.3 | Structural Multicollinearity | 114 |
| 7.4 | Examples | 115 |

**8 ANOVA for Multiple Linear Regression**   **121**
| | | |
|---|---|---:|
| 8.1 | The ANOVA Table | 121 |
| 8.2 | The General Linear $F$-Test | 122 |
| 8.3 | Lack-of-Fit Testing in the Multiple Regression Setting | 123 |
| 8.4 | Extra Sums of Squares | 124 |
| 8.5 | Partial Measures and Plots | 125 |
| 8.6 | Examples | 129 |

**9 Indicator Variables**   **137**
| | | |
|---|---|---:|
| 9.1 | Leave-One-Out Method | 137 |
| 9.2 | Coefficient Interpretations | 138 |
| 9.3 | Interactions | 140 |
| 9.4 | Coded Variables | 141 |
| 9.5 | Conjoint Analysis | 143 |
| 9.6 | Examples | 144 |

# III   Advanced Regression Diagnostic Methods          151

## 10 Influential Values, Outliers, and More Diagnostic Tests          153
10.1   More Residuals and Measures of Influence . . . . . . . . . .   154
10.2   Masking, Swamping, and Search Methods . . . . . . . . . .   163
10.3   More Diagnostic Tests . . . . . . . . . . . . . . . . . . . . .   164
10.4   Comments on Outliers and Influential Values . . . . . . . .   167
10.5   Examples . . . . . . . . . . . . . . . . . . . . . . . . . . . .   168

## 11 Measurement Errors and Instrumental Variables Regression          179
11.1   Estimation in the Presence of Measurement Errors . . . . .   180
11.2   Orthogonal and Deming Regression . . . . . . . . . . . . .   182
11.3   Instrumental Variables Regression . . . . . . . . . . . . . .   184
11.4   Structural Equation Modeling . . . . . . . . . . . . . . . .   186
11.5   Dilution . . . . . . . . . . . . . . . . . . . . . . . . . . . .   188
11.6   Examples . . . . . . . . . . . . . . . . . . . . . . . . . . . .   188

## 12 Weighted Least Squares and Robust Regression Procedures          195
12.1   Weighted Least Squares . . . . . . . . . . . . . . . . . . . .   195
12.2   Robust Regression Methods . . . . . . . . . . . . . . . . . .   197
12.3   Theil–Sen and Passing–Bablok Regression . . . . . . . . . .   201
12.4   Resistant Regression Methods . . . . . . . . . . . . . . . . .   202
12.5   Resampling Techniques for $\hat{\beta}$ . . . . . . . . . . . . . . . .   203
12.6   Examples . . . . . . . . . . . . . . . . . . . . . . . . . . . .   210

## 13 Correlated Errors and Autoregressive Structures          219
13.1   Overview of Time Series and Autoregressive Structures . .   219
13.2   Properties of the Error Terms . . . . . . . . . . . . . . . . .   221
13.3   Testing and Remedial Measures for Autocorrelation . . . .   224
13.4   Advanced Methods . . . . . . . . . . . . . . . . . . . . . . .   230
    13.4.1   ARIMA Models . . . . . . . . . . . . . . . . . . . .   230
    13.4.2   Exponential Smoothing . . . . . . . . . . . . . . . .   233
    13.4.3   Spectral Analysis . . . . . . . . . . . . . . . . . . .   234
13.5   Examples . . . . . . . . . . . . . . . . . . . . . . . . . . . .   236

## 14 Crossvalidation and Model Selection Methods          249
14.1   Crossvalidation . . . . . . . . . . . . . . . . . . . . . . . .   249
14.2   PRESS . . . . . . . . . . . . . . . . . . . . . . . . . . . . .   251
14.3   Best Subset Procedures . . . . . . . . . . . . . . . . . . . .   252
14.4   Statistics from Information Criteria . . . . . . . . . . . . .   254
14.5   Stepwise Procedures for Identifying Models . . . . . . . . .   255
14.6   Example . . . . . . . . . . . . . . . . . . . . . . . . . . . . .   256

# IV   Advanced Regression Models                          261

**15 Mixed Models and Some Regression Models for Designed Experiments**                                                        **263**
15.1   Mixed Effects Models . . . . . . . . . . . . . . . . .   263
15.2   ANCOVA . . . . . . . . . . . . . . . . . . . . . . .   265
15.3   Response Surface Regression . . . . . . . . . . . . .   268
15.4   Mixture Experiments . . . . . . . . . . . . . . . . .   273
15.5   Examples . . . . . . . . . . . . . . . . . . . . . . .   276

**16 Biased Regression, Regression Shrinkage, and Dimension Reduction**                                                        **287**
16.1   Regression Shrinkage and Penalized Regression . . . . . . .   287
16.2   Principal Components Regression . . . . . . . . . . . .   292
16.3   Partial Least Squares . . . . . . . . . . . . . . . . .   294
16.4   Other Dimension Reduction Methods and Sufficiency . . .   296
16.5   Examples . . . . . . . . . . . . . . . . . . . . . . .   301

**17 Piecewise, Nonparametric, and Local Regression Methods**   **307**
17.1   Piecewise Linear Regression . . . . . . . . . . . . . .   307
17.2   Local Regression Methods . . . . . . . . . . . . . . .   310
17.3   Splines . . . . . . . . . . . . . . . . . . . . . . . .   318
17.4   Other Nonparametric Regression Procedures . . . . . . .   324
17.5   Examples . . . . . . . . . . . . . . . . . . . . . . .   327

**18 Regression Models with Censored Data**                     **335**
18.1   Overview of Survival and Reliability Analysis . . . . . . .   335
18.2   Censored Regression Model . . . . . . . . . . . . . .   337
18.3   Survival (Reliability) Regression . . . . . . . . . . . .   339
18.4   Cox Proportional Hazards Regression . . . . . . . . . .   343
18.5   Diagnostic Procedures . . . . . . . . . . . . . . . . .   344
18.6   Truncated Regression Models . . . . . . . . . . . . . .   347
18.7   Examples . . . . . . . . . . . . . . . . . . . . . . .   350

**19 Nonlinear Regression**                                     **361**
19.1   Nonlinear Regression Models . . . . . . . . . . . . . .   361
19.2   Nonlinear Least Squares . . . . . . . . . . . . . . . .   364
       19.2.1   A Few Algorithms . . . . . . . . . . . . . . .   365
19.3   Approximate Inference Procedures . . . . . . . . . . .   367
19.4   Examples . . . . . . . . . . . . . . . . . . . . . . .   370

**20 Regression Models with Discrete Responses**               **375**
20.1   Logistic Regression . . . . . . . . . . . . . . . . . .   375
       20.1.1   Binary Logistic Regression . . . . . . . . . .   376
       20.1.2   Nominal Logistic Regression . . . . . . . . . .   383
       20.1.3   Ordinal Logistic Regression . . . . . . . . . .   384

20.2   Poisson Regression . . . . . . . . . . . . . . . . . . . .   385
20.3   Negative Binomial Regression . . . . . . . . . . . . . . .   390
20.4   Specialized Models Involving Zero Counts . . . . . . . . .   396
20.5   Examples . . . . . . . . . . . . . . . . . . . . . . . . .   398

**21 Generalized Linear Models**                                     **413**
21.1   The Generalized Linear Model and Link Functions . . . . .   413
21.2   Gamma Regression . . . . . . . . . . . . . . . . . . . . .   418
21.3   Inverse Gaussian (Normal) Regression . . . . . . . . . . .   419
21.4   Beta Regression . . . . . . . . . . . . . . . . . . . . . .   420
21.5   Generalized Estimating Equations . . . . . . . . . . . . .   422
21.6   Examples . . . . . . . . . . . . . . . . . . . . . . . . .   426

**22 Multivariate Multiple Regression**                              **439**
22.1   The Model . . . . . . . . . . . . . . . . . . . . . . . . .   439
22.2   Estimation and Statistical Regions . . . . . . . . . . . .   441
22.3   Reduced Rank Regression . . . . . . . . . . . . . . . . . .   446
22.4   Seemingly Unrelated Regressions . . . . . . . . . . . . . .   447
22.5   Examples . . . . . . . . . . . . . . . . . . . . . . . . .   449

**23 Semiparametric Regression**                                     **455**
23.1   Single-Index Models . . . . . . . . . . . . . . . . . . . .   456
23.2   (Generalized) Additive Models . . . . . . . . . . . . . . .   457
23.3   (Generalized) Partial Linear Models . . . . . . . . . . . .   458
23.4   (Generalized) Partial Linear Partial Additive Models . . . .   460
23.5   Varying-Coefficient Models . . . . . . . . . . . . . . . .   460
23.6   Projection Pursuit Regression . . . . . . . . . . . . . . .   461
23.7   Examples . . . . . . . . . . . . . . . . . . . . . . . . .   462

**24 Data Mining**                                                   **477**
24.1   Classification and Support Vector Regression . . . . . . .   478
24.2   Prediction Trees and Related Methods . . . . . . . . . . .   483
24.3   Some Ensemble Learning Methods for Regression . . . . . .   488
24.4   Neural Networks . . . . . . . . . . . . . . . . . . . . . .   492
24.5   Examples . . . . . . . . . . . . . . . . . . . . . . . . .   493

**25 Miscellaneous Topics**                                          **503**
25.1   Multilevel Regression Models . . . . . . . . . . . . . . .   503
25.2   Functional Linear Regression Analysis . . . . . . . . . . .   510
25.3   Regression Depth . . . . . . . . . . . . . . . . . . . . .   515
25.4   Mediation and Moderation Regression . . . . . . . . . . . .   517
25.5   Meta-Regression Models . . . . . . . . . . . . . . . . . .   522
25.6   Regression Methods for Analyzing Survey Data . . . . . . .   526
25.7   Regression with Missing Data and Regression Imputation .   533
25.8   Bayesian Regression . . . . . . . . . . . . . . . . . . . .   538
25.9   Quantile Regression . . . . . . . . . . . . . . . . . . . .   542

25.10 Monotone Regression . . . . . . . . . . . . . . . . . . . . . 544
25.11 Generalized Extreme Value Regression Models . . . . . . . 546
25.12 Spatial Regression . . . . . . . . . . . . . . . . . . . . . . 548
25.13 Circular Regression . . . . . . . . . . . . . . . . . . . . . 554
25.14 Rank Regression . . . . . . . . . . . . . . . . . . . . . . . 557
25.15 Mixtures of Regressions . . . . . . . . . . . . . . . . . . . 560
25.16 Copula Regression . . . . . . . . . . . . . . . . . . . . . . 563
25.17 Tensor Regression . . . . . . . . . . . . . . . . . . . . . . 565

## V  Appendices                                             571

**Appendix A  Steps for Building a Regression Model**          573

**Appendix B  Refresher on Matrices and Vector Spaces**        575

**Appendix C  Some Notes on Probability and Statistics**       579

**Bibliography**                                               583

**Index**                                                      615

# List of Examples

1993 Car Sale Data .................................................... 68
Air Passengers Data ................................................. 245
Amitriptyline Data ................................................... 449
Animal Weight Data ................................... 516, 540, 545
Annual Maximum Sea Levels Data ............................... 548
Arsenate Assay Data ......................................... 192, 212
Auditory Discrimination Data ...................................... 144
Automobile Features Data .......................................... 304
Belgian Food Expenditure Data .................................... 543
Belgian Telephone Data ............................................. 559
Biochemistry Publications Data ..................................... 405
Black Cherry Tree Data ............................................. 69
Blood Alcohol Concentration Data ................................ 188
Bone Marrow Transplant Data ..................................... 355
Boston Housing Data ............................... 473, 499, 552
Canadian Auto Insurance Data ..................................... 429
Cardiovascular Data ............................................ 238, 564
Census of Agriculture Data ......................................... 530
Cheese-Tasting Experiment Data ................................... 403
Computer Repair Data ................................... 21, 44, 79
Computer-Assisted Leaerning Data ................................ 210
Cracker Promotion Data ............................................ 278
Credit Loss Data .................................................... 431
Durable Goods Data ................................................ 356
Epilepsy Data ....................................................... 434
Ethanol Fuel Data .................................................. 561
Expenditures Data .................................................. 171
Extramarital Affairs Data ......................................... 350
Fiber Strength Data ................................................. 79
Fruit Fly Data ...................................................... 426
Gait Data ........................................................... 512
Gamma-Ray Burst Data ............................................. 327
Google Stock Data .................................................. 236
Hospital Stays Data ................................................ 409
Investment Data .................................................... 453
*James Bond* Data .................................................. 373
Job Stress Data ..................................................... 520

Kola Project Data .................................................. 103
LIDAR Data ....................................................... 331
Light Data ........................................................ 371
Mammal Sleep Data ............................................... 535
Motorette Data .................................................... 352
Mouse Liver Data ................................................. 246
NASA Data ....................................................... 567
Natural Gas Prices Data .......................................... 240
Odor Data ........................................................ 281
Pima Indian Diabetes Data ........................................ 493
Prostate Cancer Data ............................................. 301
Pulp Property Data ........................................... 45, 216
Punting Data ................................................ 168, 256
Puromycin Data .................................................. 370
Quasar Data ...................................................... 328
Radon Data ....................................................... 506
Ragweed Data .................................................... 468
Simulated Data ................................................... 131
Simulated Motorcycle Accident Data ............................... 493
Sleep Study Data ................................................. 276
Sportsfishing Survey Data ......................................... 402
Steam Output Data ................................ 20, 41, 66, 78, 146
Subarachnoid Hemorrhage Data .................................... 398
Tea Data ......................................................... 147
The 1907 Romanian Peasant Revolt Data ........................... 462
Thermal Energy Data ......................................... 99, 115, 129
Tortoise Eggs Data ........................................... 106, 119
Toy Data ......................................................... 18
Tuberclosis Vaccine Literature Data ............................... 524
U.S. Economy Data ............................................... 190
Wind Direction Data .............................................. 556
Yarn Fiber Data .................................................. 283

# Preface

The aim of this handbook is to provide a broad overview of many procedures available for performing a regression analysis. It is purposely composed with an emphasis on the *breadth* of topics covered. As a result, there is a sacrifice in terms of *depth* of coverage, especially for later topics. This strategy will be beneficial to the user of this handbook in that you will gain a broader perspective of regression tools available for a particular data problem. An earnest attempt has been made to provide relevant and helpful references to concepts treated more superficially, thus giving some direction to readers who need to dig deeper into a topic. The primary target audience of this handbook is the statistical practitioner, who will hopefully find this to be a valuable resource for their own work and research. Given the coverage of some less-than-standard regression procedures, this handbook can also be a beneficial reference for statistical researchers as well as a supplement to graduate students first learning about regression methods.

Roughly the first half of this handbook (Parts I and II) focuses on fitting regression models using ordinary least squares. Part I covers simple linear regression, which is the setting with one predictor variable (or independent variable) and one response variable (or dependent variable). Part II covers multiple linear regression, which is the setting with more than one predictor variable. Certain assumptions are made when fitting these models and various diagnostic tools are used to assess these assumptions. Throughout this handbook, we will highlight statistical inference procedures used to assess the statistical significance of the estimated model and introduce other procedures to deal with various characteristics of the dataset, such as outliers and multi-collinearity.

The second half of this handbook (Parts III and IV) mostly focuses on settings where the traditional linear regression model is not appropriate for the given data. These topics are more advanced and some are not typically presented in a traditional regression textbook. Part III includes topics about common procedures implemented when various regression assumptions are violated and determining how to select a model when faced with many predictors. In Part IV, more advanced regression models are discussed, including nonlinear regression models, count regression models, nonparametric regression models, regression models with autocorrelated data, and regression models for censored data. Such models are needed when the data are not necessarily in the form that we analyze in the first half of the handbook. As one reads through the material, it is suggested to occasionally revisit the "Steps for

Building a Regression Model" in Appendix A to help keep the bigger picture in mind when using regression modeling strategies.

Real data examples are included at the end of each chapter for nearly all concepts presented. All data analyses and visualizations are produced using R (R Core Team, 2016). A novel facet of this handbook is that we have developed a Shiny app (Chang et al., 2016) based on the R code used for all of the analyses presented. The app is located at

https://horm.as.uky.edu/

The Shiny app serves as an interactive learning tool and can help the reader better understand the presented material. The above location will also contain an errata document for this handbook. While programming with R is not a component of this handbook, all of the R code is made available through the Shiny app, thus allowing the user to verify all of the analyses. Those interested in R programming can refer to Braun and Murdoch (2008), Matloff (2011), and Wickham (2014). References for R programming in the context of regression analyses include Sheather (2009), Fox and Weisberg (2010), and Faraway (2014, 2016). Of course, there are also a plethora of question-and-answer websites, wikis, and personal blogs to which one can (cautiously) refer.

Finally, there are a number of individuals who deserve acknowledgment for helping this handbook evolve to the final published form. First and foremost, is my esteemed former colleague Bob Heckard, retired Senior Lecturer of Statistics from Penn State University. Bob graciously helped me organize my online regression course during my first semester teaching for Penn State's World Campus. The structure of the content and some examples in the first two parts of this handbook are built on portions of his lecture notes. I also acknowledge Prasanta Basak, Cedric Neumann, Jason Morton, and Durland Shumway, each of whom used this material to teach their respective online regression courses for Penn State's World Campus. They and their students provided an enormous amount of valuable feedback on this manuscript over the years. I also owe a great deal of gratitude to Glenn Johnson, who has been instrumental in organizing the online instructional material for all of the Penn State World Campus statistics courses. He, like Bob Heckard, helped organize my regression material into an online format, which in turn helped me better organize how I presented that material in this handbook. I am also very thankful to Yue Cui, who updated some of the R code to produce graphics using the ggplot2 package (Wickham, 2009). Finally, I owe much appreciation to Joshua Lambert and Kedai Cheng, who both worked tirelessly developing the Shiny app for this handbook.

*Derek S. Young*
Department of Statistics
University of Kentucky

# Part I

# Simple Linear Regression

# 1

## *Introduction*

As a scientific researcher, you will be tasked with formulating, investigating, and testing hypotheses of interest. In order to adequately explore a research hypothesis, you will need to design an experiment, collect the data, and analyze the data. During the analysis stage, the results will provide statistical evidence in support of or against your hypothesis. In some cases, the results may also be inconclusive. In this last case, you will need to assess whatever lessons were learned from the current research, implement any changes, and then proceed to run through another iteration of the research process while considering other factors, such as cost and practical constraints.

Fortunately, the field of statistics gives an excellent toolbox for guiding researchers through the experimental design, data collection, and analysis stages. A major tool in that toolbox is the area of **regression analysis**. Regression analysis is a set of techniques used to explore the relationship between at least two variables; i.e., at least one independent variable and one dependent variable. The majority of standard regression analyses implement a linear regression model, which means that the dependent variable(s) can be written in terms of a linear combination of the independent variable(s). Some reasons why this is a common approach are: (1) the linear regression model is easily understood by the majority of researchers; (2) many processes naturally follow a relationship that can be approximated by a linear relationship; and (3) statistical techniques for linear regression models are well-rooted in theory with desirable asymptotic properties (i.e., large sample properties), thus yielding tractable results.

This handbook provides a broad overview of the many regression techniques that are available to you as a scientific researcher. It is intended to serve as a solid reference that highlights the core components of these regression techniques, while keeping rigorous details at a minimum. This book is aimed at strengthening your understanding of regression strategies that are available for your data problems, which are reinforced by analyses involving real datasets.

Based on the outline in Montgomery et al. (2013), we begin by highlighting some of the primary uses of regression methods in practice:

1. **Descriptions:** A researcher may be interested in characterizing a descriptive relationship between a set of measured variables. At this point, the fewest assumptions are made while investigating such a

relationship. This relationship may or may not help to justify a possible deterministic relationship; however, there is an attempt to at least establish some sort of connection between these variables using the current sample data. *Example: A researcher may be interested in establishing a relationship between the spinal bone mineral density of individuals and other measured variables on those individuals, such as their age, height, and sex.*

2. **Coefficient Estimation:** When analyzing data, the researcher may have a theoretical, deterministic relationship in mind. This relationship could be linear or nonlinear. Regardless, the use of regression analysis can provide support for such a theory. Note that we never say that we *prove* such a theory, but rather that we only provide *support* or *evidence* for such a theory. Of particular interest will be the magnitudes and signs of the coefficients, which will yield insight into the research questions at hand. *Example: A botanist may be interested in estimating the coefficients for an established model that relates the weight of vascular plants with the amount of water they receive, the nutrients in the soil, and the amount of sunlight exposure.*

3. **Prediction:** A researcher may be concerned with predicting some response variable at given levels of other input variables. For predictions to be valid, various assumptions must be made and met. Most notably, one should not **extrapolate** beyond the range of the data since the estimation is only valid within the domain of the sampled data. *Example: A realtor has a 20-year history of the home selling prices for those properties that she sold throughout her career as well as the homes' total square footage, the years they were built, and the assessed values. She will put a new home on the market and wants to be able to predict that home's selling price given that the values of the other variables provided do not extend outside the domain of her data.*

4. **Control:** Regression models may be used for monitoring and controlling systems such that the functional relationship continues over time. If it does not, then modification of the model is necessary. *Example: A manufacturer of semiconductors continuously monitors the camber measurement on the substrate to be within certain limits. During this process, a variety of measurements in the system are recorded, such as lamination temperature, firing temperature, and lamination pressure. These inputs are always controlled within certain limits and if the camber measurement exceeds the designed limits, then the manufacturer must take corrective action.*

5. **Variable Selection or Screening:** A researcher may be faced with many independent variables and just one dependent variable. Since it is not feasible, nor necessarily informative, to model the de-

pendent variable as a function of *all* of the independent variables, a search can be conducted to identify a subset of the independent variables that explains a significant amount of the variation in the dependent variable. Historical data may also be used to help in this decision process. *Example: A wine producer may be interested in assessing how the composition of his wine relates to sensory evaluations. A score for the wine's aroma is given by a judging panel and 25 elemental concentrations are recorded from that wine. It is then desired to see which elements explain a significant amount of the variation in the aroma scores.*

# 2

## Basics of Regression Models

Statistical analyses are used to make generalizations about populations on the basis of the information in a sample. Thus, it is important to understand the distinction between a *sample* and a *population*.

- A **sample** is the collection of units (e.g., people, animals, cities, fields, whatever you study) that is actually measured or surveyed in a study.

- The **population** is the larger group of units from which the sample was selected, ideally using probability methods for the selection. A sample, which is a subset of the population, is used to estimate characteristics of the population.

For example, suppose that we measure the ages and distances at which $n = 30$ drivers can read a highway sign (we usually use $n$ to denote the sample size of the study). This collection of 30 drivers is the sample which presumably will be used to represent and estimate characteristics of the larger population of all drivers.

Different notation is used for sample and population characteristics. For example, the mean of a sample $x_1, x_2, \ldots, x_n$ is usually denoted by $\bar{x}$, whereas the mean of a population is typically denoted by the Greek letter $\mu$. An alternative notation for a population mean is $E(X)$, which is read as the "expected value of $X$."

With the observed sample, we can compute a value for $\bar{x}$. This sample value estimates the unknown value of the population mean $\mu$. It is crucial to recognize that we do not, and will not, know the exact value of $\mu$. We only know the value of its sample estimate, $\bar{x}$.

Another notational convention is that capital Roman letters (e.g., $X$) typically denote **random variables**, while lower case Roman letters (e.g., $x$) are typically **realizations** of that random variable using the data. This rule is sometimes abused when we introduce matrix notation, where capital Roman letters will be used for matrices of realizations (or the observed data).

The way we characterize a random variable and realizations of that random variable also differ. A measure computed from sample data is called a **statistic**. A measure characterizing the population is called a **parameter**. For example, $\bar{x}$ is a statistic for the sample while $\mu$ is a parameter of the population. A mnemonic to help remember this bit of terminology is to note that sample and statistic both begin with "s" while population and parameter both begin with "p."

**TABLE 2.1**
Notation for population and sample
regression coefficients

| Coefficient | Population | Sample |
|---|---|---|
| Intercept | $\beta_0$ | $\hat{\beta}_0$ |
| Slope | $\beta_1$ | $\hat{\beta}_1$ |

## 2.1 Regression Notation

A component of the simple regression model is that the mean value of the $y$-variable is a straight-line function of an $x$-variable. The two coefficients of a straight-line model are the **intercept** and the **slope**. The notation used to distinguish the sample and population versions of these coefficients is given in Table 2.1.

In statistics, the circumflex or "hat" notation above a quantity is typically used to indicate an estimate. This is the convention used in Table 2.1. Alternatively, some authors use a lower case $b$ to denote the estimate of a regression parameter $\beta$. We will use the former convention in this book.

Suppose that the regression equation relating systolic blood pressure and age for a sample of $n = 50$ individuals aged 45 to 70 years old is

$$\text{average blood pressure} = 95 + 0.80 * \text{age}.$$

Then, the sample slope is $\hat{\beta}_1 = 0.80$ and the sample intercept is $\hat{\beta}_0 = 95$. We do not know the values of $\beta_0$ and $\beta_1$, the intercept and slope, respectively, for the larger population of all individuals in this age group. For example, it would be *incorrect* to write $\beta_1 = 0.80$.

## 2.2 Population Model for Simple Linear Regression

This section discusses the theoretical regression model that we wish to estimate. Again, you will *never* know the actual population regression model in practice. If we always knew the quantities in population models, then there would be little need for the field of Statistics!

A **regression equation** describes how the *mean* value of a $y$-variable (also called the **response** or **dependent variable**) relates to specific values of the $x$-variable(s) (also called the **predictor(s)** or **independent variable(s)**) used to predict $y$. A **regression model** also incorporates a measure of uncertainty or error. A general format for a regression model is:

individual's $y$ = equation for mean + individual's deviation from mean.

Suppose that $(x_1, y_1), (x_2, y_2), \ldots, (x_n, y_n)$ are realizations of the random variable pairs $(X_1, Y_1), (X_2, Y_2), \ldots, (X_n, Y_n)$. The **simple linear regression equation** is that the mean of $Y$ is a straight-line function of $x$. This could be written as:

$$E(Y_i) = \beta_0 + \beta_1 x_i,$$

where $E(Y_i)$ is used to represent the mean value (expected value) and the subscript $i$ denotes the (hypothetical) $i^{\text{th}}$ unit in the population. To be completely pedantic, the simple regression equation should actually be written as the mean of a conditional random variable:

$$E(Y_i | X_i = x_i) = \beta_0 + \beta_1 x_i.$$

In other words, we *fix* the values of $X_i$ and then proceed to observe the values of $Y_i$.

The **simple linear regression model** for individuals in the larger population from which the sample has been taken is written as:

$$y_i = \beta_0 + \beta_1 x_i + \epsilon_i,$$

where $\epsilon_i$ is the *error* or *deviation* of $y_i$ from the line $\beta_0 + \beta_1 x_i$. We will mostly be dealing with real data, so the model is usually written in terms of the realizations, as done above. However, we could write the model in terms of the general random variables, which is $Y = \beta_0 + \beta_1 X + \epsilon$. For the purpose of statistical inference, the error terms are also assumed to be independent and identically distributed (*iid*) according to a normal distribution with mean 0 and variance $\sigma^2$. Thus, $E(\epsilon_i) = 0$ and $\text{Var}(\epsilon_i) = \sigma^2$. We later discuss how to test these assumptions and what to do when one or more of the assumptions is violated.

Recall that a random variable $Z$ is normally distributed if it has the following probability density function:

$$f_Z(z) = \frac{1}{\sqrt{2\pi}\sigma} e^{-\frac{(z-\mu)^2}{2\sigma^2}},$$

where $-\infty < \mu < +\infty$ is the mean of $Z$ and $\sigma > 0$ is the standard deviation of $Z$. Furthermore, a standard normal random variable is the special case where $Z$ has $\mu = 0$ and $\sigma = 1$. While we generally do not utilize the functional form of the normal distribution directly, it is at the root of many of our regression calculations. We will return to the normal distribution function later when we discuss the correlation model.

The term *linear* is used in the above discussion to indicate the behavior of the regression coefficients; i.e., $\beta_0$ and $\beta_1$. An example of a model that is nonlinear in the parameters is:

$$y_i = \frac{\beta_0}{\beta_1 - \beta_0} \left[ e^{-\beta_2 x_i} - e^{-\beta_1 x_i} \right] + \epsilon_i.$$

Analysis of nonlinear regression models requires more advanced techniques, which are discussed in Chapter 19. Since our current focus is on linear regression, the term *linear* may be dropped throughout earlier portions of the text.

## 2.3   Ordinary Least Squares

The first step in a regression problem is to estimate the model. The standard mathematical criterion used is **ordinary least squares**, which is short for the least sum of squared errors. The *best* sample estimate of the regression equation is that for which the observed sample has the smallest sum of squared errors; i.e., we need to find the values of $\beta_0$ and $\beta_1$ that minimize the criterion

$$S = \sum_{i=1}^{n} \epsilon_i^2 = \sum_{i=1}^{n} (y_i - (\beta_0 + \beta_1 x_i))^2$$

for the given data values $(x_1, y_1), \ldots, (x_n, y_n)$. Using calculus, we first differentiate with respect to each of the regression coefficients. This yields the system of equations

$$\frac{\partial S}{\partial \beta_0} = -2 \sum_{i=1}^{n} (y_i - \beta_0 - \beta_1 x_i)$$

$$\frac{\partial S}{\partial \beta_1} = -2 \sum_{i=1}^{n} x_i (y_i - \beta_0 - \beta_1 x_i).$$

Setting the above equal to 0 and replacing $(\beta_0, \beta_1)$ by $(\hat{\beta}_0, \hat{\beta}_1)$ yields the **normal equations**

$$\sum_{i=1}^{n} (y_i - \hat{\beta}_0 - \hat{\beta}_1 x_i) = 0$$

$$\sum_{i=1}^{n} x_i (y_i - \hat{\beta}_0 - \hat{\beta}_1 x_i) = 0.$$

The sample estimates $(\hat{\beta}_0, \hat{\beta}_1)$ in the normal equations are called the **ordinary least squares estimates**. Thus, the regression line determined by $(\hat{\beta}_0, \hat{\beta}_1)$ is called the **ordinary least squares line**. The solutions to the normal equations are:

$$\hat{\beta}_1 = \frac{\sum_{i=1}^{n} (x_i - \bar{x})(y_i - \bar{y})}{\sum_{i=1}^{n} (x_i - \bar{x})^2} \tag{2.1}$$

$$\hat{\beta}_0 = \bar{y} - \hat{\beta}_1 \bar{x}. \tag{2.2}$$

A consequence of these equations is that $\hat{\beta}_0$ and $\hat{\beta}_1$ are unbiased estimates of their population quantities; i.e., $E(\hat{\beta}_0) = \beta_0$ and $E(\hat{\beta}_1) = \beta_1$.[1] Also note that the ordinary least squares line goes through the point $(\bar{x}, \bar{y})$. Do you see why?

With the ordinary least squares estimates $(\hat{\beta}_0, \hat{\beta}_1)$ calculated, we now have the estimated simple linear regression model

$$y_i = \hat{\beta}_0 + \hat{\beta}_1 x_i + e_i,$$

from which we can calculate a few additional quantities:

- $\hat{y}_i = \hat{\beta}_0 + \hat{\beta}_1 x_i$; $\hat{y}_i$ is the **predicted value** (or **predicted fit**) of $y$ for the $i^{\text{th}}$ observation in the sample.

- $e_i = y_i - \hat{y}_i$; $e_i$ is the **observed error** (or **residual**) for the $i^{\text{th}}$ observation in the sample. This is calculated by taking the difference between the observed and predicted values of $y$ for the $i^{\text{th}}$ observation in the sample.

- $\text{SSE} = \sum_{i=1}^{n}(y_i - \hat{y}_i)^2$; SSE is the **sum of squared observed errors** for all observations in a sample of size $n$.

Thus, the *best* regression line for a sample that uses the ordinary least squares estimates in (2.1) and (2.2) minimizes the SSE and guarantees that $\sum_{i=1}^{n} e_i = 0$.

One other notational convention often used in the literature concerns the following sums of squares:

$$S_{xx} = \sum_{i=1}^{n}(x_i - \bar{x})^2$$

$$S_{yy} = \sum_{i=1}^{n}(y_i - \bar{y})^2$$

$$S_{xy} = \sum_{i=1}^{n}(x_i - \bar{x})(y_i - \bar{y}).$$

So for example, the ordinary least squares estimate for the slope can be written as:

$$\hat{\beta}_1 = \frac{S_{xy}}{S_{xx}}.$$

We will primarily use this notation only in the simple linear regression setting, as we will yield to the more succinct matrix notation for multiple linear regression starting in Chapter 6.

---

[1] A famous result is the Gauss–Markov Theorem, which states that the ordinary least squares estimates are the *best* linear unbiased estimates. This is one of the few theorems we present in this text. We formalize it later after we introduce the multiple linear regression setting and matrix notation for regression models.

## 2.4   Measuring Overall Variation from the Sample Line

We next discuss some measures of overall variation from the sample regression line.

$MSE = \frac{SSE}{n-2}$ is called the **mean squared error** for simple linear regression. MSE is the **sample variance** of the errors and estimates $\sigma^2$, the population variance for the errors. It is important to note that the divisor $n - 2$ only applies to simple linear regression. The general rule is that the divisor is $n - p$, where $p$ is the number of parameters in the regression equation. For a straight-line model, we estimate $p = 2$ regression parameters: the intercept and the slope.

The **sample standard deviation** of the errors is given by $s = \sqrt{MSE}$ and is called the **root mean square error** or **RMSE**. The value of $s$ can be interpreted (roughly) as the average absolute size of deviations of individuals from the sample regression line.

Let $\bar{y}$ be the mean of all observed $y$ values. Then, the **total sum of squares** is given by $SSTO = \sum_{i=1}^{n}(y_i - \bar{y})^2$.

### 2.4.1   $R^2$

The **coefficient of determination**

$$R^2 = \frac{SSR}{SSTO} = \frac{SSTO - SSE}{SSTO} \tag{2.3}$$

is the proportion of variation in $y$ that is explained by $x$. $R^2$ is often expressed as a percentage by multiplying the value calculated in (2.3) by 100. For example, if your calculation shows that the proportion of variation in $y$ explained by $x$ is 0.437, then it is also correct to say that 43.7% of the total variation in $y$ is explained by $x$.

While $R^2$ is a popular measure of how well the regression model fits the data (see Barrett, 1974), it should not be used solely to assess the model's adequacy without further justification. Some caveats regarding the use of $R^2$ include:

1. The value of $R^2$ is highly sensitive to the sample size.

2. The value of $R^2$ can be increased by adding more predictors to the model (as we will see when we discuss multiple linear regression). However, this can cause the unwanted situation of an increase in the MSE, which can occur with small sample sizes.

3. $R^2$ is influenced by the range of the predictors in that if the range of $X$ increases or decreases, then $R^2$ increases or decreases, respectively.

4. The magnitude of the slopes is not measured by $R^2$.

5.  $R^2$ only measures the *strength* of the linear component of a model. For example, suppose the relationship between the response and predictor is measured perfectly by $Y = \cos(X)$. Then, the value of $R^2$ will be very small even though the two variables have a perfect relationship.

6.  A high or low level of $R^2$ does not necessarily indicate one way or the other the predictability of the model.

## 2.5 Regression Through the Origin

Consider a generic assembly process. Suppose we have data on the number of items produced per hour along with the number of rejects in each of those time spans. If we have a period where no items were produced, then there are obviously 0 rejects. Such a situation may indicate deleting $\beta_0$ from the model since $\beta_0$ reflects the amount of the response (in this case, the number of rejects) when the predictor is assumed to be 0 (in this case, the number of items produced). Thus, the model to estimate becomes

$$y_i = \beta_1 x_i + \epsilon_i,$$

which is called a **regression through the origin (RTO)** model.

The estimate for $\beta_1$ when using the regression through the origin model is:

$$\hat{\beta}_{\text{RTO}} = \frac{\sum_{i=1}^n x_i y_i}{\sum_{i=1}^n x_i^2}.$$

Thus, the estimated regression equation is

$$\hat{y}_i = \hat{\beta}_{\text{RTO}} x_i,$$

Note that we no longer have to center (or "adjust") the $x_i$ and $y_i$ values by their sample means. Specifically, compare this estimate for $\beta_1$ to that of the estimate found in (2.1) for the simple linear regression model. Since there is no intercept, there is no correction factor and no adjustment for the mean; i.e., the regression line can only pivot about the point $(0, 0)$.

Generally, a regression through the origin is not recommended due to the following:

1.  Removal of $\beta_0$ is a strong assumption which forces the line to go through the point $(0, 0)$. Imposing this restriction does not give ordinary least squares as much flexibility in finding the line of best fit for the data.

2.  Generally, $\sum_{i=1}^n e_i \neq 0$. Because of this, the SSE could actually be larger than the SSTO, thus resulting in $R^2 < 0$. Since $R^2$ can

be negative, the same interpretation of this value as a measure of the strength of the linear component in the simple linear regression model cannot be used here.

If you strongly believe that a regression through the origin model is appropriate for your situation, then statistical testing can help justify your decision (to be discussed later). Moreover, if data has not been collected near $X = 0$, then forcing the regression line through the origin is likely to make for a worse-fitting model. So again, this model is not usually recommended unless there is a strong belief that it is appropriate. Eisenhauer (2003) discusses examples where regression through the origin is appropriate.

## 2.6   Distinguishing Regression from Correlation

The **(Pearson) correlation coefficient** (often denoted by $\rho$) is a bounded index (i.e., $-1 \leq \rho \leq 1$) that provides a unitless measure for the strength and direction of the association between two variables. For example, suppose we want to examine the relationship between the two random variables $X$ and $Y$ and we find that $\rho = -0.92$. This indicates that there is a strong negative relationship between these two variables; i.e., as the value of one of the variables *increases*, the value of the other variable tends to *decrease*. Thus, a value of $\rho$ close to 0 indicates that there is no association between the two variables, while values close to $-1$ or $+1$ indicate strong negative or strong positive associations, respectively.

In terms of estimation, suppose we have the samples $x_1, x_2, \ldots, x_n$ and $y_1, y_2, \ldots, y_n$. Then, the sample Pearson correlation coefficient is given by

$$
r = \frac{\sum_{i=1}^{n}(x_i - \bar{x})(y_i - \bar{y})}{\sqrt{\sum_{i=1}^{n}(x_i - \bar{x})^2 \sum_{i=1}^{n}(y_i - \bar{y})^2}}
$$
$$
= \frac{S_{xy}}{\sqrt{S_{xx}S_{yy}}},
$$

which provides an estimate of the population parameter $\rho$. Inference procedures can also be carried out on this quantity, but we will not explore those details here.

So how does the correlation coefficient connect to what we have discussed with simple linear regression? Unlike regression, correlation is treating the two variables $X$ and $Y$ as having a random bivariate structure. Correlation only establishes what kind of association *possibly* exists between these two random variables. However, regression is concerned with treating $Y$ as a random variable while fixing $X$ (i.e., $Y$ *depends* on $X$). Subsequently, a regression analysis provides a model for a cause-and-effect type of relationship, while correlation simply provides a measure of association. *Correlation does not imply causation.* Also, note that the estimated slope coefficient $(\hat{\beta}_1)$ and the estimated correlation coefficient $(r)$ are related in the following way:

$$r = \hat{\beta}_1 \sqrt{\frac{\sum_{i=1}^{n}(x_i - \bar{x})^2}{\sum_{i=1}^{n}(y_i - \bar{y})^2}}$$

$$= \hat{\beta}_1 \sqrt{\frac{S_{xx}}{S_{yy}}}.$$

The above relationship, as well as other computational and conceptual definitions of $r$, are discussed in Rodgers and Nicewander (1988). See Aldrich (1995) for a more philosophical and historical discussion about correlation and causation.

There are various correlation statistics and the Pearson correlation coefficient is perhaps the most common, especially since it is a parametric measure. If a correlation coefficient is presented in a published report and the type of coefficient is not explicitly stated, then it is likely the Pearson correlation coefficient. Some common **nonparametric** — meaning models or statistics not based on parameterized distributions — measures of association include:

- **Spearman's rank correlation coefficient** (Spearman, 1904) measures the association based on the ranks of the variables. This correlation measure is often denoted by $\theta$ and the estimate is given by

$$\hat{\theta} = \frac{\sum_{i=1}^{n}(R_i - \bar{R})(S_i - \bar{S})}{\sqrt{\sum_{i=1}^{n}(R_i - \bar{R})^2 \sum_{i=1}^{n}(S_i - \bar{S})^2}},$$

  where $R_i$ and $S_i$ are the rank of the $x_i$ and $y_i$ values, respectively. Note that this is just the estimated Pearson's correlation coefficient, but the values of the variables have been replaced by their respective ranks.

- **Hoeffding's $D$ statistic** (Hoeffding, 1948) detects more general departures from independence by measuring the distance between the joint cumulative distribution function of $X$ and $Y$ and the product of their respective marginal distribution functions. This statistic is calculated as

$$D^* = \frac{(n-2)(n-3)D_1 + D_2 - 2(n-2)D_3}{n(n-1)(n-2)(n-3)(n-4)},$$

  where $D_1 = \sum_{i=1}^{n}(Q_i - 1)(Q_i - 2)$, $D_2 = \sum_{i=1}^{n}(R_i - 1)(R_i - 2)(S_i - 1)(S_i - 2)$, and $D_3 = \sum_{i=1}^{n}(R_i - 2)(S_i - 2)(Q_i - 1)$. Here, $R_i$ and $S_i$ are defined as above and $Q_i$ is 1 plus the number of points with both $x$ and $y$ values less than the $i^{\text{th}}$ point.

- **Kendall's $\tau_b$** (Kendall, 1938) is based on the number of concordant and discordant pairs between the ordinal variables $X$ and $Y$.[2] **Concordant pairs** occur when a pair of observations are in the same direction, and **discordant pairs** occur when a pair of observations are in the opposite direction. For all $(X_i, Y_i)$ and $(X_j, Y_j)$ pairs (with $i < j$), concordance occurs if the relationship between $X_i$ and $X_j$ are in the same direction as $Y_i$ and $Y_j$, and discordance occurs if the relationships are in opposite directions. This statistic is calculated as

$$\hat{\tau}_b = \frac{\sum_{i<j} \mathbf{sgn}(x_i - x_j)\mathbf{sgn}(y_i - y_j)}{\sqrt{(T_0 - T_1)(T_0 - T_2)}},$$

where $T_0 = \frac{1}{2}n(n-1)$, $T_1 = \frac{1}{2}\sum_k t_k(t_k - 1)$, and $T_2 = \frac{1}{2}\sum_l u_l(u_l - 1)$. The $t_k$ and $u_l$ are the number of tied values in the $k^{\text{th}}$ group of tied $x$ values and the $l^{\text{th}}$ group of tied $y$ values, respectively. Note that the $\mathbf{sgn}()$ function simply means to take the sign of its argument; i.e., either "+" or "−."

- For a sample of size $n$, other measures based strictly on the number of concordant pairs ($n_c$), the number of discordant pairs ($n_d$), and the number of tied pairs ($n_t$) are:

  - **Somer's D:**
    $$D_S = \frac{n_c - n_d}{n_c + n_d + n_t}.$$

  - **Goodman–Kruskal Gamma:**
    $$\gamma_{GK} = \frac{n_c - n_d}{n_c + n_d}.$$

  - **Kendall's $\tau_{\mathbf{a}}$:**
    $$\hat{\tau}_a = \frac{2(n_c - n_d)}{n(n-1)}.$$

Comparisons between various correlation measures given above have been discussed in the context of real data applications, such as assessing the relative validity from a survey questionnaire (Masson et al., 2003) and their use in gene expression association analyses (Fujita et al., 2009). An approach for comparing any number of the same type of correlation coefficients having one variable in common is given in Meng et al. (1992).

## 2.7    Regression Effect

The **regression effect** (or **regression toward the mean**) is the phenomenon that extreme measurements (i.e., unusually large or small values)

---

[2]Ordinal variables are categorical variables where there is an implied meaning in the order of the categories. For example, if asked how you feel today on a scale of 1 to 5 with 1 being "miserable" and 5 being "great," the order has an implied meaning.

are typically followed by subsequent measurements close to the mean. Sir Francis Galton first observed the phenomenon and used the term "regression" in the context of studying seeds. He noted that the offspring of seeds did not tend to resemble the size of their parent seeds, but rather tended to be more average in size (Galton, 1885, 1886). The possibility of regression toward the mean must be considered when designing experiments and interpreting experimental, survey, and other empirical data in the physical, life, behavioral and social sciences. Otherwise, incorrect inferences might be made.

Suppose you ask a large group of people to pick a number at random from the set $\{0, 1, 2, \ldots, 10\}$. Each person's number would then be a realization from a set of independent and identically distributed random variables with a mean of, say, 5. Due to random variability, some people will have selected numbers closer to 0 or 10. Suppose we take those people who picked the numbers of 0, 1, or 2 and rerun the experiment. The mean of the new numbers randomly selected by this subgroup of people would again be expected to be near 5; i.e., the mean "regressed" back to the mean of the original group.

### 2.7.1 Regression Fallacy

The **regression fallacy** occurs when the regression effect is mistaken for a real cause or treatment effect. Some examples of statements committing a regression fallacy are as follows:

1. *A subject has a strong stomachache and visits his doctor. During the visit, the subject merely talks with his doctor and no extensive treatment plan is applied. The next day, the stomachache is gone. Therefore, the subject benefited from visiting the doctor's office.* The fact that the stomachache resolved after having, essentially, been at its worst is more easily explained by regression toward the mean. Assuming it resolved due to visiting the doctor is fallacious.

2. *An ice cream salesman at a kiosk in a mall sold more units today than he did yesterday. However, today he only worked for 5 hours instead of the usual 6 hours. Therefore, working 5 hours will result in more units sold.* Malls, like any other retailers, have day-to-day variability in the number of customers. It was likely that more customers were in the mall on the day the ice cream salesman worked 5 hours. The daily sales are likely better explained by regression towards the mean.

3. *A student basketball player kept committing fouls during a game and made no baskets. The coach decided to bench her for the remainder of that game. At their next game, she scored 15 points. Therefore, benching the player was an effective strategy in improving her performance.* Often, poor athletic performances are followed-up by better performances. In this case, she happened to have a bad game and benching her did not improve her overall athletic skills. The

performance is, again, better explained by regression toward the mean.

Another popular variant of the regression fallacy occurs when subjects are enrolled in a study on the basis of an extreme value of some measurement. When a treatment is applied to those subjects, it can then be declared effective because subsequent measurements are typically not as extreme. Similarly, it is fallacious to take individuals with extreme values from one measuring instrument (e.g., the Wechsler Adult Intelligence Scale, which is an intelligence test), reevaluate them using a different instrument (e.g., the Stanford-Binet test, which is another intelligence test), and declare the instruments to be biased relative to each other if the second instrument's measurements are not as extreme as the first instrument's measurement. The regression effect guarantees that such results must be observed in the absence of any treatment effect or bias between the instruments.

The regression effect is real and complicates the study of subjects who are initially extreme on the outcome variable, however, this does not preclude us from conducting our study. Proper experimental design can often mitigate such an issue. While we briefly discuss some experimental design topics, the breadth of those various topics is well beyond the scope of this text. Some classic texts on experimental design include Kuehl (2000), Box et al. (2005), and Montgomery (2013). For our discussion, randomization and controls are usually enough to safeguard against the regression effect involving extreme outcomes. Consider a study of subjects selected for their initially high triglyceride levels. The subjects are enrolled in a controlled diet trial to lower their levels. Regression to the mean says even the controls will show a decrease over the course of the study, but if the treatment is effective, the decrease will be greater in the treated group than in the controls. This is also something that can be assessed through the analysis of covariance, which we discuss in Chapter 15.

## 2.8    Examples

**Example 2.8.1.** *(Toy Data)*
We first analyze a toy dataset in order to illustrate some simple calculations provided thus far. The data are provided in Table 2.2.

The least squares regression line for this sample data is $\hat{y}_i = 5 + 2x_i$. A scatterplot of this data with this ordinary least squares fit overlaid is given in Figure 2.1. Notice in Table 2.2 that there are additional calculations provided based on this estimated equation. The third row gives predicted values, determined by substituting the value of $x_i$ into the estimated equation for

## TABLE 2.2
The toy data and relevant calculations

| $x_i$ | 4 | 4 | 7 | 10 | 10 |
|---|---|---|---|---|---|
| $y_i$ | 15 | 11 | 19 | 21 | 29 |
| $\hat{y}_i$ | 13 | 13 | 19 | 25 | 25 |
| $e_i$ | $15 - 13 = 2$ | $11 - 13 = -2$ | $19 - 19 = 0$ | $21 - 25 = -4$ | $29 - 25 = 4$ |

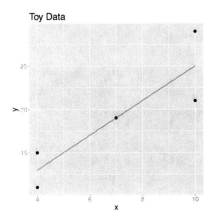

Toy Data

## FIGURE 2.1
Scatterplot of the toy data along with the ordinary least squares fit.

each $i$ (i.e., $\hat{y}_i = 5 + 2x_i$). The fourth row gives the residuals, calculated as $e_i = y_i - \hat{y}_i$, such that $i = 1, \ldots, 5$.

Relevant calculations on this dataset include:

- The sum of squared errors is SSE $= 2^2 + (-2)^2 + 0^2 + (-4)^2 + 4^2 = 40$, which is the value minimized in order to obtain the regression coefficient estimates of $\hat{\beta}_0 = 5$ and $\hat{\beta}_1 = 2$.

- MSE $= \frac{\text{SSE}}{n-2} = \frac{40}{3} = 13.33$, which estimates the unknown value of $\sigma^2$, the population variance of the errors.

- $s = \sqrt{\text{MSE}} = \sqrt{13.33} = 3.65$ is an estimate of $\sigma$, the unknown value of the population standard deviation of the errors.

- To calculate the SSTO, first calculate $\bar{y} = (15 + 11 + 19 + 21 + 29)/5 = 19$. Then

$$\text{SSTO} = (15 - 19)^2 + (11 - 19)^2 + (19 - 19)^2 + (21 - 19)^2 + (29 - 19)^2 = 184.$$

- Finally, $R^2 = \frac{\text{SSTO} - \text{SSE}}{\text{SSTO}} = \frac{184 - 40}{184} = 0.783$. So roughly 78.3% of the variation in $y$ is explained by $x$.

- We can also calculate $r = +0.8847$, which indicates a strong, positive association between the two variables. $R^2$ is often reported with a correlation analysis because it is also a measure of association and is simply the square of the sample correlation coefficient; i.e., $R^2 = r^2$.

**Example 2.8.2.** *(Steam Output Data)*
We consider a dataset of size $n = 25$ that contains observations taken at a large industrial steam plant. This dataset is from Draper and Smith (1998), which has ten variables measured for each observation. We will only focus on two of the variables:

- $Y$: pounds of steam used per month (in a coded form)

- $X$: average atmospheric temperature (in degrees Fahrenheit)

We are interested in modeling the steam used per month $(Y)$ as a function of the average atmospheric temperature $(X)$.

Figure 2.2 is a plot of the data with the least squares regression line overlaid. Some noticeable features from the scatterplot in Figure 2.2 is that the relationship appears to be linear and it appears that there is a negative slope. In other words, as the atmospheric pressure increases, then the steam used per month tends to decrease.

Next, we fit a simple linear regression line to this data using ordinary least squares. Basic output pertaining to an ordinary least squares analysis includes the following:

```
############################
Coefficients:
            Estimate Std. Error t value Pr(>|t|)
(Intercept) 13.62299    0.58146  23.429  < 2e-16 ***
temp        -0.07983    0.01052  -7.586 1.05e-07 ***
---
Signif. codes:  0 *** 0.001 ** 0.01 * 0.05 . 0.1   1

Residual standard error: 0.8901 on 23 degrees of freedom
Multiple R-squared:  0.7144,Adjusted R-squared:  0.702
F-statistic: 57.54 on 1 and 23 DF,  p-value: 1.055e-07
############################
```

The regression line fitted to these data is

$$\hat{y}_i = 13.6230 - 0.0798x_i, \tag{2.4}$$

which is overlaid on the scatterplot in Figure 2.2. Recall that we use the "hat" notation to denote estimates.

Interpretation of the estimated regression coefficients is as follows:

- The interpretation of the slope $(-0.0798)$ is that the predicted amount of steam used per month decreases, on average, by about 0.0798 pounds for every 1 degree Fahrenheit increase in temperature.

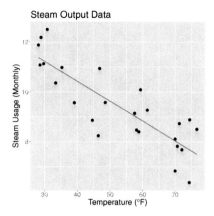

**FIGURE 2.2**
Scatterplot of steam usage versus temperature with the least squares regression line overlaid. Does a linear relationship look appropriate to you?

- The interpretation of the intercept (13.6230) is that if the temperature were 0 degrees Fahrenheit, then the average amount of steam used per month would be about 13.6230 pounds.

Some other statistics from the output worth noting are:

- The value of $s = \sqrt{\text{MSE}} = 0.8901$ tells us roughly the average difference between the $y$ values of individual observations and the predictions of $y$ based on (2.4).

- The value of $R^2$ is interpreted as the atmospheric temperature explains 71.44% of the observed variation in the steam used per month.

- We can obtain the value of the correlation coefficient $r$ by hand:

$$r = \mathbf{sgn}(\hat{\beta}_1)\sqrt{R^2}$$
$$= -\sqrt{0.7144}$$
$$= -0.8452.$$

Thus, there is a strong negative relationship between steam used per month and the atmospheric temperature.

- The other values given in the output will be discussed in later chapters.

**Example 2.8.3.** *(Computer Repair Data)*
We consider a dataset of size $n = 14$ that contains a random sample of service call records for a computer repair company. This dataset is from Chatterjee and Hadi (2012). The measured variables are:

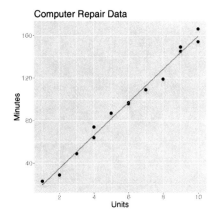

**FIGURE 2.3**
Scatterplot of service call length versus number of components repaired/replaced with the estimated regression through the origin line overlaid.

**TABLE 2.3**
Correlations for the computer repair data

| Correlation Measure | Value | |
|---|---|---|
| Pearson's | $r =$ | 0.9937 |
| Spearman's Rank | $\hat{\theta} =$ | 0.9956 |
| Hoeffding's $D$ | $D^* =$ | 0.8851 |
| Kendall's $\tau_b$ | $\hat{\tau}_b =$ | 0.9778 |
| Somer's D | $D_S =$ | 1.0000 |
| Goodman–Kruskal Gamma | $\gamma_{GK} =$ | 1.0000 |
| Kendall's $\tau_a$ | $\hat{\tau}_a =$ | 0.9560 |

- $Y$: length of service call (in minutes)

- $X$: the number of components repaired or replaced during the service call

We are interested in the relationship between these two variables.

Table 2.3 provides estimates of all of the correlation measures presented in Section 2.6. There is clearly agreement across the different measures, with Hoeffding's $D$ being the smallest at 0.8851. Thus there is a strong positive association between the length of service call and the number of components repaired or replaced. Pearson's correlation is quantifying the general association between the raw observation, Spearman's rank correlation is quantifying the general association between their respective ranks, Hoeffding's $D$ is quantifying the association between the joint distribution function of the two variables and their respective marginal distribution functions, and all of the remaining measures are quantifying the general association between the variables based on their number of concordant and discordant pairs. Note that $D_S = \gamma_{GK}$ for these data since the number of tied pairs $(n_t)$ is 0.

Next, we consider modeling the length of service call $(Y)$ as a function of the number of components repaired or replaced $(X)$. Practically speaking, if no components are repaired or replaced, then no service call is placed (which would have a length of 0 minutes). Thus, we fit a regression through the origin line to these data. Basic output pertaining to this fit is as follows:

```
##############################
Coefficients:
      Estimate Std. Error t value Pr(>|t|)
units  16.0744     0.2213   72.62   <2e-16 ***
---
Signif. codes:  0 *** 0.001 ** 0.01 * 0.05 . 0.1   1

Residual standard error: 5.502 on 13 degrees of freedom
Multiple R-squared:  0.9975,Adjusted R-squared:  0.9974
F-statistic:  5274 on 1 and 13 DF,  p-value: < 2.2e-16
##############################
```

The regression through the origin line fitted to these data is

$$\hat{y}_i = 16.0744x_i, \tag{2.5}$$

which is overlaid on the scatterplot in Figure 2.3. We interpret this estimated regression equation by stating that the predicted length of a service call increases, on average, by about 16 minutes for every increase of 1 component needing to be fixed.

Note that if using the $R^2$ quantity to compute the Pearson's correlation coefficient in Table 2.3, you cannot use the $R^2$ from the regression through the origin fit above. You must use the $R^2$ from the simple linear regression fit, which we did not report.

# 3

## Statistical Inference

**Statistical inference** concerns statistical methods for using sample data to make judgments about a larger population. Two statistical inference tools we discuss are hypothesis testing and statistical intervals. We first discuss both of these in the context of inference about the regression parameters $\beta_0$ and $\beta_1$ as well as for the Pearson correlation coefficient. We then provide a more general discussion about two groups of statistical intervals that are of interest in the regression setting. The first group (confidence intervals) is concerned with estimating the mean $E(Y)$ given a particular value or particular values of the predictor. The second group (prediction and tolerance intervals) is concerned with predicting a new response value given a particular value or particular values of the predictor.

## 3.1 Hypothesis Testing and Confidence Intervals

**Hypothesis testing** requires the formulation of a null and alternative hypothesis which specify possibilities for the value(s) of one or more population parameters. The **null hypothesis** (denoted by $H_0$) is the hypothesis being tested and is usually a statement regarding no change or difference in the situation at hand (i.e., the status quo). The **alternative hypothesis** (denoted by $H_A$ or $H_1$) is the anticipated situation should the null hypothesis be false (i.e., what we usually hope to show). Occasionally, one tests multiple alternative hypotheses against a single null hypothesis.

After establishing $H_0$ and $H_A$, a **test statistic** is then calculated using the sample data (we will introduce the general formula for a test statistic later in this chapter). The value of the test statistic is affected by the degree to which the sample supports one or the other of the two hypotheses. Two completely equivalent strategies for making a decision based on the test statistic are:

1. The **p-value** approach, which is used by all statistical software. We calculate the probability that the test statistic would be as extreme as the value found if the null hypothesis were true. We decide in favor of the alternative hypothesis when the p-value is less than the **significance level** (or **α-level**). The correct way to phrase your

decision is to state that "we reject the null hypothesis if $p < \alpha$" or "fail to reject the null hypothesis if $p \geq \alpha$." The significance level is usually set at $\alpha = 0.05$, in which case we say a result is **statistically significant** if $p < 0.05$ and a result is **marginally significant** if $0.05 \leq p < 0.10$.

2. The **critical value** (or **critical region**) approach, which is what some textbooks use; see, for example, Kutner et al. (2005). We decide in favor of the alternative hypothesis when the value of the test statistic is more extreme than a critical value. The critical region is such that if the null hypotheses were true, the probability that the test statistic ends up in the critical region is the $\alpha$-level (significance level) of the test. The $\alpha$-level, again, is usually $\alpha = 0.05$.

Once we have used one of the two methods above, we can also include a **decision rule**, which is a statement in terms of the test statistic as to which values result in rejecting or failing to reject the null hypothesis.

While we are using the term "decision" with respect to the results of our hypothesis test, we should note that the rule-of-thumb significance level of $\alpha = 0.05$ should not be viewed as a panacea for solely making scientific decisions. The results obtained from hypothesis testing merely provide support for the null or alternative and should be used in conjunction with other scientific evidence and judgment. The history of some of the controversy around this topic, as well as suggestions by the statistical community, are discussed in Wasserstein and Lazar (2016). In particular, we can emphasize estimation via statistical intervals over solely using hypothesis testing. Most common is a $100 \times (1 - \alpha)\%$ **confidence interval**, which is a statistical interval of values that is likely to include the unknown value of a population parameter. The **confidence level** is the probability that the procedure used to determine the interval will provide an interval that "captures" the population value. As an example, suppose $\alpha = 0.05$ which corresponds to the 95% confidence level. Then, the calculated confidence interval will "capture" the population value for about 95% of all random samples. The way we interpret this for the interval $(a, b)$ is by saying "with 95% confidence, we expect the true value to be between $a$ and $b$." It is *incorrect* to say "there is a 95% chance that the true value will be between $a$ and $b$." The reason is that the true value (which is unknown) is either in or out of the interval $(a, b)$, which would correspond to a probability of 1 or 0, respectively, to belonging to the interval. Thus, always make sure you use the former statement and not the latter when interpreting a confidence interval!

A general format for a confidence interval in many situations is

sample statistic $\pm$ (multiplier $\times$ standard error of the statistic).

1. The **sample statistic** is the value that estimates the parameter of interest. For example, it might be the sample slope $(\hat{\beta}_1)$ in regression.

2. The **multiplier** is determined by the confidence level and a relevant probability distribution.

3. The **standard error** of the statistic is a measure of the accuracy of the statistic as an estimate of the true population value. The "by hand" formula is different for different types of statistics. We normally rely on software to provide this value.

## Standard Errors of $\hat{\beta}_0$ and $\hat{\beta}_1$

The standard error of each estimated regression coefficient measures, roughly, the average difference between the estimate and the true, unknown population quantity. These quantities are obtained by looking at the variances of the estimated regression coefficients and are easier to write in matrix notation, which we discuss in Chapter 6. Regardless, it can be shown that the variances of $\hat{\beta}_0$ and $\hat{\beta}_1$ are

$$\text{Var}(\hat{\beta}_0) = \frac{\sigma^2 \sum_{i=1}^n x_i^2 / n}{\sum_{i=1}^n (x_i - \bar{x})^2}$$

$$= \sigma^2 \left[ \frac{1}{n} + \frac{\bar{x}^2}{\sum_{i=1}^n (x_i - \bar{x})^2} \right]$$

and

$$\text{Var}(\hat{\beta}_1) = \frac{\sigma^2}{\sum_{i=1}^n (x_i - \bar{x})^2}.$$

Then, the square root of each of the above is the corresponding standard error of that estimate.

However, $\sigma^2$ is unknown, so we use the MSE. This results in the following (estimated) standard error formulas:

$$\text{s.e.}(\hat{\beta}_0) = \sqrt{\frac{\text{MSE} \sum_{i=1}^n x_i^2 / n}{\sum_{i=1}^n (x_i - \bar{x})^2}}$$

$$= \sqrt{\text{MSE} \left[ \frac{1}{n} + \frac{\bar{x}^2}{\sum_{i=1}^n (x_i - \bar{x})^2} \right]}$$

and

$$\text{s.e.}(\hat{\beta}_1) = \sqrt{\frac{\text{MSE}}{\sum_{i=1}^n (x_i - \bar{x})^2}}.$$

To be completely pedantic, we should probably put "hats" on the estimated standard errors; i.e., $\widehat{\text{s.e.}}(\hat{\beta}_0)$ and $\widehat{\text{s.e.}}(\hat{\beta}_1)$. However, standard error quantities where $\sigma^2$ is known are rarely of interest, so we avoid the more burdensome notation and omit the "hat" as well as dropping the word "estimated."

**Hypothesis Test for the Intercept ($\beta_0$)**

This test is usually not of much interest, but does show up when one is interested in performing a regression through the origin. For a hypothesis test about the intercept, the null and alternative hypotheses are written as:

$$H_0 : \beta_0 = 0$$
$$H_A : \beta_0 \neq 0. \tag{3.1}$$

In other words, the null hypothesis is testing if the population intercept is equal to 0 versus the alternative hypothesis that the population intercept is not equal to 0. In most problems, we are not particularly interested in hypotheses about the intercept. In particular, the intercept does not give information about how the value of $Y$ changes when the value of $X$ changes, which is characterized by the slope. The test statistic for the test in (3.1) is

$$t^* = \frac{\hat{\beta}_0}{\text{s.e.}(\hat{\beta}_0)},$$

which follows a $t_{n-2}$ distribution.

**Hypothesis Test for the Slope ($\beta_1$)**

This is a test about whether or not $X$ and $Y$ are related. The slope directly tells us about the link between the mean of $Y$ with $X$. When the true population slope does not equal 0, the variables $Y$ and $X$ are linearly related. When the slope is 0, there is not a linear relationship because the mean of $Y$ does not change when the value of $X$ is changed; i.e., you just have a horizontal line. For a hypothesis test about the slope, the null and alternative hypotheses are written as:

$$H_0 : \beta_1 = 0$$
$$H_A : \beta_1 \neq 0. \tag{3.2}$$

In other words, the null hypothesis is testing if the population slope is equal to 0 versus the alternative hypothesis that the population slope is not equal to 0. The test statistic for the test in (3.2) is

$$t^* = \frac{\hat{\beta}_1}{\text{s.e.}(\hat{\beta}_1)},$$

which also follows a $t_{n-2}$ distribution.

**Confidence Interval for the Slope ($\beta_1$)**

As mentioned earlier, inference procedures regarding $\beta_0$ are rarely of interest, so we will focus our attention on $\beta_1$. However, it should be noted that most of our discussion here is applicable to procedures concerning the intercept.

A confidence interval for the unknown value of the population slope $\beta_1$ can be computed as

sample statistic $\pm$ multiplier $\times$ standard error of statistic

$$\Rightarrow \hat{\beta}_1 \pm t^*_{n-2;1-\alpha/2} \times \text{s.e.}(\hat{\beta}_1).$$

This $t^*_{n-2;1-\alpha/2}$ multiplier is not the same value as the $t^*$ value calculated for the test statistic.

For the regression through the origin setting, a confidence interval for the unknown value of the population slope $\beta_{\text{RTO}}$ can be computed as

$$\Rightarrow \hat{\beta}_{\text{RTO}} \pm t^*_{n-1;1-\alpha/2} \times \text{s.e.}(\hat{\beta}_{\text{RTO}}),$$

where

$$\text{s.e.}(\hat{\beta}_{\text{RTO}}) = \sqrt{\frac{\text{MSE}}{\sum_{i=1}^{n} x_i^2}}.$$

Note that the degrees of freedom for the $t$-multiplier are $n-1$ and *not* $n-2$.

**Bonferroni Joint Confidence Intervals for $(\beta_0, \beta_1)$**
Sometimes we may want to find the confidence interval for more than one parameter simultaneously. While we are usually only interested in a confidence interval for the slope, this section will serve as more of a foundation for this type of interval, which is more commonly used in the multiple regression setting.

Let $A$ and $B$ be two events with complements $A^C$ and $B^C$, respectively. **Bonferroni's Inequality** says that

$$P(A \cap B) \geq 1 - P(A^C) - P(B^C).$$

Now, let $A$ be the event that the confidence interval for $\beta_0$ covers $\beta_0$ and $B$ be the event that the confidence interval for $\beta_1$ covers $\beta_1$. If $P(A) = 1 - \alpha$ and $P(B) = 1 - \alpha$, then by Bonferroni's Inequality, $P(A \cap B) \geq 1 - \alpha - \alpha = 1 - 2\alpha$. Thus, to get the joint confidence interval with at least $100 \times (1 - \alpha)\%$ confidence, the Bonferroni joint confidence intervals for $\beta_0$ and $\beta_1$ are

$$\hat{\beta}_0 \pm t^*_{n-2;1-\alpha/(2q)} \times \text{s.e.}(\hat{\beta}_0) \Rightarrow \hat{\beta}_0 \pm t^*_{n-2;1-\alpha/4} \times \text{s.e.}(\hat{\beta}_0)$$

and

$$\hat{\beta}_1 \pm t^*_{n-2;1-\alpha/(2q)} \times \text{s.e.}(\hat{\beta}_1) \Rightarrow \hat{\beta}_1 \pm t^*_{n-2;1-\alpha/4} \times \text{s.e.}(\hat{\beta}_1),$$

respectively. Here, $q = p = 2$ corresponds to the number of parameters for which we are trying to jointly estimate confidence intervals. It is also easy to extend to the general case of $q = n$ joint confidence intervals by an application of Bonferroni's Inequality for $n$ sets:

$$P(A_1 \cap A_2 \cap \ldots \cap A_n) \geq 1 - \sum_{i=1}^{n} P(A_i).$$

**The Duality Principle**

Consider testing the null hypothesis $H_0 : \beta_1 = 0$ at the $\alpha = 0.05$ significance level. Suppose that $p < \alpha$, which means we reject this null hypothesis. Then, the corresponding 95% confidence interval will not include the null value of 0; i.e., it will either be entirely above 0 or entirely below 0. In other words, we are 95% confident that our interval will not include 0. Thus, we arrive at the same statistical conclusion using either a hypothesis test or constructing a confidence interval at the same $\alpha$-level. This illustrates what is called the **duality principle**, which means for a given hypothesis and $\alpha$, the test and confidence interval will lead you to the same conclusion.

**Practical Significance Versus Statistical Significance**

For statistical inference, the larger your sample size $n$, the smaller your $p$-values. This means that a larger dataset will typically yield a statistically significant predictor. However, is it an *important* effect? For example, suppose an industrial complex needs to comply with federal air quality standards. The particular air quality standards allow for a level of 500 parts per million of these particular gases. The company took a large sample of days where they recorded the air quality measurements and the amount of time that a certain purifying machine was in operation during that day. They found that a one hour increase in usage of the purifying machine only resulted in an average decrease of 0.5 parts per million (ppm) of the air quality measurement. Even though this relationship was found to be statistically significant due to the large sample size, the result might not be practically significant for reasons like the cost of running the purifying machine might be high or that the company might already be well within the established air quality limits.

Mainly, we test if a regression coefficient is equal to 0. In actuality, it is highly unlikely that a predictor which has been measured and analyzed will have no effect on the response. The effect may be minute, but it is not completely 0. One will need to defer to examining the practicality of building such a model and ask themselves if a regression coefficient is close to 0, but deemed statistically significant, does that mean anything in the context of the data? Or, if we fail to reject the null hypothesis, but a relationship between a predictor and response seems to make sense, did the size of the data affect the result? While this mainly boils down to an issue of sample size and the power of a test, it is suggested to always report confidence intervals which help with the bigger picture by giving probabilistic bounds of the estimates.

## 3.2   Power

In hypothesis testing, you can commit one of two types of errors:

- A **Type I error** is when you *reject* the null hypothesis ($H_0$) when it is actually *true* (also called a false positive).

- A **Type II error** is when you *fail to reject* the null hypothesis when it is actually *false* (also called a false negative).

Earlier in this chapter, we introduced the significance level $\alpha$, which is actually the probability of making a Type I error. Hypothesis tests are constructed to minimize the probability of committing a Type II error. In other words, we wish to maximize 1 minus the probability of committing a Type II error, which is called the **power** of a test.[1] Various factors affecting the power of a test are:

- $n$: Increasing the sample size provides more information regarding the population and thus increases power.

- $\alpha$: A larger $\alpha$ increases power because you are more likely to reject the null hypothesis with a larger $\alpha$.

- $\sigma$: A smaller $\sigma$ results in easier detection of differences, which increases power.

- population effect ($\delta$): The more similar populations are (i.e., smaller values of $\delta$), the more difficult it becomes to detect differences, thus decreasing power.

The two types of errors and their associated probabilities help us interpret $\alpha$ and $1 - \beta$. The distinction made between tests of significance and tests as decision rules between two hypotheses is understood through their respective calculations. In significance testing, we focus on a single hypothesis ($H_0$) along with a single probability (the $p$-value).[2] The goal is to measure the strength of support from the sample against the null hypothesis. Subsequent power calculations can then help assess the sensitivity of the test. If we fail to reject the null hypothesis, then we can only claim there is not sufficient support against $H_0$, *not* that it is actually true. If we are looking at this as a decision problem, then we construct a decision rule for deciding between the two hypotheses based on sample evidence. We therefore focus equally on two probabilities, which are the probabilities of the two types of errors, and decide upon one hypothesis or the other. Figure 3.1 is a decision matrix for the type of errors in hypothesis testing.

The basic thought process is to first state $H_0$ and $H_A$ in terms of a test of significance. Then, think of this as a decision problem so that the probabilities of a Type I error and Type II error are incorporated. Statisticians typically view Type I errors as more serious, so choose a significance level ($\alpha$) and

---

[1]The power of a statistical test is also written as $1 - \beta$. However, do not get this value of $\beta$ confused with the $\beta$ used to represent a regression coefficient!

[2]Multiple hypotheses can be tested, but we will only provide a superficial discussion of this topic. See Hsu (1996) and Liu (2011) for more focused discussions.

Truth About the Population

|  |  | $H_0$ True | $H_A$ True |
|---|---|---|---|
| Decision Based on Sample | Reject $H_0$ | Type I Error | Correct Decision |
|  | Fail to Reject $H_0$ | Correct Decision | Type II Error |

**FIGURE 3.1**
The two types of errors that can be made in hypothesis testing. Remember that the probability of making a Type I error is $\alpha$ and the probability of making a Type II error is $\beta$.

consider only tests with probability of a Type I error no greater than $\alpha$. Then, among all of these tests, select the one that maximizes the power; i.e., minimizes a Type II error.

A common example illustrating power is to consider a legal trial where the null hypothesis would be that the individual on trial is innocent versus the alternative hypothesis that the individual is guilty. In the legal system, an individual is *innocent until proven guilty*. Hypothesis testing must show statistically significant evidence in order to reject the null hypothesis. Similarly, the justice system must show *beyond a reasonable doubt* that the individual is guilty. It is often viewed as more serious to find someone guilty given that they are innocent (a Type I error) as opposed to finding someone not guilty who is actually guilty (a Type II error). The Type I error in this case also means that the truly guilty individual is still free, which could carry a highly variable degree of risk, while an innocent person has been convicted. This justice example illustrates a case where a Type I error is more serious, and while Type I errors usually are more serious errors, this is not true in all hypothesis tests.

In the simple linear regression setting, we are interested in testing that the slope is 0 versus the alternative that it is not equal to 0; i.e., the test in (3.2). More generally, we could also test

$$H_0 : \beta_1 = \beta_1^*$$
$$H_A : \beta_1 \neq \beta_1^*,$$

where $\beta_1^*$ is any real number. However, $\beta_1^*$ is typically 0. The power of this test is calculated by first finding the $100 \times (1 - \alpha)^{\text{th}}$ percentile of the $F_{1,n-2}$

distribution, which we write as $F_{1,n-2;1-\alpha}$. Next we calculate $F_{1,n-2;1-\alpha}(\delta)$, which is the $100 \times (1-\alpha)^{\text{th}}$ percentile of a noncentral $F_{1,n-2}$ distribution with noncentrality parameter $\delta$. This is essentially a shifted version of the $F_{1,n-2}$ distribution. The noncentrality parameter is calculated as:

$$\delta = \frac{\hat{\beta}_1^2 \sum_{i=1}^{n}(x_i - \bar{x})^2}{\text{MSE}}.$$

If $\beta_1^*$ is a number other than 0 in the hypothesis test, then we would replace $\hat{\beta}_1$ by $(\hat{\beta}_1 - \beta_1^*)$ in the equation for $\delta$. Finally, power is simply the probability that the calculated $F_{1,n-2;1-\alpha}(\delta)$ value is greater than the calculated $F_{1,n-2;1-\alpha}$ value under the $F_{1,n-2}(\delta)$ distribution. Details on these power calculations are found in Dupont and Plummer, Jr. (1998).

We can also test

$$H_0 : \beta_0 = \beta_0^*$$
$$H_A : \beta_0 \neq \beta_0^*,$$

where $\beta_0^*$ is any real number. We can find the power in a similar manner, but where the noncentrality parameter is defined as:

$$\delta = \frac{(b_0 - \beta_0^*)^2 \sum_{i=1}^{n}(x_i - \bar{x})^2}{\text{MSE}}.$$

Luckily, statistical software will calculate these quantities for us!

There is also a relationship between the $t$ distribution and $F$ distribution which can be exploited here. If the random variable $Z$ is distributed as $t_n$, then $Z^2$ is distributed as $F_{1,n}$. Because of this relationship, we can calculate the power in a similar manner using the $t$ distribution and noncentral $t$ distribution. However, the noncentral $t$ distribution has noncentrality parameter $\delta^* = \sqrt{\delta}$, such that $\delta$ is as defined above.

## 3.3    Inference on the Correlation Model

Earlier we introduced the (Pearson) correlation coefficient $\rho$. Now we will formally develop this quantity in the context of a probability model. Suppose that we no longer have the $X$ values at fixed constants. Instead we have two random variables, say, $Y_1$ and $Y_2$. For example, these random variables could be the height and weight of a person or they could be the daily temperature and humidity levels. The objective is to develop a formal probability model to perform inferences on the correlation coefficient.

The **correlation model** assumes that the random variables $Y_1$ and $Y_2$ are jointly normally distributed. Their joint distribution is the bivariate normal distribution:

$$f_{Y_1,Y_2}(y_1, y_2) = \frac{1}{2\pi\sigma_1\sigma_2\sqrt{1-\rho^2}} \exp\left\{ -\frac{1}{2(1-\rho^2)}\left[ \left(\frac{y_1 - \mu_1}{\sigma_1}\right)^2 \right. \right.$$
$$\left. \left. - 2\rho\left(\frac{y_1 - \mu_1}{\sigma_1}\right)\left(\frac{y_2 - \mu_2}{\sigma_2}\right) + \left(\frac{y_2 - \mu_2}{\sigma_2}\right)^2 \right] \right\},$$

where $-\infty < \mu_j < +\infty$ and $\sigma_j > 0$ are the mean and standard deviation, respectively, of $Y_i$, for $i = 1, 2$, and $-1 \leq \rho \leq +1$ is the coefficient of correlation between $Y_1$ and $Y_2$. Note that if $Y_1$ and $Y_2$ are jointly normally distributed, then you can integrate out the other variable in the bivariate normal distribution above, which will yield the marginal distributions of $Y_1$ and $Y_2$. These marginal distributions are both normal with the respective parameters stated above.

When the population is bivariate normal, it is often of interest to test the following hypothesis:

$$H_0 : \rho = 0$$
$$H_A : \rho \neq 0.$$

This test is of interest because if $\rho = 0$, then $Y_1$ and $Y_2$ are independent of each other. The test statistic for this test is

$$t^* = \frac{r\sqrt{n-2}}{\sqrt{1-r^2}},$$

which is distributed according to a $t_{n-2}$ distribution. Thus we can carry out inference just the way we described testing for the regression parameters.

A confidence interval for $\rho$, however, is more complicated since the sampling distribution of $r$ is complex; see Hotelling (1953) for a detailed derivation of this distribution. We proceed to conduct inference using **Fisher's z-transformation**:

$$z = \frac{1}{2}\ln\left(\frac{1+r}{1-r}\right) = \text{arctanh}(r).$$

When $n \geq 25$, then $z$ is approximately normally distributed with mean $\zeta = \text{arctanh}(\rho)$ and standard error $\left(\sqrt{n-3}\right)^{-1}$. Thus, the standardized statistic

$$\sqrt{n-3}\,(z - \text{arctanh}(\rho))$$

is approximately a standard normal random variable. This yields the following (approximate) $100 \times (1-\alpha)\%$ confidence interval for $\zeta$:

$$z' \pm z^*_{1-\alpha/2}\sqrt{\frac{1}{n-3}},$$

where $z^*_{1-\alpha/2}$ is the $100 \times (1 - \alpha/2)^{\text{th}}$ percentile of the standard normal distribution. We can then transform these limits back to limits for $\rho$ using the definition of $\rho = \tanh(\varsigma)$. Then, the interpretation of this interval would be similar to how we have interpreted other confidence intervals.

Note that we have the stipulation that $n \geq 25$. The above approximation can be used for $n < 25$, but you will likely have very wide intervals. In addition to the exact procedures discussed in Hotelling (1953), there are also approximate procedures for calculating these intervals, such as by simulation (Tuğran et al., 2015).

## 3.4 Intervals for a Mean Response

**Confidence Intervals**
A $100 \times (1 - \alpha)\%$ confidence interval for $E(Y|X = x_h)$ (i.e., the mean of the random variable $Y$) estimates the mean value of $Y$ for individuals that have the particular value of $x_h$. In other words, this confidence interval estimates the location of the line at the specific location $x_h$.

The corresponding fit at the specified level $x_h$ is $\hat{y}_h = b_0 + b_1 x_h$. The standard error of the fit at $x_h$ is given by the formula:

$$\text{s.e.}(\hat{y}_h) = \sqrt{\text{MSE}\left(\frac{1}{n} + \frac{(x_h - \bar{x})^2}{\sum_{i=1}^{n}(x_i - \bar{x})^2}\right)}. \tag{3.3}$$

Formula (3.3) applies only to simple linear regression. The answer describes the accuracy of a particular $\hat{y}_h$ as an estimate of $E(Y|X = x_h)$. It is also important to note that the subscript $i$ is associated with the index of the observed dataset, while the subscript $h$ is used to denote *any* possible level of $x$. Thus, a $100 \times (1 - \alpha)\%$ confidence interval for $E(Y)$ at $x_h$ is calculated by

$$\hat{y}_h \pm t^*_{n-2;1-\alpha/2}\text{s.e.}(\hat{y}_h). \tag{3.4}$$

Note that the equivalent formula to (3.3) for the regression through the origin setting is

$$\text{s.e.}(\hat{y}_h) = \sqrt{\text{MSE}\frac{x_h^2}{\sum_{i=1}^{n} x_i^2}}. \tag{3.5}$$

Any of the statistical intervals we discuss can be calculated for the regression through the origin setting by using (3.5) in place of (3.3).

**Bonferroni Joint Confidence Intervals**
**Bonferroni** $100 \times (1 - \alpha)\%$ **joint confidence intervals** for $E(Y|X = x_h)$ at $q$ different values of $x_h$ (i.e., $x_{h_1}, x_{h_2}, \ldots, x_{h_q}$) estimate the mean value of $Y$ for individuals with these $q$ different values of $x$. In other words, the

confidence intervals estimate the location of the line at $q$ specific locations of $x$. Bonferroni $100 \times (1 - \alpha)\%$ joint confidence intervals for $E(Y)$ at the values $x_{h_1}, x_{h_2}, \ldots, x_{h_q}$ are calculated by

$$\hat{y}_{h_i} \pm t^*_{n-2;1-\alpha/(2q)} \text{s.e.}(\hat{y}_{h_i}), \tag{3.6}$$

where $i = 1, 2, \ldots, q$ and the percentile of the $t^*$ multiplier has been adjusted by $q$.

One note about terminology. When we construct joint statistical intervals like those in (3.6), such intervals are sometimes called familywise or simultaneous intervals. When we construct individual statistical intervals like those in (3.4), such intervals are called pointwise intervals in a regression setting.

**Working–Hotelling Confidence Bands**
A **Working–Hotelling** $100 \times (1 - \alpha)\%$ **confidence band** for $E(Y)$ at all possible values of $X = x_h$ estimates the mean value of $Y$ for all different values of $x_h$. This confidence band contains the entire regression line (for all values of $X$) with confidence level $1 - \alpha$. A Working–Hotelling $100 \times (1 - \alpha)\%$ confidence band for $E(Y)$ at all possible values of $X = x_h$ is calculated by

$$\hat{y}_h \pm \sqrt{2F^*_{2,n-2;1-\alpha}} \text{s.e.}(\hat{y}_h), \tag{3.7}$$

where $F^*_{2,n-2;1-\alpha}$ is the multiplier from a $F_{2,n-2}$ distribution.

## 3.5 Intervals for a New Observation

**Prediction Intervals**
A $100 \times (1 - \alpha)\%$ **prediction interval** for $Y = y$ estimates the value of $y$ for an individual observation with a particular value of $x$. Equivalently, a prediction interval estimates the range of a future value for a response variable at a particular value of $x$ given a specified confidence level. Since our data is just one sample, it is reasonable to consider taking another sample, which would likely yield different values for $\hat{\beta}_0$ and $\hat{\beta}_1$. This illustrates the sampling variability in the slope and intercept of the regression line as well as the variability of the observations about the regression line. Since it will not always be feasible to get another sample of data, a prediction interval will allow us to assess the limits for the response at any given value of $x$; i.e., $x_h$. You can think of a prediction interval as pertaining to some characteristic of interest in a *future* sample of data from the population.

A prediction interval for $Y = y_h$ at $x_h$ is calculated as

$$\hat{y}_h \pm t^*_{n-2;1-\alpha/2}\sqrt{\text{MSE} + [\text{s.e.}(\hat{y}_h)]^2}. \tag{3.8}$$

If you look at the formulas for both the confidence intervals (3.4) and prediction intervals (3.8), you will see that a prediction interval will always be wider than a confidence interval at the same $\alpha$-level. This is because a prediction interval considers a larger source of possible variation than do confidence intervals. Confidence intervals generally bound the estimate of a true parameter (e.g., a mean) around an average value. Thus, you are bounding a statistic and NOT an individual value.

It is also possible to construct a $100 \times (1 - \alpha)\%$ prediction interval for the mean $\bar{Y} = \bar{y}_h$ when $m$ new observations are drawn at the level $x_h$. These are calculated as

$$\hat{y}_h \pm t^*_{n-2;1-\alpha/2}\sqrt{\text{MSE}/m + [\text{s.e.}(\hat{y}_h)]^2}. \tag{3.9}$$

Note that the case of $m = 1$ reduces to the original prediction interval in (3.8).

**Bonferroni Joint Prediction Intervals**
**Bonferroni** $100 \times (1 - \alpha)\%$ **joint prediction intervals** for $Y = y$ estimate the values of $y$ for $q$ different values of $x$. Equivalently, these intervals estimate the range of a future value for a response variable at a particular value of $x$ given a specified confidence level. Bonferroni $100 \times (1 - \alpha)\%$ joint prediction intervals for $Y = y_{h_i}$ at $x_{h_i}$ for $i = 1, 2, \ldots, q$ are calculated as

$$\hat{y}_{h_i} \pm t^*_{n-2;1-\alpha/(2q)}\sqrt{\text{MSE} + [\text{s.e.}(\hat{y}_{h_i})]^2}. \tag{3.10}$$

**Scheffé Joint Prediction Intervals**
**Scheffé** $100 \times (1 - \alpha)\%$ **joint prediction intervals** for $Y = y$ estimate the values of $y$ for $q$ different values of $x$. Equivalently, these intervals estimate the range a future value for a response variable at a particular value of $x$ given a specified confidence level. Scheffé joint $100 \times (1 - \alpha)\%$ prediction intervals for $Y = y_{h_i}$ at $x_{h_i}$ for $i = 1, 2, \ldots, q$ are calculated as

$$\hat{y}_{h_i} \pm \sqrt{qF^*_{q,n-2;1-\alpha}(\text{MSE} + [\text{s.e.}(\hat{y}_{h_i})]^2)}. \tag{3.11}$$

Scheffé joint prediction intervals accomplish the same goal as Bonferroni joint prediction intervals, but they have a slightly different formula. Regarding which interval to choose, you would want the procedure that produces the tighter interval at the same confidence level.

**Pointwise Tolerance Intervals**
A $[100 \times (1 - \alpha)\%]/[100 \times P\%]$ **tolerance interval** for $y$ estimates the value of $y$ for an individual with a particular predictor value $x_h$. Equivalently, for a given predictor value of $x_h$, tolerance intervals capture a specified proportion of individual values $(P)$ in the entire population with a given degree of confidence $(1 - \alpha)$. As we can see, tolerance intervals, like prediction intervals, are concerned with probabilistic bounds of individual data points rather than of statistics.

To construct a one-sided tolerance interval for $y$ at $x_h$, first set the values of $\alpha$ and $P$. Next, calculate

$$
n^* = \left[ \frac{1}{n} + \frac{(x_h - \bar{x})^2}{\sum_{i=1}^{n}(x_i - \bar{x})^2} \right]^{-1}
$$
$$
= \frac{\text{MSE}}{\text{s.e.}(\hat{y}_h)^2}.
$$

Finally, find $k_1$ (which is called a $k$-**factor**) such that:

$$
\text{P}(\sqrt{n^*} k_1 \leq t^*_{n-2;1-\alpha}(\delta)) = 1 - \alpha,
$$

where $t^*_{n-2;1-\alpha}(\delta)$ is from a noncentral $t$ distribution with $n - 2$ degrees of freedom and noncentrality parameter $\delta = \sqrt{n^*} z_P^*$.[3] Again, $z_P^*$ is the $100 \times P^{\text{th}}$ percentile from a standard normal distribution. Finally, the upper and lower one-sided tolerance intervals are calculated as

$$
(-\infty, \hat{y}_h + k_1 \sqrt{\text{MSE}}) \tag{3.12}
$$

and

$$
(\hat{y}_h - k_1 \sqrt{\text{MSE}}, +\infty), \tag{3.13}
$$

respectively.

A two-sided tolerance interval is given by

$$
\hat{y}_h \pm k_2 \sqrt{\text{MSE}}. \tag{3.14}
$$

For two-sided tolerance intervals, it is incorrect to take the intersection of the two one-sided tolerance intervals just presented, although sometimes it yields a decent approximation. $k_2$ is the solution to the integral equation

$$
\sqrt{\frac{2}{\pi n^*}} \int_0^{\infty} \text{P}\left( \chi^2_{n-2} > \frac{(n-2)\chi^2_{1;P}(z^2)}{k_2^2} \right) e^{-\frac{z^2}{2n^*}} dz = 1 - \alpha, \tag{3.15}
$$

where $\chi^2_{1;P}$ and $\chi^2_{1;P}(z^2)$ are the $100 \times P^{\text{th}}$ percentile, respectively, of a chi-square distribution with 1 degree of freedom and a noncentral chi-square distribution with 1 degree of freedom and noncentrality parameter $z^2$. Due to the complicated form of (3.15), $k_2$ is often approximated with the fairly accurate formula due to Wallis (1946):

$$
k_2^* = \sqrt{\frac{(n-2)\chi^2_{1;P}(1/n^*_h)}{\chi^2_{n-2;\alpha}}}, \tag{3.16}
$$

Thus, we would use $k_2^*$ in (3.14) for ease of computation, which would make

---

[3] Analogous to the discussion about the noncentral $F$ distribution, a noncentral $t$ distribution is basically a $t$ distribution shifted from 0 by its noncentrality parameter.

it an *approximate* (yet fairly accurate) $[100 \times (1 - \alpha)\%]/[100 \times P\%]$ tolerance interval. More details on regression tolerance intervals are given in Chapter 3 of Krishnamoorthy and Mathew (2009) and Young (2013).

### Calibration and Regulation Intervals

Consider the calibration of a venturi that measures the flow rate of a chemical process. Let $X$ denote the actual flow rate and $Y$ denote a reading on the venturi. In this calibration study, the flow rate is controlled at $n$ levels of $X$. Then, the corresponding $Y$ readings on the venturi are observed. Suppose we assume the simple linear regression model with the standard assumptions on the error terms. In the future, the experimenter may be interested in estimating the flow rate from a particular venturi reading.

The method of **calibration** (or **inverse regression**) involves predicting a new value of a predictor that produces a given response value. Suppose we observe a series of $m$ responses $y_{h_1}, \ldots, y_{h_m}$, which are associated with the same, but unknown, predictor value $x_h$. Conditioning the simple linear regression model on $Y = \bar{y}_h = \sum_{j=1}^{m} y_{h_j}$ (i.e., the *given* mean of the response values), the predicted value for $X = x_h$ (i.e., the *unknown* predictor value) is

$$\hat{x}_h = \frac{\bar{y}_h - \hat{\beta}_0}{\hat{\beta}_1},$$

where $\hat{\beta}_1 \neq 0$. Calibration is concerned with finding an interval for the value $X = x_h$.

Basically, estimation and construction of intervals in the regression setting assume the predictor values are fixed while the response is random. However, calibration switches these roles. What results is that the statistical intervals for calibration need to be solved numerically, but various formulas for approximate intervals exist in the literature. The theory behind statistical calibration is developed in Scheffé (1973).

An exact $100 \times (1 - \alpha)\%$ **calibration confidence interval** for $X = x_h$ (see Draper and Smith, 1998) is calculated as

$$\frac{(\hat{x}_h - g\bar{x}) \pm t^*_{n+m-3, 1-\alpha/2} \sqrt{\frac{\text{MSE}}{\hat{\beta}_1^2} \left[ (1 - g) \left( \frac{1}{m} + \frac{1}{n} \right) + \frac{(\hat{x}_h - \bar{x})^2}{S_{xx}} \right]}}{1 - g}, \tag{3.17}$$

where

$$g = \frac{\text{MSE}(t^*_{n-2; 1-\alpha/2})^2}{\hat{\beta}_1^2 S_{xx}}.$$

Notice that the interval in (3.17) is not centered at $\hat{x}_h$, partly because the quantity which describes the "standard error" of $\hat{x}_h$ is also a function of $\hat{x}_h$. Compare this to the traditional confidence and prediction intervals where the standard error of $\hat{y}_h$ does not depend on $\hat{y}_h$. However, estimates for calibration intervals which are centered at $\hat{x}_h$ can also be used, but do not have as much theoretical development as the intervals presented above. One approach is to use a large-sample approach as discussed in Greenwell and Schubert-Kabban (2014). The formulas for this large-sample centered calibration interval is

$$\hat{x}_h \pm t^*_{n+m-3;1-\alpha/2} \text{s.e.}(\hat{x}_h), \tag{3.18}$$

where

$$\text{s.e.}(\hat{x}_h) = \sqrt{\frac{\text{MSE}}{\hat{\beta}_1^2}\left[\frac{1}{m} + \frac{1}{n} + \frac{(\hat{x}_h - \bar{x})^2}{S_{xx}}\right]}.$$

The above is equivalent to setting $g = 0$ in (3.17).

In calibration, we infer the value of $x_h$ that corresponds to an observed value (or mean of observed values) for the response $y_h$ (or $\bar{y}_h$). Suppose we want to infer the value of $x_h$ for a specified value of the mean response, which we treat as a known parameter; i.e., $\mu_h$. Then,

$$\hat{x}_h = \frac{\mu_h - \hat{\beta}_0}{\hat{\beta}_1}$$

and the resulting interval, which is called a **regulation interval** (see Graybill and Iyer, 1994), is

$$\frac{(\hat{x}_h - g\bar{x}) \pm t^*_{n-2,1-\alpha/2}\sqrt{\frac{\text{MSE}}{\hat{\beta}_1^2}\left[\frac{(1-g)}{n} + \frac{(\hat{x}_h - \bar{x})^2}{S_{xx}}\right]}}{1-g}, \tag{3.19}$$

which is the same formula as (3.17), but with $1/m$ set to 0 and the degrees of freedom changed from $n + m - 3$ to $n - 2$. Greenwell and Schubert-Kabban (2014), analogously, established the large-sample centered regulation interval,

$$\hat{x}_h \pm t^*_{n-2;1-\alpha/2} \text{s.e.}(\hat{x}_h), \tag{3.20}$$

where

$$\text{s.e.}(\hat{x}_h) = \sqrt{\frac{\text{MSE}}{\hat{\beta}_1^2}\left[\frac{1}{n} + \frac{(\hat{x}_h - \bar{x})^2}{S_{xx}}\right]}.$$

A couple of final notes. First is that the relationship between calibration and regulation intervals is akin to the relationship between prediction and confidence intervals, respectively. Second is that the method for constructing tolerance intervals can actually be inverted to provide an estimate for calibration intervals. See Chapter 3 of Krishnamoorthy and Mathew (2009) for details.

## 3.6 Examples

**Example 3.6.1.** *(Steam Output Data, cont'd)*
Recall for the steam output data in Example 2.8.2 that we obtained the following output for the simple linear regression fit:

```
############################
Coefficients:
            Estimate Std. Error t value Pr(>|t|)
(Intercept) 13.62299    0.58146  23.429  < 2e-16 ***
temp        -0.07983    0.01052  -7.586 1.05e-07 ***
---
Signif. codes:  0 *** 0.001 ** 0.01 * 0.05 . 0.1   1

Residual standard error: 0.8901 on 23 degrees of freedom
Multiple R-squared:  0.7144,Adjusted R-squared:  0.702
F-statistic: 57.54 on 1 and 23 DF,  p-value: 1.055e-07
############################
```

In the above output, the standard errors for the estimated regression parameters are s.e.$(\hat{\beta}_0) = 0.582$ and s.e.$(\hat{\beta}_1) = 0.011$. These measure, roughly, the average difference between the estimates and the unknown population quantities.

To test if the pounds of steam used per month is linearly related to the average atmospheric temperature, we use the hypothesis test about the slope:

$$
\begin{aligned}
H_0 &: \beta_1 = 0 \\
H_A &: \beta_1 \neq 0.
\end{aligned}
\tag{3.21}
$$

The test statistic for this test is $t^* = -7.586$ and has a $p$-value of $1.05 \times 10^{-7}$. Using a significance level of $\alpha = 0.05$, there is statistically significant evidence that the steam used per month is linearly related to the average atmospheric temperature; i.e., that $\beta_1$ is significantly different from 0. The power of this test, as well as for the test of $\beta_0 = 0$, is given below:

```
############################
                ncp       power
Intercept 1.675783e+06 1.0000000
Slope        5.754279e+01 0.9999999
############################
```

Clearly, both tests are very powerful since there is both a relatively small MSE and a relatively large population effect to detect. Note that if we were to drastically reduce the population effect, for example, to test $\beta_0 = 13.6$ and $\beta_1 = -0.05$, we would get much smaller power values for our test:

```
##############################
            ncp       power
Intercept 4.772250 0.5528894
Slope     8.034176 0.7747259
##############################
```

In this example, $n = 25$ and $df = n - 2 = 23$. The multiplier for a 95% confidence interval is $t^*_{23;0.975} = 2.0687$. Thus, a 95% confidence interval for $\beta_1$, the true population slope, is:

$$-0.0798 \pm (2.0687 \times 0.0105) \Rightarrow (-0.1016, -0.0581).$$

The 95% confidence interval does not include 0, which matches the result of the corresponding hypothesis test at $\alpha = 0.05$. The interpretation is that with 95% confidence, we can say the mean steam used per month decreases somewhere between 0.0581 and 0.1016 pounds per each degree increase in temperature. Remember, it is incorrect to say with 95% probability that the mean steam used per month decreases somewhere between these two limits!

For Bonferroni 95% joint confidence intervals, the multiplier is $t_{23;1-0.05/4} = t_{23;0.9875} = 2.3979$, where the joint confidence intervals for $\beta_0$ and $\beta_1$ are, respectively,

$$13.6230 \pm (2.3979 \times 0.5815) \Rightarrow (12.2287, 15.0173)$$

and

$$-0.0798 \pm (2.3979 \times 0.0105) \Rightarrow (-0.1051, -0.0546).$$

The interpretation is that with 95% joint confidence, we can say that the true population intercept and slope terms are in the intervals $(12.2287, 15.0173)$ and $(-0.1051, -0.0546)$, respectively.

### TABLE 3.1

Statistical intervals for the given levels of $x$, where $q = 2$ for the joint intervals

| | $x_h$ | 40 | 50 |
|---|---|---|---|
| | $\hat{y}_h$ | 10.4298 | 9.6316 |
| 95% CI | | $(9.9706, 10.8890)$ | $(9.2590, 10.0042)$ |
| Bonferroni 95% CI | | $(9.8976, 10.9621)$ | $(9.1997, 10.0634)$ |
| Working–Hotelling 95% CB | | $(9.8491, 11.0106)$ | $(9.1603, 10.1028)$ |
| 95% PI | | $(8.5321, 12.3276)$ | $(7.7529, 11.5102)$ |
| Bonferroni 95% PI | | $(8.2301, 12.6296)$ | $(7.4539, 11.8092)$ |
| Scheffé 95% PI | | $(8.0299, 12.8299)$ | $(7.2557, 12.0075)$ |
| Upper 95%/99% TI | | $(-\infty, 13.2998)$ | $(-\infty, 12.4617)$ |
| Lower 95%/99% TI | | $(7.5599, +\infty)$ | $(6.8014, +\infty)$ |
| 95%/99% TI | | $(7.3010, 13.5587)$ | $(6.5323, 12.7308)$ |

Suppose we wish to calculate 95% confidence intervals, 95% prediction intervals, and 95%/99% tolerance intervals for the values of $x_{h_1} = 40$ and $x_{h_2} = 50$. Table 3.1 provides the various intervals from our earlier discussion. The interpretations of these intervals are as follows:

- **95% CI:** With 95% confidence, the mean steam usage per month is between 9.9706 and 10.8890 for an atmospheric temperature of 40°F. With 95% confidence, the mean steam usage per month is between 9.2590 and 10.0042 for an atmospheric temperature of 50°F.

- **Bonferroni 95% CI:** With 95% joint confidence, the mean steam usage per month is between 9.8976 and 10.9621 and between 9.1997 and 10.0634 for atmospheric temperatures of 40°F and 50°F, respectively.

- **Working–Hotelling 95% CB:** With 95% joint confidence, the mean steam usage per month is between 9.8491 and 11.0106 and between 9.1603 and 10.1028 for atmospheric temperatures of 40°F and 50°F, respectively. Note that this is the same interpretation as the Bonferroni confidence intervals, but the Working-Hotelling limits are part of a region (specifically, a hyperbolic region) in which the entire regression line lies.

- **95% PI:** With 95% confidence, the mean steam usage per month is between 8.5321 and 12.3276 for a future observation with an atmospheric temperature of 40°F. With 95% confidence, the mean steam usage per month is between 7.7529 and 11.5102 for a future observation with an atmospheric temperature of 50°F. Notice that these are wider than the respective 95% confidence intervals.

- **Bonferroni 95% PI:** With 95% joint confidence, the mean steam usage per month is between 8.2301 and 12.6296 and between 7.4539 and 11.8092 for future observations with atmospheric temperatures of 40°F and 50°F, respectively. Again, notice that these are wider than the respective Bonferroni 95% confidence intervals.

- **Scheffé 95% PI:** With 95% joint confidence, the mean steam usage per month is between 8.0299 and 12.8299 and between 7.2557 and 12.0075 for future observations with atmospheric temperatures of 40°F and 50°F, respectively. These have the same interpretation as the Bonferroni 95% prediction intervals, but we would report the results from the procedure that yields the tighter set of intervals. In this case, the Bonferroni procedure is better.

- **Upper 95%/99% TI:** With 95% confidence, we expect 99% of all measurements taken at an atmospheric temperature of 40°F to be less than 13.2998. With 95% confidence, we expect 99% of all measurements taken at an atmospheric temperature of 50°F to be less than 12.4617.

- **Lower 95%/99% TI:** With 95% confidence, we expect 99% of all measurements taken at an atmospheric temperature of 40°F to be greater than

7.5599. With 95% confidence, we expect 99% of all measurements taken at an atmospheric temperature of 50°F to be greater than 6.8014.

- **95%/99% TI:** With 95% confidence, we expect 99% of all measurements taken at an atmospheric temperature of 40°F to be between 7.3010 and 13.5587. With 95% confidence, we expect 99% of all measurements taken at an atmospheric temperature of 50°F to be between 6.5323 and 12.7308.

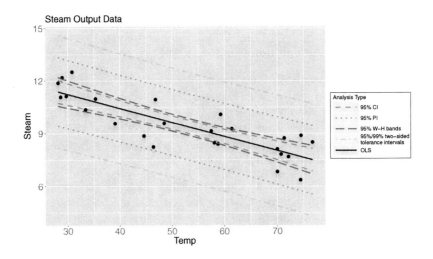

### FIGURE 3.2

Scatterplot of the steam output data with the ordinary least squares fit (solid blue line), pointwise 95% confidence intervals (dashed gold lines), pointwise 95% prediction intervals (dotted light blue lines), 95% Working–Hotelling bands (dashed green lines), and pointwise 95%/99% two-sided tolerance intervals (dotted-dashed yellow lines). Colors are in e-book version.

Finally, Figure 3.2 shows some of the statistical intervals calculated for this problem overlaid on a plot of the steam output dataset. The statistical intervals are centered at the estimated regression line given by the solid blue line. Then, the intervals plotted, in increasing width, are the pointwise 95% confidence intervals, 95% Working–Hotelling bands, pointwise 95% prediction intervals, and pointwise 95%/99% tolerance intervals.

**Example 3.6.2.** *(Computer Repair Data, cont'd)*
Let us justify the use of a regression through the origin model in Example 2.8.3. First, fit a simple linear regression model:

```
##############################
Coefficients:
            Estimate Std. Error t value Pr(>|t|)
```

```
(Intercept)     4.162      3.355    1.24    0.239
units          15.509      0.505   30.71 8.92e-13 ***
---
Signif. codes:  0 *** 0.001 ** 0.01 * 0.05 . 0.1   1

Residual standard error: 5.392 on 12 degrees of freedom
Multiple R-squared:  0.9874,Adjusted R-squared:  0.9864
F-statistic: 943.2 on 1 and 12 DF,  p-value: 8.916e-13
############################
```

Next, use the hypothesis test about the intercept:

$$H_0 : \beta_0 = 0$$
$$H_A : \beta_0 \neq 0. \tag{3.22}$$

The test statistic for this test is $t^* = 1.24$ and has a $p$-value of 0.239. Using a significance level of $\alpha = 0.05$, we would claim that the intercept term is not significantly different from 0.

Using the regression through the origin fit in Example 2.8.3, we can also calculate a 95% confidence interval for the slope. Using that fit, MSE $= 0.2213$ and the $t$-multiplier is $t^*_{13;0.975} = 2.1604$. Therefore, the 95% confidence interval for the slope is $(15.5963, 16.5526)$.

**Example 3.6.3.** *(Pulp Property Data)*
Santos et al. (2012) studied the pulp properties of wood density of the Australian blackwood tree. Two of the variables studied were:

- $Y$: the percentage of pulp yield

- $X$: the Kappa number, which is a measurement of standard potassium permanganate solution that the pulp will consume

The researchers had one sample of size $n = 7$ and the data are given in Figure 3.3. We are interested in determining if there are relationships between the percentage of pulp yield and the Kappa number.

The estimate and test for the Pearson correlation $\rho$ is given below:

```
############################
Pearson's product-moment correlation

data:  pulp and Kappa
t = -3.7183, df = 5, p-value = 0.01374
alternative hypothesis: true correlation is not equal to 0
95 percent confidence interval:
 -0.9785339 -0.2930078
sample estimates:
      cor
-0.8569731
############################
```

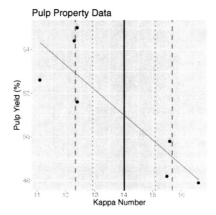

**FIGURE 3.3**
Scatterplot of the pulp property data along with the ordinary least squares fit, inverted $x_h$ value (solid black line), the centered 90% calibration interval (dashed red lines), and the centered 90% regulation interval (dotted green lines). Colors are in e-book version.

The observed correlation between the percentage of pulp yield and the Kappa number is $r = -0.8570$, which indicates a strong, negative relationship between the two variables. The test statistic is $t^* = -3.7183$, which has a $p$-value of 0.0137. Thus, there is a statistically significant correlation between the two variables. The confidence interval reported above is calculated using the Fisher's $z$-transformation, where $z = \operatorname{arctanh}(r) = -1.2818$. Thus, we are 95% confident that the true correlation between the percentage of pulp yield and the Kappa number is between $-0.9785$ and $-0.2930$.

Next, we fit a simple linear regression model to these data:

```
#############################
Coefficients:
             Estimate Std. Error t value Pr(>|t|)
(Intercept)  66.9850     4.2465   15.774 1.86e-05 ***
Kappa        -1.1407     0.3068   -3.718   0.0137 *
---
Signif. codes:  0 *** 0.001 ** 0.01 * 0.05 . 0.1   1

Residual standard error: 1.605 on 5 degrees of freedom
Multiple R-squared:  0.7344,Adjusted R-squared:  0.6813
F-statistic: 13.83 on 1 and 5 DF,  p-value: 0.01374
#############################
```

Thus, the regression line fitted to these data is:

$$\hat{y}_i = 66.9850 - 1.1407x_i,$$

**TABLE 3.2**

Calibration and regulation intervals for the pulp property data

|  | Calibration | Regulation |
|---|---|---|
| Uncentered | $(12.2046, 15.9923)$ | $(12.8460, 15.4405)$ |
| Centered | $(12.3361, 15.6901)$ | $(12.9283, 15.0978)$ |

which is overlaid on the scatterplot in Figure 3.3.

Suppose we want to predict the Kappa number for a tree with a sample of pulp percentages of $y_{h_1} = 51$, $y_{h_2} = 50$, and $y_{h_3} = 52$, along with a 90% calibration interval. In this case, $\bar{y}_h = 51$. Suppose that we are also interested in a 90% regulation interval with $\mu_h = 51$. The exact and centered versions of both types of intervals are summarized in Table 3.2.

Let us interpret the centered versions of both intervals (the uncentered versions are interpreted analogously):

- **Calibration:** With 90% confidence, the true Kappa number $x_h$ corresponding to the pulp yield percentages of 51, 50, and 52 is between 12.3361 and 15.6901. See the dashed red lines in Figure 3.3.

- **Regulation:** With 90% confidence, the true Kappa number $x_h$ corresponding to a mean pulp yield percentage of $\mu_h = 51$ is between 12.9283 and 15.0978. See the dotted green lines in Figure 3.3.

By construction, the point estimate is $\hat{x}_h = 14.0131$ for both sets of intervals calculated above. See the solid black line in Figure 3.3.

# 4

# Regression Assumptions and Residual Diagnostics

When we fit a simple linear regression model to our sample data and use the estimated model to make predictions and statistical inferences about a larger population, we make several assumptions that may or may not be correct for the data at hand. The key theoretical assumptions about the linear regression model are:

1. **Linearity.** The equation that is used for the connection between the expected value of the $Y$ (dependent) variable and the different levels of the $X$ (independent) variable describes the actual pattern of the data. In other words, we use a straight-line equation because we assume the "average" pattern in the data is indeed linear.

2. **Normality.** The errors are normally distributed with a mean of 0.

3. **Homoskedasticity.** The errors have the same theoretical variance, $\sigma^2$, regardless of the values of $X$ and, thus, regardless of the expected value of $Y$. For a straight line, this means that the vertical variation of data points around the line has about the same magnitude everywhere.

4. **Independence.** The errors are independent of each other (i.e., that they are a random sample) and are independent of any time order in the data. A discussion assessing dependency requires an introduction to time series analysis, which is presented in Chapter 13.

This chapter presents various regression diagnostic procedures to assess these assumptions. Figure 4.1 shows how all of these assumptions translate to the simple linear regression setting. The values for the dependent variable are distributed normally, with the mean value falling on the regression line. Moreover, the same standard deviation holds at all values of the independent variable. Thus, the distribution curves (shown in red at three different values of the independent variable) all look the same, but they are just translated along the $x$-axis based on the regression relationship. Note that the only assumption that is not visualized here is the assumption of independence, which will usually be satisfied if there is no temporal component and the experimenter did a proper job of designing the experiment.

It is standard to write for the assumptions about the errors that the $\epsilon_i$ are

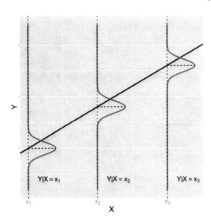

**FIGURE 4.1**
This figure shows data fit with a simple linear regression line. The distribution of the errors is shown at three different values of the predictor.

*iid* $\mathcal{N}(0, \sigma^2)$, where *iid* stands for independent and identically distributed, and $\mathcal{N}(0, \sigma^2)$ is a normal distribution with mean 0 and variance $\sigma^2$. Also, these assumptions play a role when we discuss **robustness**, which is the ability to produce estimators not heavily affected by small departures from the model assumptions.

There is one thing to keep in mind when assessing regression assumptions — it is just as much an art as it is a science. When dealing with real datasets, it is somewhat subjective as to what you are willing to claim as far as satisfying a certain assumption. If you look at datasets in textbooks (including this one), they are often chosen because they provide ideal plots or can clearly demonstrate some concept. However, real datasets encountered in practice will require greater scrutiny. It is important to be able to justify your decisions based on the statistical tools you have at your disposal.

## 4.1 Consequences of Invalid Assumptions

- **Linearity.** Using the wrong equation (such as using a straight line for curved data) is very serious. Predicted values will be wrong in a biased manner, meaning that predicted values will systematically miss the true pattern of the expected value of $Y$ as related to $X$.

- **Normality.** If the errors do not have a normal distribution, it usually is not particularly serious. Simulation results have shown that regression inferences tend to be robust with respect to normality (or nonnormality of the errors). In practice, the residuals may appear to be nonnormal when the wrong regression equation has been used. With that stated, if the error distribution is significantly nonnormal, then inferential procedures can be very misleading; e.g., statistical intervals may be too wide or too narrow. It is not possible to check the assumption that the overall mean of the errors is equal to 0 because the least squares process causes the residuals to sum to 0. However, if the wrong equation is used and the predicted values are biased, the sample residuals will be patterned so that they may not average 0 at specific values of $X$.

- **Homoskedasticity.** The principal consequence of nonconstant variance — i.e., where the variance is not the same at each level of $X$ — are prediction intervals for individual $Y$ values will be wrong because they are determined assuming constant variance. There is a small effect on the validity of $t$-test and $F$-test results, but generally regression inferences are robust with regard to the variance issue.

- **Independence.** As noted earlier, this assumption will usually be satisfied if there is no temporal component and the experimenter did a proper job of designing the experiment. However, if there is a time-ordering of the observations, then there is the possibility of correlation between consecutive errors. If present, then this is indicative of a badly misspecified model and any subsequent inference procedures will be severely misleading.

## 4.2    Diagnosing Validity of Assumptions

Visualization plots and diagnostic measures used to detect types of disagreement between the observed data and an assumed regression model have a long history. Central to many of these methods are residuals based on the fitted model. As will be shown, there are more than just the raw residuals; i.e., the difference between the observed values and the fitted values. We briefly outline the types of strategies that can be used to diagnose validity of each of the four assumptions. We will then discuss these procedures in greater detail. More procedures are presented throughout this text, but appear under topics with which they are more aligned. Deeper treatment of visualization plots and diagnostic measures in regression is given in Cook and Weisberg (1982, 1994).

*Diagnosing Whether the Right Type of Equation Was Used*

1. Examine a **plot of residuals versus fitted values** (predicted values). A curved pattern for the residuals versus fitted values plot indicates that the wrong type of equation has been used.

2. Use **goodness-of-fit** measures. For example, $R^2$ can be used as a rough goodness-of-fit measure, but by no means should be used to solely determine the appropriateness of the model fit. The plot of residuals versus fitted values can also be used for assessing goodness-of-fit. Other measures, like model selection criteria, are discussed in Chapter 14.

3. If the regression has repeated measurements at the same $x_i$ values, then you can perform a formal **lack-of-fit test** (also called a **pure error test**) in which the null hypothesis is that the type of equation used as the regression equation is correct. Failure to reject this null hypothesis is a good thing since it means that the regression equation is okay. We will discuss this test later, after we introduce analysis of variance.

*Diagnosing Whether the Errors Have a Normal Distribution*

1. Examine a **histogram of the residuals** to see if it appears to be bell-shaped, such as the residuals from the simulated data given in Figure 4.2(a). The difficulty is that the shape of a histogram may be difficult to judge unless the sample size is large.

2. Examine a **normal probability plot of the residuals**. Essentially, the ordered (standardized) residuals are plotted against theoretical expected values for a sample from a standard normal curve population. A straight-line pattern for a normal probability plot (NPP) indicates that the assumption of normality is reasonable, such as the NPP given in Figure 4.2(b).

3. Do a hypothesis test in which the null hypothesis is that the errors have a normal distribution. Failure to reject this null hypothesis is a good result. It means that it is reasonable to assume that the errors have a normal distribution. Goodness-of-fit tests are also used for this purpose. We discuss some testing procedures later in this chapter.

*Diagnosing Whether or Not the Variance is Constant*

1. Examine a plot of residuals versus fitted values. Obvious differences in the vertical spread of the residuals indicate nonconstant variance. The most typical pattern for nonconstant variance is a plot of residuals versus fitted values with a pattern that resembles a sideways cone.

2. Do a hypothesis test with the null hypothesis that the variance of the errors is the same for all values of the predictor variable(s). There are various statistical tests that can be used, such as the modified Levene test and Bartlett's test. These tests are often secondary because nonconstant variance tends to be obvious from the

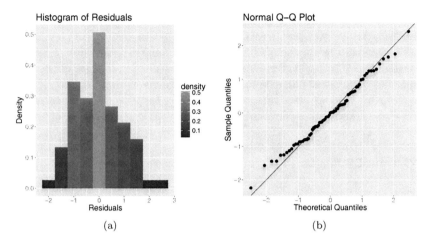

**FIGURE 4.2**
(a) Histogram of the residuals from a fitted simple linear regression model with normal errors. (b) A normal probability plot (NPP) or quantile-quantile (Q-Q) plot of the residuals.

plot of residuals versus fitted values plot. However, it is good practice to include results from these tests along with the visualizations produced.

*Diagnosing Independence of the Error Terms*

1.  Examine a plot of the residuals versus their order. Trends in this plot are indicative of a correlated error structure and, hence, dependency.

2.  If the observations in the dataset represent successive time periods at which they were recorded and you suspect a temporal component to the study, then time series can be used for analyzing and remedying the situation. Again, these procedures are discussed in Chapter 13.

## 4.3 Plots of Residuals Versus Fitted Values

**Predicted Values, Fitted Values, and Residuals**
A predicted value for an individual is $\hat{y}_i = \hat{\beta}_0 + \hat{\beta}_1 x_i$, where $x_i$ is a specified value for the predictor variable of interest. **Fit** and **fitted value** are both synonyms for predicted value. A **(raw) residual** for an individual is $e_i =$

$y_i - \hat{y}_i$. This is the difference between the actual and predicted values for an individual. **Error** and **disturbance** are both synonyms for the deviation of the observed value from the unobserved true population value of interest; e.g., the population mean. Thus, the residuals are estimating the errors, which are population quantities. For example, suppose that a sample regression line is $\hat{y}_i = 5 + 3x_i$ and an observation has $x_i = 4$ and $y_i = 20$. The fit is $\hat{y}_i = 5 + 3(4) = 17$ and the residual is $e_i = 20 - 17 = 3$.

**Internally Studentized Residuals**
The **internally Studentized residuals**[1] for the $i^{\text{th}}$ observation is given by

$$r_i = \frac{e_i}{s\sqrt{(1 - h_{i,i})}},$$

where $s = \sqrt{\text{MSE}}$ is the standard deviation of the residuals and

$$h_{i,i} = \frac{1}{n} + \frac{(x_i - \bar{x})^2}{\sum_{j=1}^{n}(x_j - \bar{x})^2},$$

which is a measure of leverage discussed in Chapter 10. We use $r_i$ because while the variance of the true errors $(\epsilon_i)$ is $\sigma^2$, the variance of the computed residuals $(e_i)$ is not. Using $r_i$ now gives us residuals with constant variance. A value of $|r_i| > 3$ usually indicates the value can be considered an outlier. Other methods can be used in outlier detection, which are discussed later.

Plots of residuals versus fitted values are constructed by plotting all pairs of $r_i$ and $\hat{y}_i$ for the observed sample, with residuals on the vertical axis. The idea is that properties of the residuals should be the same for all different values of the predicted values. Specifically, as we move across the plot (across predicted values), the average of the residuals should always be about 0 and the vertical variation in residuals should maintain about the same magnitude. Note that plots of $e_i$ and $\hat{y}_i$ can also be used and typically appear similar to those based on the Studentized residuals.

## 4.3.1   Ideal Appearance of Plots

The usual assumptions for regression imply that the pattern of deviations (errors) from the regression line should be similar regardless of the value of the predicted value (and value of the $X$ variable). The consequence is that a plot of residuals versus fitted values (or residuals versus an $X$ variable) ideally has a random "zero correlation" appearance.

---

[1]There is also the notion of externally Studentized residuals discussed in Chapter 10. Typically, when authors use plots and perform an analysis of "Studentized residuals," they are doing so using the internally Studentized residuals. We will usually drop the adverb "internally" and make it clear from the context which type of Studentized residual we are discussing.

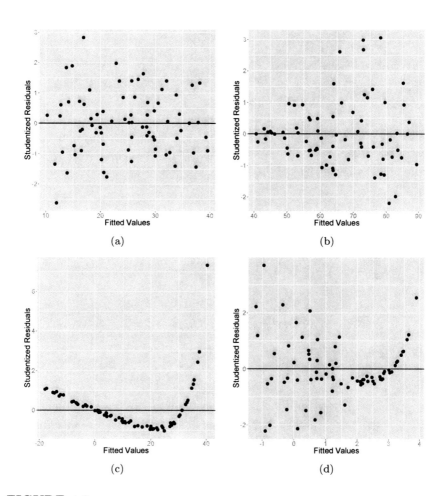

**FIGURE 4.3**

(a) Ideal scatterplot of residuals versus fitted values showing constant variance. (b) A fit demonstrating nonconstant variance. (c) A nonlinear fit. (d) A nonlinear fit with nonconstant variance.

Figure 4.3(a) is a scatterplot of residuals versus fitted values from simulated data that more or less has the ideal appearance for this type of scatterplot. The other scatterplots of Figure 4.3 show situations requiring further remediation.

- The residuals are plotted on the vertical axis and the fitted values are on the horizontal. This is the usual convention.

- Notice that the residuals are averaging about 0 all the way across the plot. This is a good thing — the regression line is describing the center of the data in all locations.

- We are interested in the vertical range as we move across the horizontal axis. In Figure 4.3(a), we see roughly the same type of residual variation (vertical variation in the plot) for all predicted values. This means that deviations from the regression line are about the same as we move across the fitted values.

### 4.3.2 Difficulties Possibly Seen in the Plots

Three primary difficulties may show up in plots of residuals versus fitted values:

1. Outliers in the data.

2. The regression equation for the average does not have the right form.

3. The residual variance is not constant.

**Difficulty 1: Outliers in the $Y$ Values**
An unusual value for the $Y$ variable will often lead to a large residual. Thus, an outlier may show up as an extreme point in the vertical direction of the residuals versus fitted values plot. Figure 4.4 gives a plot of data simulated from a simple linear regression model. However, the point at $(6, 35)$ is an outlier, especially since the residual at this point is so large. This is also confirmed by observing the plot of Studentized residuals versus fitted values, which is not shown here.

**Difficulty 2: Wrong Mathematical Form of the Regression Equation**
A curved appearance in a plot of residuals versus fitted values indicates that we used a regression equation that does not match the curvature of the data. Thus, we have misspecified our model. Figures 4.3(c) and 4.3(d) show the case of nonlinearity in the residual plots. When this happens, there is often a pattern to the data similar to that of an exponential or trigonometric function.

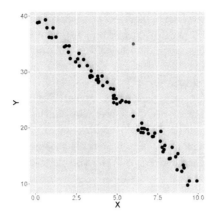

**FIGURE 4.4**
A simulated dataset with an outlier.

### Difficulty 3: Nonconstant Residual Variation
Many times, nonconstant residual variance leads to a sideways cone or funnel shape for the plot of residuals versus fitted values. Figures 4.3(b) and 4.3(d) show plots of such residuals versus fitted values. Nonconstant variance is noted by how the data "fans out" as the predicted value increases. In other words, the residual variance (i.e., the vertical range in the plot) is increasing as the size of the predicted value increases. This typically has a minimal impact on regression estimates, but is a feature of the data that needs to be taken into account when reporting the accuracy of the predictions, especially for inferential purposes.

---

## 4.4    Data Transformations

Transformations of the variables are used in regression to describe curvature and sometimes are also used to adjust for nonconstant variance in the errors and the response variable. Below are some general guidelines when considering transformations, but a more expansive discussion on data transformation strategies in regression settings can be found in Carroll and Ruppert (1988).

### What to Try?
When there is curvature in the data, there might possibly be some theory in the literature of the subject matter that suggests an appropriate equation. Or, you might have to use trial-and-error data exploration to determine a model that fits the data. In the trial-and-error approach, you might try polynomial

models or transformations of $X$ and/or $Y$, such as square root, logarithmic, or reciprocal. One of these will often end up improving the overall fit, but note that interpretations of any quantities will be on the transformed variable(s).

## Transform $X$ or Transform $Y$?

In the data exploration approach, if you transform $Y$, you will change the variance of the $Y$ and the errors. You may wish to try common transformations of the $Y$ (e.g., $\log(Y)$, $\sqrt{Y}$, or $Y^{-1}$) when there is nonconstant variance and possible curvature to the data. Try transformations of the $X$ (e.g., $X^{-1}$, $X^2$, or $X^3$) when the data are curved, but the variance looks to be constant in the original scale of $Y$. Sometimes it will be just as much art as it is science!

## Why Might Logarithms Work?

Logarithms often are used because they are connected to common exponential growth and power curve relationships. The relationships discussed below are easily verified using the algebra of logarithms. The **exponential growth equation** for variables $Y$ and $X$ may be written as

$$Y = \alpha \times 10^{\gamma X},$$

where $\alpha$ and $\gamma$ are parameters to be estimated. Taking logarithms on both sides of the exponential growth equation gives

$$\log_{10} Y = \log_{10} \alpha + \gamma X.$$

Thus, an equivalent way to express exponential growth is that the logarithm of $Y$ is a straight-line function of $X$.

A general **power curve** equation is

$$Y = \alpha X^{\gamma},$$

where, again, $\alpha$ and $\gamma$ are parameters to be estimated. Taking logarithms on both sides of the exponential growth equation gives[2]

$$\log Y = \log \alpha + \gamma \log X.$$

Thus, an equivalent way to write a power curve equation is that the logarithm of $Y$ is a straight-line function of the logarithm of $X$. This regression equation is sometimes referred to as a **log-log regression equation**.

## Power and Box–Cox Transformations

It is often difficult to determine which transformation on $Y$ to use. One approach is to use the family of **power transformations** on $Y$ such that

---

[2]Note that here $\log()$ refers to the natural logarithm. In Statistics, $\log()$ and $\ln()$ are used interchangeably for the natural logarithm. If any other base is ever used, then the appropriate subscript will be used; e.g., $\log_{10}$ where the base is 10.

$Y' = Y^\lambda$, where $\lambda$ is a parameter to be determined using the data. We have already alluded to some transformations that are part of this family, e.g., $\lambda = 2$ ($Y' = Y^2$), $\lambda = 0.5$ ($Y' = \sqrt{Y}$), and $\lambda = -1$ ($Y' = 1/Y$). Note that, by convention, $\lambda = 0$ is defined to be $Y' = \log Y$. For a sample $(x_1, y_1), \ldots, (x_n, y_n)$, the normal error regression model with a power transformation is

$$y_i^\lambda = \beta_0 + \beta_1 x_i + \epsilon_i, \tag{4.1}$$

where $i = 1, \ldots, n$.

**Box–Cox transformations** (Box and Cox, 1964) are determined from the family of power transformations on $Y$ such that $Y' = Y^\lambda$, but are done so using the method of maximum likelihood (which we discuss later) to estimate $\lambda$. A standard approach is to perform a simple search over a range of reasonable candidate values for $\lambda$; e.g., $\lambda = -4.0, -3.5, -3.0, \ldots, 3.0, 3.5, 4.0$. For each $\lambda$ value, the $y_i^\lambda$ observations are standardized so that the analysis using the SSEs does not depend on $\lambda$. The standardization is

$$w_i = \begin{cases} \frac{(y_i^\lambda - 1)}{\lambda C^{\lambda-1}}, & \lambda \neq 0; \\ C(\log y_i), & \lambda = 0, \end{cases}$$

where $C = \prod_{i=1}^n y_i^{1/n}$. Once the $w_i$ have been calculated for a given $\lambda$, they are then regressed on the $x_i$ and the SSE is retained. We then select the $\lambda$ for which the SSE is a minimum and call it $\hat{\lambda}$. This is then what we call the maximum likelihood estimate for $\lambda$.

## 4.5  Tests for Normality

As mentioned earlier, a hypothesis test can be performed in which the null hypothesis is that the errors have a normal distribution. Failure to reject this null hypothesis is a good result. It means that it is reasonable to assume that the errors have a normal distribution. Typically, assessment of the appropriate residual plots is sufficient to diagnose deviations from normality. However, more rigorous and formal quantification of normality may be requested. So this section provides a discussion of some common testing procedures (of which there are many) for normality. For each test discussed below, the formal hypothesis test is written as:

$H_0$ : the errors follow a normal distribution

$H_A$ : the errors do not follow a normal distribution.

## Anderson–Darling Test

The **Anderson–Darling test** (Anderson and Darling, 1954) measures the area between the fitted line (based on the chosen distribution) and the non-parametric step function (based on the plot points). The statistic is a squared distance that is weighted more heavily in the tails of the distribution. Smaller Anderson–Darling values indicate that the distribution fits the data better. The test statistic is given by:

$$A^2 = -n - \sum_{i=1}^{n} \frac{2i-1}{n} [\log F(e_{(i)}) + \log(1 - F(e_{(n+1-i)}))],$$

where $F(\cdot)$ is the cumulative distribution of the normal distribution and $e_{(1)} \leq e_{(2)} \leq \ldots \leq e_{(n)}$ are the ordered values of the residuals. The test statistic is compared against the critical values from a normal distribution in order to determine the $p$-value.

## Kolmogorov–Smirnov Test

The **Kolmogorov–Smirnov test** (see Chapter 10 of DeGroot and Schervish, 2012, for discussion) compares the empirical cumulative distribution function of your sample data with the distribution expected if the data were normal. If this observed difference is sufficiently large, the test will reject the null hypothesis of population normality. The test statistic is given by:

$$D = \max(D^+, D^-),$$

where

$$D^+ = \max_i(i/n - F(e_{(i)}))$$
$$D^- = \max_i(F(e_{(i)}) - (i-1)/n),$$

where $e_{(i)}$ pertains to the $i^{\text{th}}$ largest value of the error terms. The test statistic is compared against the critical values from a normal distribution in order to determine the $p$-value.

## Shapiro–Wilk Test

The **Shapiro–Wilk test** (Shapiro and Wilk, 1965) uses the test statistic

$$W = \frac{\left(\sum_{i=1}^{n} a_i e_{(i)}\right)^2}{\sum_{i=1}^{n}(e_i - \bar{e})^2},$$

where the $a_i$ values are calculated using the means, variances, and covariances of the $e_{(i)}$, the form of which requires more of technical discussion beyond what we present. $W$ is compared against tabulated values of this statistic's distribution. Small values of $W$ will lead to rejection of the null hypothesis.

**Ryan–Joiner Test**

The **Ryan–Joiner test** (Ryan and Joiner, 1976) can be viewed as a simpler alternative to the Shapiro–Wilk test. The test statistic is actually a correlation coefficient calculated by

$$R_p = \frac{\sum_{i=1}^{n} e_{(i)} z_{(i)}}{\sqrt{s^2(n-1) \sum_{i=1}^{n} z_{(i)}^2}},$$

where the $z_{(i)}$ values are the $z$-score values (i.e., normal values) of the corresponding $e_{(i)}$ values and $s^2$ is the sample variance. Values of $R_p$ closer to 1 indicate that the errors are normally distributed.

**Chi-Square Goodness-of-Fit Test**

The **chi-square goodness-of-fit test** is used for the more general hypothesis:

$$H_0 : \text{the data follow a specified distribution}$$
$$H_A : \text{the data do not follow the specified distribution.}$$

To construct the test statistic, the data are binned into $k$-bins that partition the range of the observed data. The test statistic is then defined as

$$X^2 = \sum_{j=1}^{k} \frac{(O_j - E_j)^2}{E_j},$$

where $O_j$ and $E_j$ are the observed frequency of our data and expected frequency, respectively, for bin $j = 1, \ldots, k$. This test is asymptotically $\chi^2_{k-3}$ for the simple regression setting because three degrees of freedom are lost due to estimation of the parameters $\beta_0$, $\beta_1$, and $\sigma$. In a regression context, we would create bins by partitioning the $y$-axis. We would then count the number of observations and fitted values that fall in each bin to determine the $O_j$ and $E_j$ values.

A major drawback with this test is that it depends on the binwidths, which greatly varies the power of the test. Typically, one of the other tests listed in this section would be performed. However, the use of this chi-square goodness-of-fit test and a slightly-modified version are often used in the field of astrostatistics for regression problems; see Babu and Feigelson (2006).

## 4.5.1    Skewness and Kurtosis

Suppose we have the residuals $e_1, \ldots, e_n$ from our simple linear regression fit. We have already introduced the notion of the mean and variance of the residuals. More broadly, for a univariate dataset, the mean and variance of the data are based on the first and second moments of the data, respectively. In particular, the $r$**th** (**sample**) **moment** of the residuals is given by

$$m_r' = \frac{\sum_{i=1}^{n} e_i^r}{n}.$$

Moreover, the $r^{\text{th}}$ **central moment** of the residuals is given by

$$m_r = \frac{\sum_{i=1}^{n}(e_i - \bar{e})^r}{n},$$

where $\bar{e}$ is the mean of the residuals, which will be 0 when using an ordinary least squares fit to the data. In other words, the $r^{\text{th}}$ moment and $r^{\text{th}}$ central moment are the same for our setting. Also note that the mean and variance are given by $m_1$ and $(n-1)m_2/n$, respectively.

Two other moment estimators can be used to assess the degree of nonnormality - both of which can be estimated in a variety of ways. We just present one formulation for each of these estimators, but it should be noted that there are a number of variations to each of these estimators in the literature. So you need to be clear about the formulation being used.

The first of these moment estimators is **skewness**, which is a statistic that measures the direction and degree of asymmetry of your data. In other words, it characterizes how long of a tail the data exhibits. A value of 0 indicates a symmetrical distribution, a positive value indicates skewness to the right, and a negative value indicates skewness to the left. Values between $-3$ and $+3$ are typical values of samples from a normal distribution. The formulation we present for the sample skewness is

$$g = \frac{m_3}{m_2^{3/2}}.$$

Note that this particular estimate of the skewness, like the others found in the literature, has a degree of bias to it when estimating the "true" population skewness.

The other moment estimator is **kurtosis**, which is a statistic that measures the heaviness of the tails of a distribution. Unimodal distributions with kurtosis *less* than 3 are called **platykurtic** and tend to exhibit peaks that are broader than those in a normal distribution. Unimodal distributions with kurtosis *more* than 3 are called **leptokurtic** and tend to exhibit sharper peaks in the center of the distribution relative to those in a normal distribution. If the kurtosis statistic equals 3 and the skewness is 0, then the distribution is normal. The formulation we present for the sample kurtosis is

$$k = \frac{m_4}{m_2^2}.$$

Be forewarned that this statistic is an unreliable estimator of kurtosis for small sample sizes and, as with skewness, there are other formulations in the literature that can be calculated.

There are formal tests using these statistics with how they pertain to a normal distribution. However, simply using the rules-of-thumb stated above with respect to the raw residuals is generally acceptable. While these statistics are not commonly reported with assessing the normality of the residuals, they can provide further justification or clarity in a setting where it appears that the normality assumption is violated.

## 4.6 Tests for Constant Error Variance

There are various tests that may be performed on the residuals for testing if they have constant variance. The following should not be used solely in determining if the residuals do indeed have constant variance. Oftentimes, it is sufficient to produce a powerful visualization (e.g., residuals versus fitted values plot) to make this assessment. However, the tests we discuss can provide an added layer of justification to your analysis. Some of the procedures require you to partition the residuals into a specified number of groups, say, $k \geq 2$ groups of sizes $n_1, \ldots, n_k$ such that $\sum_{i=1}^{k} n_i = n$. For these procedures, the sample variance of group $i$ is given by:

$$s_i^2 = \frac{\sum_{j=1}^{n_i} (e_{i,j} - \bar{e}_{i,\cdot})^2}{n_i - 1},$$

where $e_{i,j}$ is the $j^{\text{th}}$ residual from group $i$. Moreover, the pooled variance is given by:

$$s_p^2 = \frac{\sum_{i=1}^{k} (n_i - 1) s_i^2}{n - k}.$$

### $F$-Test

Suppose we partition the residuals of observations into two groups — one consisting of the residuals associated with the lowest predictor values and one consisting of the residuals associated with the highest predictor values. Treating these two groups as if they could (potentially) represent two different populations, we can test

$$H_0 : \sigma_1^2 = \sigma_2^2$$
$$H_A : \sigma_1^2 \neq \sigma_2^2$$

using the $F$-statistic $F^* = s_1^2 / s_2^2$. This test statistic is distributed according to a $F_{n_1-1,n_2-1}$ distribution. Thus, if $F^* \geq F_{n_1-1,n_2-1;1-\alpha}$, then reject the null hypothesis and conclude that there is statistically significant evidence that the variance is not constant.

### Levene Tests

Nonconstant variance can also be tested using the **Levene tests** (Levene, 1960). These tests do not require the error terms to be drawn from a normal distribution. The tests are constructed by grouping the residuals into $k$ groups according to the values of $X$. It is typically recommended that each group has at least 25 observations. Usually two groups are used, but we develop this test from the more general perspective of $k$ groups.

Begin by letting group 1 consist of the residuals associated with the $n_1$ lowest values of the predictor. Then, let group 2 consist of the residuals associated with the $n_2$ lowest remaining values of the predictor. Continue on in this manner until you have partitioned the residuals into $k$ groups. The objective is to perform the following hypothesis test:

$$H_0 : \text{the variance is constant}$$
$$H_A : \text{the variance is not constant.}$$

While hypothesis tests are usually constructed to *reject* the null hypothesis, this is a case where we actually hope we *fail to reject* the null hypothesis. The test statistic for the above is computed as follows:

$$L^2 = \frac{(n-k)\sum_{i=1}^{k} n_i(\bar{d}_{i,\cdot} - \bar{d}_{\cdot,\cdot})^2}{(k-1)\sum_{i=1}^{k}\sum_{j=1}^{n_i}(d_{i,j} - \bar{d}_{i,\cdot})^2},$$

where $d_{i,j}$ is one of the following quantities:

- $d_{i,j} = |e_{i,j} - \bar{e}_{i,\cdot}|$, where $\bar{e}_{i,\cdot}$ is the mean of the $i^{\text{th}}$ group of residuals. This is how Levene (1960) originally defined the test.

- $d_{i,j} = |e_{i,j} - \tilde{e}_{i,\cdot}|$, where $\tilde{e}_{i,\cdot}$ is the median of the $i^{\text{th}}$ group of residuals.

- $d_{i,j} = |e_{i,j} - \check{e}_{i,\cdot;\gamma}|$, where $\check{e}_{i,\cdot}$ is the $100 \times \gamma\%$ trimmed mean of the $i^{\text{th}}$ group of residuals.

The $100 \times \gamma\%$ **trimmed mean** removes the $100 \times \gamma\%$ smallest and $100 \times \gamma\%$ largest values (rounded to the nearest integer) and then calculates the mean of the remaining values. Typically for this test, $\gamma = 0.10$.

Tests based on the second and third quantities for $d_{i,j}$ are due to Brown and Forsythe (1974) and referred to as **modified Levene tests** or **Brown–Forsythe tests**. In each case, $L$ is approximately distributed according to a $t_{n-k}$ distribution. Similarly, $L^2$ is approximately distributed according to a $F_{1,n-k}$ distribution.

So why are there multiple ways to define $d_{i,j}$? The choice of how you define $d_{i,j}$ determines the robustness and power of the Levene test. Robustness in this setting is the ability of the test to *not* falsely detect unequal variances when the underlying distribution of the residuals is nonnormal and the variance is constant. Power in this setting is the ability of the test to detect that the variance is not constant when, in fact, the variance is not constant.

The modified Levene test based on the median of the residuals is usually the preferred test used in textbooks and statistical packages. Since the Levene test does not make any distributional assumptions, the median will provide a more informative measure of the center of the data — compared to the mean or trimmed mean — if the distribution of the data is skewed. But if the data are (approximately) symmetric, then all versions of the Levene test will provide similar results.

One additional note is that $d_{i,j}$ can also be formulated by the squared versions of the quantities, namely:

- $d_{i,j} = (e_{i,j} - \bar{e}_{i,\cdot})^2$;

- $d_{i,j} = (e_{i,j} - \tilde{e}_{i,.})^2;$

- $d_{i,j} = (e_{i,j} - \breve{e}_{i,\cdot;\gamma})^2.$

The testing is still done similarly, it is just that this version utilizes an $L_2$-norm, which results in larger residuals having a (potentially) bigger effect on the calculated $d_{i,j}$. The earlier definitions of $d_{i,j}$ using the absolute values (i.e., $L_1$-norm) are typically the quantities used.

**Bartlett's Test**
**Bartlett's test** can also be used to test for constant variance. The objective is to again perform the following hypothesis test:

$$H_0 : \text{the variance is constant}$$
$$H_A : \text{the variance is not constant.}$$

Bartlett's test is highly sensitive to the normality assumption, so if the residuals do not appear normal, even after transformations, then this test should not be used. Instead, the Levene test is the alternative to Bartlett's test that is less sensitive to departures from normality.

This test is carried out similarly to the Levene test. Once you have partitioned the residuals into $k$ groups, the following test statistic can be constructed:

$$B = \frac{(n-k)\log(s_p^2) - \sum_{i=1}^{k}(n_i - 1)\log(s_i^2)}{1 + \left[\frac{1}{3(k-1)}\left(\left(\sum_{i=1}^{k}\frac{1}{n_i-1}\right) - \frac{1}{n-k}\right)\right]}. \tag{4.2}$$

The test statistic $B$ is approximately distributed according to a $\chi_{k-1}^2$ distribution, so if $B \geq \chi_{k-1;1-\alpha}^2$, then reject the null hypothesis and conclude that there is statistically significant evidence that the variance is not constant. The test statistic in (4.2) is actually an approximation to a test statistic having a more complicated functional form, which follows what is called **Bartlett's distribution**. The details of Bartlett's test are given in Dyer and Keating (1980).

**Breusch–Pagan Test**
The **Breusch–Pagan test** (Breusch and Pagan, 1979) is considered as an alternative to the Levene tests. Whereas the Levene tests are distribution-free tests, the Breusch–Pagan test assumes that the error terms are normally distributed, with $E(\epsilon_i) = 0$ and $Var(\epsilon_i) = \sigma_i^2$; i.e., nonconstant variance. For a sample $(x_1, y_1), \ldots, (x_n, y_n)$, the $\sigma_i^2$ values depend on the $x_i$ values in the following way:

$$\log \sigma_i^2 = \gamma_0 + \gamma_1 x_i,$$

where $i = 1, \ldots, n$.

We are interested in testing the null hypothesis of constant variance versus the alternative hypothesis of nonconstant variance. Specifically, the hypothesis test is formulated as:

$$H_0 : \gamma_1 = 0$$
$$H_A : \gamma_1 \neq 0.$$

This test is carried out by first regressing the residuals on the predictor; i.e., regressing $e_i$ on $x_i$. The sum of squares resulting from this analysis is denoted by $SSR^*$, which provides a measure of the dependency of the error term on the predictor. The test statistic is given by

$$X^{2*} = \frac{SSR^*/2}{(SSE/n)^2},$$

where SSE is from the regression analysis of the response on the predictor. The test statistic follows a $\chi_1^2$ distribution, so larger values of $X^{2*}$ support the conclusion that the error variance is not constant.

## 4.7    Examples

**Example 4.7.1.** *(Steam Output Data, cont'd)*
We have already performed statistical tests on the steam output dataset to assess the significance of the slope and intercept terms as well as construction of various statistical intervals. However, all of these procedures are technically invalid if we do not demonstrate that the underlying assumptions are met.

Figure 4.5(a) gives a histogram of the residuals. While the histogram is not completely symmetric and bell-shaped, it still provides a reasonable indication of normality. The important thing is to not see very erratically-shaped (e.g., heavily-skewed) histograms. The skewness and kurtosis are $-0.149$ and $2.039$, respectively, which provide reasonably good indications of a symmetric distribution and not too heavy/thin tails with respect to the normal distribution.

Figure 4.5(b) provides a normal probability plot that indicates the residuals lie on a straight line. Thus, the assumption of normality appears to be met. Table 4.1 provides a summary of the various normality tests that were presented. Clearly all tests have $p$-values much greater than 0.05, suggesting that we fail to reject the null hypothesis of normality. Recall that the Ryan–Joiner test is actually a correlation measure. The value for these data is 0.9789, which is close to 1. Thus, it is indicative that normality is an appropriate assumption.

Finally, Figures 4.5(c) and 4.5(d) show the residual plots which demonstrate random scatter about 0 with no noticeable trends. In fact, there is not much difference between the two plots, which is often the case. Table 4.2 provides results from some of the constant variance tests that we discussed. All

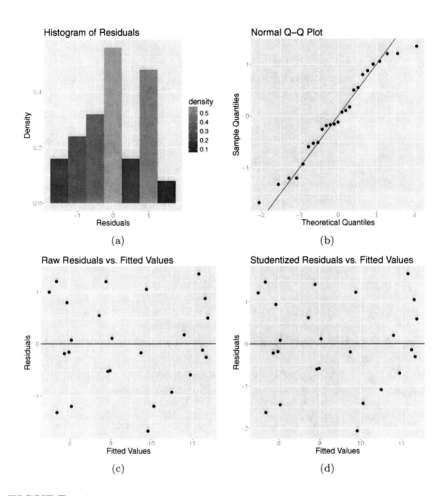

**FIGURE 4.5**
Various plots pertaining to the linear regression fit for the steam output data:
(a) histogram of the raw residuals, (b) NPP of the raw residuals, (c) scatterplot of the raw residuals versus the fitted values, and (d) scatterplot of the
Studentized residuals versus the fitted values.

**TABLE 4.1**

Results from different normality tests

| | Test Statistic | $p$-Value |
|---|---|---|
| Anderson–Darling | 0.3051 | 0.5429 |
| Kolmogorov–Smirnov | 0.0892 | 0.9782 |
| Shapiro–Wilk | 0.9596 | 0.4064 |
| Ryan–Joiner | 0.9789 | N/A |
| $\chi^2$ GOF (4 Bins) | 0.8825 | 0.3475 |
| $\chi^2$ GOF (5 Bins) | 1.7000 | 0.4274 |

**TABLE 4.2**

Results from different tests for constant variance

| | Test Statistic | $p$-Value |
|---|---|---|
| $F$-Test | 0.7305 | 0.6948 |
| Bartlett's Test | 0.2687 | 0.6042 |
| Breusch–Pagan Test | 0.1940 | 0.6596 |

of these have very high $p$-values. Thus, the assumption of constant variance appears to be valid.

**Example 4.7.2.** *(1993 Car Sale Data)*
Lock (1993) presented a dataset that contained information on $n = 93$ cars for the 1993 model year. While 26 variables are available, we are interested in determining if the vehicle's engine size is predictive of the miles per gallon (MPG) for city driving. In particular, we want to assess the appropriateness of a linear relationship between the following two variables:

- $Y$: vehicle's city MPG

- $X$: vehicle's engine size (in liters)

Figure 4.6(a) is a scatterplot of these data with the estimated simple linear regression line overlaid. Figure 4.6(b) is a scatterplot of the Studentized residuals versus the fitted values. Clearly there is curvature exhibited in the plot of the Studentized residuals. Based on the shape of the data, we next fit a log-log regression equation. A log-log plot of the data with the fitted regression line is given in Figure 4.6(c), while a scatterplot of the Studentized residuals versus the fitted values is given in Figure 4.6(d). This clearly shows a much-improved fit. While it still looks like there could be issues with constant variance (especially due to a potential outlier for the fitted value around 3.4), Levene's test suggests that the constant variance is appropriate. The results for Levene's test assuming the three different measures of the center of the groups are given in Table 4.3. All three tests have high $p$-values, thus indicating that the assumption of constant error variance is appropriate.

**TABLE 4.3**
Results from Levene's test for different group centers

|  | Test Statistic | $p$-Value |
|---|---|---|
| Mean | 1.3303 | 0.2518 |
| Median | 1.9520 | 0.1658 |
| 10% Trimmed Mean | 1.9025 | 0.1712 |

**Example 4.7.3.** *(Black Cherry Tree Data)*
Chen et al. (2002) provided an extensive study about the utility of Box–Cox transformations in linear regression settings. One dataset they analyzed pertained to various dimensional measurements on $n = 31$ black cherry trees in the Allegheny National Forest. The original data are from Ryan et al. (1976). We focus on a simple linear regression relationship between the following two measured variables:

• $Y$: the tree's volume

• $X$: the tree's diameter

Figure 4.7(a) is a scatterplot of the black cherry tree data. Figure 4.7(b) is a Q-Q plot of the Studentized residuals for the ordinary least squares fit. While this Q-Q plot is not too bad, there does appear to be some room for improvement. Namely, the extreme quantiles appear to moderately deviate from the $y = x$ line, plus there is a bit more variability of the other quantiles about this line. The Box–Cox procedure is run to determine an appropriate transformation.

Figure 4.8 shows a plot of the **profile loglikelihood** versus the different $\lambda$ values in the family of Box–Cox transformations under consideration. While we do not explicitly develop the profile loglikelihood here (see Davison, 2003, Chapter 4), it basically involves rewriting the likelihood function[3] into the parameters of interest by taking the supremum of the likelihood function over the parameters not directly of interest, called **nuisance parameters**. For our purposes, it is a function of the SSE of the estimated simple linear regression model using the given value of $\lambda$. The value that maximizes the profile loglikelihood is $\hat{\lambda} = 0.400$, which is given by the solid red line in Figure 4.8. A 95% confidence interval can also be calculated based on the $\chi_1^2$ distribution, which is $(0.100, 0.600)$. Using the Box–Cox transformation with $\hat{\lambda} = 0.400$, the transformed data and new estimated regression line are given in Figure 4.7(c). This appears to be a better fit compared to the untransformed data. This is confirmed with the corresponding Q-Q plot in Figure 4.7(d).

---

[3]See Appendix C for a definition of the likelihood function.

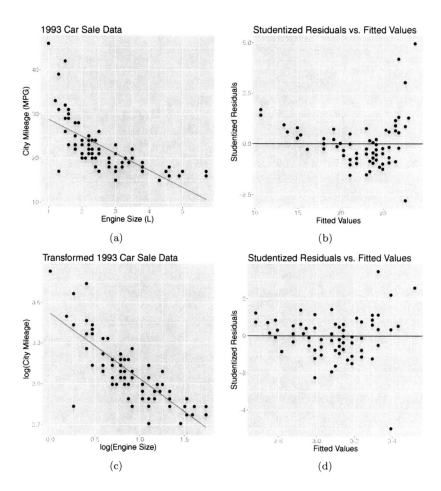

**FIGURE 4.6**
Various plots pertaining to the linear regression fit for the 1993 car sale data:
(a) scatterplot of city mileage versus engine size, (b) scatterplot of the Studentized residuals versus fitted values, (c) scatterplot of the log of city mileage versus the log of engine size, and (d) scatterplot of the Studentized residuals versus the fitted values for this log-log fit.

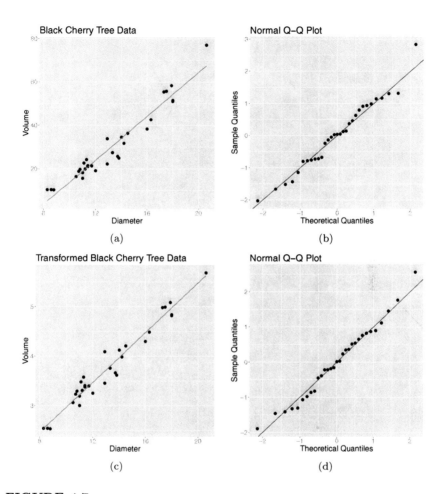

**FIGURE 4.7**
(a) Scatterplot of the black cherry tree data with estimated linear regression line. (b) Q-Q plot of the Studentized residuals for the fit. (c) Scatterplot of the Box–Cox transformed black cherry tree data with estimated linear regression line. (d) Q-Q plot of the Studentized residuals for the fit using the Box–Cox transformation.

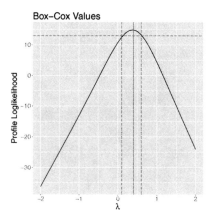

**FIGURE 4.8**
A plot of the profile loglikelihood, the optimal $\lambda$ value (solid vertical line) and
the 95% confidence interval (vertical dashed lines).

# 5

## ANOVA for Simple Linear Regression

An **analysis of variance** (**ANOVA**) table for regression displays quantities that measure how much of the variability in the response variable is explained and how much is not explained by the regression relationship with the predictor variable. The table also gives the construction and value of the mean squared error (MSE) and a significance test of whether the variables are related in the sampled population.

The use of the ANOVA appears in various capacities throughout this handbook. In this chapter, the ANOVA table is presented for testing in the context of a regression model with a single predictor. In Chapter 8, it is shown that the ANOVA table is easily extended to the setting with more than one predictor variable. In Chapter 15, the notion of ANOVA-type models (expressed as regression models) for common experimental designs is presented. The theory of ANOVA is extensive and only superficial aspects are presented here. Most regression and experimental design textbooks develop various aspects of the ANOVA theory, but the classic text on the topic is Scheffé (1959).

## 5.1 Constructing the ANOVA Table

An underlying conceptual idea for the construction of the ANOVA table is:

overall variation in $y$ = regression variation + error variation.

More formally, we can partition our estimated simple linear regression model to reflect the above concept:

$$\begin{aligned} y_i &= \hat{\beta}_0 + \hat{\beta}_1 x_i + e_i \\ &= \hat{y}_i + e_i \\ &= \hat{y}_i + (y_i - \hat{y}_i) \\ \Rightarrow (y_i - \bar{y}) &= (\hat{y}_i - \bar{y}) + (y_i - \hat{y}_i). \end{aligned}$$

The above relationship is also illustrated in Figure 5.1, where the different colored lines represent the different sources of error.

The quantities in the ANOVA table, some of which we have introduced, are the following:

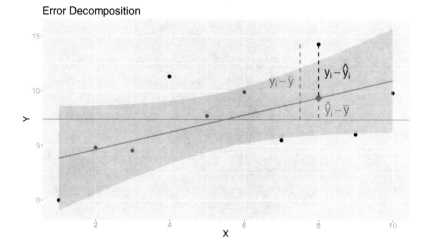

**FIGURE 5.1**
How the estimated regression model is decomposed into its various error
sources. These quantities are the basis of an ANOVA table, which is the
platform for our statistical testing and variance component analysis. A 95%
Working–Hotelling confidence band is also plotted for reference.

- The total sum of squares is

$$\text{SSTO} = \sum_{i=1}^{n} (y_i - \bar{y})^2,$$

  which is the sum of squared deviations from the overall mean of $y$. The value
  $n - 1$ is the **total degrees of freedom** $(\text{df}_T)$. SSTO is a measure of the
  overall variation in the observed $y$-values.

- The error sum of squares is

$$\text{SSE} = \sum_{i=1}^{n} (y_i - \hat{y}_i)^2,$$

  which is the sum of squared residuals. SSE is a measure of the variation
  in the observed $y$-values that is not explained by the regression. For simple
  linear regression, the value $n - 2$ is the **error degrees of freedom** $(\text{df}_E)$.

- The mean squared error is

$$\text{MSE} = \frac{\text{SSE}}{\text{df}_E} = \frac{\text{SSE}}{n - 2},$$

  which estimates $\sigma^2$, the variance of the errors.

- The **sum of squares due to the regression** is

$$\text{SSR} = \text{SSTO} - \text{SSE} = \sum_{i=1}^{n}(\hat{y}_i - \bar{y})^2,$$

and it is a measure of the total variation in the modeled values; i.e., it measures how well the simple linear regression model represents the data being modeled. Also, for the **regression degrees of freedom** ($\text{df}_R$), we have $\text{df}_R = \text{df}_T - \text{df}_E$. For simple linear regression, $\text{df}_R = (n-1) - (n-2) = 1$.

- The **mean square for the regression** is

$$\text{MSR} = \frac{\text{SSR}}{\text{df}_R} = \frac{\text{SSR}}{1},$$

which, roughly, is the variance of the modeled responses relative to the mean of the observed responses.

**TABLE 5.1**
ANOVA table for simple linear regression

| Source | df | SS | MS | F |
|---|---|---|---|---|
| **Regression** | 1 | $\sum_{i=1}^{n}(\hat{y}_i - \bar{y})^2$ | MSR | MSR/MSE |
| **Error** | $n-2$ | $\sum_{i=1}^{n}(y_i - \hat{y}_i)^2$ | MSE | |
| **Total** | $n-1$ | $\sum_{i=1}^{n}(y_i - \bar{y})^2$ | | |

The uses of the ANOVA table in Table 5.1 include:

- The error line tells the value for the estimated error variance, MSE.

- The "Total" and "Error" lines give the SS values used in the calculation of $R^2$.

- The $F$-statistic can be used to test whether the $Y$ and $X$ are related.

Specifically, in the simple linear regression model, $F^* = \text{MSR}/\text{MSE}$ is a test statistic for:

$$H_0 : \beta_1 = 0$$
$$H_A : \beta_1 \neq 0$$

and the corresponding degrees of freedom are 1 and $n-2$. Statistical software will report a $p$-value for this test statistic. The same decision rules apply here as when we discussed statistical inference.

Suppose that we take many random samples of size $n$ from some population at the same predictor values, $X_1, \ldots, X_n$. We then estimate the regression line and determine the MSR and MSE for each dataset. If we were to take the average of each of these quantities, we would obtain what are called the **expected mean squares**, which can be shown to be

$$E(\text{MSR}) = \sigma^2 + \beta_1^2 \sum_{i=1}^{n} (X_i - \bar{X})^2$$

and

$$E(\text{MSE}) = \sigma^2.$$

These two expected values are connected to how we obtain the $F$-test and its interpretation. Basically, the intuition is that if the null hypothesis is true (i.e., $\beta_1 = 0$), then the sampling distribution of the MSR and the MSE are the same. Thus, MSR/MSE $= 1$. If the null hypothesis is false (i.e., $\beta_1 \neq 0$), then the sampling distribution of the MSR and the MSE are very different. Thus, MSR/MSE $> 1$. These expressions are not always reported in the ANOVA table since they tend to be secondary to the analysis at hand.

As a special case of the ANOVA table in Table 5.1, consider the regression through the origin setting. When performing a regression through the origin, we force $\bar{y} = 0$ to get the uncorrected sums of squares totals in Table 5.2. By forcing $\bar{y} = 0$, the sums of squares can now be shown to be additive. Table 5.2 gives the ANOVA table for the regression through the origin. Notice that (generally) $\sum_{i=1}^{n} e_i \neq 0$. Because of this, the SSE could actually be larger than the SSTO in Table 5.1, hence we use the uncorrected values in Table 5.2. Note that by forcing $\bar{y} = 0$, we are not asserting that the mean value of our response is 0, but are doing so to get additivity in the ANOVA table. Note also that the expected mean squares for the regression through the origin setting can be shown to be

$$E(\text{MSR}) = \sigma^2 + \beta_1^2 \sum_{i=1}^{n} X_i^2$$

and

$$E(\text{MSE}) = \sigma^2.$$

**TABLE 5.2**
ANOVA table for regression through the origin

| Source | df | SS | MS | F |
|--------|-----|-----|------|--------|
| Regression | 1 | $\sum_{i=1}^{n} \hat{y}_i^2$ | MSR | MSR/MSE |
| Error | $n-1$ | $\sum_{i=1}^{n} (y_i - \hat{y}_i)^2$ | MSE | |
| Total | $n$ | $\sum_{i=1}^{n} y_i^2$ | | |

## 5.2 Formal Lack of Fit

It was just shown that every regression divides up the variation into two segments:

$$\text{SSTO} = \text{SSR} + \text{SSE}.$$

If the regression also has repeated measurements with the same levels of $X$ (called **replicates**), then a lack-of-fit term can be estimated to detect non-linearities in the data. A lack-of-fit test cannot be used with data which has a continuous spread of $x$ values with no replicates, such as what might come from an existing process dataset or a purely observational study. The formal lack-of-fit test is written as:

$$H_0 : \text{there is no linear lack of fit}$$
$$H_A : \text{there is a linear lack of fit}.$$

To perform a lack-of-fit test, first partition the SSE into two quantities: lack of fit and pure error. The **pure error sum of squares (SSPE)** is found by considering only those observations that are replicates. The $x$ values are treated as the levels of the factor in a one-way ANOVA. With $x$ held constant, the SSE from this analysis measures the underlying variation in $y$. Thus, it is called pure error. Subtracting the pure error sum of squares from the SSE of the linear regression measures the amount of nonlinearity in the data, called the **lack-of-fit sum of squares (SSLOF)**.

Formally, let $j = 1, \ldots, m < n$ be the index for the unique levels of the predictors and $i = 1, \ldots, n_j$ be the index for the $i^{\text{th}}$ replicate of the $j^{\text{th}}$ level. For example, if we have the set of six predictor values $\{0,0,2,4,4,4\}$, then there are $m = 3$ unique predictor levels such that $n_1 = 2$, $n_2 = 1$, and $n_3 = 3$. Notice that $\sum_{j=1}^{m} n_j = n$. We assume that the random variable $Y_{i,j}$ can be written as

$$Y_{i,j} = \mu_j + \epsilon_{i,j},$$

such that the $\mu_j$ are the mean responses at the $m$ unique levels of the predictor and $\epsilon_{i,j}$ are again assumed to be *iid* $\mathcal{N}(0, \sigma^2)$. Then,

$$\text{SSE} = \sum_{j=1}^{m} \sum_{i=1}^{n_j} (y_{i,j} - \hat{y}_j)^2,$$

where $y_{i,j}$ is the $i^{\text{th}}$ observed response at the $j^{\text{th}}$ predictor level and $\hat{y}_j$ is the expected response at the $j^{\text{th}}$ predictor level. Then,

$$\text{SSE} = \text{SSPE} + \text{SSLOF}$$
$$= \sum_{j=1}^{m} \sum_{i=1}^{n_j} (y_{i,j} - \bar{y}_j)^2 + \sum_{j=1}^{m} \sum_{i=1}^{n_j} (\bar{y}_j - \hat{y}_{i,j})^2$$

An $F$-statistic can be constructed from these two values that will test the statistical significance of the lack of fit:

$$F_{LOF}^* = \frac{\frac{\text{SSLOF}}{\text{df}_{LOF}}}{\frac{\text{SSPE}}{\text{df}_{PE}}} = \frac{\text{MSLOF}}{\text{MSPE}},$$

where $\text{df}_{LOF} = m - 2$ and $\text{df}_{PE} = n - m$ are the degrees of freedom for the SSLOF and SSPE, respectively. $F_{LOF}^*$ follows an $F_{\text{df}_{LOF}, \text{df}_{PE}}$ distribution. It can further be shown that the expected mean squares for the lack-of-fit and pure error terms are

$$E(\text{MSLOF}) = \sigma^2 + \frac{\sum_{j=1}^{m} (\mu_j - (\beta_0 + \beta_1 X_j))^2}{m - 2}$$

and

$$E(\text{MSPE}) = \sigma^2,$$

respectively.

The ANOVA table including lack-of-fit and pure error is given in Table 5.3. The actual formulas for the sum of squares have been omitted to simplify the table, but are all found earlier in this chapter.

**TABLE 5.3**
ANOVA table for simple linear regression with a lack-of-fit test

| Source | df | SS | MS | F |
|--------|-----|-------|-------|-----------|
| **Regression** | 1 | SSR | MSR | MSR/MSE |
| **Error** | $n - 2$ | SSE | MSE | |
| **Lack of Fit** | $m - 2$ | SSLOF | MSLOF | MSLOF/MSPE |
| **Pure Error** | $n - m$ | SSPE | MSPE | |
| **Total** | $n - 1$ | SSTO | | |

## 5.3   Examples

**Example 5.3.1.** *(Steam Output Data, cont'd)*
Let us construct the ANOVA table with the simple linear regression model fit to the steam output data:

```
#############################
Analysis of Variance Table

Response: steam
          Df Sum Sq Mean Sq F value    Pr(>F)
```

```
Regression  1 45.592  45.592  57.543 1.055e-07 ***
Residuals  23 18.223   0.792
Total      24 63.816
---
Signif. codes:  0 *** 0.001 ** 0.01 * 0.05 . 0.1   1
###########################
```

The value of the $F$-statistic is $F^* = 45.592/0.792 = 57.543$ and the $p$-value for this $F$-statistic is $p = 1.055 \times 10^{-7}$. Thus, we reject the null hypothesis that the two variables are not related at the $\alpha = 0.05$ level and conclude that there is a significant linear relationship between steam output and atmospheric pressure. This is the same result that we obtained earlier with the $t$-test for $H_0 : \beta_1 = 0$.

**Example 5.3.2.** *(Computer Repair Data, cont'd)*
Let us construct the ANOVA table with the regression through the origin model fit to the computer repair data:

```
Analysis of Variance Table

Response: minutes
           Df Sum Sq Mean Sq F value     Pr(>F)
Regression  1 159683  159683  5274.4 < 2.2e-16 ***
Residuals  13    394      30
Total      14 160077
---
Signif. codes:  0 *** 0.001 ** 0.01 * 0.05 . 0.1   1
```

Recall for these data that $n = 14$, which equals the $df_T$ for the regression through the origin ANOVA. The value of the $F$-statistic is $F^* = 159683/30 = 5274.4$ and the $p$-value for this $F$-statistic is $p < 2.2 \times 10^{-6}$. Thus, we reject the null hypothesis that the two variables are not related at the $\alpha = 0.05$ level and conclude that there is a significant linear relationship between the length of the service call and the number of components repaired or replaced during the call. This, again, is the same result that we obtained earlier with the $t$-test for $H_0 : \beta_1 = 0$ in the regression through the origin model.

**Example 5.3.3.** *(Fiber Strength Data)*
Ndaro et al. (2007) conducted a study about the strength of a particular type of fiber based on the amount of pressure applied. The experiment consisted of five replicates at each of six different water pressure levels for a total sample size of $n = 30$. The measured variables are:

- $Y$: tensile strength of fiber (measured in N/5 cm)

- $X$: amount of water pressure applied (measured in bars); the unique levels are 60, 80, 100, 120, 150, and 200

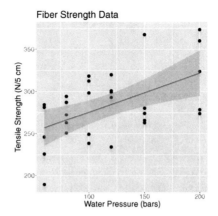

**FIGURE 5.2**
Scatterplot of tensile strength versus water pressure applied with the estimated simple linear regression line overlaid and 95% Working–Hotelling confidence band.

Figure 5.2 provides a scatterplot of these data with the estimated simple linear
regression line overlaid. For the sake of this example, we will assume that all
of the assumptions have been met.

    Since there are $m = 6$ unique predictor levels — each with 5 replicates —
we proceed to test if there is a linear lack of fit. The following is the ANOVA
table with a lack-of-fit test for the fiber data:

```
#############################
Analysis of Variance Table

Response: tensile
              Df Sum Sq Mean Sq F value   Pr(>F)
Regression     1  13868 13868.4 11.0868 0.002447 **
Residuals     28  35025  1250.9
 Lack of fit   4   1729   432.3  0.3116 0.867366
 Pure Error   24  33296  1387.3
Total         29  48893
---
Signif. codes:  0 *** 0.001 ** 0.01 * 0.05 . 0.1   1
#############################
```

The value of the $F$-statistic for the lack-of-fit test is $F^*_{LOF} = 432.3/13287.3 =$
0.3116 and the $p$-value for this $F$-statistic is $p = 0.867$. Thus, we fail to reject
the null hypothesis and conclude that there is no significant linear lack of fit
at the $\alpha = 0.05$ level. Thus, we can proceed to test if the tensile strength of
the fiber and the water pressure level applied are, indeed, linearly related. The

$F$-statistic for this test is $F^* = 13868.4/1250.9 = 11.0868$, which has a $p$-value of $p = 0.0024$. Thus, we reject the null hypothesis that the two variables are not related at the $\alpha = 0.05$ level and conclude that there is a significant linear relationship between tensile strength and water pressure applied.

# Part II

# Multiple Linear Regression

# 6

# Multiple Linear Regression Models and Inference

A **multiple linear regression model** is a linear model that describes how a response variable relates to two or more predictor variables or transformations of those predictor variables. For example, suppose that a researcher is studying factors that might affect systolic blood pressures for women aged 45 to 65 years old. The response variable is systolic blood pressure $(Y)$. Suppose that two predictor variables of interest are age $(X_1)$ and body mass index $(X_2)$. The general structure of a multiple linear regression model for this situation would be

$$Y = \beta_0 + \beta_1 X_1 + \beta_2 X_2 + \epsilon.$$

- The equation $\beta_0 + \beta_1 X_1 + \beta_2 X_2$ describes the mean value of blood pressure for specific values of age and body mass index.

- The error term, $\epsilon$, characterizes the differences between individual values of blood pressure and their expected values of blood pressure.

One note to reiterate about terminology. A linear model is one that is linear in the regression coefficients, meaning that each coefficient multiplies a predictor variable or a transformation of those predictor variables. For example,

$$Y = \beta_0 + \beta_1 X + \beta_2 X^2 + \epsilon$$

is a multiple linear regression model even though it describes a quadratic, curved, relationship between $Y$ and a single $X$.

## 6.1   About the Model

### Model Notation

- The population multiple linear regression model that relates a response variable $Y$ to $p-1$ predictor variables $X_1, X_2, \ldots, X_{p-1}$ is written as

$$Y = \beta_0 + \beta_1 X_1 + \beta_2 X_2 + \ldots + \beta_{p-1} X_{p-1} + \epsilon. \tag{6.1}$$

- Let $(x_{1,1}, \ldots, x_{1,p-1}, y_1), \ldots, (x_{n,1}, \ldots, x_{n,p-1}, y_n)$ be the realizations of the $p$ variables for our sample of size $n$. Then the multiple linear regression model is written as

$$y_i = \beta_0 + \beta_1 x_{i,1} + \beta_2 x_{i,2} + \ldots + \beta_{p-1} x_{i,p-1} + \epsilon_i. \qquad (6.2)$$

- We assume that the $\epsilon_i$ are *iid* $\mathcal{N}(0, \sigma^2)$. This is the same assumption made in simple linear regression.

- The subscript $i$ refers to the $i^{\text{th}}$ individual or unit in the population. In the notation for each predictor, the subscript following $i$ corresponds to the predictor variable. For example $x_{i,2}$ is the observed value of the second predictor for observation $i$.

## Estimates of the Model Parameters

- The estimates of the regression coefficients are the values that minimize the SSE for the sample. The exact formula for this will be given shortly.

- $\hat{\beta}_j$ is used to represent a sample estimate of the regression parameter $\beta_j$, $j = 0, 1, \ldots, p-1$.

- MSE $= \frac{\text{SSE}}{n-p}$ estimates $\sigma^2$, the variance of the errors. In the formula, $n$ is the sample size, $p$ is the number of regression parameters in the model, and SSE is the sum of squared errors. Notice that for simple linear regression $p = 2$. Thus, we get the formula for MSE that we introduced in that context of one predictor.

- In the case of two predictors, the estimated regression equation is a plane as opposed to a line in the simple linear regression setting. For more than two predictors, the estimated regression equation is a hyperplane.

## Predicted Values and Residuals

- A predicted value is calculated as

$$\hat{y}_i = \hat{\beta}_0 + \hat{\beta}_1 x_{i,1} + \hat{\beta}_2 x_{i,2} + \ldots + \hat{\beta}_{p-1} x_{i,p-1},$$

where the $\hat{\beta}$ values are estimated from ordinary least squares using statistical software.

- A **residual** term is calculated as $e_i = y_i - \hat{y}_i$; i.e., the difference between an observed and a predicted value of the response variable $Y$.

- A **plot of residuals versus predicted values** ideally should resemble a horizontal random band. Departures from this form indicates difficulties with the model and/or data. This holds the same utility as it did in the simple linear regression setting.

- Other residual analyses can be done exactly as we did in simple linear regression. For instance, we might wish to examine a normal probability plot (NPP) of the residuals. Additional plots to consider are plots of residuals versus each predictor variable separately. This might help us identify sources of curvature or nonconstant variance.

### Interaction Terms

- An **interaction** term is when there is a coupling or combined effect of two or more independent variables.

- Suppose we have a response variable, $Y$, and two predictors, $X_1$ and $X_2$. Then, the regression model with an interaction term is written as

$$Y = \beta_0 + \beta_1 X_1 + \beta_2 X_2 + \beta_3 X_1 X_2 + \epsilon.$$

Suppose you also have a third predictor, $X_3$. Then, the regression model with all the interaction terms is written as

$$Y = \beta_0 + \beta_1 X_1 + \beta_2 X_2 + \beta_3 X_3 + \beta_4 X_1 X_2 + \beta_5 X_1 X_3 \\ + \beta_6 X_2 X_3 + \beta_7 X_1 X_2 X_3 + \epsilon.$$

In a model with more predictors, you can imagine how much the model grows by adding interactions. Just make sure that you have enough observations to cover the degrees of freedom used in estimating the corresponding regression parameters!

- For each observation, their value of the interaction is found by multiplying the recorded values of the predictor variables in the interaction.

- In models with interaction terms, the significance of the interaction term should always be assessed first before proceeding with significance testing of the main variables.

- If one of the main variables is removed from the model, then the model should *not* include any interaction terms involving that variable.

- At least one of the variables in an interaction term is typically an indicator variable, which is discussed in Chapter 9.

## 6.2 Matrix Notation in Regression

There are two main reasons for using matrices in regression. First, the notation simplifies the writing of the model. Second, and more importantly, matrix formulas provide the means by which statistical software calculates the estimated coefficients and their standard errors, as well as the set of predicted values for the observed sample. If necessary, a review of matrices and some of their basic properties can be found in Appendix B.

In matrix (vector) notation, the theoretical multiple linear regression model for the population is written as

$$Y = \beta_0 + \beta_1 X_1 + \beta_2 X_2 + \ldots + \beta_{p-1} X_{p-1} + \epsilon$$
$$\underline{X}^{\mathrm{T}} \beta + \epsilon, \tag{6.3}$$

where $\underline{X} = (1, X_1, X_2, \ldots, X_{p-1})^{\mathrm{T}}$ is a $p$-dimensional vector of the predictor variables augmented with a 1 in the first position for the intercept term. More commonly, we will directly utilize the multiple linear regression model written in terms of the observed sample $(x_{1,1}, \ldots, x_{1,p-1}, y_1), \ldots, (x_{n,1}, \ldots, x_{n,p-1}, y_n)$:

$$\mathbf{Y} = \mathbf{X}\beta + \epsilon. \tag{6.4}$$

The four different model components in (6.4) are:

1. $\mathbf{Y}$ is an $n$-dimensional column vector that vertically lists the response values:

$$\mathbf{Y} = \begin{pmatrix} y_1 \\ y_2 \\ \vdots \\ y_n \end{pmatrix}.$$

2. The $\mathbf{X}$ **matrix** is a matrix in which each row gives the predictor for a different observation. The first column equals 1 for all observations, unless doing a regression through the origin. Each column after the first gives the data for a different predictor variable. There is a column for each variable, including any added interactions, transformations, indicators, and so on. The abstract formulation is:

$$\mathbf{X} = \begin{pmatrix} \mathbf{x}_1^{\mathrm{T}} \\ \mathbf{x}_2^{\mathrm{T}} \\ \vdots \\ \mathbf{x}_n^{\mathrm{T}} \end{pmatrix} = \begin{pmatrix} 1 & x_{1,1} & \cdots & x_{1,p-1} \\ 1 & x_{2,1} & \cdots & x_{2,p-1} \\ \vdots & \vdots & \ddots & \vdots \\ 1 & x_{n,1} & \cdots & x_{n,p-1} \end{pmatrix}.$$

In the subscript, the first value is the observation number and the second number is the variable number. The first column is always a column of 1s. The $\mathbf{X}$ matrix has $n$ rows and $p$ columns.

3. $\beta$ is a $p$-dimensional column vector listing the coefficients:

$$\beta = \begin{pmatrix} \beta_0 \\ \beta_1 \\ \vdots \\ \beta_{p-1} \end{pmatrix}.$$

Notice the subscript for $\beta$. As an example, for simple linear regression, $\beta = (\beta_0, \beta_1)^T$. The $\beta$ vector will contain symbols, not numbers, as it gives the population parameters.

4. $\epsilon$ is an $n$-dimensional column vector listing the errors:

$$\epsilon = \begin{pmatrix} \epsilon_1 \\ \epsilon_2 \\ \vdots \\ \epsilon_n \end{pmatrix}.$$

Again, we will not have numerical values for the $\epsilon$ vector.

There is a slight disconnect between how we wrote the population model in (6.3) and the model with our sample data in (6.4). In (6.3), we underlined $X$ (i.e., $\underline{X}$) to represent our vector of predictor variables. In the $\mathbf{X}$ matrix, we used boldface to represent a vector of observation $i$ (i.e., $\mathbf{x}_i$). We attempt to always use boldface quantities to represent vectors or matrices, but for the population multiple linear regression model, we will usually underline quantities to represent vectors to avoid multiple definitions for a particular quantity. By underlining $X$ in (6.3), we avoid using $\mathbf{X}$ to represent both a vector of (random) variables *and* the matrix of observed predictor variables

As an example, suppose that data for a response variable and two predictor variables is as given in Table 6.1. For the model

$$y_i = \beta_0 + \beta_1 x_{i,1} + \beta_2 x_{i,2} + \beta_3 x_{i,1} x_{i,2} + \epsilon_i,$$

$\mathbf{Y}$, $\mathbf{X}$, $\beta$, and $\epsilon$ are as follows:

$$\mathbf{Y} = \begin{pmatrix} 6 \\ 5 \\ 10 \\ 12 \\ 14 \\ 18 \end{pmatrix}, \mathbf{X} = \begin{pmatrix} 1 & 1 & 1 & 1 \\ 1 & 1 & 2 & 2 \\ 1 & 3 & 1 & 3 \\ 1 & 5 & 1 & 5 \\ 1 & 3 & 2 & 6 \\ 1 & 5 & 2 & 10 \end{pmatrix}, \beta = \begin{pmatrix} \beta_0 \\ \beta_1 \\ \beta_2 \\ \beta_3 \end{pmatrix}, \epsilon = \begin{pmatrix} \epsilon_1 \\ \epsilon_2 \\ \epsilon_3 \\ \epsilon_4 \\ \epsilon_5 \\ \epsilon_6 \end{pmatrix}.$$

Notice that the first column of the $\mathbf{X}$ matrix equals 1 for all rows (observations), the second column gives the values of $x_{i,1}$, the third column lists the values of $x_{i,2}$, and the fourth column gives the values of the interaction values $x_{i,1} x_{i,2}$. For the model, we do not know the values of $\beta$ or $\epsilon$, thus we can only list the symbols for these vectors.

Finally, using calculus rules for matrices, it can be derived that the ordinary least squares estimates of $\beta$ is calculated using the matrix formula

**TABLE 6.1**

A sample dataset

| $y_i$ | 6 | 5 | 10 | 12 | 14 | 18 |
|---|---|---|---|---|---|---|
| $x_{i,1}$ | 1 | 1 | 3 | 5 | 3 | 5 |
| $x_{i,2}$ | 1 | 2 | 1 | 1 | 2 | 2 |

$$\hat{\boldsymbol{\beta}} = (\mathbf{X}^T\mathbf{X})^{-1}\mathbf{X}^T\mathbf{Y},$$

which minimizes the sum of squared errors

$$\begin{aligned}
||\mathbf{e}||^2 &= \mathbf{e}^T\mathbf{e} \\
&= (\mathbf{Y} - \hat{\mathbf{Y}})^T(\mathbf{Y} - \hat{\mathbf{Y}}) \\
&= (\mathbf{Y} - \mathbf{X}\hat{\boldsymbol{\beta}})^T(\mathbf{Y} - \mathbf{X}\hat{\boldsymbol{\beta}}),
\end{aligned}$$

where

$$\hat{\boldsymbol{\beta}} = \begin{pmatrix} \hat{\beta}_0 \\ \hat{\beta}_1 \\ \vdots \\ \hat{\beta}_{p-1} \end{pmatrix}.$$

As in the simple linear regression case, these regression coefficient estimators are unbiased; i.e., $E(\hat{\boldsymbol{\beta}}) = \boldsymbol{\beta}$. The formula above is used by statistical software to perform calculations.

An important theorem in regression analysis (and statistics in general) is the **Gauss–Markov Theorem**, which we alluded to earlier. Since we have introduced the proper matrix notation, we now formalize this very important result.

**Theorem 1.** (Gauss–Markov Theorem) Suppose that we have the linear regression model

$$\mathbf{Y} = \mathbf{X}\boldsymbol{\beta} + \boldsymbol{\epsilon},$$

where $E(\epsilon_i|\mathbf{x}_i) = 0$ and $E(\epsilon_i|\mathbf{x}_i) = \sigma^2$ for all $i = 1, \ldots, n$. Then

$$\hat{\boldsymbol{\beta}} = (\mathbf{X}^T\mathbf{X})^{-1}\mathbf{X}^T\mathbf{Y}$$

is an unbiased estimator of $\boldsymbol{\beta}$ and has the smallest variance of all other unbiased estimates of $\boldsymbol{\beta}$. Any estimator which is unbiased and has smaller variance than any other unbiased estimators is called a **best linear unbiased estimator** or **BLUE**.

An important note regarding the matrix expressions introduced above is that

$$\begin{aligned}
\hat{\mathbf{Y}} &= \mathbf{X}\hat{\boldsymbol{\beta}} \\
&= \mathbf{X}(\mathbf{X}^T\mathbf{X})^{-1}\mathbf{X}^T\mathbf{Y} \\
&= \mathbf{H}\mathbf{Y}
\end{aligned}$$

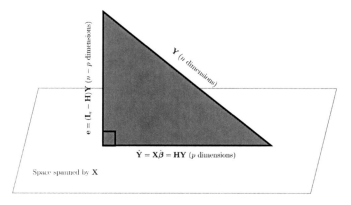

**FIGURE 6.1**
Geometric representation of the estimation of $\beta$ via ordinary least squares.

and

$$\mathbf{e} = \mathbf{Y} - \hat{\mathbf{Y}}$$
$$= \mathbf{Y} - \mathbf{HY}$$
$$= (\mathbf{I}_n - \mathbf{H})\mathbf{Y},$$

where $\mathbf{H} = \mathbf{X}(\mathbf{X}^{\mathrm{T}}\mathbf{X})^{-1}\mathbf{X}^{\mathrm{T}}$ is the $n \times n$ **hat matrix** and $\mathbf{I}_n$ is the $n \times n$ identity matrix. $\mathbf{H}$ is important for several reasons as it appears often in regression formulas. One important implication of $\mathbf{H}$ is that it is a projection matrix, meaning that it projects the response vector, $\mathbf{Y}$, as a linear combination of the columns of the $\mathbf{X}$ matrix in order to obtain the vector of fitted values, $\hat{\mathbf{Y}}$. Also, the diagonal of this matrix contains the $h_{j,j}$ values we introduced earlier in the context of Studentized residuals, which is important when discussing leverage.

Figure 6.1 provides a geometric representation of the estimation of $\beta$ via ordinary least squares. The data vector $\mathbf{Y}$ is an $n$-dimensional vector that extends out from the model space, which is the region represented as being spanned by $\mathbf{X}$ in this graphic. $\mathbf{Y}$ is then projected orthogonally onto the model space using the hat matrix $\mathbf{H}$. Hence, the vector of fitted values is $\hat{\mathbf{Y}} = \mathbf{HY} = \mathbf{X}\hat{\beta}$, which is $p$-dimensional. Moreover, the difference between these two vectors is simply $\mathbf{e} = (\mathbf{I}_n - \mathbf{H})\mathbf{Y}$, which is $(n - p)$-dimensional.

## 6.3 Variance–Covariance Matrix and Correlation Matrix of $\hat{\beta}$

Two important characteristics of the sample multiple regression coefficients are their standard errors and their correlations with each other. The **variance–covariance matrix** of the sample coefficients $\hat{\beta}$ is a symmetric $p \times p$ square matrix. Remember that $p$ is the number of regression parameters in the model, including the intercept.

The rows and the columns of the variance–covariance matrix are in coefficient order (first row is information about $\hat{\beta}_0$, second is about $\hat{\beta}_1$, and so on).

- The *diagonal values* (from top left to bottom right) are the variances of the sample coefficients, written as $\text{Var}(\hat{\beta}_i)$. The standard error of a coefficient is the square root of its variance.

- An *off-diagonal value* is the covariance between two coefficient estimates, written as $\text{Cov}(\hat{\beta}_i, \hat{\beta}_j)$.

- The *correlation* between two coefficient estimates can be determined using the following relationship: *correlation = covariance divided by product of standard deviations*, written as $\text{Corr}(\hat{\beta}_i, \hat{\beta}_j)$.

In regression, the theoretical variance–covariance matrix of the sample coefficients is

$$V(\hat{\beta}) = \sigma^2 (\mathbf{X}^\mathsf{T}\mathbf{X})^{-1}.$$

Recall, the MSE estimates $\sigma^2$, so the **estimated variance–covariance matrix** of the estimated regression parameters is calculated as

$$\hat{V}(\hat{\beta}) = \text{MSE}(\mathbf{X}^\mathsf{T}\mathbf{X})^{-1}.$$

We can also calculate the **correlation matrix of** $\hat{\beta}$ (denoted by $\mathbf{r}_{\hat{\beta}}$):

$$\mathbf{r}_{\hat{\beta}} = \begin{pmatrix} \dfrac{\text{Var}(\hat{\beta}_0)}{\sqrt{\text{Var}(\hat{\beta}_0)\text{Var}(\hat{\beta}_0)}} & \dfrac{\text{Cov}(\hat{\beta}_0,\hat{\beta}_1)}{\sqrt{\text{Var}(\hat{\beta}_0)\text{Var}(\hat{\beta}_1)}} & \cdots & \dfrac{\text{Cov}(\hat{\beta}_0,\hat{\beta}_{p-1})}{\sqrt{\text{Var}(\hat{\beta}_0)\text{Var}(\hat{\beta}_{p-1})}} \\ \dfrac{\text{Cov}(\hat{\beta}_1,\hat{\beta}_0)}{\sqrt{\text{Var}(\hat{\beta}_1)\text{Var}(\hat{\beta}_0)}} & \dfrac{\text{Var}(\hat{\beta}_1)}{\sqrt{\text{Var}(\hat{\beta}_1)\text{Var}(\hat{\beta}_1)}} & \cdots & \dfrac{\text{Cov}(\hat{\beta}_1,\hat{\beta}_{p-1})}{\sqrt{\text{Var}(\hat{\beta}_1)\text{Var}(\hat{\beta}_{p-1})}} \\ \vdots & \vdots & \ddots & \vdots \\ \dfrac{\text{Cov}(\hat{\beta}_{p-1},\hat{\beta}_0)}{\sqrt{\text{Var}(\hat{\beta}_{p-1})\text{Var}(\hat{\beta}_0)}} & \dfrac{\text{Cov}(\hat{\beta}_{p-1},\hat{\beta}_1)}{\sqrt{\text{Var}(\hat{\beta}_{p-1})\text{Var}(\hat{\beta}_1)}} & \cdots & \dfrac{\text{Var}(\hat{\beta}_{p-1})}{\sqrt{\text{Var}(\hat{\beta}_{p-1})\text{Var}(\hat{\beta}_{p-1})}} \end{pmatrix}$$

$$= \begin{pmatrix} 1 & \text{Corr}(\hat{\beta}_0, \hat{\beta}_1) & \cdots & \text{Corr}(\hat{\beta}_0, \hat{\beta}_{p-1}) \\ \text{Corr}(\hat{\beta}_1, \hat{\beta}_0) & 1 & \cdots & \text{Corr}(\hat{\beta}_1, \hat{\beta}_{p-1}) \\ \vdots & \vdots & \ddots & \vdots \\ \text{Corr}(\hat{\beta}_{p-1}, \hat{\beta}_1) & \text{Corr}(\hat{\beta}_{p-1}, \hat{\beta}_2) & \cdots & 1 \end{pmatrix}. \tag{6.5}$$

$r_{\hat{\beta}}$ is an estimate of the population correlation matrix $\rho_{\hat{\beta}}$. For example, $\text{Corr}(\hat{\beta}_1, \hat{\beta}_2)$, is interpreted as the average change in $Y$ for each unit change in $X_1$ and each unit change in $X_2$. The direction of the "changes" will depend on the signs of $\hat{\beta}_1$ and $\hat{\beta}_1$. Note that we usually do not care about correlations concerning the intercept, $\hat{\beta}_0$, since we usually want to provide an interpretation concerning the predictor variables.

If all of the predictor variables are uncorrelated with each other, then all covariances between pairs of sample coefficients that multiply those predictor variables will equal 0. This means that the estimate of one regression coefficient is not affected by the presence of the other predictor variables. Many experiments are designed to achieve this property, but achieving it with real data is often a greater challenge.

The correlation matrix presented in (6.5) should *not* be confused with the correlation matrix, $\mathbf{r}$, constructed for each pairwise combination of the variables $Y, X_1, X_2, \ldots, X_{p-1}$:

$$\mathbf{r} = \begin{pmatrix} 1 & \text{Corr}(Y, X_1) & \cdots & \text{Corr}(Y, X_{p-1}) \\ \text{Corr}(X_1, Y) & 1 & \cdots & \text{Corr}(X_1, X_{p-1}) \\ \vdots & \vdots & \ddots & \vdots \\ \text{Corr}(X_{p-1}, Y) & \text{Corr}(X_{p-1}, X_1) & \cdots & 1 \end{pmatrix}.$$

Note that all of the diagonal entries are 1 because the correlation between a variable and itself is a perfect (positive) association. This correlation matrix is what most statistical software reports and it does not always report $r_{\hat{\beta}}$. The interpretation of each entry in $\mathbf{r}$ is identical to the Pearson correlation coefficient interpretation presented earlier. Specifically, it provides the strength and direction of the association between the variables corresponding to the row and column of the respective entry.

$100 \times (1 - \alpha)\%$ confidence intervals are also readily available for $\beta$:

$$\hat{\beta}_j \pm t^*_{n-p;1-\alpha/2} \sqrt{\hat{V}(\hat{\beta})_{j,j}},$$

where $\hat{V}(\hat{\beta})_{j,j}$ is the $j^{\text{th}}$ diagonal element of the estimated variance–covariance matrix of the sample beta coefficients; i.e., the (estimated) standard error. Furthermore, the Bonferroni joint $100 \times (1 - \alpha)\%$ confidence intervals are:

$$\hat{\beta}_j \pm t^*_{n-p;1-\alpha/(2p)} \sqrt{\hat{V}(\hat{\beta})_{j,j}},$$

for $j = 0, 1, 2, \ldots, (p - 1)$.

## 6.4   Testing the Contribution of Individual Predictor Variables

Within a multiple linear regression model, we may want to know whether a particular predictor variable is making a useful contribution to the model. That is, given the presence of the other predictor variables in the model, does a particular $X$-variable help us predict or explain $Y$? For example, suppose that we have three predictor variables in the model. The multiple linear regression model is

$$Y = \beta_0 + \beta_1 X_1 + \beta_2 X_2 + \beta_3 X_3 + \epsilon. \tag{6.6}$$

To determine if $X_2$ is a useful predictor variable in this model, we would test

$$H_0 : \beta_2 = 0$$
$$H_A : \beta_2 \neq 0.$$

If the null hypothesis above were the case, then a change in the value of $X_2$ would not change $Y$, so $Y$ and $X_2$ are not related. Also, we would still be left with variables $X_1$ and $X_3$ being present in the model. When we fail to reject the null hypothesis above, we do not need variable $X_2$ in the model *given* that variables $X_1$ and $X_3$ will remain in the model. This is the major subtlety in the interpretation of a slope in multiple linear regression. Correlations among the predictors can change the slope values dramatically from what they would be in separate simple linear regressions.

For the multiple linear regression model in (6.3), we test the effect of $X_j$ on $Y$ *given* that all other predictors are in the model by carrying out a test on the corresponding regression coefficient $\beta_j$; i.e.,

$$H_0 : \beta_j = 0$$
$$H_A : \beta_j \neq 0$$

for $j = 1, \ldots, p - 1$. Statistical software will report $p$-values for each of the above tests. The $p$-value is based on a $t$-statistic calculated as

$$t^* = \frac{\hat{\beta}_j - 0}{\text{s.e.}(\hat{\beta}_j)} = \frac{\hat{\beta}_j}{\text{s.e.}(\hat{\beta}_j)},$$

which is distributed according to a $t_{n-p}$ distribution. Note that more generally we can test whether or not $\beta_j$ equals a non-zero quantity $\beta^*$; however, this is rarely of interest and the hypothesized value is usually just 0.

## 6.5   Statistical Intervals

The statistical intervals for estimating the mean or predicting new observations in the simple linear regression case is easily extended to the multiple regression case. Here, it is only necessary to present the formulas.

First, let use define the vector of given predictors as

$$
\mathbf{x}_h = \begin{pmatrix} 1 \\ x_{h,1} \\ x_{h,2} \\ \vdots \\ x_{h,p-1} \end{pmatrix}.
$$

We are interested in either intervals for $E(Y|\mathbf{X} = \mathbf{x}_h)$ or intervals for the value of a new response $y_h$ given that the observation has the particular value $\mathbf{x}_h$. First we define the standard error of the fit at $\mathbf{x}_h$ given by:

$$
\text{s.e.}(\hat{y}_h) = \sqrt{\text{MSE}(\mathbf{x}_h^T(\mathbf{X}^T\mathbf{X})^{-1}\mathbf{x}_h)}.
$$

Now, we can give the formulas for the various intervals:

- $100 \times (1 - \alpha)\%$ **Confidence Interval:**

$$
\hat{y}_h \pm t^*_{n-p;1-\alpha/2}\text{s.e.}(\hat{y}_h).
$$

- **Bonferroni Joint** $100 \times (1 - \alpha)\%$ **Confidence Intervals:**

$$
\hat{y}_{h_i} \pm t^*_{n-p;1-\alpha/(2q)}\text{s.e.}(\hat{y}_{h_i}),
$$

for $i = 1, 2, \ldots, q$.

- $100 \times (1 - \alpha)\%$ **Working–Hotelling Confidence Band:**

$$
\hat{y}_h \pm \sqrt{pF^*_{p,n-p;1-\alpha}}\text{s.e.}(\hat{y}_h).
$$

- $100 \times (1 - \alpha)\%$ **Prediction Interval:**

$$
\hat{y}_h \pm t^*_{n-p;1-\alpha/2}\sqrt{\text{MSE}/m + [\text{s.e.}(\hat{y}_h)]^2},
$$

where $m = 1$ corresponds to a prediction interval for a new observation at a given $\mathbf{x}_h$ and $m > 1$ corresponds to the mean of $m$ new observations calculated at the same $\mathbf{x}_h$.

- **Bonferroni Joint** $100 \times (1 - \alpha)\%$ **Prediction Intervals:**

$$
\hat{y}_{h_i} \pm t^*_{n-p;1-\alpha/(2q)}\sqrt{\text{MSE} + [\text{s.e.}(\hat{y}_{h_i})]^2},
$$

for $i = 1, 2, \ldots, q$.

- **Scheffé Joint** $100 \times (1 - \alpha)\%$ **Prediction Intervals:**

$$\hat{y}_{h_i} \pm \sqrt{qF^*_{q,n-p;1-\alpha}(\text{MSE} + [\text{s.e.}(\hat{y}_h)]^2)},$$

for $i = 1, 2, \ldots, q$.

- $[100 \times (1 - \alpha)\%]/[100 \times P\%]$ **Tolerance Intervals:**

  - *One-Sided Intervals:*

  $$(-\infty, \hat{y}_h + K_{\alpha,P}\sqrt{\text{MSE}})$$

  and

  $$(\hat{y}_h - K_{\alpha,P}\sqrt{\text{MSE}}, \infty)$$

  are the upper and lower one-sided tolerance intervals, respectively, where $K_{\alpha,P}$ is found similarly as in the simple linear regression setting, but with $n^* = (\mathbf{x}_h^{\text{T}}(\mathbf{X}^{\text{T}}\mathbf{X})^{-1}\mathbf{x}_h)^{-1}$.

  - *Two-Sided Interval:*

  $$\hat{y}_h \pm K_{\alpha/2,P/2}\sqrt{\text{MSE}},$$

  where $K_{\alpha/2,P/2}$ is found similarly as in the simple linear regression setting, but with $n^*$ as given above and $f = n - p$, where $p$ is the dimension of $\mathbf{x}_h$.

Note that calibration and regulation in the context of multiple predictors are more challenging problems. Some approximations and numerical approaches are presented in Brown (1994).

## 6.6  Polynomial Regression

In our earlier discussions on multiple linear regression, we mentioned that ways to check the assumption of linearity is by looking for curvature in various residual plots, just as in simple linear regression. For example, we can look at the plot of residuals versus the fitted values or a scatterplot of the response value versus each predictor. Sometimes, a plot of the response versus a predictor may also show some curvature in that relationship. Such plots may suggest there is a nonlinear relationship. If we believe there is a nonlinear relationship between the response and predictor(s), then one way to account for it is through a **polynomial regression** model:

$$Y = \beta_0 + \beta_1 X + \beta_2 X^2 + \ldots + \beta_q X^q + \epsilon, \tag{6.7}$$

where $q$ is called the **degree** or **order** of the polynomial. For lower degrees, the relationship has a specific name (i.e., $q = 2$ is called **quadratic**, $q = 3$ is called cubic, $q = 4$ is called **quartic**, and so on). As for a bit of semantics, it was noted in Chapter 2 that nonlinear regression (which we discuss in Chapter 19) refers to the nonlinear behavior of the coefficients, which are linear in polynomial regression. Thus, polynomial regression is still considered linear regression.

In order to estimate the polynomial regression model in (6.7), we only need the response variable $(Y)$ and the predictor variable $(X)$. However, polynomial regression models may have other predictor variables in them as well, which could lead to interaction terms. So as you can see, equation (6.7) is a relatively simple model, but you can imagine how the model can grow depending on your situation.

For the most part, we implement the same analysis procedures as in multiple linear regression. For example, consider the data in Figure 6.2(a), which were generated according to a quadratic regression model; i.e., $q = 2$. We have overlaid both the true mean curve used to generate these data as well as the ordinary least squares fit assuming a simple linear regression model. Clearly a simple linear regression model is not appropriate here. We are also able to assess this lack of fit using our standard regression visualizations:

- Figure 6.2(b) is a histogram of the Studentized residuals, which clearly has a peak broader than a regular normal distribution as well as extreme values to the left.

- Figure 6.2(c) is a scatterplot of the Studentized residuals versus the fitted values. Clearly the residuals do not exhibit a random pattern around the horizontal line of 0, but rather a strong curved pattern is noticeable. This plot alone would suggest that there is something wrong with the model being used and especially indicate the use of a higher-degree model.

- Figure 6.2(d) is a normal probability plot of the Studentized residuals, which reveals extreme values at both ends that deviate substantially from the 0-1 (i.e., $y = x$) line.

Some general guidelines to keep in mind when estimating a polynomial regression model are:

- The fitted model is more reliable when it is built on a larger sample size $n$.

- Do not extrapolate beyond the limits of your observed values.

- Consider how large the size of the predictor(s) will be when incorporating higher degree terms, as this may cause overflow.

- Do not go strictly by low $p$-values to incorporate a higher degree term, but rather just use these to support your model only if the plot looks reasonable. This is a situation where you need to determine practical significance versus statistical significance.

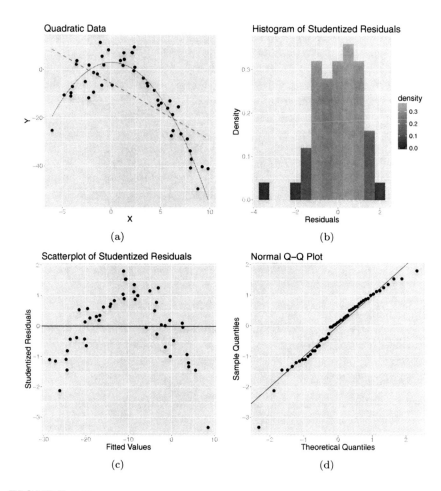

**FIGURE 6.2**
(a) Scatterplot of the quadratic data with the true quadratic regression line
(solid red line) and the fitted simple linear regression line (dashed blue line).
(b) Histogram of the Studentized residuals. (c) Residual plot for the simple
linear regression fit. (d) Normal probability plot for the Studentized residuals.
Colors are in e-book version.

- In general, you should obey the **hierarchy principle**, which says that if your model includes $X^q$ and $X^q$ is shown to be a statistically significant predictor of $Y$, then your model should also include each $X^k$ for all $k < q$, whether or not the coefficients for these lower-order terms are statistically significant.

## 6.7 Examples

**Example 6.7.1.** *(Thermal Energy Data)*
Heat flux is usually measured as part of solar thermal energy tests. An engineer wants to determine how total heat flux is predicted by a number of other measured variables. The data are from $n = 29$ homes used to test solar thermal energy and were reported and analyzed in Montgomery et al. (2013). The variables are:

- $Y$: total heat flux (in kilowatts)

- $X_1$: the east focal point (in inches)

- $X_2$: the south focal point (in inches)

- $X_3$: the north focal point (in inches)

- $X_4$: the insolation value (in watts/m$^2$)

- $X_5$: the time of day (converted to continuous values)

For now, we are only interested in assessing the relationship between total heat flux ($Y$) and the location predictors $X_1$, $X_2$, and $X_3$. First, we look at the correlation matrix **r**:

```
#############################
             heat.flux        east        south        north
heat.flux    1.0000000   0.1023521    0.1120914  -0.8488370
east         0.1023521   1.0000000   -0.3285435  -0.1172981
south        0.1120914  -0.3285435    1.0000000   0.2874106
north       -0.8488370  -0.1172981    0.2874106   1.0000000
#############################
```

The correlations are interpreted just as before. For example, the total heat flux and east focal point have a very weak, positive association (0.1024), while the total heat flux and north focal point have a strong, negative association ($-0.8488$).

The regression model of interest is

$$y_i = \beta_0 + \beta_1 x_{i,1} + \beta_2 x_{i,2} + \beta_3 x_{i,3} + \epsilon_i.$$

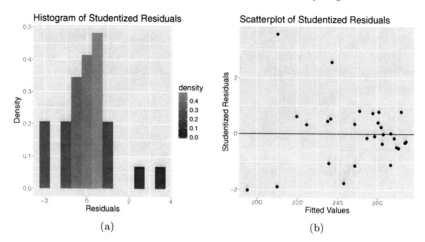

(a)                                                          (b)

**FIGURE 6.3**

(a) Histogram of the residuals for the thermal energy dataset with the three-predictor model. (b) Plot of the residuals for the thermal energy dataset with the three-predictor model.

Figure 6.3(a) is a histogram of the Studentized residuals. While the shape is not completely bell-shaped, it is not suggestive of any severe departures from normality. Figure 6.3(b) gives a plot of the Studentized residuals versus the fitted values. Again, the values appear to be randomly scattered about 0, suggesting constant variance.

The following provides the $t$-tests for the individual regression coefficients:

```
#############################
Coefficients:
            Estimate Std. Error t value Pr(>|t|)
(Intercept) 389.1659   66.0937    5.888 3.83e-06 ***
east          2.1247    1.2145    1.750   0.0925 .
south         5.3185    0.9629    5.523 9.69e-06 ***
north       -24.1324    1.8685  -12.915 1.46e-12 ***
---
Signif. codes:  0 *** 0.001 ** 0.01 * 0.05 . 0.1   1

Residual standard error: 8.598 on 25 degrees of freedom
Multiple R-squared:  0.8741,Adjusted R-squared:  0.859
F-statistic: 57.87 on 3 and 25 DF,  p-value: 2.167e-11
#############################
```

At the $\alpha = 0.05$ significance level, both north and south appear to be statistically significant predictors of heat flux. However, east is not (with a $p$-value

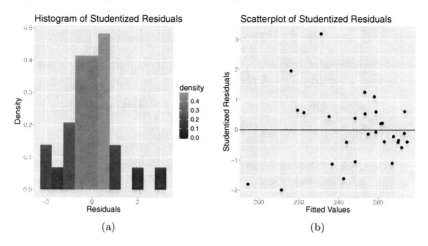

(a)                                    (b)

**FIGURE 6.4**
(a) Histogram of the residuals for the thermal energy dataset with the two-predictor model. (b) Plot of the residuals for the thermal energy dataset with the two-predictor model.

of 0.0925). While we could claim this is a marginally significant predictor, we will rerun the analysis by dropping the east predictor.

The following provides the $t$-tests for the individual regression coefficients for the newly suggested model:

```
##############################
Coefficients:
             Estimate Std. Error t value Pr(>|t|)
(Intercept) 483.6703    39.5671  12.224 2.78e-12 ***
south         4.7963     0.9511   5.043 3.00e-05 ***
north       -24.2150     1.9405 -12.479 1.75e-12 ***
---
Signif. codes:  0 *** 0.001 ** 0.01 * 0.05 . 0.1   1

Residual standard error: 8.932 on 26 degrees of freedom
Multiple R-squared:  0.8587,Adjusted R-squared:  0.8478
F-statistic: 79.01 on 2 and 26 DF,  p-value: 8.938e-12
##############################
```

The residual plots (Figure 6.4) based on this new fit appear very similar. Some things to note from this final analysis are:

- The estimated multiple linear regression equation is

$$\hat{y}_i = 483.67 + 4.80x_{i,1} - 24.22x_{i,2},$$

where we have reordered the index for the predictors such that $X_1$ is for the south focal point and $X_2$ is for the north focal point.

- To use our estimated equation for prediction, we substitute specified values for the two directions; i.e., north and south.

- We can interpret the coefficients in the same way that we do for a straight-line model, but we have to add the constraint that values of other variables remain constant.

  - When the north position is held constant, the average flux temperature for a home increases by 4.80 degrees for each 1 unit increase in the south position.

  - When the south position is held constant, the average flux temperature for a home decreases by 24.22 degrees for each 1 unit increase in the north position.

- The value of $R^2 = 0.8587$ means that the model (the two $x$-variables) explains 85.87% of the observed variation in a home's flux temperature.

- The value $\sqrt{\text{MSE}} = 8.932$ is the estimated standard deviation of the residuals. Roughly, it is the average absolute size of a residual.

The variance–covariance matrix of $\hat{\beta}$, $\hat{V}(\hat{\beta})$, is:

```
#############################
              (Intercept)       south          north
(Intercept)   1565.55372  -23.2796882  -44.0479176
south          -23.27969    0.9045889   -0.5304571
north          -44.04792   -0.5304571    3.7656828
#############################
```

The square root of the diagonal entries of the above matrix give us s.e.$(\hat{\beta}_0) = 39.57$, s.e.$(\hat{\beta}_1) = 0.95$, and s.e.$(\hat{\beta}_2) = 1.94$. These match the standard errors from the testing output given earlier.

We can also calculate the correlation matrix of $\hat{\beta}$, $r(\hat{\beta})$,:

```
#############################
              (Intercept)       south          north
(Intercept)    1.0000000  -0.6186109  -0.5736798
south          -0.6186109   1.0000000  -0.2874106
north          -0.5736798  -0.2874106   1.0000000
#############################
```

For example, $\text{Corr}(b_1, b_2) = -0.2874$, which implies there is a fairly low, negative correlation between the average change in flux for each unit increase in the south position and each unit increase in the north position. Therefore, the presence of the north position only slightly affects the estimate of the south's

**TABLE 6.2**
Statistical intervals for the given levels of $(x_{h,1}, x_{h,2})$, where $q = 2$ for the joint intervals

| $(x_{h,1}, x_{h,2})$ | $(35, 17)$ | $(40, 16)$ |
|---|---|---|
| $\hat{y}_h$ | 239.88 | 288.08 |
| 90% CI | $(236.69, 243.08)$ | $(279.38, 296.78)$ |
| Bonferroni 90% CI | $(235.68, 244.09)$ | $(276.62, 299.54)$ |
| Working–Hotelling 90% CB | $(235.86, 243.91)$ | $(277.12, 299.04)$ |
| 90% PI | $(224.32, 255.45)$ | $(270.54, 305.63)$ |
| Bonferroni 90% PI | $(219.38, 260.39)$ | $(264.97, 311.20)$ |
| Scheffé 90% PI | $(220.28, 259.49)$ | $(265.98, 310.18)$ |
| Upper 90%/99% TI | $(-\infty, 266.09)$ | $(-\infty, 317.47)$ |
| Lower 90%/99% TI | $(213.68, +\infty)$ | $(258.69, +\infty)$ |
| 90%/99% TI | $(211.08, 268.69)$ | $(256.24, 319.92)$ |

regression coefficient. The consequence is that it is fairly easy to separate the individual effects of these two variables.

Next we calculate and interpret the 95% confidence intervals for the slope coefficients:

- The 95% confidence interval for $\beta_1$ is $(2.841, 6.751)$. The interpretation is that we are 95% confident that the average increase in heat flux is between 2.841 and 6.751 for each unit increase in the south focal point at a fixed value of the north focal point.

- The 95% confidence interval for $\beta_2$ is $(-28.204, -20.226)$. The interpretation is that we are 95% confident that the average decrease in heat flux is between 20.226 and 28.204 for each unit increase in the north focal point at a fixed value of the south focal point.

Suppose the engineer is interested in the two pairs of values, $(x_{1,h}, x_{2,h})$, for the south and north focal points: $(35, 17)$ and $(40, 16)$. The statistical intervals discussed in Section 6.5 can each be calculated. In particular, the engineer wishes to calculate 90% confidence intervals, Bonferroni Joint 90% confidence intervals, 90% Working–Hotelling confidence bands, 90% prediction intervals, Bonferroni Joint 90% prediction intervals, Scheffé joint 90% prediction intervals, and 90%/99% one-sided and two-sided tolerance intervals. These are reported in Table 6.2 and hold similar interpretations as those reported in Example 3.6.1.

**Example 6.7.2.** *(Kola Project Data)*
The Kola Project ran from 1993–1998 and involved extensive geological surveys of Finland, Norway, and Russia. The entire published dataset (Reimann et al., 1998) consists of 605 geochemical samples measured on 111 variables. For our analysis, we will only look at a subset of the data pertaining to those

samples having a lithologic classification[1] of "1." The sample size of this subset is $n = 131$. The variables are:

- $Y$: the measure of chromium in the sample analyzed by instrumental neutron activation analysis (Cr_INAA)

- $X_1$: the measure of chromium in the sample measured by aqua regia extraction (Cr)

- $X_2$: the measure of cobalt in the sample measured by aqua regia extraction (Co)

The investigators are interested in modeling the geological composition variable Cr_INAA ($Y$) as a function of Cr ($X_1$) and Co ($X_2$). A 3D scatterplot of this data with the least squares plane is provided in Figure 6.5(a). In this 3D plot, observations above the plane (i.e., observations with positive residuals) are given by green points and observations below the plane (i.e., observations with negative residuals) are given by red points. The output for fitting a multiple linear regression model to this data is below:

```
#############################
Coefficients:
            Estimate Std. Error t value Pr(>|t|)
(Intercept)  53.3483    11.6908   4.563 1.17e-05 ***
Cr            1.8577     0.2324   7.994 6.66e-13 ***
Co            2.1808     1.7530   1.244    0.216
---
Signif. codes:  0 *** 0.001 ** 0.01 * 0.05 . 0.1   1

Residual standard error: 74.76 on 128 degrees of freedom
Multiple R-squared:  0.544,Adjusted R-squared:  0.5369
F-statistic: 76.36 on 2 and 128 DF,  p-value: < 2.2e-16
#############################
```

Note that Co is not statistically significant. However, the scatterplot in Figure 6.5(a) clearly shows that the data is skewed to the right for each of the variables; i.e., the bulk of the data is clustered near the lower-end of values for each variable while there are fewer values as you increase along a given axis. In fact, a plot of the standardized residuals against the fitted values (Figure 6.6(a)) indicates that a transformation is needed.

   Since the data appears skewed to the right for each of the variables, a log transformation on Cr_INAA, Cr, and Co will be taken. This is consistent with the Box–Cox transformation analysis performed in Reimann et al. (2002). The scatterplot in Figure 6.5(b) shows the results from this transformations along with the new least squares plane. Clearly, the transformation has done a

---

[1]The lithologic classification of a rock or geophysical sample is basically a description of that sample's physical characteristics that are visible on the surface.

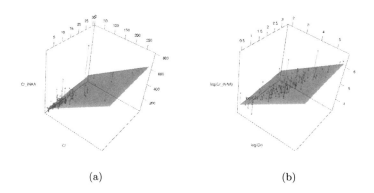

(a)          (b)

**FIGURE 6.5**
3D scatterplots of (a) the Kola dataset with the least squares plane and (b)
the Kola dataset where the logarithm of each variable has been taken.

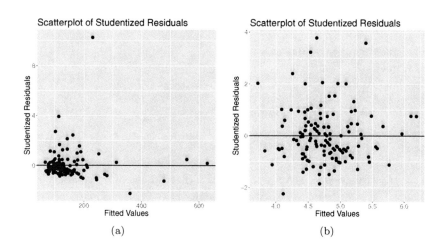

(a)          (b)

**FIGURE 6.6**
The Studentized residuals versus the fitted values for (a) the raw Kola dataset
and (b) the log-transformed Kola dataset.

better job linearizing the relationship. The output for fitting a multiple linear regression model to this transformed data is below:

```
############################
Coefficients:
            Estimate Std. Error t value Pr(>|t|)
(Intercept)  2.65109    0.17630  15.037  < 2e-16 ***
ln.Cr        0.57873    0.08415   6.877 2.42e-10 ***
ln.Co        0.08587    0.09639   0.891    0.375
---
Signif. codes:  0 *** 0.001 ** 0.01 * 0.05 . 0.1   1

Residual standard error: 0.3784 on 128 degrees of freedom
Multiple R-squared:  0.5732,Adjusted R-squared:  0.5665
F-statistic: 85.94 on 2 and 128 DF,  p-value: < 2.2e-16
############################
```

There is also a noted improvement in the plot of the standardized residuals versus the fitted values (Figure 6.6(b)). Notice that the log transformation of Co is not statistically significant as it has a high $p$-value (0.375).

After omitting the log transformation of Co from our analysis, a simple linear regression model is fit to the data. Figure 6.7 provides a scatterplot of the data and a plot of the Studentized residuals against the fitted values. These plots, combined with the following simple linear regression output, indicate a highly statistically significant relationship between the log transformation of Cr_INAA and the log transformation of Cr.

```
############################
Coefficients:
            Estimate Std. Error t value Pr(>|t|)
(Intercept)  2.60459    0.16826  15.48   <2e-16 ***
ln.Cr        0.63974    0.04887  13.09   <2e-16 ***
---
Signif. codes:  0 *** 0.001 ** 0.01 * 0.05 . 0.1   1

Residual standard error: 0.3781 on 129 degrees of freedom
Multiple R-squared:  0.5705,Adjusted R-squared:  0.5672
F-statistic: 171.4 on 1 and 129 DF,  p-value: < 2.2e-16
############################
```

**Example 6.7.3.** *(Tortoise Eggs Data)*
This dataset of size $n = 18$ contains measurements from a study on the number of eggs in female gopher tortoises in southern Florida (Ashton et al., 2007). The number of eggs in each tortoise — called the clutch size — was obtained using X-rays. The variables are:

- $Y$: number of eggs (clutch size)

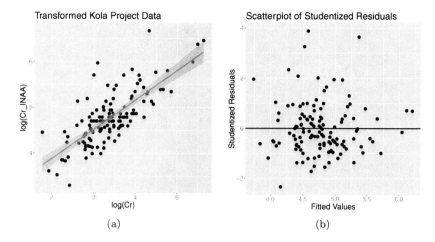

**FIGURE 6.7**
(a) Scatterplot of the Kola dataset where the logarithm of Cr_INAA has been regressed on the logarithm of Cr. (b) Plot of the Studentized residuals for this simple linear regression fit.

- $X$: carapace length (in millimeters)

Figure 6.8 gives scatterplots of the data with a simple linear regression fit and a quadratic regression fit overlaid. Obviously the quadratic fit appears better.

Here we have the linear fit results:

```
#############################
Coefficients:
            Estimate Std. Error t value Pr(>|t|)
(Intercept) -0.43532   17.34992  -0.025    0.980
length       0.02759    0.05631   0.490    0.631

Residual standard error: 3.411 on 16 degrees of freedom
Multiple R-squared:  0.01478,Adjusted R-squared:  -0.0468
F-statistic:  0.24 on 1 and 16 DF,  p-value: 0.6308
#############################
```

Here we have the quadratic fit results:

```
#############################
Coefficients:
             Estimate Std. Error t value Pr(>|t|)
(Intercept) -8.999e+02  2.703e+02  -3.329  0.00457 **
length       5.857e+00  1.750e+00   3.347  0.00441 **
I(length^2) -9.425e-03  2.829e-03  -3.332  0.00455 **
```

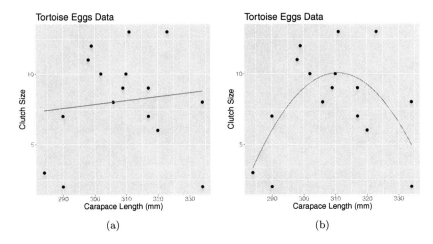

(a)                                                    (b)

**FIGURE 6.8**
The tortoise eggs dataset with (a) a linear fit and (b) a quadratic fit.

```
---
Signif. codes:  0 *** 0.001 ** 0.01 * 0.05 . 0.1   1

Residual standard error: 2.671 on 15 degrees of freedom
Multiple R-squared:  0.4338,Adjusted R-squared:  0.3583
F-statistic: 5.747 on 2 and 15 DF,  p-value: 0.01403
#############################
```

We see that the carapace length in the linear fit is not significant ($p$-value
of 0.631). However, the carapace length and its square are both significant
predictors in the quadratic regression fit ($p$-values of 0.004 and 0.005, respectively). Thus, our model should include a quadratic term.

# 7

## Multicollinearity

Recall that the columns of a matrix are linearly dependent if one column can be expressed as a linear combination of the other columns. Suppose we have the multiple linear regression model with $p - 1$ predictors:

$$\mathbf{Y} = \mathbf{X}\boldsymbol{\beta} + \boldsymbol{\epsilon}.$$

A matrix theorem states that if there is a linear dependence among the columns of $\mathbf{X}$, then $(\mathbf{X}^T\mathbf{X})^{-1}$ does not exist. This means that we cannot determine an estimate of $\boldsymbol{\beta}$ since the formula for determining the estimate involves $(\mathbf{X}^T\mathbf{X})^{-1}$.

In multiple regression, the term **multicollinearity** refers to the linear relationships among the predictor variables. For example, perfect multicollinearity would involve some subset of $q \leq p - 1$ predictors from our multiple linear regression model having the following relationship:

$$\alpha_0 \mathbf{1}_n + \sum_{j^*=1}^{q} \alpha_{j^*} \mathbf{X}_{j^*} = \mathbf{0},$$

where the $j^*$ represents the index relative to the $q$ predictors for which there is multicollinearity. We refer to the special case when $q = 2$ as **collinearity**. Practically speaking, the use of the term multicollinearity implies that the predictor variables are correlated with each other, so when they are not correlated with each other, we might claim that there is no multicollinearity.

## 7.1 Sources and Effects of Multicollinearity

There are various potential sources for multicollinearity, such as the following:

- During the data collection phase, an investigator may have drawn the data from such a narrow subspace of the independent variables that multicollinearity appears.

- Physical constraints, such as design limits, may also impact the range of some of the independent variables.

- Model specification, such as defining more variables than observations or specifying too many higher-ordered terms/interactions (i.e., polynomial terms).

- A large number of outliers can lead to multicollinearity.

When there is no multicollinearity among the predictor variables, the effects of the individual predictors can be estimated independently of each other, although we will still use a fitted multiple regression model. When multicollinearity is present, the estimated coefficients are correlated (confounded) with each other. This creates difficulty when we attempt to interpret how individual predictor variables affect the response variable.

Along with this correlation, multicollinearity has a multitude of other ramifications on our analysis, including:

- inaccurate regression coefficient estimates,

- inflated standard errors of the regression coefficient estimates,

- deflated $t$-tests and wider confidence intervals for significance testing of the regression coefficients,

- false nonsignificance determined by the $p$-values, and

- degradation of model predictability.

In designed experiments with multiple predictor variables, researchers usually choose the value of each predictor (or design variable) so that there is no multicollinearity. The **X** matrix in designed experiments is also usually called the **design matrix**. In observational studies and sample surveys, it is nearly always the case that variables in the **X** matrix will have some correlation.

## 7.2 Detecting and Remedying Multicollinearity

We introduce three primary ways for detecting multicollinearity, two of which are fairly straightforward to implement, while the third method is actually a variety of measures based on the eigenvalues and eigenvectors of the standardized design matrix. These methods, as well as other methods that are essentially variations on what we present, are discussed in Belsley (1991).

**Method 1: Pairwise Scatterplots**
For the first method, we can visually inspect the data by doing pairwise scatterplots of the independent variables. For $p - 1$ independent variables, you would need to inspect all $\binom{p-1}{2}$ pairwise scatterplots. You will be looking for

any plots that seem to indicate a linear relationship between pairs of independent variables.

## Method 2: VIF

Second, we can use a measure of multicollinearity called the **variance inflation factor** (**VIF**). This is defined as

$$VIF_j = \frac{1}{1 - R_j^2},$$

where $R_j^2$ is the coefficient of determination obtained by regressing $\mathbf{x}_j$ on the remaining independent variables. A common rule of thumb is that if $VIF_j = 1$, then there is no multicollinearity, if $1 < VIF_j < 5$, then there is possibly some moderate multicollinearity, and if $VIF_j \geq 5$, then there is a strong indication of a problem with multicollinearity. Most of the time, we will aim for values as close to 1 as possible and that usually will be sufficient. The bottom line is that the higher the $VIF$, the more likely multicollinearity is an issue.

Sometimes, the tolerance is also reported. The **tolerance** is simply the inverse of the $VIF$ (i.e., $Tol_j = VIF_j^{-1}$). In this case, the lower the $Tol$, the more likely multicollinearity is an issue. Note that this tolerance quantity is not related to the notion of tolerance intervals discussed in Chapter 3.

If multicollinearity is suspected after doing the above, then a couple of things can be done. First, reassess the choice of model and determine if there are any unnecessary terms that can be removed. You may wish to start by removing the one you most suspect first, because this will then drive down the $VIF$s of the remaining variables.

Next, check for outliers and assess the impact some of the observations with higher residuals have on the analysis. Remove some (or all) of the suspected outliers and see how that changes the pairwise scatterplots and $VIF$ values.

You can also standardize the variables, which involves simply subtracting each variable by its mean and dividing by its standard deviation. Thus, the **standardized X matrix** is given by:

$$\mathbf{X}^* = \frac{1}{\sqrt{n-1}} \begin{pmatrix} \frac{X_{1,1}-\bar{X}_1}{s_{X_1}} & \frac{X_{1,2}-\bar{X}_2}{s_{X_2}} & \cdots & \frac{X_{1,p-1}-\bar{X}_{p-1}}{s_{X_{p-1}}} \\ \frac{X_{2,1}-\bar{X}_1}{s_{X_1}} & \frac{X_{2,2}-\bar{X}_2}{s_{X_2}} & \cdots & \frac{X_{2,p-1}-\bar{X}_{p-1}}{s_{X_{p-1}}} \\ \vdots & \vdots & \ddots & \vdots \\ \frac{X_{n,1}-\bar{X}_1}{s_{X_1}} & \frac{X_{n,2}-\bar{X}_2}{s_{X_2}} & \cdots & \frac{X_{n,p-1}-\bar{X}_{p-1}}{s_{X_{p-1}}} \end{pmatrix},$$

which is an $n \times (p-1)$ matrix, and the **standardized Y vector** is given by:

$$\mathbf{Y}^* = \frac{1}{\sqrt{n-1}} \begin{pmatrix} \frac{Y_1-\bar{Y}}{s_Y} \\ \frac{Y_2-\bar{Y}}{s_Y} \\ \vdots \\ \frac{Y_n-\bar{Y}}{s_Y} \end{pmatrix},$$

which is still an $n$-dimensional vector. Here,

$$s_{X_j} = \sqrt{\frac{\sum_{i=1}^{n}(X_{i,j} - \bar{X}_j)^2}{n-1}}$$

for $j = 1, 2, \ldots, (p-1)$ and

$$s_Y = \sqrt{\frac{\sum_{i=1}^{n}(Y_i - \bar{Y})^2}{n-1}}.$$

Notice that we have removed the column of 1s in forming $\mathbf{X}^*$, effectively reducing the column dimension of the original $\mathbf{X}$ matrix by 1. Because of this, we can no longer estimate an intercept term, which may be an important part of the analysis. Thus, proceed with this method only if you believe the intercept term adds little value to explaining the science behind your regression model!

When using the standardized variables, the regression model of interest becomes:

$$\mathbf{Y}^* = \mathbf{X}^*\boldsymbol{\beta}^* + \boldsymbol{\epsilon}^*,$$

where $\boldsymbol{\beta}^*$ is now a $(p-1)$-dimensional vector of standardized regression coefficients and $\boldsymbol{\epsilon}^*$ is an $n$-dimensional vector of errors pertaining to this standardized model. Thus, the ordinary least squares estimates are

$$\hat{\boldsymbol{\beta}}^* = (\mathbf{X}^{*\mathrm{T}}\mathbf{X}^*)^{-1}\mathbf{X}^{*\mathrm{T}}\mathbf{Y}^*$$

$$= \mathbf{r}_{XX}^{-1}\mathbf{r}_{XY},$$

where $\mathbf{r}_{XX}$ is the $(p-1)\times(p-1)$ correlation matrix of the predictors and $\mathbf{r}_{XY}$ is the $(p-1)$-dimensional vector of correlation coefficients between the predictors and the response. Because $\hat{\boldsymbol{\beta}}^*$ is a function of correlations, this method is called a **correlation transformation**. Sometimes it may be enough to simply center the variables by their respective means in order to decrease the $VIF$s. Note the relationship between the quantities introduced above and the correlation matrix $\mathbf{r}$ from earlier:

$$\mathbf{r} = \begin{pmatrix} 1 & \mathbf{r}_{XY}^{\mathrm{T}} \\ \mathbf{r}_{XY} & \mathbf{r}_{XX} \end{pmatrix}.$$

**Method 3: Eigenvalue Methods**
Finally, the third method for identifying potential multicollinearity concerns a variety of measures utilizing eigenvalues and eigenvectors. Note that the eigenvalue $\lambda_j$ and the corresponding $(p-1)$-dimensional orthonormal eigenvectors $\xi_j$ are solutions to the system of equations:

$$\mathbf{X}^{*\mathrm{T}}\mathbf{X}^*\xi_j = \lambda_j\xi_j,$$

for $j = 1, \ldots, (p-1)$. Since the $\lambda_j$'s are normalized, it follows that

$$\xi_j^{\mathrm{T}}\mathbf{X}^{*\mathrm{T}}\mathbf{X}^*\xi_j = \lambda_j.$$

Therefore, if $\lambda_j \approx 0$, then $\mathbf{X}^* \xi_j \approx 0$; i.e., the columns of $\mathbf{X}^*$ are approximately linearly dependent. Thus, since the sum of the eigenvalues must equal the number of predictors (i.e., $(p-1)$), then very small $\lambda_j$'s (say, near 0.05) are indicative of multicollinearity. Another criterion commonly used is to declare multicollinearity is present when $\sum_{j=1}^{p-1} \lambda_j^{-1} > 5(p-1)$. Moreover, the entries of the corresponding $\xi_j$'s indicate the nature of the linear dependencies; i.e., large elements of the eigenvectors identify the predictor variables that comprise the multicollinearity.

A measure of the overall multicollinearity of the variables can be obtained by computing what is called the **condition number** of the correlation matrix $\mathbf{r}$, which is defined as $\sqrt{\lambda_{(p-1)}/\lambda_{(1)}}$, such that $\lambda_{(1)}$ and $\lambda_{(p-1)}$ are the minimum and maximum eigenvalues, respectively. Obviously this quantity is always greater than 1, so a large number is indicative of multicollinearity. Empirical evidence suggests that a value less than 15 typically means weak multicollinearity, values between 15 and 30 is evidence of moderate multicollinearity, while anything over 30 is evidence of strong multicollinearity.

Condition numbers for the individual eigenvalues can also be calculated. This is accomplished by taking $c_j = \sqrt{\lambda_{(p-1)}/\lambda_j}$ for each $j = 1, \ldots, (p-1)$. When data is centered and scaled, then $c_j \le 100$ indicates no multicollinearity, $100 < c_j < 1000$ indicates moderate multicollinearity, while $c_j \ge 1000$ indicates strong multicollinearity for predictor $\mathbf{x}_j$. When the data is only scaled (i.e., for regression through the origin models), then multicollinearity will always be worse. Thus, more relaxed limits are usually used. For example, a common rule-of-thumb is to use 5 times the limits mentioned above; namely, $c_j \le 500$ indicates no multicollinearity, $500 < c_j < 5000$ indicates moderate multicollinearity, while $c_j \ge 5000$ indicates strong multicollinearity involving predictor $\mathbf{x}_j$.

Finally, we can summarize the proportion of variance contributed by each regression coefficient to each component index. In other words, let

$$\phi_{jk} = \xi_{jk}^2/\lambda_j \quad \text{and} \quad \phi_k = \sum_{j=1}^{p-1} \phi_{jk},$$

where $\xi_{jk}$ is the $k^{\text{th}}$ element of the eigenvector $\xi_j$ and $k = 1, \ldots, p-1$. Then, the **variance–decomposition proportion** is defined as

$$\pi_{jk} = \phi_{jk}/\phi_k,$$

where for large condition indices, large values of $\phi_{jk}$ for two or more variables are indicative that these variables are causing collinearity problems. Belsley (1991) suggests large values of $\phi_{jk}$ are anything above 0.50.

It should be noted that there are many heuristic ways other than those described above to assess multicollinearity with eigenvalues and eigenvectors.[1]

---

[1] For example, one such technique involves taking the square eigenvector relative to the square eigenvalue and then seeing what percentage each quantity in this $(p-1)$-dimensional vector explains of the total variation for the corresponding regression coefficient.

Moreover, it should be noted that some observations can have an undue influence on these various measures of multicollinearity. These observations are called **collinearity–influential observations** and care should be taken with how these observations are handled. You can typically use some of the residual diagnostic measures, such as those discussed in Chapter 10, for identifying potential collinearity-influential observations since there is no established or agreed-upon method for classifying such observations.

Finally, there are also some more advanced regression procedures that can be performed in the presence of multicollinearity. Such methods include principal components regression, weighted regression, and ridge regression. These methods are discussed later.

## 7.3 Structural Multicollinearity

Our discussion of multicollinearity thus far has been on data that were collected from, perhaps, a poorly-designed experiment where two or more of the predictor variables have a linear relationship. Thus, the multicollinearity is characterized by the data. We can also have **structural multicollinearity**, which is a mathematical artifact caused by creating new predictors that are functions of other predictors. For example, in polynomial regression, we create powers of one or more of our predictors. Thus, structural multicollinearity results from the structure of your underlying model and not something that is, say, indicative of a poorly-designed experiment.

The typical way to handle structural multicollinearity is to center the predictor variables. For simplicity, let us use the $q^{\text{th}}$-degree polynomial regression model, which we will write as

$$\mathbf{Y} = \mathbf{X}_q \boldsymbol{\beta} + \boldsymbol{\epsilon},$$

where

$$\mathbf{X}_q = \begin{pmatrix} 1 & x_1 & x_1^2 & \cdots & x_1^q \\ 1 & x_2 & x_2^2 & \cdots & x_2^q \\ \vdots & \vdots & & \ddots & \vdots \\ 1 & x_n & x_n^2 & \cdots & x_n^q \end{pmatrix}.$$

Note in the above matrix that the subscript for $x_i$ refers to observation $i$, just like in the simple linear regression setting. Thus, the **centered X matrix** is given by:

$$\mathbf{X}_q^\star = \begin{pmatrix} 1 & (x_1 - \bar{x}) & (x_1 - \bar{x})^2 & \cdots & (x_1 - \bar{x})^q \\ 1 & (x_2 - \bar{x}) & (x_2 - \bar{x})^2 & \cdots & (x_2 - \bar{x})^q \\ \vdots & \vdots & & \ddots & \vdots \\ 1 & (x_n - \bar{x}) & (x_n - \bar{x})^2 & \cdots & (x_n - \bar{x})^q \end{pmatrix}.$$

Note that the column for the intercept is left unchanged. We then fit the centered model

$$Y = X_q^\star \beta^\star + \epsilon,$$

where the interpretation of the estimated values for $\beta^\star$ take on a slightly different meaning. Specifically,

- $\hat{\beta}_0^\star$ is the predicted response when the predictor value equals the sample mean of the $n$ values of the predictor.

- $\hat{\beta}_1^\star$ is the estimated slope of the tangent line at the predictor mean.

- $\hat{\beta}_2^\star$ indicates the direction of the curve; i.e., the curve is concave up when $\hat{\beta}_2^\star > 0$ and concave down when $\hat{\beta}_2^\star < 0$.

- $\hat{\beta}_3^\star, \hat{\beta}_4^\star, \ldots$ would all indicate relative changes in the response with respect to the centered value it is multiplied by.

## 7.4 Examples

**Example 7.4.1.** *(Thermal Energy Data, cont'd)*
Let us return to the heat flux dataset in Example 6.7.1. We let our model include the east, south, and north focal points, but also incorporate time and insolation as predictors. We will contrive these data by creating an artificial predictor variable, which is simply half of the east focal point measurements. Let us assume these were measurements intended to represent the west focal point. Figure 7.1 provides a scatterplot matrix of all the pairwise scatterplots for the six predictors and the response. Clearly, we see a perfect correlation between the east and west predictors. In practice, one could choose whichever of these two predictors they would like to retain. But since we contrived this example, we omit the west focal point. Note that if we try to compute $(X^T X)^{-1}$, the computer program will return an error message stating that a singularity has occurred since the inverse does not exist.

We next run a multiple regression analysis using the remaining predictors:

```
##############################
Coefficients:
             Estimate Std. Error t value Pr(>|t|)
(Intercept) 325.43612   96.12721   3.385  0.00255 **
east          2.55198    1.24824   2.044  0.05252 .
north       -22.94947    2.70360  -8.488 1.53e-08 ***
south         3.80019    1.46114   2.601  0.01598 *
time          2.41748    1.80829   1.337  0.19433
insolation    0.06753    0.02899   2.329  0.02900 *
```

---
```
Signif. codes:  0 '***' 0.001 '**' 0.01 '*' 0.05 '.' 0.1 ' ' 1

Residual standard error: 8.039 on 23 degrees of freedom
Multiple R-Squared: 0.8988,     Adjusted R-squared: 0.8768
F-statistic: 40.84 on 5 and 23 DF,  p-value: 1.077e-10
##############################
```

We see that time is not a statistically significant predictor of heat flux and, in fact, east has become marginally significant.

However, let us now look at the $VIF$ values:

```
##############################
     east       north       south       time insolation
  1.355448    2.612066    3.175970   5.370059   2.319035
##############################
```

Notice that the $VIF$ for time is fairly high (about 5.37). This is a somewhat high value and should be investigated further. We can confirm this potential problem with multicollinearity by also taking the eigenvalue approach. First, the eigenvalues are given below:

```
##############################
[1] 8.459534 6.130335 4.413890 2.865005 1.778934
##############################
```

While none of them are very small, the smallest value of 1.78 could be indicative of some effects due to multicollinearity. A better assessment can be obtained by the condition numbers. Below, we see that the largest condition number is 72.73, which indicates evidence of strong multicollinearity. In fact, three of the condition numbers are above the rule-of-thumb value of 30. We further see that for the largest condition number, south, time, and insolation could be the biggest contributors to the multicollinearity problem since they all have variance–decomposition proportions greater than 0.50.

```
##############################
Condition
Index Variance Decomposition Proportions
          east  north south time  insolation
1    1.000 0.000 0.000 0.000 0.000 0.000
2   14.092 0.003 0.001 0.000 0.091 0.105
3   33.103 0.304 0.080 0.013 0.149 0.258
4   55.899 0.668 0.895 0.001 0.126 0.005
5   72.733 0.025 0.023 0.985 0.634 0.632
##############################
```

Looking back at the pairwise scatterplots given in Figure 7.1 (we will only focus on the plots involving the time variable since that is the variable we

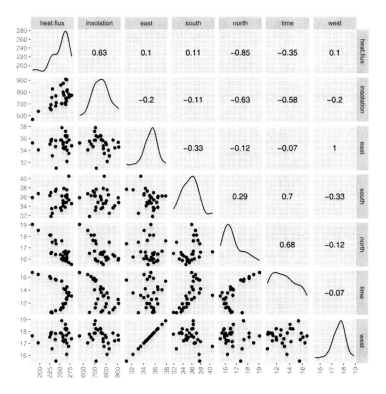

**FIGURE 7.1**

Matrix of pairwise scatterplots of the thermal energy data, including the made-up west focal point variable. The diagonal of this matrix gives a density estimate of the respective variable, while the strictly upper triangular portion of this matrix gives the correlations between the column and row variables.

are investigating), there appears to be a noticeable linear trend between time and the south focal point. There also appears to be some sort of curvilinear trend between time and the north focal point as well as between time and insolation. These plots, combined with the $VIF$ for time and assessment based on eigenvalue methods, suggests to look at a model without the time variable.

After removing the time variable, we obtain the new $VIF$ values:

```
##############################
     east      north     south insolation
 1.277792   1.942421  1.206057   1.925791
##############################
```

Notice how removal of the time variable has sharply decreased the $VIF$ values for the other variables. If one wishes to report the tolerances, simply take the inverse of the above $VIF$ values:

```
##############################
      east        north        south insolation
  0.7825999   0.5148215   0.8291484   0.5192672
##############################
```

We also see improvements with the eigenvalues:

```
##############################
[1] 6.997676 6.119343 4.172404 2.859583
##############################
```

There is further improvement in the condition numbers, although the largest suggests that some collinearity might still exist between the north and south focal point variables:

```
##############################
Condition
Index Variance Decomposition Proportions
          east  north south insolation
1    1.000 0.000 0.000 0.000 0.001
2   18.506 0.007 0.050 0.010 0.665
3   37.809 0.715 0.009 0.362 0.016
4   53.511 0.278 0.941 0.628 0.319
##############################
```

Regardless, we proceed to include both the north and south focal points as predictors in our model.

The regression coefficient estimates are:

```
##############################
Coefficients:
               Estimate Std. Error t value Pr(>|t|)
(Intercept) 270.21013   88.21060   3.063  0.00534 **
east          2.95141    1.23167   2.396  0.02471 *
north       -21.11940    2.36936  -8.914 4.42e-09 ***
south         5.33861    0.91506   5.834 5.13e-06 ***
insolation    0.05156    0.02685   1.920  0.06676 .
---
Signif. codes:  0 '***' 0.001 '**' 0.01 '*' 0.05 '.' 0.1 ' ' 1

Residual standard error: 8.17 on 24 degrees of freedom
Multiple R-Squared: 0.8909,    Adjusted R-squared: 0.8727
F-statistic: 48.99 on 4 and 24 DF,  p-value: 3.327e-11
##############################
```

Notice that now east is statistically significant while insolation is marginally significant. If we proceeded to drop insolation from the model, then we would be back to the analysis we did in Example 6.7.1. This illustrates how dropping

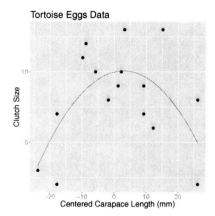

Centered Carapace Length (mm)

**FIGURE 7.2**
Plot of the tortoise eggs dataset where length has been centered.

or adding a predictor to a model can change the significance of other predictors. We will return to this when we discuss stepwise regression methods.

**Example 7.4.2.** *(Tortoise Eggs Data, cont'd)*
In Example 6.7.3, we fit a quadratic regression model to the data obtained from the study of tortoise eggs. Let us look at the $VIF$ value between the linear and quadratic length term:

```
#############################
  length  length2
1575.807  1575.807
#############################
```

Clearly this $VIF$ value is high due to the induced structural collinearity. Let us now center the length variable and recalculate the $VIF$:

```
#############################
 length.ctr length2.ctr
   1.036423    1.036423
#############################
```

This clearly shows a drastic improvement in the $VIF$ value.
Let us next look at the estimates when fitting the model using the centered predictor:

```
#############################
Coefficients:
            Estimate Std. Error t value Pr(>|t|)
(Intercept) 9.976718   0.853600  11.688 6.19e-09 ***
```

```
length.ctr    0.055623    0.044883    1.239   0.23428
length2.ctr  -0.009425    0.002829   -3.332   0.00455 **
---
Signif. codes:  0 *** 0.001 ** 0.01 * 0.05 . 0.1   1

Residual standard error: 2.671 on 15 degrees of freedom
Multiple R-squared:   0.4338, Adjusted R-squared:   0.3583
F-statistic: 5.747 on 2 and 15 DF,  p-value: 0.01403
##############################
```

$\hat{\beta}_1^\star$ is noticeably different from $\hat{\beta}_1$ in the original untransformed fit, but the other two estimates are very similar. The interpretations of the estimates in this example are:

- The predicted response is $\hat{\beta}_0^\star = 9.9767$ when the value of length equals the sample mean of the recorded length values (307.7778).

- The estimated slope of the tangent line at the sample mean of length (307.7778) is $\hat{\beta}_1^\star = 0.0556$.

- $\hat{\beta}_2^\star = -0.0094 < 0$, so the curve is concave down.

Figure 7.2 shows the estimated regression function using the transformed predictor values, which still appears to do a good job characterizing the trend.

# 8

# ANOVA for Multiple Linear Regression

As in simple linear regression, the ANOVA table for a multiple linear regression model displays quantities that measure how much of the variability in the response variable is explained and how much is not explained by the predictor variables. The calculation of the quantities involved is nearly identical to what we did in simple linear regression. The main difference has to do with the degrees of freedom quantities.

## 8.1   The ANOVA Table

The basic structure of the ANOVA table for multiple linear regression is given in Table 8.1. The explanations — nearly identical to those presented in the simple linear regression setting — are as follows:

- The total sum of squares is SSTO $= \sum_{i=1}^{n}(y_i - \bar{y})^2$, which is the sum of squared deviations from the overall mean of $y$ and $\mathrm{df}_T = n - 1$. SSTO is a measure of the overall variation in the $y$-values. In matrix notation,

$$\mathbf{SSTO} = ||\mathbf{Y} - \bar{Y}\mathbf{1}_n||^2.$$

- The **error sum of squares** is SSE $= \sum_{i=1}^{n}(y_i - \hat{y}_i)^2$, which is the sum of squared residuals. SSE is a measure of the variation in $y$ that is not explained by the regression. For multiple linear regression, $\mathrm{df}_E = n - p$, where $p$ is the dimension of $\boldsymbol{\beta}$ in the model, including the intercept $\beta_0$. In matrix notation,

$$\mathbf{SSE} = ||\mathbf{Y} - \hat{\mathbf{Y}}||^2.$$

- The mean squared error is MSE $= \frac{\text{SSE}}{\mathrm{df}_E} = \frac{\text{SSE}}{n-p}$, which estimates $\sigma^2$, the variance of the errors.

- The sum of squares due to the regression is SSR $=$ SSTO $-$ SSE $= \sum_{i=1}^{n}(\hat{y}_i - \bar{y})^2$, and it is a measure of the total variation in $y$ that can be explained by the regression with the identified predictor variables. Also, $\mathrm{df}_R = \mathrm{df}_T - \mathrm{df}_E$. For multiple regression, $\mathrm{df}_R = (n-1) - (n-p) = p - 1$. In matrix notation,

$$\mathbf{SSR} = ||\hat{\mathbf{Y}} - \bar{Y}\mathbf{1}_n||^2.$$

**TABLE 8.1**

ANOVA table for any linear regression

| Source | df | SS | MS | F |
|--------|-----|-----|-----|-----|
| **Regression** | $p-1$ | $\sum_{i=1}^{n}(\hat{y}_i - \bar{y})^2$ | MSR | MSR/MSE |
| **Error** | $n-p$ | $\sum_{i=1}^{n}(y_i - \hat{y}_i)^2$ | MSE | |
| **Total** | $n-1$ | $\sum_{i=1}^{n}(y_i - \bar{y})^2$ | | |

- The **mean square for the regression** is MSR $= \frac{\text{SSR}}{\text{df}_R} = \frac{\text{SSR}}{p-1}$.

The $F$-statistic in the ANOVA given in Table 8.1 can be used to test whether the response variable is related to one or more of the predictor variables in the model. Specifically, $F^* = \text{MSR}/\text{MSE}$ is a test statistic for

$$H_0 : \beta_1 = \beta_2 = \ldots = \beta_{p-1} = 0$$
$$H_A : \text{at least one } \beta_i \neq 0 \text{ for } i = 1, \ldots, p-1.$$

Thus, $F^* \sim F_{p-1,n-p}$. The null hypothesis means that the response variable is not related to any of the predictor variables in the model. The alternative hypothesis means that the response variable is related to at least one of the predictor variables in the model. The usual decision rule also applies here in that if $p < 0.05$, reject the null hypothesis. If that is our decision, we conclude that the response variable is related to at least one of the predictor variables in the model.

The MSE is the estimate of the error variance $\sigma^2$. Thus $s = \sqrt{\text{MSE}}$ estimates the standard deviation of the errors. Moreover, as in simple regression, $R^2 = \frac{\text{SSTO-SSE}}{\text{SSTO}}$, but here it is called the **coefficient of multiple determination**. $R^2$ is interpreted as the proportion of variation in the observed response values that is explained by the model; i.e., by the entire set of predictor variables in the model.

## 8.2   The General Linear $F$-Test

The **general linear $F$-test procedure** is used to test any null hypothesis that, if true, still leaves us with a linear model (linear in the $\beta$s). The most common application is to test whether a particular set of the $\beta$ coefficients are all equal 0. As an example, suppose we have a response variable $(Y)$ and 5 predictor variables $(X_1, X_2, X_3, X_4, X_5)$. We might wish to test

$$H_0 : \beta_1 = \beta_3 = \beta_4 = 0$$
$$H_A : \text{at least one of } \{\beta_1, \beta_3, \beta_4\} \neq 0.$$

The purpose of testing a hypothesis like this is to determine whether we could eliminate variables $X_1$, $X_3$, and $X_4$ from a multiple regression model, an action implied by the statistical truth of the null hypothesis.

**TABLE 8.2**

ANOVA table for multiple linear regression which includes a lack-of-fit test

| Source | df | SS | MS | F |
|--------|-----|------|-------|-----------|
| Regression | $p-1$ | SSR | MSR | MSR/MSE |
| Error | $n-p$ | SSE | MSE | |
| Lack of Fit | $m-p$ | SSLOF | MSLOF | MSLOF/MSPE |
| Pure Error | $n-m$ | SSPE | MSPE | |
| Total | $n-1$ | SSTO | | |

The **full model** is the multiple regression model that includes all the variables under consideration. The **reduced model** is the regression model that would result if the null hypothesis were true. The **general linear $F$-statistic** is

$$F^* = \frac{\frac{\text{SSE(reduced)} - \text{SSE(full)}}{\text{df}_E(\text{reduced}) - \text{df}_E(\text{full})}}{\text{MSE(full)}}.$$

Here, $F^* \sim F_{df_1, df_2}$, where the degrees of freedom are $df_1 = \text{df}_E(\text{reduced}) - \text{df}_E(\text{full})$ and $df_2 = \text{df}_E(\text{full})$.

To summarize, the general linear $F$-test is used in settings where there are many predictors and it is desirable to see if only one or a subset of the predictors can adequately perform the task of estimating the mean response and prediction of new observations. Sometimes the full and reduced sums of squares are referred to as extra sums of squares.

## 8.3 Lack-of-Fit Testing in the Multiple Regression Setting

Formal lack-of-fit testing can also be performed in the multiple regression setting; however, the ability to obtain replicates tends to be more difficult the more predictors there are in the model. Note that the corresponding ANOVA table (Table 8.2) is similar to that introduced for the simple linear regression setting. However, now we have the notion of $p$ regression parameters and the number of replicates $(m)$ refers to the number of unique $\mathbf{x}_i$. In other words, the predictor vectors for two or more observations must have the same values for all of their entries to be considered a replicate. For example, suppose we have 3 predictors for our model. The observations $(40, 10, 12)$ and $(40, 10, 7)$ are unique levels, whereas the observations $(10, 5, 13)$ and $(10, 5, 13)$ would constitute a replicate.

Formal lack-of-fit testing in multiple linear regression can be difficult due to sparse data, unless the experiment was designed properly to achieve replicates. However, other methods can be employed for lack-of-fit testing when you do not have replicates. Such methods involve **data subsetting**. The basic approach is to establish criteria by introducing indicator variables, which in turn creates coded variables, which are discussed in Chapter 9. By coding the variables, you can artificially create replicates and then you can proceed with lack-of-fit testing. Another approach with data subsetting is to look at central regions of the data (i.e., observations where the leverage is less than $(1.1) * p/n$) and treat this as a reduced dataset. Then compare this reduced fit to the full fit (i.e., the fit with all of the data), for which the formulas for a lack-of-fit test can be employed. Be forewarned that these methods should only be used as exploratory methods and they heavily depend on the data subsetting method used.

## 8.4   Extra Sums of Squares

**Extra sums of squares** measure the marginal reduction in the SSE when including additional predictor variables in a regression model *given* that other predictors are already specified in the model. In probability theory, we write $A|B$ which means that event $A$ happens *given* that event $B$ happens (the vertical bar means *given*). We also utilize this notation when writing extra sums of squares. For example, suppose we are considering two predictors, $X_1$ and $X_2$. The SSE when both variables are in the model is smaller than when only one of the predictors is in the model. This is because when both variables are in the model, they both explain additional variability in $Y$, which drives down the SSE compared to when only one of the variables is in the model. This difference is what we call the extra sums of squares. Specifically,

$$\text{SSR}(X_1|X_2) = \text{SSE}(X_2) - \text{SSE}(X_1, X_2)$$

measures the marginal effect of adding $X_1$ to the model *given* that $X_2$ is already in the model. An equivalent expression is

$$\text{SSR}(X_1|X_2) = \text{SSR}(X_1, X_2) - \text{SSR}(X_2),$$

which can be viewed as the marginal increase in the regression sum of squares. Notice for the second formulation that the corresponding degrees of freedom is $(3-1) - (2-1) = 1$ (because the degrees of freedom for $\text{SSR}(X_1, X_2)$ is $(3-1) = 2$ and the degrees of freedom for $\text{SSR}(X_2)$ is $(2-1) = 1$). Thus,

$$\text{MSR}(X_1|X_2) = \frac{\text{SSR}(X_1|X_2)}{1}.$$

When more predictors are available, then there are a vast array of possible decompositions of the SSR into extra sums of squares. One generic formulation is that if you have $p$ predictors, then

$$
\begin{aligned}
\text{SSR}(X_1,\ldots,X_j,\ldots,X_p) ={}& \text{SSR}(X_j) + \text{SSR}(X_1|X_j) + \ldots \\
&+ \text{SSR}(X_{j-1}|X_1,\ldots,X_{j-2},X_j) \\
&+ \text{SSR}(X_{j+1}|X_1,\ldots,X_j) + \ldots \\
&+ \text{SSR}(X_p|X_1,\ldots,X_{p-1}).
\end{aligned}
$$

In the above, $j$ is just being used to indicate any one of the $p$ predictors. $X_j$ could also just as easily be a set of predictor variables. You can also calculate the marginal increase in the regression sum of squares when adding more than one predictor. One generic formulation is

$$
\begin{aligned}
\text{SSR}(X_1,\ldots,X_j|X_{j+1},\ldots,X_p) ={}& \text{SSR}(X_1|X_{j+1},\ldots,X_p) \\
&+ \text{SSR}(X_2|X_1,X_{j+1},\ldots,X_p) + \ldots \\
&+ \text{SSR}(X_{j-1}|X_1,\ldots,X_{j-2},X_{j+1},\ldots,X_p) \\
&+ \text{SSR}(X_j|X_1,\ldots,X_{j-1},X_{j+1},\ldots,X_p).
\end{aligned}
$$

Again, you could imagine many such possibilities.

The primary use of extra sums of squares is in the testing of whether or not certain predictors can be dropped from the model; i.e., the general linear $F$-test. Furthermore, they can also be used to calculate a version of $R^2$ for such models, called the partial $R^2$. A more detailed treatment of extra sums of squares and related measures can be found in Chapter 6 of Draper and Smith (1998).

## 8.5 Partial Measures and Plots

Here we introduce a few different types of "partial" measures that allow us to assess the contribution of individual predictors in the overall model.

### Partial $R^2$

Suppose we have set up a general linear $F$-test. We might be interested in seeing what percent of the variation in the response *cannot* be explained by the predictors in the reduced model (i.e., the model specified by $H_0$), but *can* be explained by the rest of the predictors in the full model. If we obtain a large percentage, then it is likely we would want to specify some or all of the remaining predictors to be in the final model since they explain so much variation.

The way we formally define this percentage is with the **partial $R^2$** or the **coefficient of partial determination**. Suppose, again, that we have three predictors: $X_1$, $X_2$, and $X_3$. For the corresponding multiple linear regression model (with response $Y$), we wish to know what percent of the variation is explained by $X_2$ and $X_3$ that is not explained by $X_1$. In other words, given $X_1$, what additional percent of the variation can be explained by $X_2$ and $X_3$? Note that here the full model will include all three predictors, while the reduced model will only include $X_1$. We next calculate the relevant ANOVA tables for the full and reduced models. The partial $R^2$ is then calculated as follows:

$$
\begin{aligned}
R^2_{Y,2,3|1} &= \frac{\text{SSR}(X_2, X_3|X_1)}{\text{SSE}(X_1)} \\
&= \frac{\text{SSE}(X_1) - \text{SSE}(X_1, X_2, X_3)}{\text{SSE}(X_1)} \\
&= \frac{\text{SSE(reduced)} - \text{SSE(full)}}{\text{SSE(reduced)}}.
\end{aligned}
$$

The above gives us the proportion of variation explained by $X_2$ and $X_3$ that cannot be explained by $X_1$. Note that the last line of the above equation is just demonstrating that the partial $R^2$ has a similar form to the traditional $R^2$.

More generally, consider partitioning the predictors $X_1, X_2, \ldots, X_p$ into two groups, $A$ and $B$, containing $u$ and $(p - u)$ predictors, respectively. The proportion of variation explained by the predictors in group $B$ that cannot be explained by the predictors in group $A$ is given by

$$
\begin{aligned}
R^2_{Y,B|A} &= \frac{\text{SSR}(B|A)}{\text{SSE}(A)} \\
&= \frac{\text{SSE}(A) - \text{SSE}(A, B)}{\text{SSE}(A)}.
\end{aligned}
$$

These partial $R^2$ values can also be used to calculate the power for the corresponding general linear $F$-test. The power is the probability that the calculated $F_{u,n-p-1;1-\alpha}(\delta)$ value is greater than the calculated $F_{u,n-p-1;1-\alpha}$ value under the $F_{u,n-p-1}(\delta)$ distribution, where

$$
\delta = n \times \left( \frac{R^2_{Y,A,B} - R^2_{Y,B}}{1 - R^2_{Y,A,B}} \right)
$$

is the noncentrality parameter.

**Partial Correlation**
Partial $R^2$ values between the response and a single predictor conditioned on one or more predictors can be used to define a **coefficient of partial correlation**. Specifically, suppose we have a response variable $Y$ and $p - 1$ predictors $X_1, \ldots, X_{p-1}$. The partial correlation coefficient between $Y$ and a predictor variable $X_j$ is

$$
r_{Y,j|1,\ldots,j-1,j+1,\ldots,p-1} = \textbf{sgn}(\hat{\beta}_j) \sqrt{R^2_{Y,j|1,\ldots,j-1,j+1,\ldots,p-1}},
$$

where $\hat{\beta}_j$ is the estimate of the slope coefficient for $X_j$ in the full multiple linear regression model. The above quantity closely mirrors the traditional correlation coefficient. Specifically, it describes the joint behavior between $Y$ and $X_j$, but given that $X_1, \ldots, X_{j-1}, X_{j+1}, \ldots, X_{p-1}$ are held constant.

The coefficient of partial correlation can be used as a criterion in stepwise procedures for adding predictors in a model; see Chapter 14. Coefficients of partial correlation can also be expressed in terms of simple correlation coefficients or other partial correlation coefficients depending on the specific correlation of interest. For example, suppose again that we have three predictors: $X_1$, $X_2$, and $X_3$. Then we can calculate the following coefficients of partial correlations:

$$r_{Y,2|1} = \frac{r_{Y,2} - r_{1,2}r_{Y,1}}{\sqrt{(1 - r_{1,2}^2)(1 - r_{Y,1}^2)}}$$

$$r_{Y,3|1,2} = \frac{r_{Y,3|2} - r_{1,3|2}r_{Y,1|2}}{\sqrt{(1 - r_{1,3|2}^2)(1 - r_{Y,1|2}^2)}},$$

where extensions are straightforward. Note that a partial correlation coefficient between $Y$ and a predictor can differ substantially from their simple correlation as well as possibly have different signs. Another way to calculate coefficients of partial correlation is shown below.

## Partial Regression Plots

Next we establish a way to visually assess the relationship of a given predictor to the response when accounting for all of the other predictors in the multiple linear regression model. Suppose we have $p - 1$ predictors $X_1, \ldots, X_{p-1}$ and that we are trying to assess each predictor's relationship with a response $Y$ given that the other predictors are already in the model. Let $r_{Y_{[j]}}$ denote the residuals that result from regressing $Y$ on all of the predictors *except* $X_j$. Moreover, let $r_{X_{[j]}}$ denote the residuals that result from regressing $X_j$ on all of the remaining $p - 2$ predictors. A **partial regression plot** (also referred to as an **added variable plot**, **adjusted variable plot**, or **individual coefficients plot**) is constructed by plotting $r_{Y_{[j]}}$ on the $y$-axis and $r_{X_{[j]}}$ on the $x$-axis.[1] Then, the relationship between these two sets of residuals is examined to provide insight into $X_j$'s contribution to the response given the other $p - 2$ predictors are in the model. This is a helpful exploratory measure if you are not quite sure about what type of relationship (e.g., linear, quadratic, logarithmic, etc.) that the response may have with a particular predictor.

For a simple linear regression of $r_{Y_{[j]}}$ on $r_{X_{[j]}}$, the ordinary least squares line goes through the origin. This is because the ordinary least squares line goes through the point $(\bar{r}_{X_{[j]}}, \bar{r}_{Y_{[j]}})$ and each of these means is 0. Thus, a regression through the origin can be used for this step of the assessment. Moreover, the slope from this regression through the origin fit is equal to the

---

[1] Note that you can produce $p - 1$ partial regression plots, one for each predictor.

slope for $X_j$ if it were included in the full model where $Y$ is regressed on *all* $p-1$ predictors. Finally, the correlation between $r_{Y_{[j]}}$ and $r_{X_{[j]}}$ yields the corresponding coefficient of partial correlation; i.e.,

$$r_{Y,j|1,\ldots,j-1,j+1,\ldots,p-1} = \mathrm{Corr}(r_{Y_{[j]}}, r_{X_{[j]}}).$$

**Partial Leverage Plots**
**Partial leverage** is used to measure the contribution of each predictor variable to the leverage of each observation. Recall that $h_{i,i}$ is the $i^{\mathrm{th}}$ diagonal entry of the hat matrix. Then the partial leverage measures how $h_{i,i}$ changes as a variable is added to the regression model and is computed as:

$$(h_j^*)_i = \frac{r^2_{X_{[j]},i}}{\sum_{k=1}^{n} r^2_{X_{[j]},k}}.$$

A **partial leverage plot** is then constructed by plotting the partial leverage values versus the index of the observations or versus $X_j$. The latter will almost always have a distinct U-shaped pattern. Thus, a rule-of-thumb criterion is that a high partial leverage value is typically on that is greater than $3p/n$.

**Partial Residual Plots**
Finally, **partial residuals** are residuals that have not been adjusted for a particular predictor variable, say, $X_{j^*}$. These can be used to check the assumption of linearity for each predictor. Suppose, we partition the $\mathbf{X}$ matrix such that $\mathbf{X} = (\mathbf{X}_{-j^*}, \underline{X}_{j^*})$. For this formulation, $\mathbf{X}_{-j^*}$ is the same as the $\mathbf{X}$ matrix, but with the vector of observations for the predictor $X_{j^*}$ omitted; i.e., this vector of values is $\underline{X}_{j^*}$. Similarly, let us partition the vector of estimated regression coefficients as $\hat{\boldsymbol{\beta}} = (\hat{\boldsymbol{\beta}}_{-j}^{\mathrm{T}}, \hat{\beta}_{j^*})^{\mathrm{T}}$. Here, $\hat{\beta}_{j^*}$ is the estimated regression coefficient corresponding to $X_{j^*}$ in the full model. Then, the set of partial residuals for the predictor $X_{j^*}$ would be

$$\begin{aligned}
\mathbf{e}_j^* &= \mathbf{Y} - \mathbf{X}_{j^*}\hat{\boldsymbol{\beta}}_{-j^*} \\
&= \mathbf{Y} - \hat{\mathbf{Y}} + \hat{\mathbf{Y}} - \mathbf{X}_{j^*}\hat{\boldsymbol{\beta}}_{-j^*} \\
&= \mathbf{e} - (\mathbf{X}_{j^*}\hat{\boldsymbol{\beta}}_{-j^*} - \mathbf{X}\hat{\boldsymbol{\beta}}) \\
&= \mathbf{e} + \hat{\beta}_{j^*}\underline{X}_{j^*}.
\end{aligned}$$

Note in the above that $\hat{\beta}_{j^*}$ is just a univariate quantity, so $\hat{\beta}_{j^*}\underline{X}_{j^*}$ is still an $n$-dimensional vector. Finally, a plot of $e_j^*$ versus $X_j^*$ has slope $\hat{\beta}_{j^*}$. This plot is called a **partial residual plot**.[2] The more the data deviates from a straight-line fit for the partial residual plot, the greater the evidence that a higher-ordered term or transformation on this predictor variable is necessary. Note also that the vector $\mathbf{e}$ would provide the residuals if a straight-line fit were made to these data.

---

[2] Plots of $\hat{\beta}_{j^*}X_{j^*}$ versus $X_{j^*}$ are called **component-plus-residual plots**.

## 8.6 Examples

**Example 8.6.1.** *(Thermal Energy Data, cont'd)*
We continue with the thermal energy data in Example 6.7.1. We will again look at the full model which includes all of the predictor variables, even though we have subsequently determined which predictors should likely be removed from the model. The full model with the five predictors is

$$\mathbf{Y} = \mathbf{X}\beta + \epsilon,$$

where $\beta$ is a 6-dimensional vector of the regression parameters, $\mathbf{Y}$ is a 29-dimensional response vector, $\mathbf{X}$ is a $(29 \times 5)$-dimensional design matrix, and $\epsilon$ is a 29-dimensional error vector.

The ANOVA for the model with all of the predictors is as follows:

```
#############################
Analysis of Variance Table

Response: heat.flux
            Df  Sum Sq Mean Sq F value    Pr(>F)
Regression   5 13195.5 2639.11  40.837 1.077e-10 ***
Residuals   23  1486.4   64.63
Total       28 14681.9
---
Signif. codes:  0 *** 0.001 ** 0.01 * 0.05 . 0.1   1
#############################
```

Now, using the results from earlier, which indicate a model including only the north and south focal points as predictors, let us test the following hypothesis:

$$H_0 : \beta_1 = \beta_2 = \beta_5 = 0$$
$$H_A : \text{at least one of } \{\beta_1, \beta_2, \beta_5\} \neq 0.$$

In other words, we only want our model to include the south $(X_3)$ and north $(X_4)$ focal points. We see that MSE(full) = 64.63, SSE(full) = 1486.40, and $\mathrm{df}_E(\text{full}) = 23$.

Next we calculate the ANOVA table for the above null hypothesis:

```
#############################
Analysis of Variance Table

Response: heat.flux
            Df  Sum Sq Mean Sq F value    Pr(>F)
Regression   2 12607.6  6303.8  79.013 8.938e-12 ***
Residuals   26  2074.3    79.8
Total       28 14681.9
```

```
---
Signif. codes:  0 *** 0.001 ** 0.01 * 0.05 . 0.1   1
#############################
```

The ANOVA for this analysis shows that SSE(reduced) $= 2074.33$ and $\text{df}_E(\text{reduced}) = 26$. Thus, the $F$-statistic is:

$$F^* = \frac{\frac{2074.33 - 1486.40}{26 - 23}}{64.63} = 3.0325,$$

which follows an $F_{3,23}$ distribution. The output for this analysis is also given below:

```
#############################
Analysis of Variance Table

Model 1: heat.flux ~ north + south
Model 2: heat.flux ~ insolation + east + south + north + time
  Res.Df    RSS Df Sum of Sq      F  Pr(>F)
1     26 2074.3
2     23 1486.4  3    587.94 3.0325 0.04978 *
---
Signif. codes:  0 *** 0.001 ** 0.01 * 0.05 . 0.1   1
#############################
```

The $p$-value is 0.0498. Thus, we just barely claim statistical significance and conclude that at least one of the other predictors (insolation, east focal point, and time) is a statistically significant predictor of heat flux.

We can also calculate the power of this $F$-test by using the partial $R^2$ values. Specifically,

$$R^2_{Y,1,2,3,4,5} = 0.8987602$$
$$R^2_{Y,3,4} = 0.8587154$$
$$\delta = (29)\frac{0.8987602 - 0.8587154}{1 - 0.8987602}$$
$$= 11.47078.$$

Therefore, $P(F^* > 3.027998)$ under an $F_{3,23}(11.47078)$ distribution gives the power, which is as follows:

```
#############################
          Power
F-Test 0.7455648
#############################
```

Notice that this is not extremely powerful; i.e., we usually hope to attain a power of at least 0.80. Thus, the probability of committing a Type II error is somewhat high.

We can also calculate the partial $R^2$ for this testing situation:

$$R^2_{Y,1,2,5|3,4} = \frac{\text{SSE}(X_3, X_4) - \text{SSE}(X_1, X_2, X_3, X_4, X_5)}{\text{SSE}(X_3, X_4)}$$

$$= \frac{2074.33 - 1486.4}{2074.33}$$

$$= 0.2834,$$

which implicitly uses the definitions of extra sums of squares. The above partial $R^2$ values means that insolation, the east focal point, and time explain about 28.34% of the variation in heat flux that could not be explained by the north and south focal points.

**Example 8.6.2.** *(Simulated Data)*
In order to concisely demonstrate different trends that can be observed in partial regression, partial residual, and partial leverage plots, we use simulated data so that we know the true underlying model. Suppose we have a response variable, $Y$, and two predictors, $X_1$ and $X_2$. Let us consider three models:

1. $Y = 9 + 7X_1 + \epsilon$;
2. $Y = 9 + 7X_1 - 4X_2 + \epsilon$; and
3. $Y = 9 + 7X_1 - 4X_2 + 7X_2^2 + \epsilon$.

In each model, $\epsilon$ is assumed to follow a normal distribution with $\mu = 0$ and $\sigma = 5$. Note that Model 3 is a quadratic regression model which falls under the polynomial regression framework discussed in Chapter 6. Figure 8.1 shows the partial regression plots for $X_1$ and $X_2$ for each of these three settings. In Figure 8.1(a), we see a strong linear relationship between $Y$ and $X_1$ when $X_2$ is in the model, but this is not present between $Y$ and $X_2$ when $X_1$ is in the model (Figure 8.1(b)). Figures 8.1(c) and 8.1(d) show that there is a strong linear relationship between $Y$ and $X_1$ when $X_2$ is in the model as well as between $Y$ and $X_2$ when $X_1$ is in the model. Finally, Figure 8.1(e) shows that there is a linear relationship between $Y$ and $X_1$ when $X_2$ is in the model, but there is an indication of a quadratic (i.e., curvilinear) relationship between $Y$ and $X_2$ when $X_1$ is in the model.

Figure 8.2 gives the partial leverage plots for $X_1$ and $X_2$, which are the same for each of the three models. Both of the plots are fairly similar and this is pretty standard for the partial leverage plots. These plots are the same for all three models since the same generated values of $X_1$ and $X_2$ we used in all three data settings. If we had generated different $X_1$ and $X_2$ values for each model, we would have different partial leverage plots. We are looking for any partial leverage values greater than $3p/n$. For Model 1, Model 2, and Model 3, this threshold is 0.12, 0.18, and 0.24, respectively. For each model, no partial leverage values exceed the respective threshold, which is obviously a feature of how we generated these data.

For Model 2, the fitted model when $Y$ is regressed on $X_1$ and $X_2$ is $\hat{Y} = 9.92 + 6.95X_1 - 4.20X_2$ as shown in the following output:

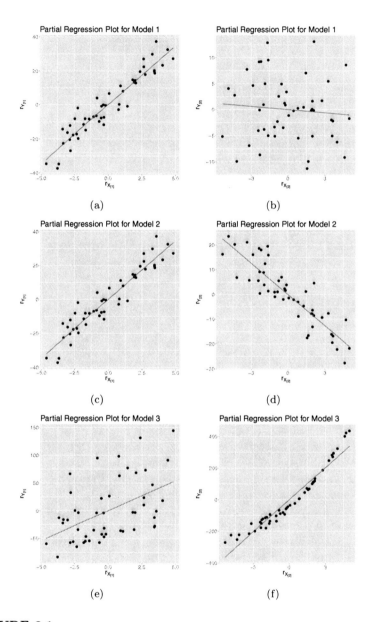

**FIGURE 8.1**
Partial regression plots displaying (a) $r_{Y_{[1]}}$ versus $r_{X_{[1]}}$ and (b) $r_{Y_{[2]}}$ versus $r_{X_{[2]}}$ for Model 1, (c) $r_{Y_{[1]}}$ versus $r_{X_{[1]}}$ and (d) $r_{Y_{[2]}}$ versus $r_{X_{[2]}}$ for Model 2, and (e) $r_{Y_{[1]}}$ versus $r_{X_{[1]}}$ and (f) $r_{Y_{[2]}}$ versus $r_{X_{[2]}}$ for Model 3.

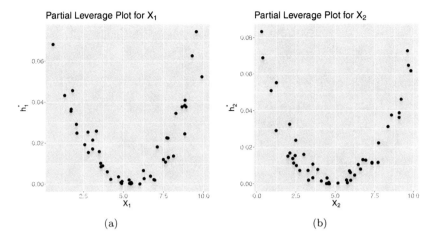

(a)                                                          (b)

**FIGURE 8.2**
Partial leverage plots of (a) $X_1$ and (b) $X_2$. Note that these are the same for all three models since the same generated values of $X_1$ and $X_2$ we used in all three data settings.

```
#############################
Coefficients:
             Estimate Std. Error t value Pr(>|t|)
(Intercept)   9.9174     2.4202    4.098 0.000164 ***
X1            6.9483     0.3427   20.274  < 2e-16 ***
X2           -4.2044     0.3265  -12.879  < 2e-16 ***
---
Signif. codes:  0 '***' 0.001 '**' 0.01 '*' 0.05 '.' 0.1 ' ' 1

Residual standard error: 6.058 on 47 degrees of freedom
Multiple R-squared: 0.915,      Adjusted R-squared: 0.9114
F-statistic:   253 on 2 and 47 DF,  p-value: < 2.2e-16
#############################
```

The slope from the regression through the origin fit of $r_{Y_{[1]}}$ on $r_{X_{[1]}}$ is 6.95 while the slope from the regression through the origin fit of $r_{Y_{[2]}}$ on $r_{X_{[2]}}$ is -4.20. The output from both of these fits is given below:

```
#############################
Coefficients:
          Estimate Std. Error t value Pr(>|t|)
r_X.1      6.9483     0.3357    20.7    <2e-16 ***
---
Signif. codes:  0 '***' 0.001 '**' 0.01 '*' 0.05 '.' 0.1 ' ' 1
```

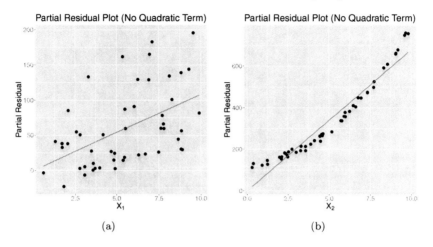

**FIGURE 8.3**
Partial residuals versus (a) $X_1$ and (b) $X_2$ for Model (2), but where the data were generated according to Model (3).

```
Residual standard error: 5.933 on 49 degrees of freedom
Multiple R-squared: 0.8974,     Adjusted R-squared: 0.8953
F-statistic: 428.5 on 1 and 49 DF,  p-value: < 2.2e-16
-----------------------------------------------------------
Coefficients:
          Estimate Std. Error t value Pr(>|t|)
r_X.2      -4.2044     0.3197  -13.15   <2e-16 ***
---
Signif. codes:  0 '***' 0.001 '**' 0.01 '*' 0.05 '.' 0.1 ' ' 1

Residual standard error: 5.933 on 49 degrees of freedom
Multiple R-squared: 0.7792,     Adjusted R-squared: 0.7747
F-statistic: 172.9 on 1 and 49 DF,  p-value: < 2.2e-16
#############################
```

Note that while the slopes from these partial leverage regression routines are the same as their respective slopes in the full multiple linear regression routine, the other statistics regarding hypothesis testing are *not* the same since, fundamentally, different assumptions are made for the respective tests.

Let us next consider the data generated from Model 3, but where we start by fitting Model 2. Figure 8.3 gives the partial residual plots pertaining to both predictors. Figure 8.3(a) seems to indicate that there is a linear relationship between $Y$ and $X_1$, but clearly Figure 8.3(b) shows a curvilinear relationship between $Y$ and $X_2$. Thus, we proceed to fit a quadratic term; i.e., we fit Model 3.

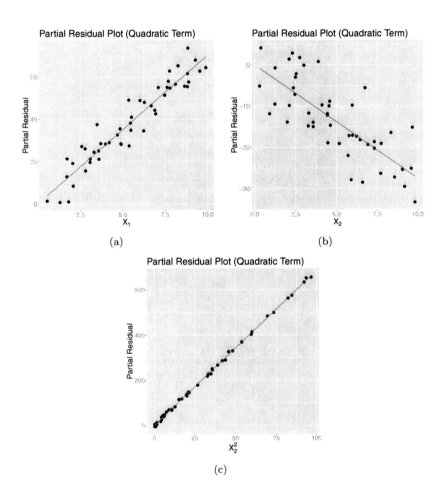

**FIGURE 8.4**
Partial residuals versus (a) $X_1$, (b) $X_2$, and (c) $X_2^2$ for Model (3).

Figure 8.4 gives the partial residual plots pertaining to each term in Model 3. All three partial residual plots clearly indicate that $Y$ has a linear relationship with each of the predictors. Moreover, each figure is fitted with a regression through the origin line, where the slope is determined using the respective slope from the estimated regression equation for the full model:

```
##############################
Coefficients:
            Estimate Std. Error t value Pr(>|t|)
(Intercept)   6.7567     3.7062   1.823   0.0748 .
X1            7.0269     0.3489  20.143   <2e-16 ***
X2           -2.7520     1.3329  -2.065   0.0446 *
I(X2^2)       6.8588     0.1256  54.598   <2e-16 ***
---
Signif. codes:  0 *** 0.001 ** 0.01 * 0.05 . 0.1   1

Residual standard error: 6.041 on 46 degrees of freedom
Multiple R-squared:  0.9991,Adjusted R-squared:  0.999
F-statistic: 1.692e+04 on 3 and 46 DF,  p-value: < 2.2e-16
##############################
```

# 9

## Indicator Variables

We next discuss how to include categorical predictor variables in a regression model. A **categorical variable** is a variable for which the possible outcomes are nameable characteristics, groups, or treatments. Some examples are gender (male or female), highest educational degree attained (secondary school, college undergraduate, or college graduate), and political affiliation of the subject (Democrat, Republican, or Other).

We use indicator variables to incorporate categorical predictor variables into a regression model. An **indicator variable** equals 1 when an observation is in a particular group and equals 0 when an observation is not in that group. An interaction between an indicator variable and a quantitative variable exists if the slope between the response and the quantitative variable depends upon the specific value present for the indicator variable.

## 9.1   Leave-One-Out Method

When a categorical predictor variable has $k$ categories, it is possible to define $k$ indicator variables. However, as will be explained shortly, we should only use $k - 1$ of them as predictor variables in the regression model.

Let us consider an example where we are analyzing data for a clinical trial done to compare the effectiveness of three different medications used to treat high blood pressure. Suppose that $n = 90$ participants are randomly assigned into one of the three different medication treatment groups, thus yielding three groups of 30 patients each. The response variable is the reduction in diastolic blood pressure over a three-month period. In addition to the treatment variables, two other variables are measured: age $(X_1)$ and body mass index $(X_2)$.

We are examining three different treatments so we start by defining the following three indicator variables for the treatment:

$$X_3 = 1 \text{ if patient used treatment 1, 0 otherwise;}$$
$$X_4 = 1 \text{ if patient used treatment 2, 0 otherwise;}$$
$$X_5 = 1 \text{ if patient used treatment 3, 0 otherwise.}$$

On the surface, it seems that our model should be the following over-parameterized model:

$$y_i = \beta_0 + \beta_1 x_{i,1} + \beta_2 x_{i,2} + \beta_3 x_{i,3} + \beta_4 x_{i,4} + \beta_5 x_{i,5} + \epsilon_i. \qquad (9.1)$$

The difficulty with this model is that the $\mathbf{X}$ matrix has a linear dependency, so we cannot estimate the individual coefficients. Technically, this is because there will be an infinite number of solutions for $\boldsymbol{\beta}$. The dependency stems from the fact that $x_{i,3} + x_{i,4} + x_{i,5} = 1$ for all observations since each patient uses one (and only one) of the treatments. In the $\mathbf{X}$ matrix, the linear dependency is that the sum of the last three columns will equal the first column (all 1s). This scenario leads to perfect multicollinearity.

A solution for avoiding this difficulty is the **leave-one-out method**, which involves defining $k-1$ indicators to describe the differences among $k$ categories. For the overall fit of the model, it does not matter which set of $k-1$ indicators we use. The choice of which $k-1$ indicator variables we use, however, does affect the interpretation of the coefficients that multiply the specific indicators in the model. So one must be cognizant of these interpretations in the context of their data problem.

In our example with three treatments and three possible indicator variables, we might leave out the third indicator giving us this model:

$$y_i = \beta_0 + \beta_1 x_{i,1} + \beta_2 x_{i,2} + \beta_3 x_{i,3} + \beta_4 x_{i,4} + \epsilon_i. \qquad (9.2)$$

For the overall fit of the model, it would work equally well to leave out the first indicator and include the other two or to leave out the second and include the first and third.

## 9.2 Coefficient Interpretations

The interpretation of the coefficients that multiply indicator variables requires some care. The interpretation for the individual $\beta$ values in the leave-one-out method is that it measures the difference between the group defined by the indicator in the model and the group defined by the indicator that was left. Usually, a control or placebo group is the one that is "left out," in which case this is referred to as a **treatment contrast**.

Let us consider our example again. We are predicting decreases in blood pressure in response to age $(X_1)$, body mass $(X_2)$, and which of three different treatments a person used. The variables $X_3$ and $X_4$ are indicators of the treatment as defined above. The model we examine is

$$y_i = \beta_0 + \beta_1 x_{i,1} + \beta_2 x_{i,2} + \beta_3 x_{i,3} + \beta_4 x_{i,4} + \epsilon_i.$$

To further our understanding, look at each treatment separately by substituting the appropriately defined values of the two indicators into the equation:

- For *treatment 1*, $X_3 = 1$ and $X_4 = 0$, by definition. This leads to

$$y_i = \beta_0 + \beta_1 x_{i,1} + \beta_2 x_{i,2} + \beta_3(1) + \beta_4(0) + \epsilon_i$$
$$= \beta_0 + \beta_1 x_{i,1} + \beta_2 x_{i,2} + \beta_3 + \epsilon_i.$$

- For *treatment 2*, $X_3 = 0$ and $X_4 = 1$, by definition. This leads to

$$y_i = \beta_0 + \beta_1 x_{i,1} + \beta_2 x_{i,2} + \beta_3(0) + \beta_4(1) + \epsilon_i$$
$$= \beta_0 + \beta_1 x_{i,1} + \beta_2 x_{i,2} + \beta_4 + \epsilon_i.$$

- For *treatment 3*, $X_3 = 0$ and $X_4 = 0$, by definition. This leads to

$$y_i = \beta_0 + \beta_1 x_{i,1} + \beta_2 x_{i,2} + \beta_3(0) + \beta_4(0) + \epsilon_i$$
$$= \beta_0 + \beta_1 x_{i,1} + \beta_2 x_{i,2} + \epsilon_i.$$

Now compare the three equations above to each other. The only difference between the equations for treatments 1 and 3 is the coefficient $\beta_3$. The only difference between the equations for treatments 2 and 3 is the coefficient $\beta_4$. This leads to the following meanings for the coefficients:

- $\beta_3$ : the difference in mean response for treatment 1 versus treatment 3, assuming the same value of age and the same value of body mass.

- $\beta_4$ : the difference in mean response for treatment 2 versus treatment 3, assuming the same value of age and the same value of body mass.

Here the coefficients are measuring differences from the third treatment, thus highlighting how we interpret the regression coefficients in the leave-one-out method. It is worth emphasizing that the coefficient multiplying an indicator variable in the model does not retain the meaning implied by the definition of the indicator. A common mistake is to incorrectly state that a coefficient measures the difference between that group and the other groups. It is also incorrect to state that only a coefficient multiplying an indicator measures the effect of being in that group. An effect has to involve a comparison, which with the leave-one-out method is a comparison to the group associated with the omitted indicator.

One additional note is that, in theory, a linear dependence results in an infinite number of suitable solutions for the $\beta$ coefficients. With the leave-one-out method, we are picking one with a particular meaning and then the resulting coefficients measure differences from the specified group, thus avoiding an infinite number of solutions. In experimental design, it is more common to parameterize in such a way that a coefficient measures how a group differs from an overall average.

To test the overall significance of a categorical predictor variable, we use a general linear $F$-test procedure. We form the reduced model by dropping the indicator variables from the model. More technically, the null hypothesis is that the coefficients multiplying the indicator all equal 0.

For our example with three treatments of high blood pressure and additional predictors of age and body mass, an overall test of treatment differences uses the following:

- Full model is: $y_i = \beta_0 + \beta_1 x_{i,1} + \beta_2 x_{i,2} + \beta_3 x_{i,3} + \beta_4 x_{i,4} + \epsilon_i$.

- Null hypothesis is: $H_0 : \beta_3 = \beta_4 = 0$.

- Reduced model is: $y_i = \beta_0 + \beta_1 x_{i,1} + \beta_2 x_{i,2} + \epsilon_i$.

Recalling that we have a sample of size $n = 90$, the resulting test statistic would then be $F^* \sim F_{2,85}$.

## 9.3   Interactions

An interaction between a categorical predictor and a quantitative predictor is characterized by including a new variable that is the product of the two predictors. For example, suppose we thought there could be an interaction between the body mass variable ($X_2$) and the treatment variable (characterized by $X_3$ and $X_4$). This would mean that treatment differences in blood pressure reduction further depend on the specific value of body mass. The model we would use is:

$$y_i = \beta_0 + \beta_1 x_{i,1} + \beta_2 x_{i,2} + \beta_3 x_{i,3} + \beta_4 x_{i,4} + \beta_5 x_{i,2} x_{i,3} + \beta_6 x_{i,2} x_{i,4} + \epsilon_i.$$

Just as with the interpretation of indicator variables, care needs to be taken with the interpretation of coefficients when interaction terms are present. For example, $\beta_5$ would characterize the additional change in blood pressure for each unit increase in body mass when considering treatment 1. In other words, when considering treatment 1, the average change in blood pressure for each unit increase in body mass is $(\beta_2 + \beta_5)$.

To test whether there is an interaction, the null hypothesis is $H_0 : \beta_5 = \beta_6 = 0$. We use the general linear $F$-test procedure to carry out the test. The full model is the interaction model given above. The reduced model is now:

$$y_i = \beta_0 + \beta_1 x_{i,1} + \beta_2 x_{i,2} + \beta_3 x_{i,3} + \beta_4 x_{i,4} + \epsilon_i.$$

A visual way to assess if there is an interaction is by using an interaction plot. An **interaction plot** is created by plotting the response versus the quantitative predictor and connecting the successive values according to the grouping of the observations. Recall that an interaction between factors occurs when the change in response from lower levels to higher levels of one factor is not quite the same as going from lower levels to higher levels of another factor.

Interaction plots allow us to compare the relative strength of the effects across factors. What results is one of three possible trends:

- The lines could be (nearly) parallel, which indicates no interaction. This means that the change in the response from lower levels to higher levels for each factor is roughly the same.

- The lines intersect within the scope of the study, which indicates an interaction. This means that the change in the response from lower levels to higher levels of one factor is noticeably different than the change in another factor. This type of interaction is called a **disordinal interaction**.

- The lines do not intersect within the scope of the study, but the trends indicate that if we were to extend the levels of our factors, then we may see an interaction. This type of interaction is called an **ordinal interaction**.

Figure 9.1 illustrates each type of interaction plot using a mock dataset involving three different levels of a categorical variable. Figure 9.1(a) illustrates the case where no interaction is present because the change in the mean response is similar for each treatment level as continuous predictor value increases; i.e., the lines are parallel. Figure 9.1(b) illustrates an interaction because as the continuous predictor value increase, the change in the mean response is noticeably different depending on which level of the categorical predictor is used; i.e., the lines cross. Figure 9.1(c) illustrates an ordinal interaction where no interaction is present within the scope of the range of the continuous predictor values, but if these trends continued for higher values, then we might see an interaction; i.e., the lines might cross.

It should also be noted that just because lines cross, it does not necessarily imply the interaction is statistically significant. Lines that appear nearly parallel, yet cross at some point, may not yield a statistically significant interaction term. If two lines cross, the more different the slopes appear and the more data that is available, then the more likely the interaction term will be significant.

## 9.4 Coded Variables

In the early days when computing power was limited, coding of the variables accomplished simplifying the linear algebra and thus allowing least squares solutions to be solved manually. Many methods exist for coding data, such as:

- Converting variables to two values; e.g., $\{-1, 1\}$ or $\{0, 1\}$.

- Converting variables to three values; e.g., $\{-1, 0, 1\}$.

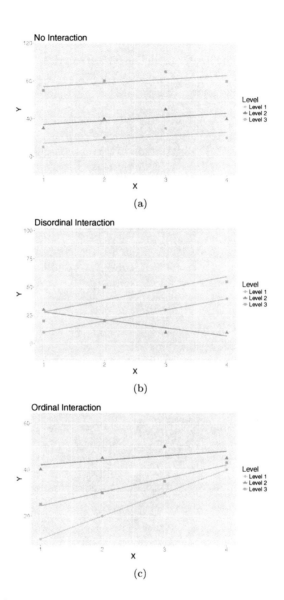

**FIGURE 9.1**
(a) A plot of no interactions among the groups — notice how the lines are parallel. (b) A plot of a disordinal interaction among the groups — notice how the lines intersect. (c) A plot of an ordinal interaction among the groups — notice how the lines do not intersect, but if we were to extrapolate beyond the predictor limits, then the lines would likely cross.

- Coding continuous variables to reflect only important digits; e.g., if the costs of various business programs range from \$100,000 to \$150,000, coding can be done by dividing through by \$100,000, resulting in the range being from 1 to 1.5.

The purpose of coding is to simplify the calculation of $(\mathbf{X}^T\mathbf{X})^{-1}$ in the various regression equations, which was especially important when this had to be done by hand. It is important to note that the above methods are just a few possibilities and that there are no specific guidelines or rules of thumb for when to code data. Moreover, such coding might be imposed from the onset, such as when designing an experiment.

Today, $(\mathbf{X}^T\mathbf{X})^{-1}$ is easily calculated with computers, but there may be rounding error in the linear algebra manipulations if the difference in the magnitude of the predictors is large. Good statistical programs assess the probability of such errors, which would warrant using coded variables. If coding variables, one should be aware of different magnitudes of the parameter estimates relative to the original data. The intercept term can change dramatically, but we are more concerned with any drastic changes in the slope estimates. We need to protect against additional errors due to the varying magnitudes of the regression parameters. This can be assessed by comparing plots of the actual data and the coded data and noting significant discrepancies.

## 9.5 Conjoint Analysis

An application where many indicator variables (or binary predictors) is used is **conjoint analysis**, which is a marketing tool developed by psychometricians that attempts to capture a respondent's preference given the particular levels of a categorical variable. The $\mathbf{X}$ matrix is called a **dummy matrix** as it consists of only 1s and 0s, which is propagated based on the leave-one-out method. The response is then regressed on the indicators using ordinary least squares and researchers attempt to quantify items like identification of different market segments, predict profitability, or predict the impact of a new competitor.

For a conjoint analysis, we attempt to characterize both a respondent's preference and the larger target population's preference. We have $m$ **attributes** of interest (categorical variables), which have $k_j$ unique levels, $j = 1, \ldots, m$. Each combination of attribute levels yields a **profile** and a respondent provides their rating of a profile on some sort of scale; e.g., on a scale of 1 to 100 or simply ranking the different profiles. This implies that each respondent provides a response (called the **utility**) for each of the $k \leq \prod_{j=1}^{m} k_j$ profiles. Note that strict equality occurs if all possible profiles are of interest, while choosing $k$ less than the product is perfectly acceptable if only a subset of profiles are truly of interest.

Let us write these responses as $\mathbf{y}_i^{\mathrm{T}} = (y_{i,1}, y_{i,2}, \ldots, y_{i,m})$ for respondent $i = 1, \ldots, n$. Thus, there are

$$p = 1 + \sum_{j=1}^{m} (k_j - 1)$$

predictors in a particular conjoint regression model. We can either estimate a respondent-level model:

$$y_{i,j} = \beta_0^{(i)} + \sum_{j^*=1}^{m} \mathbf{x}_{j^*}^{\mathrm{T}} \boldsymbol{\beta}_{j^*}^{(i)} + \epsilon_{i,j} \tag{9.3}$$

or we can estimate a population-level model based on the entire sample:

$$y_{i,j} = \beta_0 + \sum_{j^*=1}^{m} \mathbf{x}_{j^*}^{\mathrm{T}} \boldsymbol{\beta}_{j^*} + \epsilon_{i,j}. \tag{9.4}$$

In each case, the $\mathbf{x}_{j^*}$ are the dummy vectors for attribute $j$. The difference between the models is that (9.3) only uses the data from respondent $i$ for estimation (which yields a separate $\boldsymbol{\beta}^{(i)}$ vector for each respondent) and that (9.4) uses all of the data from the $n$ respondents for estimation (which yields a single $\boldsymbol{\beta}$ vector for the sampled population). Refer to Rao (2014) for more details on conjoint analysis.

Suppose we are interested in consumer preferences about coffee. For example, we might be interested in the attributes of blend (light, medium, or bold), aroma (weak or strong), and caffeine level (decaffeinated or caffeinated). Thus, there are $3 \times 2 \times 2 = 12$ profiles to which a respondent would provide a rating. The basic layout for this study is provided in Table 9.1.

---

## 9.6    Examples

**Example 9.6.1.** *(Auditory Discrimination Data)*
Hendrix et al. (1982) analyzed data from a study to assess auditory differences between environmental sounds given several other factors. The experiment was designed to study auditory discrimination with respect to sex, cultural status, and a particular treatment program. For the purposes of illustration, we only look at a subset of these data. We look at those subjects who were female and exposed to an additional experimental treatment, for a total sample size of $n = 20$. The measured variables considered for our analysis are:

- $Y$: gain in auditory score between the pre- and post-test administration of the treatment

**TABLE 9.1**

Profiles for the coffee example

| Profile | Attributes | | | Rating | | |
|---------|-------|-------|----------|-------------|-------------|-----|
| | **Blend** | **Aroma** | **Caffeine** | $y_1$ | $y_2$ | $\cdots$ |
| 1 | light | weak | decaffeinated | $y_{1,1}$ | $y_{2,1}$ | $\cdots$ |
| 2 | light | weak | caffeinated | $y_{1,2}$ | $y_{2,2}$ | $\cdots$ |
| 3 | light | strong | decaffeinated | $y_{1,3}$ | $y_{2,3}$ | $\cdots$ |
| 4 | light | strong | caffeinated | $y_{1,4}$ | $y_{2,4}$ | $\cdots$ |
| 5 | medium | weak | decaffeinated | $y_{1,5}$ | $y_{2,5}$ | $\cdots$ |
| 6 | medium | weak | caffeinated | $y_{1,6}$ | $y_{2,6}$ | $\cdots$ |
| 7 | medium | strong | decaffeinated | $y_{1,7}$ | $y_{2,7}$ | $\cdots$ |
| 8 | medium | strong | caffeinated | $y_{1,8}$ | $y_{2,8}$ | $\cdots$ |
| 9 | bold | weak | decaffeinated | $y_{1,9}$ | $y_{2,9}$ | $\cdots$ |
| 10 | bold | weak | caffeinated | $y_{1,10}$ | $y_{2,10}$ | $\cdots$ |
| 11 | bold | strong | decaffeinated | $y_{1,11}$ | $y_{2,11}$ | $\cdots$ |
| 12 | bold | strong | caffeinated | $y_{1,12}$ | $y_{2,12}$ | $\cdots$ |

- $X_1$: pre-test score

- $X_2$: cultural status of individuals, where 1 means the subject is from a "culturally-nondeprived" group and 0 means the subject is from a "culturally-deprived" group

Suppose we wish to estimate the average gain in auditory score. We suspect a possible interaction between the pre-test score and the cultural status of the female subjects. Thus, we consider the multiple regression model

$$y_i = \beta_0 + \beta_1 x_{i,1} + \beta_2 x_{i,2} + \beta_3 x_{i,1} x_{i,2} + \epsilon_i.$$

First we fit the above model and assess the significance of the interaction term.

```
##############################
Coefficients:
                  Estimate Std. Error t value Pr(>|t|)
(Intercept)        18.0837     5.9653   3.031 0.007940 **
pre.test           -0.2193     0.1219  -1.800 0.090796 .
Culture1           46.4076     8.2060   5.655 3.58e-05 ***
pre.test:Culture1  -0.6873     0.1566  -4.389 0.000457 ***
---
Signif. codes:  0 *** 0.001 ** 0.01 * 0.05 . 0.1   1

Residual standard error: 4.063 on 16 degrees of freedom
Multiple R-squared:  0.8624,Adjusted R-squared:  0.8366
F-statistic: 33.43 on 3 and 16 DF,  p-value: 4.018e-07
##############################
```

The above gives the $t$-tests for these predictors. Notice that the pre-test predictor is marginally significant, while the cultural status and interaction terms

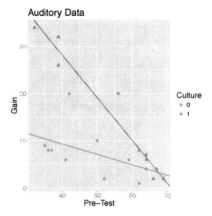

**FIGURE 9.2**
Interaction plot of the auditory discrimination data. Clearly the two cultural status levels demonstrate different relationships between the auditory score and the pre-test score.

are both significant. Since the interaction term is significant, we retain the pre-test score as a predictor. Thus, the estimated regression equation is

$$\hat{y}_i = 18.08 - 0.22x_{i,1} + 46.41x_{i,2} - 0.69x_{i,1}x_{i,2} + \epsilon_i,$$

where the average gain in auditory score for each unit increase in pre-test score is $(-0.22 - 0.69) = -0.91$ for culturally-nondeprived females and $-0.22$ for culturally-deprived females.

An interaction plot can also be used to justify use of an interaction term. Figure 9.2 provides the interaction plot for this dataset. This plot clearly indicates an (ordinal) interaction. A test of this interaction term yields $p = 0.0005$ (see the earlier output).

**Example 9.6.2.** *(Steam Output Data, cont'd)*
Consider coding the steam output data by rounding the temperature to the nearest ten value. For example, a temperature of 57.5 degrees would be rounded up to 60 degrees while a temperature of 72.8 degrees would be rounded down to 70 degrees. While you would probably not utilize coding on such an easy dataset where magnitude is not an issue, it is utilized here just for illustrative purposes.

Figure 9.3 compares the scatterplots of this dataset with the original temperature value and the coded temperature value. The plots look comparable, suggesting that coding could be used here. Recall that the estimated regression equation for the original data was $\hat{y}_i = 13.6230 - 0.0798x_i$. The estimated regression equation for the coded data is $\hat{y}_i = 13.6760 - 0.0805x_i$, which is comparable to the uncoded fit.

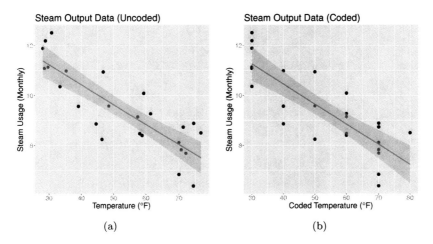

**FIGURE 9.3**
Comparing scatterplots of the steam output data with (a) the original temperature and (b) the coded temperature. A line of best fit for each is also shown.

**Example 9.6.3.** *(Tea Data)*
Bąk and Bartłomowicz (2012) studied consumer preferences regarding various attributes of tea. $m = 4$ attributes of the tea were studied: price (low, average, high), variety (black, green, red), kind (bags, granulated, leaf), and aroma (yes, no). A total of $k = 13$ profiles were studied with $n = 100$ respondents. Using the leave-one-out method for the predictor variables, the variables in our model are as follows:

- **Y**: a utility vector of length 13 giving a respondent's score to the profiles

- $X_1$: a price indicator where 1 means low and 0 otherwise

- $X_2$: a price indicator where 1 means average and 0 otherwise

- $X_3$: a variety indicator where 1 means black and 0 otherwise

- $X_4$: a variety indicator where 1 means green and 0 otherwise

- $X_5$: a kind indicator where 1 means bags and 0 otherwise

- $X_6$: a kind indicator where 1 means granulated and 0 otherwise

- $X_7$: an aroma indicator where 1 means yes and 0 means no

Table 9.2 summarizes the 13 profiles based on the indicators above. Moreover, the last column in the table gives the scores for the first respondent. Suppose we want to build a respondent-level model for this respondent to

**TABLE 9.2**

Profiles for the tea data

| Profile | Attributes | | | | $y_1$ |
|---|---|---|---|---|---|
| | Price | Variety | Kind | Aroma | |
| 1 | high | black | bags | yes | 8 |
| 2 | low | green | bags | yes | 1 |
| 3 | average | green | granulated | yes | 1 |
| 4 | average | black | leaf | yes | 3 |
| 5 | high | red | leaf | yes | 9 |
| 6 | average | black | bags | no | 2 |
| 7 | high | green | bags | no | 7 |
| 8 | average | red | bags | no | 2 |
| 9 | high | black | granulated | no | 2 |
| 10 | low | red | granulated | no | 2 |
| 11 | low | black | leaf | no | 2 |
| 12 | average | green | leaf | no | 3 |
| 13 | high | green | leaf | no | 4 |

characterize their preferences. Below are the estimates for those coefficients in that model:

```
#############################
Coefficients:
            Estimate Std. Error t value Pr(>|t|)
(Intercept)   3.3937     0.5439   6.240  0.00155 **
price1       -1.5172     0.7944  -1.910  0.11440
price2       -1.1414     0.6889  -1.657  0.15844
variety1     -0.4747     0.6889  -0.689  0.52141
variety2     -0.6747     0.6889  -0.979  0.37234
kind1         0.6586     0.6889   0.956  0.38293
kind2        -1.5172     0.7944  -1.910  0.11440
aroma1        0.6293     0.5093   1.236  0.27150
---
Signif. codes:  0 *** 0.001 ** 0.01 * 0.05 . 0.1   1

Residual standard error: 1.78 on 5 degrees of freedom
Multiple R-squared:  0.8184,Adjusted R-squared:  0.5642
F-statistic:  3.22 on 7 and 5 DF,  p-value: 0.1082
#############################
```

Even though none of them are significant, they would still be retained in a conjoint regression model simply for the purpose of characterizing this respondent's preferences.

We can then proceed to build the population-level model based on the entire sample of $n = 100$:

```
#############################
```

```
Coefficients:
              Estimate Std. Error t value Pr(>|t|)
(Intercept)    3.55336    0.09068  39.184  < 2e-16 ***
price1         0.24023    0.13245   1.814    0.070 .
price2        -0.14311    0.11485  -1.246    0.213
variety1       0.61489    0.11485   5.354 1.02e-07 ***
variety2       0.03489    0.11485   0.304    0.761
kind1          0.13689    0.11485   1.192    0.234
kind2         -0.88977    0.13245  -6.718 2.76e-11 ***
aroma1         0.41078    0.08492   4.837 1.48e-06 ***
---
Signif. codes:  0 *** 0.001 ** 0.01 * 0.05 . 0.1   1

Residual standard error: 2.967 on 1292 degrees of freedom
Multiple R-squared:  0.09003, Adjusted R-squared:  0.0851
F-statistic: 18.26 on 7 and 1292 DF,  p-value: < 2.2e-16
###############################
```

Here, we can identify those levels that are significant for the population's overall preferences, such as the average difference in score between granulated and leaf teabags and aroma. Moreover, one could compare the general signs of the coefficients in a respondent-level model with those in this population-level model, thus characterizing how the respondent differs from the overall preferences of the population.

# Part III

# Advanced Regression Diagnostic Methods

# 10

## Influential Values, Outliers, and More Diagnostic Tests

A concern in residual diagnostics is the influence of a single data value on the estimated regression equation. For example, the estimated regression equation may be too dependent on a single, possibly outlying, data value. Several statistics discussed in this chapter measure what happens when a data value is not included in the data during the estimation process.

Relevant notation for our discussion, some of which is a recap from previous chapters, is as follows:

- $y_i$ = observed value of $y$ for the $i^{\text{th}}$ observation.

- $\hat{y}_i$ = predicted value of $y$ for the $i^{\text{th}}$ observation, assuming all data is used during estimation.

- $\hat{y}_{i(i)}$ = predicted value of $y$ for the $i^{\text{th}}$ observation, when the $i^{\text{th}}$ observation is not included in the dataset during estimation.

- $\hat{y}_{j(i)}$ = predicted value of $y$ for the $j^{\text{th}}$ observation, when the $i^{\text{th}}$ observation is not included in the dataset during estimation.

- $\mathbf{X}_{(i)}$ = the $\mathbf{X}$ matrix, but with the entry for the $i^{\text{th}}$ observation omitted; i.e., the $i^{\text{th}}$ row is omitted from $\mathbf{X}$.

- $\hat{\boldsymbol{\beta}}_{(i)}$ = the estimated value of $\boldsymbol{\beta}$ calculated when the $i^{\text{th}}$ observation is not included in the dataset during estimation.

- $\text{MSE}_{(i)}$ = the MSE calculated when the $i^{\text{th}}$ observation is omitted from the analysis, which is given by

$$\text{MSE}_{(i)} = \frac{\text{SSE} - \frac{e_i^2}{1 - h_{i,i}}}{n - p - 1},$$

where $h_{i,i}$ is the leverage value and is defined below.

As will be shown, there are numerous measures that can be reported when determining if an observation could be considered as an outlier or influential. While only a couple of these would be reported when doing an analysis, it is helpful to be cognizant of the various measurements available when trying to make the case one way or the other about removing an observation.

Most textbooks on regression modeling will devote at least a chapter to the identification of outliers and influential values. A few references devoted primarily to this topic include Barnett and Lewis (1994), Rousseeuw and Leroy (2003), and Aggarwal (2013). Some of the guiding discussions in this chapter have also been "influenced" by the online handbook published by the National Institute of Standards and Technology (NIST, 2011).

## 10.1   More Residuals and Measures of Influence

**Leverage**
One of the assumptions we introduced when building a regression model is that the error terms, $\epsilon_1, \ldots, \epsilon_n$, are independent. However, the computed residuals are not. While this lack of independence is often not a concern when constructing diagnostic plots and graphs, it is a concern when developing theoretical tests.

Assume that $\text{Var}(\epsilon_i) = \sigma^2$ for all $i = 1, 2, \ldots, n$. Recall that the error vector is given by $\mathbf{e} = \mathbf{Y} - \hat{\mathbf{Y}}$. $\text{Var}(e_i)$ is *not* $\sigma^2$, but rather given by the diagonal elements of the variance–covariance matrix of $\mathbf{e}$:

$$V(\mathbf{e}) = \sigma^2(\mathbf{I}_n - \mathbf{X}(\mathbf{X}^T\mathbf{X})^{-1}\mathbf{X}^T)$$
$$= \sigma^2(\mathbf{I}_n - \mathbf{H}),$$

where $\mathbf{H}$ is the **hat matrix**. Thus, $\text{Var}(e_i) = \sigma^2(1 - h_{i,i})$, where $h_{i,i}$ is the $i^{\text{th}}$ diagonal element of the hat matrix. However, $\sigma^2$ is rarely known, so we use the corresponding estimates from the diagonal of the sample variance–covariance matrix of $\mathbf{e}$:

$$\hat{V}(\mathbf{e}) = \text{MSE}(\mathbf{I}_n - \mathbf{H}).$$

$\mathbf{H}$ provides a measure that captures an observation's remoteness in the **predictor space** (sometimes called the **X-space** or **covariate space**), which is referred to as **leverage**. Typically, high leverage values will be noted when an observation has a high or low value for one or more of the predictors relative to the other observations. A common rule of thumb is to consider any observations with $h_{i,i} > 3p/n$ (where $p$ is the number of parameters and $n$ is the sample size) as influential and thus label them as high leverage observations. It is important to note that a high leverage does not necessarily imply an observation is bad. It could simply be due to chance variation, measurement error, or both. High leverage observations simply exert additional influence on the final estimated equation, so one should try and ensure that the observation is indeed correct.

## Standardized Residuals

**Standardized residuals** (also called **normalized residuals** or **semistudentized residuals**) for the $i^{\text{th}}$ observation are simply given by

$$e_i^* = \frac{e_i - \bar{e}}{\sqrt{\text{MSE}}} = \frac{e_i}{\sqrt{\text{MSE}}},$$

which are helpful for identifying outliers when plotted. A general rule of thumb is that an observation might possibly be an outlier if it has a standardized residual value of $|e_i^*| > 3$. However, the residuals may have standard deviations that differ greatly and require a standardization that takes this into account.

## Studentized Residuals[1]

As we saw earlier, Studentized residuals for the $i^{\text{th}}$ observation are given by

$$r_i = \frac{e_i}{\sqrt{\text{MSE}(1 - h_{i,i})}}.$$

We use $r_i$ because while the variance of the true errors $(\epsilon_i)$ is $\sigma^2$, the variance of the computed residuals $(e_i)$ is not. Using $r_i$ produces residuals with constant variance, which is not attained with standardized residuals. A general rule of thumb is that an observation might possibly be an outlier if it has a Studentized residual value of $|r_i| > 3$. Typically it is safe to simply use the Studentized residuals instead of the standardized residuals as there is usually no practical difference.

## Deleted Residuals[2]

The **deleted residual** for the $i^{\text{th}}$ point is the difference between the actual $y$ and the predicted $y$ when the prediction is based on an equation estimated from the other $n - 1$ data points. The formula is

$$d_i = y_i - \hat{y}_{i(i)}$$
$$= \frac{e_i}{1 - h_{i,i}}.$$

However, the **Studentized deleted residuals** (sometimes called the **R Student residuals**) are typically used in determining if an observation is an outlier:

$$t_i = \frac{d_i}{s_{d_i}}$$
$$= \frac{e_i}{\sqrt{\text{MSE}_{(i)}(1 - h_{i,i})}},$$

---

[1]Studentized residuals are also called **internally Studentized residuals** because the MSE has within it $e_i$. So $e_i$ is stated explicitly in the numerator, but also concealed in the denominator of $r_i$.

[2]Deleted residuals are also called **externally Studentized residuals** because MSE$_{(i)}$ has excluded $e_i$. So $e_i$ is stated explicitly in the numerator, but excluded in the denominator of $t_i$.

where
$$s_{d_i}^2 = \frac{\text{MSE}_{(i)}}{1 - h_{i,i}}.$$

The Studentized deleted residuals follow a $t_{df_E - 1}$ distribution. A $p$-value can be calculated for each $i$, say $p_i$, and if $p_i < \alpha$, then observation $i$ can be considered an outlier. However, again, a common rule of thumb is to simply consider any observation with $|t_i| > 3$ as a possible outlier.

### Recursive Residuals

Let $\mathbf{X}$ and $\mathbf{y}$ be partitioned according to the first $p \le r \le n$ observations in our dataset, where

$$\mathbf{X}_r = \begin{pmatrix} \mathbf{x}_1^{\text{T}} \\ \mathbf{x}_2^{\text{T}} \\ \vdots \\ \mathbf{x}_r^{\text{T}} \end{pmatrix} \quad \text{and} \quad \mathbf{y}_r = \begin{pmatrix} y_1 \\ y_2 \\ \vdots \\ y_r \end{pmatrix},$$

respectively. Assuming that $\mathbf{X}_r^{\text{T}} \mathbf{X}$ is nonsingular and letting $\hat{\boldsymbol{\beta}}_r$ be the least squares estimator based on $\mathbf{X}_r$ and $\mathbf{y}_r$, we can then define the **recursive residuals** (Brown et al., 1975) as

$$w_r = \frac{y_r - \mathbf{x}_r^{\text{T}} \hat{\boldsymbol{\beta}}_{r-1}}{\sqrt{1 + \mathbf{x}_r^{\text{T}} \left( \mathbf{X}_{r-1}^{\text{T}} \mathbf{X}_{r-1} \right)^{-1} \mathbf{x}_r}}, \tag{10.1}$$

where $r = p + 1, \ldots, n$. Since the $\epsilon_i$ are $iid$ $\mathcal{N}(0, \sigma^2)$, it then follows that the $w_i$ are also $iid$ $\mathcal{N}(0, \sigma^2)$. Thus, the recursive residuals characterize the effect of successively deleting points from a dataset while maintaining the property of independence.

### Grubbs' Test

**Grubbs' test** (Grubbs, 1969), also called the **maximum normed residual test**, can be used to detect outliers in univariate datasets when normality is assumed. Provided that the residuals from a regression fit appear to meet the assumption of normality, then Grubbs' test may be used. Grubbs' test detects one outlier at a time and is iterated until no further outliers are detected. Formally, it concerns the (two-sided) test

$$H_0 : \text{there are no outliers in the dataset}$$
$$H_A : \text{there is at least one outlier in the dataset.}$$

The test statistic is defined as

$$G = \frac{\max_{i=1,\ldots,n} |e_i - \bar{e}|}{\sqrt{\text{MSE}}}$$

and rejection of the null hypothesis of no outliers occurs at the significance level $\alpha$ when

$$G > \frac{n-1}{\sqrt{n}} \sqrt{\frac{F_{1,n-2;\alpha/(2n)}}{n-2+F_{1,n-2;\alpha/(2n)}}}.$$

Individual tests of whether the minimum residual is an outlier or the maximum residual is an outlier can also be performed by calculating

$$G_{\min} = \frac{\bar{e} - e_{(1)}}{\sqrt{\text{MSE}}}$$

and

$$G_{\max} = \frac{e_{(n)} - \bar{e}}{\sqrt{\text{MSE}}},$$

respectively. In the above, $e_{(1)}$ and $e_{(n)}$ are the minimum and maximum residuals, respectively. Since these are one-sided tests, the critical region specified earlier requires replacing $F_{1,n-2;\alpha/(2n)}$ by $F_{1,n-2;\alpha/n}$.

A caveat with using Grubbs' test is that it should not be used with small datasets (say $n < 10$) as it typically classifies a disproportionate number of observations as outliers. Furthermore, even when normality appears appropriate for the residuals, outliers can often appear in large datasets and should never be automatically discarded, regardless if testing or residual plots seem to indicate otherwise. Care needs to be taken in that sound scientific judgment and statistical results are working harmoniously.

## Chi-Square Outlier Test

A **chi-square outlier test** was presented in Dixon (1951), which is distinct from the traditional chi-square goodness-of-fit test. The chi-square outlier test is for the following hypotheses:

$$H_0 : \text{there are no outliers in the dataset}$$
$$H_A : \text{there is one outlier in the dataset.}$$

The test statistic can be constructed similarly to Grubbs' test depending on if you want to test whether the minimum residual is an outlier or the maximum residual is an outlier:

$$X_{\min}^* = \frac{(e_{(1)} - \bar{e})^2}{\text{MSE}},$$

or

$$X_{\max}^* = \frac{(e_{(n)} - \bar{e})^2}{\text{MSE}},$$

both of which follow a $\chi_1^2$ distribution. If the null hypothesis is rejected, then the observation corresponding to the test statistic used above would be labeled an outlier. This test is typically not used as much because it has been shown to be less powerful than some of the other tests presented in this chapter.

## Tietjen–Moore Test

The **Tietjen–Moore test** (Tietjen and Moore, 1972) is basically a generalized version of the Grubbs' test. It is used to detect multiple outliers (specifically, $k$ outliers) in a dataset that is approximately normally distributed. When $k = 1$, then we have exactly one form of the Grubbs' test.

Formally, this test is written as

$$H_0 : \text{there are no outliers in the dataset}$$
$$H_A : \text{there are exactly } k \text{ outliers in the dataset.}$$

First, begin by ordering the residuals from the fitted regression model. So $e_{(m)}$ will denote the $m^{\text{th}}$ largest residual. The test statistic for the $k$ *largest* observations is

$$L_k = \frac{\sum_{i=1}^{n-k} (e_{(i)} - \bar{e}_k)^2}{\sum_{i=1}^{n} (e_{(i)} - \bar{e})^2},$$

where $\bar{e}_k$ is the sample mean with the $k$ largest observations removed. Similarly, the test statistic for the $k$ *smallest* observations is

$$L_{k'} = \frac{\sum_{i=k+1}^{n} (e_{(i)} - \bar{e}_{k'})^2}{\sum_{i=1}^{n} (e_{(i)} - \bar{e})^2},$$

where $\bar{e}_{k'}$ is the sample mean with the $k$ smallest observations removed.

Finally, to test for outliers in both tails simultaneously, compute the absolute residuals

$$a_i = |e_i - \bar{e}|$$

and then let $a_{(i)}$ denote the sorted absolute residuals. The test statistic for this setting is

$$E_k = \frac{\sum_{i=k+1}^{n} (a_{(i)} - \bar{a}_k)^2}{\sum_{i=1}^{n} (a_{(i)} - \bar{a})^2},$$

where, again, the subscript $k$ is used to indicate that the sample mean has been calculated with the $k$ largest observations removed.

A $p$-value is obtained via simulation. First generate $B$ sets of $n$ random variates from a standard normal distribution and compute the Tietjen–Moore test statistic. $B$ should be chosen sufficiently large, say, greater than $10,000$. The value of the Tietjen–Moore statistic obtained from the data are compared to this reference distribution and the proportion of test statistics that are more extreme than the one calculated from the data gives you the simulated $p$-value.

## Dixon's Q Test

The **Dixon's Q test** (Dixon, 1951) can be used to test whether one (and only one) observation can be considered an outlier. This test is not to be iterated in the manner as done with Grubbs' test. Formally, the test is

$$H_0 : \text{the extreme value in question is not an outlier}$$
$$H_A : \text{the extreme value in question is an outlier.}$$

The test can be applied to either $e_{(1)}$ or $e_{(n)}$, depending on which value is in question. The respective test statistics are

$$Q_{(1)} = \frac{e_{(2)} - e_{(1)}}{e_{(n)} - e_{(1)}}$$

and

$$Q_{(n)} = \frac{e_{(n)} - e_{(n-1)}}{e_{(n)} - e_{(1)}}.$$

The respective $p$-values are $P(Q_{(1)} > Q_{1-\alpha})$ and $P(Q_{(n)} > Q_{1-\alpha})$, where $Q_{1-\alpha}$ determines the critical region at the $\alpha$ significance level. Tabulated values for $Q_{1-\alpha}$ exist only for small values of $n$, but other values can be determined through numerical methods.

**TABLE 10.1**

Critical values for Dixon's Q test

| $n$ | $Q_{0.80}$ | $Q_{0.90}$ | $Q_{0.95}$ | $Q_{0.99}$ |
|---|---|---|---|---|
| 3 | 0.886 | 0.941 | 0.970 | 0.994 |
| 4 | 0.679 | 0.765 | 0.829 | 0.926 |
| 5 | 0.557 | 0.642 | 0.710 | 0.821 |
| 6 | 0.482 | 0.560 | 0.625 | 0.740 |
| 7 | 0.434 | 0.507 | 0.568 | 0.680 |
| 8 | 0.399 | 0.468 | 0.526 | 0.634 |
| 9 | 0.370 | 0.437 | 0.493 | 0.598 |
| 10 | 0.349 | 0.412 | 0.466 | 0.568 |
| 20 | 0.252 | 0.300 | 0.342 | 0.425 |
| 30 | 0.215 | 0.260 | 0.298 | 0.372 |

Table 10.1 provides tabulated values for $n = 1, \ldots, 10, 20, 30$ for four levels of $\alpha$. An interesting mathematical note, the $p$-values for $n = 3$ and $n = 4$ can be calculated in closed-form using the following equations:

$$p = \begin{cases} 1 - \frac{6}{\pi}\arctan\left(\frac{2Q-1}{\sqrt{3}}\right), & n = 3; \\ -8 + \frac{12}{\pi}\left[\arctan(\sqrt{4Q^2 - 4Q + 3}) + \arctan\left(\frac{\sqrt{3Q^2-4Q+4}}{Q}\right)\right], & n = 4. \end{cases}$$

In the formulas above, $Q$ is either $Q_{(n)}$ or $Q_{(1)}$, depending on the extreme value of interest. Most likely, you would never be interested in determining if a value is an outlier for such small datasets since you have so little data to work with in the first place.

**Generalized ESD Test**

The **generalized extreme Studentized deviate (ESD) test** (Rosner, 1983) can also be used to test if one or more observations can be considered as outliers. The Grubbs' and Tietjen–Moore tests require that the suspected number of outliers, $k$, be specified exactly, which can lead to erroneous conclusions if $k$ is misspecified. The generalized ESD test only requires specification of an upper bound, say $c$, for the suspected number of outliers.

The generalized ESD test proceeds sequentially with $c$ separate tests: a test for one outlier, a test for two outliers, and so on up to $c$ outliers. Formally, the test is

$$H_0 : \text{there are no outliers in the dataset}$$
$$H_A : \text{there are up to } c \text{ outliers in the dataset.}$$

Since this test is performed sequentially, first calculate the test statistic

$$C_1 = \frac{\max |e_i - \bar{e}|}{\sqrt{\text{MSE}}}.$$

Next, remove the observation that maximizes $|e_i - \bar{e}|$ and then recompute the above statistic with $n-1$ observations, which requires the regression model to be refit. We then repeat this process until $c$ observations have been removed, thus yielding a sequence of test statistics $C_1, C_2, \ldots, C_c$.

The following $c$ critical values can be calculated for each test statistic:

$$C_i^* = \frac{t_{n-i-p;q(n-i)}}{\sqrt{(n-i-p+t_{n-i-p;q}^2)(n-i+1)}},$$

where $i = 1, 2, \ldots, c$ and $q = 1 - \frac{\alpha}{2(n-i+1)}$, such that $\alpha$ is the significance level. The number of outliers, $k^*$, is determined by finding the following:

$$k^* = \underset{i=1,\ldots,c}{\arg\max}\{C_i > C_{i^*}\}.$$

This test is considered fairly accurate for sample sizes of $n \geq 15$. Note that the generalized ESD test makes appropriate adjustments for the critical values based on the number of outliers being tested, whereas the sequential application of Grubbs' test does not.

**DFITs**
The **unstandardized difference in fits residual** for the $i^{\text{th}}$ observation is the difference between the usual predicted value and the predicted value when the $i^{\text{th}}$ point is excluded when estimating the equation. The calculation is as follows:

$$f_i = \hat{y}_i - \hat{y}_{i(i)}. \tag{10.2}$$

However, the standardized version of $f_i$, simply called the **difference in fits** or **DFITs**, is what is typically used. This formula is:

$$\text{DFIT}_i = \frac{f_i}{\sqrt{\text{MSE}_{(i)} h_{i,i}}}$$

$$= e_i \sqrt{\frac{n-p-1}{\text{SSE}(1-h_{i,i}) - e_i^2}} \Big/ \sqrt{\frac{1-h_{i,i}}{h_{i,i}}}.$$

This statistic represents the number of estimated standard errors that the fitted value changes if the $i^{\text{th}}$ observation is omitted from the dataset. A common rule-of-thumb is to consider any observation with $|\text{DFIT}_i| > 1$ as "large" and, thus, influential.

## Cook's $D_i$

Whereas $\text{DFIT}_i$ attempts to measure the influence of observation $i$ on its fitted value, **Cook's $D_i$**, called **Cook's distance** (Cook, 1977), measures how much the $i^{\text{th}}$ point influences the complete set of predicted values for all observations. A value of Cook's $D_i$ is computed for each data point and the formula is

$$D_i = \frac{(\hat{\boldsymbol{\beta}} - \hat{\boldsymbol{\beta}}_{(i)})^{\text{T}} \mathbf{X}^{\text{T}} \mathbf{X} (\hat{\boldsymbol{\beta}} - \hat{\boldsymbol{\beta}}_{(i)})}{p \times \text{MSE}}$$
$$= \frac{\sum_{j=1}^{n} (\hat{y}_j - \hat{y}_{j(i)})^2}{p \times \text{MSE}}$$
$$= \frac{e_i^2 h_{i,i}}{\text{MSE} \times p(1 - h_{i,i})^2}.$$

Notice that the sum is over all observations. The numerator is the total squared difference between two sets of predicted values, that is, the difference between when all data is used and when the $i^{\text{th}}$ observation is excluded.

A large value indicates that the observation has a large influence on the results. Some rules-of-thumb for considering an observation influential are when $D_i > F_{p,n-p;0.50}$, $D_i > 1$, and $D_i > 3/n$. However, many analysts just look at a plot of Cook's $D_i$ values to see if any points are obviously greater than the others.

## Hadi's Influence Measure

**Hadi's influence measure** (Hadi, 1992) of the $i^{\text{th}}$ observation is due to influential observations being outliers in the response variable, the predictors, or both. Hadi's influence of the $i^{\text{th}}$ observation can be measured by

$$H_i = \frac{h_{i,i}}{1 - h_{i,i}} + \left(\frac{p+1}{1 - h_{i,i}}\right) \frac{e_i^{*2}}{1 - e_i^{*2}},$$

$i = 1, \ldots, n$. The first term on the right-hand side of the expression for $H_i$ is called the **potential function**, which measures outlyingness in the predictor space. The second term is the **residual function**, which measures outlyingness in the response variable. It can be seen that observations will have large values of $H_i$ if they are outliers in the response (large values of $e_i$), the predictor variables (large values of $h_{i,i}$), or both. $H_i$ is not a measure that focuses on a specific regression result, but rather it can be thought of as an *overall* general measure of influence that depicts observations that are influential on at least one regression result. Finally, note that $D_i$ and $\text{DFIT}_i$ are multiplicative functions of the residuals and leverage values, whereas $H_i$ is an additive function.

As we mentioned earlier, the functional form of $H_i$ consists of the potential function and what we termed the residual function. This breakdown of $H_i$ into the potential function and what we termed the residual function enables us to utilize a simple graph that aids in classifying observations as high leverage points, outliers, or both. The graph is called the **potential-residual plot** because it is a scatterplot where we plot the value of the potential function on the $y$-axis and the value of the residual function on the $x$-axis.[3] The intent with this plot is to look for values that appear to be outlying along either axis or both axes. Such values should be flagged as influential and investigated as to whether or not there is a special cause to those more extreme values. Some practical discussions and examples using Hadi's influence measure and the potential-residual plot is found in Chapter 4 of Chatterjee and Hadi (2012).

### DFBETAS

The **difference in betas** (or **DFBETAS**) criterion measures the standardized change in a regression coefficient when observation $i$ is omitted. For $k = 1, 2, \ldots, p$, the formula for this criterion is

$$\text{DFBETAS}_{(k-1)(i)} = \frac{\hat{\beta}_{k-1} - \hat{\beta}_{(k-1)(i)}}{\sqrt{\text{MSE}_{(i)} c_{k,k}}},$$

where $\hat{\beta}_{(k-1)(i)}$ is the $k^{\text{th}}$ entry of $\hat{\boldsymbol{\beta}}_{(i)}$ and $c_{k,k}$ is the $k^{\text{th}}$ diagonal entry of $(\mathbf{X}^{\text{T}}\mathbf{X})^{-1}$. A value of $|\text{DFBETAS}_{(k-1)(i)}| > 2/\sqrt{n}$ indicates that the $i^{\text{th}}$ observation is possibly influencing the estimate of $\beta_{k-1}$.

### CovRatio

Belsley et al. (1980) proposed the **covariance ratio** (or **CovRatio**) value, which measures the effect of an observation on the determinant of the variance–covariance matrix of the regression parameter estimates. The formula is given by

$$C_i = \frac{|\text{MSE}_{(i)}(\mathbf{X}_{(i)}^{\text{T}}\mathbf{X}_{(i)})^{-1}|}{|\text{MSE}(\mathbf{X}^{\text{T}}\mathbf{X})^{-1}|}.$$

Belsley et al. (1980) suggest that values of $C_i$ near 1 indicate that the $i^{\text{th}}$ observation has little effect on the precision of the estimates while observations with $|C_i - 1| \geq 3p/n$ suggest need for further investigation.

---

[3]Similarly, we can construct a **leverage-residual plot**, where we plot $h_{i,i}$ on the $y$-axis and $e_i^{*2}$ on the $x$-axis.

## 10.2 Masking, Swamping, and Search Methods

In the previous section, we discussed a variety of tests that are used to determine outliers. However, we must also be concerned about misclassifying observations as outliers when, in fact, they are not as well as not classifying observations as outliers when, in fact, they are. We briefly discuss these two errors in this section.

**Masking** is when *too few* outliers are specified in the test, oftentimes because of the presence of a cluster of outliers. For example, if testing for a single outlier when there are actually two or more outliers, then these additional outliers may influence the value of the test statistic being used. Hence, we may declare that no points are outliers. This is akin to a "false negative".

**Swamping** is when *too many* outliers are specified in the test. For example, if testing for two or more outliers when there is actually only a single outlier, both points may be declared outliers. In fact, many tests often declare either all or none of the tested points as outliers. This is akin to a "false positive."

A way to handle the possibility of masking and swamping, is to use formal statistical tests as well as graphics, much like we do when testing for normality or constant variance. Graphics can aid in identifying when masking or swamping may be present. Masking and swamping are also a reason that many tests require an exact number of outliers be specified in the hypothesis as well as result in the failure of applying single outlier tests sequentially.

There are also search methods, such as those discussed in Mavridis and Moustaki (2009), that can be used to help protect against masking and swamping. The first algorithm is called the **backward search method**:

---

Backward Search Method

1. Start with the whole dataset.

2. Compute a measure that quantifies each observation's "outlyingness."

3. Omit the observation that is most extreme with respect to the measure chosen in Step 2.

4. Repeat this process until a specified number of observations is omitted. You can simultaneously monitor the impact of deleting these observations throughout the search to help guide your choice of the number to omit.

---

The second algorithm is called the **forward search method**:

---

**Forward Search Method**

1. Start with an outlier-free subset of the data, which is called a **basic set**.

2. Using some measure (such as a distance metric), order the observations that are not in the basic set (i.e., the **non-basic set**) according to their closeness to the basic set.

3. Add the closest observation from the non-basic set to the basic set.

4. Repeat this process until all observations are included.

5. Monitor the search throughout the iterations.

---

In both of the above search methods, the last step makes a note to "monitor the search." What this means is to search for sharp changes in various statistical measures to help you identify a threshold where it is most likely that all of the remaining observations are outliers. Such statistical measures that can be used are changes in the parameter estimates, $t$-tests, goodness-of-fit tests, and the monitoring of residual diagnostics, like those discussed in this chapter.

More on the effects of masking and swamping is discussed in Bendre (1989). A theoretical treatment about the foundations of masking and swamping is given in Serfling and Wang (2014).

## 10.3 More Diagnostic Tests

In this section we present some miscellaneous and, perhaps less frequently used, diagnostic tests.

**Glejser's Test**

**Glejser's test** (Glejser, 1969) is constructed in a manner similar to the Breusch–Pagan test for common variance. Glejser's test takes the absolute value of the residuals from an ordinary least squares fit and regresses it on the $j^{\text{th}}$ predictor variable using one of the following models:

$$|e_i| = \gamma_0 + \gamma_1 x_{i,j} + u_i \tag{10.3}$$

$$|e_i| = \gamma_0 + \gamma_1 \sqrt{x_{i,j}} + u_i \tag{10.4}$$

$$|e_i| = \gamma_0 + \gamma_1 x_{i,j}^{-1} + u_i. \tag{10.5}$$

Then, choose Equation (10.3), (10.4), or (10.5) based on which one produces the highest $R^2$ with the lowest standard errors and perform the following hypothesis test

$$H_0 : \gamma_1 = 0$$
$$H_0 : \gamma_1 \neq 0.$$

Rejecting the above null hypothesis is equivalent to rejecting the null hypothesis of constant variance.

## White's Test
**White's test** (White, 1980) is another alternative to the Glejser test and Breusch–Pagan test. Suppose we have the standard multiple linear regression model with $p - 1$ predictors. To perform White's test, take the squared residuals from the ordinary least squares fit for the regression model of interest. Then, regress those squared residuals on all of the original predictors along with their squares and cross-products; i.e., a total of $m = 2(p - 1) + \binom{p-1}{2}$ predictors. Then, take the $R^2$ from the fit of this auxiliary regression model and compute the Lagrange multiplier test statistic, which is

$$L^* = nR^2.$$

Then, $L^* \sim \chi^2_{m-1}$. A small $p$-value based on $L^*$ means to reject the null hypothesis of constant variance.

## Goldfeld–Quandt Test
The **Goldfeld–Quandt test** (Goldfeld and Quandt, 1965) is a variation on the $F$-test presented in Chapter 4 for testing common variance. In this case, we partition the observations into two groups (possibly of different sample sizes) based on an identified predictor, say $X_j$. Thus, one group consists of the observations with the lowest values of $X_j$ (say $n_1$ observations) and the other group consists of the observations with the higher values of $X_j$ (say $n_2 = n - n_1$ observations). Then, fit a separate linear regression model to each of these subsets. Let $\text{MSE}_1$ and $\text{MSE}_2$ be the mean square error values for the respective fits. Then, the test statistic for testing constant variance is

$$G^* = \frac{\text{MSE}_1}{\text{MSE}_2},$$

which is distributed according to a $F_{n_1-p,n_2-p}$ distribution. A small $p$-value based on $G^*$ means to reject the null hypothesis of constant variance.

## Ramsey's RESET
Consider the fitted values from the multiple linear regression model, i.e.,

$$\hat{\mathbf{Y}} = \mathbf{X}\hat{\beta}.$$

We can assess the misspecification of the above linear functional form by including nonlinear combinations of the fitted values as predictors. This is known

as **Ramsey's** **regression** **equation** **specification** **error** **test** (**RESET**)
(Ramsey, 1969). In Ramsey's RESET, we test if $(\mathbf{X}\beta)^2, (\mathbf{X}\beta)^3, \cdots, (\mathbf{X}\beta)^q$
has any power in explaining the variability in $Y$. This is accomplished by first
fitting the following auxiliary regression equation:

$$\mathbf{Y} = \mathbf{X}\alpha + \sum_{k=2}^{q} \gamma_{k-1} \hat{\mathbf{Y}}^k + \boldsymbol{\omega}^*,$$

where $\boldsymbol{\omega}^*$ has been used to denote a vector of *iid* normal errors for the above
model and to distinguish it from the errors in the original multiple linear
regression model. Then, a general linear $F$-test is performed on

$$H_0 : \gamma_1 = \ldots = \gamma_{q-1} = 0$$
$$H_A : \text{at least one } \gamma_k \neq 0.$$

If $p - 1$ predictors are in the original multiple linear regression model, then
the test statistic for the above test follows a $F_{q-1, n-p-q+1}$ distribution.

**Rainbow Test**
Utts (1982) introduced the **rainbow test** for lack-of-fit in regression. The
approach compares a model fit over low leverage points (i.e., the "center" of
the sample) with a fit over the entire set of data. Suppose the size of the
subsample of observations with low leverage points is $m < n$, where $m$ is
usually taken to be $[0.5n]$. Suppose we are interested in testing if there is a
lack-of-fit with the multiple linear regression model with $p - 1$ predictors. Let
$\text{SSE}_{Full}$ and $\text{SSE}_{Central}$ be the error sums of squares for the fits to the full
dataset and the central dataset, respectively. The rainbow test then defines
the error sum of squares for the lack-of-fit as

$$\text{SSE}_{LOF} = \text{SSE}_{Full} - \text{SSE}_{Central}.$$

Hence, the test statistic for the rainbow test is

$$F^* = \frac{\text{SSE}_{LOF}/(n-m)}{\text{SSE}_{Central}/(m-p)},$$

which follows a $F_{n-m, m-p}$ distribution.

**Harvey–Collier Test**
The **Harvey–Collier test** (Harvey and Collier, 1977) is used to detect model
misspecification in a regression equation. The test is based on the recursive
residuals in Equation (10.1). The recursive residuals are fitting the model using
the first $i - 1$ observations and then the $i^{\text{th}}$ observation is predicted from that
fit and appropriately standardized. If the model is correctly specified, the
recursive residuals will have mean 0. If the ordering variable has an influence

on the model fit, then the recursive residuals will have a mean different from 0. The test statistic for the Harvey–Collier test is

$$W = \left[ \frac{n-p}{n-p-1} \sum_{r=p+1}^{n} (w_r - \bar{w})^2 \right]^{-1/2} \sum_{r=p+1}^{n} w_r,$$

where $\bar{w} = (n-p)^{-1} \sum_{r=p+1}^{n} w_r$ is the mean of the recursive residuals. It can be shown that $W$ follows a $t_{n-p-1}$ distribution.

---

## 10.4   Comments on Outliers and Influential Values

Many measures were presented to assess if an observation is an outlier or influential. Classifying an observation as such is as much art as it is science. An observation which is indicated as being highly influential is not necessarily bad. One should think twice before removing such a value and assess what the overall impact is of that observation on the regression estimates. The tools presented in this chapter will aid you in that decision. Outliers need to also be handled with care. As an investigator, you need to assess if that observation occurred due to chance variation or if there is something fundamentally wrong with that recorded observation. If it is the latter reason, then there is a strong case for dropping that observation from the analysis. Moreover, if an observation appears to be highly influential *and* an extreme outlier, then this could provide sufficient justification to omit the observation from the analysis altogether.

Regarding outliers, Iglewicz and Hoaglin (1993) delineated three basic issues:

1. **Outlier labeling** is the process of flagging potential outliers for further investigation. For example, the outliers may be due to data entry error, erroneously collected data, or use of a misspecified model.

2. **Outlier accommodation** is done by using robust statistical techniques that will not be affected by outliers. So if we cannot determine the appropriate status of potential outliers, then we want our statistical analysis to more accurately accommodate these observations. The use of robust regression techniques is discussed later.

3. **Outlier identification** pertains to the formal tests that we can run to determine if observations are outliers, such as with the Dixon's Q test and the generalized ESD test.

The take-home message from this chapter is to be diplomatic in your decision to remove an observation. Even if you omit an influential observation or an outlier, chances are that when you rerun the analysis other observations will be designated as influential or as an outlier. This is especially true with large datasets. Thus, do not *torture* the data by overusing the diagnostic measures presented in this chapter.

## 10.5    Examples

**Example 10.5.1.** *(Punting Data)*
Leg strength is a critical component for a punter to be successful in American football. Renner, Jr. (1983) studied various physical characteristics and abilities for $n = 13$ football punters. Each punter punted a football ten times and the average distance and hang time were recorded. We are interested in modeling the average hang time as a function of the various physical characteristics recorded by the investigator. The variables are:

- $Y$: mean hang time (in seconds)

- $X_1$: right leg strength (in pounds)

- $X_2$: left leg strength (in pounds)

- $X_3$: right leg flexibility (in degrees)

- $X_4$: left leg flexibility (in degrees)

- $X_5$: overall leg strength (in foot-pounds)

We begin by fitting a multiple linear regression model to these data:

```
#############################
Coefficients:
              Estimate Std. Error t value Pr(>|t|)
(Intercept)  0.6449613  0.9856374   0.654   0.5338
RStr         0.0011040  0.0068466   0.161   0.8764
LStr         0.0121468  0.0072592   1.673   0.1382
RFlex       -0.0002985  0.0217420  -0.014   0.9894
LFlex        0.0069159  0.0111723   0.619   0.5555
OStr         0.0038897  0.0019581   1.986   0.0873 .
---
Signif. codes:  0 *** 0.001 ** 0.01 * 0.05 . 0.1   1

Residual standard error: 0.2198 on 7 degrees of freedom
Multiple R-squared:  0.8821,Adjusted R-squared:  0.7979
F-statistic: 10.47 on 5 and 7 DF,  p-value: 0.003782
#############################
```

Even though only one of the predictors is found marginally significant in the estimated model, we will proceed to use the model for the purpose of illustrating certain calculations.

Figure 10.1 gives six different diagnostic plots for six different measures discussed in this chapter. Figures 10.1(a), 10.1(b), and 10.1(c) give plots of the raw residuals, standardized residuals, and Studentized residuals, respectively, versus the fitted values. All three plots look similar, but it looks like there could be some curvature in these plots as well as a possible outlier for the observation with a fitted value around 4.25. This observation also appears as potentially influential as it is the most extreme point on the Q-Q plot (Figure 10.1(d)), the Cook's distance plot (Figure 10.1(e)) and the plot of DFITs (Figure 10.1(f)). All of theses values are reported in Table 10.2 as well as the values for the CovRatio ($C_i$) in the last column.

The value in question on the plots in Figure 10.1 is observation 13 in Table 10.2. While it looks like a potential outlier and/or influential observation on the plots, the values in Table 10.2 indicate that most of the measures are below the rule-of-thumb thresholds stated earlier in the chapter. We could potentially drop this point and rerun the analysis, but without any indication that this was truly an aberrant observation, we retain it in the dataset.

**TABLE 10.2**
Values of different diagnostic measures for the punting data

| $i$ | $y_i$ | $\hat{y}_i$ | $h_{i,i}$ | $e_i$ | $e_i^*$ | $r_i$ | DFIT$_i$ | D$_i$ | $C_i$ |
|---|---|---|---|---|---|---|---|---|---|
| 1 | 4.75 | 4.53 | 0.35 | 0.22 | 1.22 | 1.27 | 0.94 | 0.13 | 0.93 |
| 2 | 4.07 | 3.75 | 0.18 | 0.32 | 1.59 | 1.84 | 0.88 | 0.10 | 0.21 |
| 3 | 4.04 | 4.02 | 0.91 | 0.02 | 0.38 | 0.35 | 1.13 | 0.24 | 25.20 |
| 4 | 4.18 | 4.14 | 0.28 | 0.04 | 0.19 | 0.18 | 0.11 | 0.00 | 3.38 |
| 5 | 4.35 | 4.30 | 0.50 | 0.05 | 0.30 | 0.28 | 0.28 | 0.01 | 4.68 |
| 6 | 4.16 | 4.22 | 0.53 | -0.06 | -0.38 | -0.36 | -0.38 | 0.03 | 4.71 |
| 7 | 4.43 | 4.57 | 0.48 | -0.14 | -0.90 | -0.89 | -0.85 | 0.12 | 2.32 |
| 8 | 3.20 | 3.23 | 0.40 | -0.03 | -0.17 | -0.16 | -0.13 | 0.00 | 4.08 |
| 9 | 3.02 | 3.19 | 0.55 | -0.17 | -1.14 | -1.17 | -1.30 | 0.27 | 1.63 |
| 10 | 3.64 | 3.57 | 0.45 | 0.07 | 0.40 | 0.38 | 0.34 | 0.02 | 4.02 |
| 11 | 3.68 | 3.62 | 0.67 | 0.06 | 0.44 | 0.42 | 0.60 | 0.07 | 6.48 |
| 12 | 3.60 | 3.60 | 0.27 | 0.00 | -0.01 | -0.01 | -0.01 | 0.00 | 3.44 |
| 13 | 3.85 | 4.21 | 0.42 | -0.36 | -2.14 | -3.38 | -2.89 | 0.56 | 0.01 |

We will also test separately if the observations with the maximum and minimum raw residuals are outliers using Dixon's test. The observation with the maximum residual is observation 1, which has the following test results:

```
##############################
Dixon test for outliers

data:  resid(out)
Q.7 = 0.37557, p-value = 0.4933
```

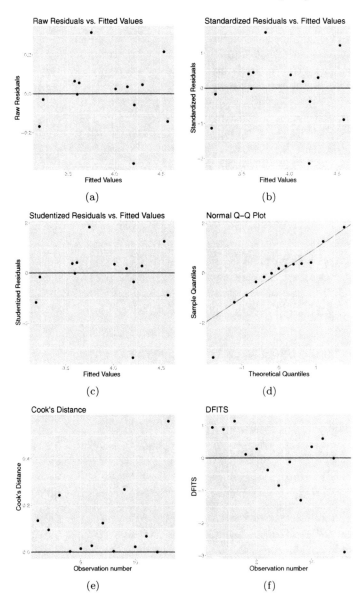

**FIGURE 10.1**
(a) Scatterplot of the raw residuals versus the fitted values. (b) Scatterplot of the standardized residuals versus the fitted values. (c) Scatterplot of the Studentized residuals versus the fitted values. (d) A normal probability plot (NPP) or quantile-quantile (Q-Q) plot of the residuals. (e) Plot of Cook's distance versus the observation number. (f) Plot of the DFITs versus the observation number.

```
alternative hypothesis: lowest value -0.3581 is an outlier
#############################
```

The *p*-value indicates a non-significant result and, thus, we would not claim that this observation is an outlier. The observation with the minimum residual is observation 13, which is the influential observation noted above. Below are the corresponding test results:

```
#############################
Dixon test for outliers

data:  resid(out)
Q.2 = 0.51739, p-value = 0.1053
alternative hypothesis: highest value 0.3153 is an outlier
#############################
```

The *p*-value, again, indicates a non-significant result and, thus, we would not claim this observation is an outlier.

**Example 10.5.2.** *(Expenditures Data)*
Greene (1993) provided data on the per capita expenditure on public schools and the per capita income for the 50 United States and Washington, DC. These data were analyzed in Kleiber and Zeileis (2008) for the purpose of illustrating various regression diagnostics, much like we will do with our analysis. The variables are:

- $Y$: per capita expenditure on public schools (based on U.S. dollars)

- $X$: per capita income (based on U.S. dollars)

However, the value for per capita expenditure for the state of Wisconsin is missing, so this dataset consists of only $n = 50$ complete observations. We begin by fitting a simple linear regression model to these data:

```
#############################
Coefficients:
            Estimate Std. Error t value Pr(>|t|)
(Intercept) -151.26509   64.12183  -2.359   0.0224 *
Income         0.06894    0.00835   8.256 9.05e-11 ***
---
Signif. codes:  0 *** 0.001 ** 0.01 * 0.05 . 0.1   1

Residual standard error: 61.41 on 48 degrees of freedom
  (1 observation deleted due to missingness)
Multiple R-squared:  0.5868,Adjusted R-squared:  0.5782
F-statistic: 68.16 on 1 and 48 DF,  p-value: 9.055e-11
#############################
```

A plot of these data is given in Figure 10.2(a), which clearly shows that a linear regression fit is appropriate, however, there are some interesting features worth discussing.

First is that a potentially influential point shows up on the Studentized residuals plot (Figure 10.2(b)), the Hadi's influence measure plot (Figure 10.2(d)), and the DFBETAS plots (Figures 10.2(e) and 10.2(f)). The recursive residuals plot in Figure 10.2(c) shows that those few observations with higher fitted values in this dataset tend to have a bigger effect on the fit and thus give the recursive residuals the appearance of a decreasing trend.

The first thing we do is to test separately if the observations with the maximum and minimum raw residuals are outliers using Grubbs' test. The observation with the maximum residual is the state of Alaska, which has the following test results:

```
##############################
Grubbs test for one outlier

data:  resid(out.pub)
G.Alaska = 3.68890, U = 0.71662, p-value = 0.00173
alternative hypothesis: highest value 224.2100 is an outlier
##############################
```

The $p$-value indicates a highly significant result and, thus, we could claim Alaska as an outlier. The observation with the minimum residual is the state of Nevada, which has the following test results:

```
##############################
Grubbs test for one outlier

data:  resid(out.pub)
G.Nevada = 1.84910, U = 0.92879, p-value = 1
alternative hypothesis: lowest value -112.3903 is an outlier
##############################
```

The $p$-value indicates a highly non-significant result and, thus, we could not claim Nevada as an outlier.

We can also conduct the chi-square outlier test analogously to the Grubbs' tests above. In this case, we again find that Alaska is a significant outlier:

```
##############################
chi-squared test for outlier

data:  resid(out.pub)
X-squared.Alaska = 13.608, p-value = 0.0002252
alternative hypothesis: highest value 224.2100 is an outlier
##############################
```

However, this test finds that Nevada is also a significant outlier:

```
#############################
chi-squared test for outlier

data:   resid(out.pub)
X-squared.Nevada = 3.4193, p-value = 0.06444
alternative hypothesis: lowest value -112.3903 is an outlier
#############################
```

Given that the Grubbs' test is more powerful, we would likely not claim that Nevada is an outlier in these data.

We next use the Tietjen–Moore test to test for $k = 1, \ldots, 9$ outliers in the dataset. The output below gives the test statistic ($E_k$) and the (simulated) critical value at the $\alpha = 0.05$ level. If the test statistic is less than the critical value, then we reject the null hypothesis of no outliers in favor of the alternative of exactly $k$ outliers. These tests indicate that up to 8 values could be considered as outliers, which is a very high percentage of the data (16%). Thus, this procedure is likely producing too conservative of results.

```
#############################
          T Talpha.5.          Sig
1 0.7166202 0.7980108    Significant
2 0.6510669 0.6804948    Significant
3 0.5776389 0.5970196    Significant
4 0.5131566 0.5274330    Significant
5 0.4559929 0.4696364    Significant
6 0.4044778 0.4154624    Significant
7 0.3543785 0.3730985    Significant
8 0.3251761 0.3325733    Significant
9 0.3016139 0.2972028 Not Significant
#############################
```

Next we merely consider the same values of $k$ as upper bounds and run the generalized ESD test. Using this approach, we find that only $k = 1$ produces a significant result, which is more consistent with what we found using the Grubbs' test. Thus, assuming $k > 1$ for the number of outliers in this dataset is likely not reasonable.

```
#############################
  No.Outliers Test.Stat Critical.Val
1          1   3.688889     3.128247
2          2   2.237705     3.120128
3          3   2.122274     3.111796
4          4   2.241822     3.103243
5          5   2.214463     3.094456
6          6   2.204626     3.085425
7          7   2.281448     3.076135
8          8   1.964458     3.066572
```

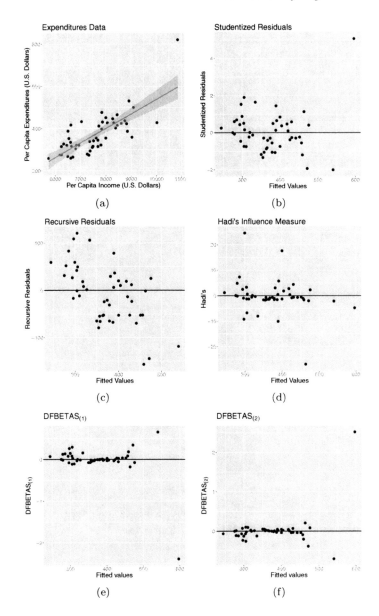

**FIGURE 10.2**
(a) Scatterplot of the expenditures data. (b) Scatterplot of the Studentized residuals versus the fitted values. (c) Scatterplot of the recursive residuals versus the fitted values. (d) Scatterplot of Hadi's influence measure versus the fitted values. (e) Scatterplots of the DFBETAS$_{(1)}$ values versus the fitted values. (f) Scatterplots of the DFBETAS$_{(2)}$ values versus the fitted values.

```
9          9  1.860030      3.056723
##############################
```

We next turn to the various diagnostic tests discussed in Section 10.3. First we conduct Glejser's test for constant variance. We use the same three functional forms presented in Equations (10.3)–(10.5). The $R^2$ values for these three models are, respectively, 0.1916, 0.1753, and 0.1347. Therefore, we would go with the model in Equation (10.3) which is a linear function of the untransformed version of the per capita income predictor. Below is the fit for that model:

```
##############################
Coefficients:
             Estimate Std. Error t value Pr(>|t|)
(Intercept) -80.445951  37.699808  -2.134  0.03799 *
Income        0.016556   0.004909   3.372  0.00148 **
---
Signif. codes:  0 *** 0.001 ** 0.01 * 0.05 . 0.1   1

Residual standard error: 36.11 on 48 degrees of freedom
Multiple R-squared:  0.1916,Adjusted R-squared:  0.1747
F-statistic: 11.37 on 1 and 48 DF,  p-value: 0.00148
##############################
```

The test statistic shows that $\gamma_1$, which is the coefficient for the per capita income predictor in the above output, is statistically significant with a $p$-value of 0.0015. Thus, we would reject the null hypothesis of constant variance based on Glejser's test.

Next we present the output for White's test:

```
##############################
White's Test for Heteroskedasticity:
====================================

 No Cross Terms

 H0: Homoskedasticity
 H1: Heteroskedasticity

 Test Statistic:
 9.9631

 Degrees of Freedom:
 12

 P-value:
 0.6192
```

```
##############################
```

For this test, we obtain a large $p$-value (0.6192), which means we would fail to reject the null hypothesis of constant variance. The same decision is reached using the Goldfeld–Quandt test:

```
##############################
Goldfeld-Quandt test

data:  out.pub
GQ = 0.80022, df1 = 23, df2 = 23, p-value = 0.7013
##############################
```

As can be seen, the Goldfeld–Quandt test yielded a $p$-value of 0.7013. Since the Goldfeld–Quandt and White tests both returned highly non-significant results, it is reasonable to conclude that the variance is constant.

Finally, we assess possible model misspecification using a simple linear regression model for the expenditures data. First we run Ramsey's RESET where we include quadratic and cubic terms of the fitted values:

```
##############################
RESET test

data:  out.pub
RESET = 6.5658, df1 = 2, df2 = 46, p-value = 0.003102
##############################
```

Ramsey's RESET produces a significant result, which indicates that there is some detectable misspecification in modeling the relationship between per capita expenditures as a function of per capita income. Specifically, a significant amount of additional variability in the per capita expenditures can be explained by these higher-order terms.

We next run the rainbow test to assess if the model fit for the "central" 50% of the states (based on per capita income) differs from the fit to the whole dataset. The results are below:

```
##############################
Rainbow test

data:  out.pub
Rain = 1.4762, df1 = 25, df2 = 23, p-value = 0.1756
##############################
```

Here, the results are not statistically significant. Thus, there does not appear to be any lack-of-fit out in the tails of the per capita income values.

Finally, we run the Harvey–Collier test:

```
##############################
Harvey-Collier test

data:  out.pub
HC = 0.80688, df = 47, p-value = 0.4238
##############################
```

This agrees with the rainbow test in that there is no statistically significant model misspecification detected. Since the rainbow and Harvey–Collier tests are in fairly strong agreement, it is reasonable to claim that using a simple linear regression model for the relationship between per capita expenditures and per capita income is adequate.

# 11

## Measurement Errors and Instrumental Variables Regression

The linear regression model $\mathbf{Y} = \mathbf{X}\boldsymbol{\beta} + \boldsymbol{\epsilon}$ reflects that $\mathbf{Y}$ is being measured with an error term $\boldsymbol{\epsilon}$. However, some or all of the predictors in $\mathbf{X}$ may also be measured with some error, which is not what the $\boldsymbol{\epsilon}$ term reflects. For example, suppose you wish to model an individual's heart rate after they have jogged a certain distance. While an individual may be assigned to jog 500 feet, the exact distance they jog could be 501 feet and 3 inches. Clearly, such situations show that it may be difficult to know or measure the values of the predictor *exactly*.[1]

Measurement error models share some similarities with the traditional linear regression models, such as the true relationship between our response variable and our set of predictor variables is well-approximated by a linear regression with negligible model error. By negligible model error, we mean that the model used is not biased with respect to the true underlying relationship. However, measurement error models assume that the response variable *and* one or more of the predictor variables are subject to additive measurement errors. For this initial discussion, $X_j$ generically denotes a predictor variable, $W_j$ is the observed variable, and $U_j$ is the measurement error, where $j = 1, \ldots, p - 1$ is the number of predictors under consideration. The measurement error can be characterized in two primary ways:

1. The **classical measurement error model** is

$$W_j = X_j + U_j,$$

where $W_j$ is an unbiased measure of $X_j$ so that $\mathrm{E}(X_j|U_j) = 0$. Typically, we reasonably assume that $U_j|X_j \sim \mathcal{N}(0, \sigma_{u_j}^2)$.

---

[1] When the effect of the response $Y$ may be measured at values of the $p-1$ predictors (i.e., $X_1, \ldots, X_{p-1}$) that were chosen by the experimenter, then this is called a **type I regression problem**. When $Y$ and $X_1, \ldots, X_{p-1}$ are all chosen *at random* in the $p$-dimensional space, then this is called a **type II regression problem** (Crow et al., 1960). Typically, a common assumption for the type II regression problem is that all of the variables are normally distributed. As such, the shape that forms from the observations will often be more like an oval compared to data drawn under a type I regression problem, depending on how strong the correlation is with each predictor and the response. The type of regression problem you have will ultimately impact various residual diagnostic measures and statistical intervals. One note about terminology: the type I and type II regression problems are not related to Type I and Type II errors in inference.

2. The **Berkson model** (Berkson, 1950) alters the classical measurement error model in such a way that now we assume that

$$X_j = W_j + U_j,$$

where $E(U_j|W_j) = 0$. Thus, the true predictor will have more variability than the observed predictor.

These models are generically referred to as **measurement error models** or **errors-in-variables models**. The classical error model and Berkson model seem somewhat similar, but they do have some important differences, such as those that arise in power calculations. These differences are not discussed here. Refer to the texts by Fuller (1987) and Carroll et al. (2006) for details.

Carroll et al. (2006) also highlight an important defining characteristic about the unobserved true values of $X_j$:

1. In a **functional measurement error model**, the sequence of $X_{i,j}$, $i = 1, \ldots, n$, are treated as unknown fixed constants or parameters.

2. In a **structural measurement error model**, the $X_{i,j}$ are assumed to be random variables with a specified distribution.

Moreover, the errors associated with the response and each of the predictors are typically assumed to be jointly normally distributed with mean 0 and a particular variance–covariance matrix, which can include a correlation coefficient. An additional measurement error may be assumed in the response, but this is typically just absorbed into $\epsilon$ unless a correlation structure is present. For the purpose of our discussion, we only focus on the functional measurement error model case where the random error and measurement error terms are uncorrelated. The structural measurement error model settings require further modifications to the procedures discussed in this chapter.

## 11.1    Estimation in the Presence of Measurement Errors

Suppose we have a dataset of size $n$ where $\mathbf{Y}$ is the response vector and $\mathbf{X}$ is the $n \times (p - 1)$ matrix of predictors. Thus far, we have primarily focused on fitting the multiple linear regression model

$$\mathbf{Y} = \beta_0 \mathbf{1}_n + \mathbf{X}^{\mathsf{T}} \boldsymbol{\beta}_- + \boldsymbol{\epsilon}, \tag{11.1}$$

where $\boldsymbol{\epsilon}$ is a vector of *iid* normal random errors with mean 0 and variance $\sigma^2$. Note that we have used a slightly different formulation above by separating out the intercept term. In particular, the intercept term has been decoupled from $\mathbf{X}$ and $\boldsymbol{\beta}_-$ is defined as the same vector of regression coefficients as $\boldsymbol{\beta}$ in our regular model, but with $\beta_0$ omitted from the first entry. This allows for easier notation in what follows.

Suppose instead that we observe the matrix of predictors $\mathbf{W}$, which has measurement error based on the relationship for the classical measurement error model:

$$\mathbf{W} = \mathbf{X} + \mathbf{U}.$$

Then $\mathbf{W}$ and $\mathbf{U}$ are $n \times (p-1)$ matrices of observed predictors and measurement errors, respectively. We assume that the "true" values (i.e., the unobserved $\mathbf{X}$) have known population mean $\boldsymbol{\mu}_X = (\mu_{X_1}, \ldots, \mu_{X_{p-1}})^{\mathrm{T}}$ and known population variance–covariance matrix $\Sigma_X$. Therefore, the mean vector is $(p-1)$-dimensional and the variance–covariance matrix has $(p-1) \times (p-1)$ dimensions. Note that no distributional assumption is being made about $X_1, \ldots, X_{p-1}$. The interpretations for $\boldsymbol{\mu}_X$ and $\Sigma_X$ are simply that they are the center and spread, respectively, of the nominal design points. As stated at the beginning of this chapter, making a distributional assumption is possible, but results in needing a more complex estimation procedure for the measurement error model.

We further assume that the predictors are independent of one another, so $\Sigma_X = \mathrm{diag}(\sigma^2_{X_1}, \ldots, \sigma^2_{X_{p-1}})$. Moreover, the corresponding measurement errors, say, $U_1, \ldots, U_{p-1}$, are assumed to have respective means of 0 and are independent of each other. Therefore, we have $\Sigma_U = \mathrm{diag}(\sigma^2_{U_1}, \ldots, \sigma^2_{U_{p-1}})$.

With the above in place, we now end up estimating

$$\mathbf{Y} = \beta^*_0 \mathbf{1}_n + \mathbf{W}^{\mathrm{T}} \boldsymbol{\beta}^*_- + \boldsymbol{\epsilon}^*. \tag{11.2}$$

The new vector of error terms, $\boldsymbol{\epsilon}^* = (\epsilon^*_1, \ldots, \epsilon^*_n)^{\mathrm{T}}$, is composed of the old error term and the measurement error term. The $\epsilon^*_i$s are assumed to be independent with mean 0 and variance $\sigma^{*2}$ and thus we can use ordinary least squares for estimating the parameters of the model based on the observed data. In addition, we assume that the original errors and the measurement errors are independent, both with mean 0 and variances $\sigma^2$ and $\sigma^2_{X_j}$, $j = 1, \ldots, p-1$, respectively. Moreover, the original errors and measurement errors are each assumed to be independent across the observations.

However, the ordinary least squares estimates of (11.2) are biased because of the following relationships:

$$\boldsymbol{\beta}^* \equiv \begin{pmatrix} \beta^*_0 \\ \boldsymbol{\beta}^*_- \end{pmatrix} = \begin{pmatrix} \beta^*_0 \\ \beta^*_1 \\ \vdots \\ \beta^*_{p-1} \end{pmatrix} = \begin{pmatrix} \beta_0 + \sum_{j=1}^{p-1} \beta_j \mu_{X_j} (1 - \lambda_j) \\ \Sigma_X (\Sigma_X + \Sigma_U)^{-1} \boldsymbol{\beta}_- \end{pmatrix}$$

and

$$\sigma^{*2} = \boldsymbol{\beta}^{\mathrm{T}}_- \Sigma_{\mathbf{X}} (\mathbf{I}_{p-1} - \Sigma_X (\Sigma_X + \Sigma_U)^{-1}) \boldsymbol{\beta}_- + \sigma^2,$$

where

$$\lambda_j = \frac{\sigma^2_{X_j}}{\sigma^2_{X_j} + \sigma^2_{U_j}} \tag{11.3}$$

is known as the **reliability ratio** (Fuller, 1987). Note that in the simple linear regression setting, the above formulas reduce to:

$$\beta_0^* = \beta_0 + (1 - \lambda)\beta_1,$$
$$\beta_1^* = \lambda\beta_1,$$

and

$$\sigma^{*2} = \beta_1^2 \sigma_X^2 (1 - \lambda) + \sigma^2,$$

where the reliability ratio is for the single predictor; i.e., $\lambda = \sigma_X^2/(\sigma_X^2 + \sigma_U^2)$.

There are other estimators for this model which possess various asymptotic properties (such as consistency and unbiasedness) as well as estimators when assuming a correlation structure in the errors. Again, we refer to Fuller (1987) and Carroll et al. (2006) for such discussions.

## 11.2   Orthogonal and Deming Regression

When all of the variables have measurement error, there may be another dilemma: which variable is the response and which variables are the predictors? A well-planned experiment should have already addressed the roles of each variable; however, there may be times that their roles are not clear. Sometimes, it may even be arbitrary as to which role each variable is designated. One way to fit a regression hyperplane to such data is by **orthogonal regression**, which is sometimes called **total least squares**. Orthogonal regression is used in fields such as astronomy, signal processing, and electrical engineering, where they build regression relationships based on quantities with known measurement errors.

In ordinary least squares, we estimate the regression coefficients by minimizing the sum of the squared vertical distances from the hyperplane of interest to the data points; i.e., the residuals. However, orthogonal regression minimizes the sum of the squared orthogonal distances from the hyperplane of interest to the data points. The orthogonal regression line is calculated through a relatively easily implemented algorithm. But first, we must establish a few calculations.

Assume that we have $n$ data points, $\mathbf{Z}_1, \ldots, \mathbf{Z}_n$, each of length $p$. You can think of these vectors as being the vectors of predictors in the ordinary least squares setting, but with the response value appended to each vector; i.e., $\mathbf{Z}_i = (\mathbf{X}_i^{\mathrm{T}}, Y_i)^{\mathrm{T}}$. Thus, we are not distinguishing the response from the predictors.

Orthogonal regression fits a hyperplane passing through the point

$$\bar{\mathbf{Z}} = \begin{pmatrix} \bar{Z}_1 \\ \bar{Z}_2 \\ \vdots \\ \bar{Z}_p \end{pmatrix} = \begin{pmatrix} (1/n) \sum_{i=1}^{n} Z_{i,1} \\ (1/n) \sum_{i=1}^{n} Z_{i,2} \\ \vdots \\ (1/n) \sum_{i=1}^{n} Z_{i,p} \end{pmatrix}$$

with a normal vector $\boldsymbol{\beta}^{\perp} = (\beta_1^{\perp}, \dots, \beta_p^{\perp})^{\mathrm{T}}$ for the equation

$$(\mathbf{Z} - \bar{\mathbf{Z}})^{\mathrm{T}} \boldsymbol{\beta}^{\perp} = 0, \tag{11.4}$$

which minimizes the sum of the squared distances from the data points to that hyperplane,

$$\operatorname*{argmin}_{\boldsymbol{\beta}^{\perp}} \sum_{i=1}^{n} \frac{[(\mathbf{Z}_i - \bar{\mathbf{Z}})^{\mathrm{T}} \boldsymbol{\beta}^{\perp}]^2}{(\boldsymbol{\beta}^{\perp})^{\mathrm{T}} \boldsymbol{\beta}^{\perp}}. \tag{11.5}$$

Next, form the matrix

$$\mathbf{M} = \begin{pmatrix} (\mathbf{Z}_1 - \bar{\mathbf{Z}})^{\mathrm{T}} \\ \vdots \\ (\mathbf{Z}_n - \bar{\mathbf{Z}})^{\mathrm{T}} \end{pmatrix} = \begin{pmatrix} Z_{1,1} - \bar{Z}_1 & \cdots & Z_{1,p} - \bar{Z}_p \\ \vdots & \ddots & \vdots \\ Z_{n,1} - \bar{Z}_1 & \cdots & Z_{n,p} - \bar{Z}_p \end{pmatrix}.$$

Then, compute the smallest singular value of $\mathbf{M}$. That is, compute the smallest eigenvalue of $\mathbf{M}^{\mathrm{T}}\mathbf{M}$ and one eigenvector $\boldsymbol{\beta}^{\perp}$ in the corresponding eigenspace of $\mathbf{M}^{\mathrm{T}}\mathbf{M}$. Thus, the hyperplane with equation (11.4) minimizes the sum of the squared distances from the data points to that hyperplane; i.e., provides the minimum sought in (11.5).

An easier way to think about a total least squares fit is by using principal components. Suppose that $\psi$ is the vector pertaining to the last principal component. Then, $\psi = \boldsymbol{\beta}^{\perp}$, which is the solution to (11.4). More on total least squares is found in Markovsky and Van Huffel (2007).

Also, one can obtain confidence intervals for the orthogonal regression coefficients. Theoretical confidence intervals have been constructed, but they do not have well-understood properties in the literature. An alternative way to proceed is with bootstrapping; see Chapter 12. Bootstrapping is an advisable alternative in this setting if you wish to report confidence intervals for your estimates.

Let us next consider the single predictor setting of the model in (11.2). Note that in the single predictor setup, we observe the pairs $(x_i, y_i)$, where the true values are $(x_i^*, y_i^*)$. These are related as follows:

$$x_i = x_i^* + u_i$$
$$y_i = y_i^* + \epsilon_i,$$

where, again, the $\epsilon_i$ and $u_i$ are independent. We wish to estimate the parameters for

$$y_i^* = \beta_0^* + \beta_1^* x_i^* + \epsilon_i^*.$$

However, assume that the ratio of the variances of $\epsilon$ and $u$, given by

$$\eta = \frac{\sigma^2}{\sigma_U^2},$$

is known. In practice, $\eta$ might be estimated from related data sources or from established assumptions. This specific errors-in-variables model is called the **Deming regression**, named after the renowned statistician W. Edwards Deming. The least squares estimates of the Deming regression model are

$$\hat{\beta}_1^* = \frac{S_{yy} - \eta S_{xx} + \sqrt{(S_{yy} - \eta S_{xx})^2 + 4\eta S_{xy}^2}}{2S_{xy}},$$

$$\hat{\beta}_0^* = \bar{y} - \hat{\beta}_1 \bar{x},$$

$$\hat{x}_i^* = x_i + \frac{\hat{\beta}_1^*}{\hat{\beta}_1^{*2} + \eta}(y_i - \hat{\beta}_0^* - \hat{\beta}_1^* x_i).$$

When $\eta = 1$, the above formulas reduce to the closed-form total least squares solution to (11.5) for the case of a single predictor variable.

---

## 11.3    Instrumental Variables Regression

**Instrumental variables regression** is used in the estimation of causal relationships when controlled experiments are not feasible. Roughly speaking, when estimating the causal effect of some independent variable $X$ (which is correlated with the error $\epsilon$ and called an **exogenous variable**) on a dependent variable $Y$ (which is uncorrelated with the error and called an **endogenous variable**), an **instrumental variable** $Z$ is one such that it affects $Y$ only through its effect on $X$. Thus, in order for an instrumental variables regression to work properly, you must have $\text{Cov}(Z, \epsilon) = 0$ and $\text{Cov}(X, \epsilon) \neq 0$. An example is modeling survival time of individuals with heart attacks (the endogenous variable) as a function of cardiac catheterization treatment (the exogenous variable), but where distance of the nearest catheterization hospital is correlated with receiving the catheterization treatment (the instrumental variable).

Suppose we consider the classic linear regression model

$$\mathbf{Y} = \mathbf{X}\boldsymbol{\beta} + \boldsymbol{\epsilon},$$

consisting of $n$ observations and $p - 1$ predictors. However, suppose that $\mathbf{X}$ and $\epsilon$ are correlated, in which case we say that $\mathbf{X}$ is **contaminated**. Then, the columns of the $n \times q$ matrix $\mathbf{Z}$ consist of instruments such that $q \geq p$. Instrumental variables have two properties:

1. They are uncorrelated with the error terms ($\epsilon$), in which case the instruments are said to be **clean**.

2. The matrix of correlations between the variables in $\mathbf{X}$ and the variables in $\mathbf{Z}$ are of maximum rank; i.e., their rank is $p$.

We proceed with estimation by first premultiplying the classic linear regression model be $\mathbf{Z}^{\mathrm{T}}$, which gives

$$\mathbf{Z}^{\mathrm{T}}\mathbf{Y} = \mathbf{Z}^{\mathrm{T}}\mathbf{X}\beta + \mathbf{Z}^{\mathrm{T}}\epsilon.$$

We next look at the expectation of the model conditioned on $\mathbf{Z}$:

$$E[\mathbf{Z}^{\mathrm{T}}\mathbf{Y}|\mathbf{Z}] = E[\mathbf{Z}^{\mathrm{T}}\mathbf{X}\beta|\mathbf{Z}] + E[\mathbf{Z}^{\mathrm{T}}\epsilon|\mathbf{Z}].$$

Since $\mathbf{Z}$ and $\epsilon$ are uncorrelated, the last term on the right is 0. Therefore, while skipping over some of the details regarding asymptotic properties, we are interested in finding the solution to

$$\mathbf{Z}^{\mathrm{T}}\mathbf{Y} = \mathbf{Z}^{\mathrm{T}}\mathbf{X}\beta,$$

which is a system of $p$ equations needing to be solved for $p$ unknowns.

At first glance, it appears that ordinary least squares is a good method for estimating $\beta$. When the number of columns of $\mathbf{X}$ equals the number of columns of $\mathbf{Z}$ (i.e., $p = q$), then ordinary least squares is appropriate. In this case, the ordinary least squares estimate for instrumental variables regression is

$$\hat{\beta}_{\mathrm{IV}} = (\mathbf{Z}^{\mathrm{T}}\mathbf{X})^{-1}\mathbf{Z}^{\mathrm{T}}\mathbf{Y},$$

which is uniquely identified. However, if $p < q$, then an estimated for instrumental variables regression is overly identified and $\beta$ is not uniquely determined. So the method of **two-stage least squares** (or **2SLS**) is used:

**Stage 1:** Regress each column of $\mathbf{X}$ on $\mathbf{Z}$ to get the predicted values of $\mathbf{X}$:

$$\begin{aligned}\hat{\mathbf{X}} &= \mathbf{Z}(\mathbf{Z}^{\mathrm{T}}\mathbf{Z})^{-1}\mathbf{Z}^{\mathrm{T}}\mathbf{X} \\ &= P_{\mathbf{Z}}\mathbf{X},\end{aligned}$$

where $P_{\mathbf{Z}}\mathbf{X}$ is an idempotent matrix.

**Stage 2:** Regress $\mathbf{Y}$ on $\hat{\mathbf{X}}$. The ordinary least squares estimate for this case yields the instrumental variables regression estimate:

$$\begin{aligned}\hat{\beta}_{\mathrm{IV}} &= (\hat{\mathbf{X}}^{\mathrm{T}}\hat{\mathbf{X}})^{-1}\hat{\mathbf{X}}^{\mathrm{T}}\mathbf{Y} \\ &- (\mathbf{X}^{\mathrm{T}}P_{\mathbf{Z}}^{\mathrm{T}}P_{\mathbf{Z}}\mathbf{X})^{-1}\mathbf{X}^{\mathrm{T}}P_{\mathbf{Z}}^{\mathrm{T}}\mathbf{Y} \\ &= (\mathbf{X}^{\mathrm{T}}P_{\mathbf{Z}}\mathbf{X})^{-1}\mathbf{X}^{\mathrm{T}}P_{\mathbf{Z}}\mathbf{Y}.\end{aligned}$$

If the error terms are correlated, then a generalized least squares estimator can be used, which is discussed in greater detail in Chapter 13. Suppose that the variance–covariance matrix of $\epsilon$ is $\Omega$. If you provide an estimate of $\Omega$ (either from historical information or empirical results), then the method of seemingly unrelated regressions (or SUR) can be used for estimation. SUR is discussed in greater detail in Chapter 22, but for now, it is enough to note that SUR uses ordinary least squares in the context of a multivariate multiple regression problem where the assumption of $\text{Cov}(\epsilon) = \hat{\Omega}$ is made. Then, an instrumental variables regression estimation routine can be accommodated by first performing SUR and then the same steps from 2SLS. This procedure is called **three-stage least squares** (or **3SLS**) and yields the estimate:

$$\hat{\beta}_{\text{IV}} = [\hat{\mathbf{X}}^{\text{T}}\mathbf{Z}(\mathbf{Z}^{\text{T}}\hat{\Omega}\mathbf{Z})^{-1}\mathbf{Z}^{\text{T}}\hat{\mathbf{X}}]^{-1}\hat{\mathbf{X}}^{\text{T}}\mathbf{Z}(\mathbf{Z}^{\text{T}}\hat{\Omega}\mathbf{Z})^{-1}\mathbf{Z}^{\text{T}}\mathbf{Y}.$$

More on some of the details discussed above and instrumental variables regression can be found in Scott and Holt (1982), Staiger and Stock (1997), and Chapter 12 of Stock and Watson (2015).

One final note is that there can be serious problems with instrumental variables regression for finite sample sizes. Almost the entire theory surrounding instrumental variables regression is based on asymptotic properties. If using this method, then be aware of this limitation. Of course, the next most logical question is "how large does $n$ need to be?" There is no clear answer for this, but you should have either additional statistical justification or knowledge of the study which indicates the estimate you obtain appears reasonable.

---

## 11.4   Structural Equation Modeling

A **structural equation model** (or **SEM**) is a set of linear equations used to specify phenomena in terms of their presumed cause-and-effect variables, similar to instrumental variables regression. However, the main difference with SEMs is that they generally allow for **latent variables**, which are variables that are not directly observed. SEMs are commonly applied in areas such as the social sciences (e.g., in studying the relationship between socio-economic status and achievement) and economics (e.g., in studying supply and demand).

A widely used method for specifying, fitting, and evaluating SEMs is with the **LISREL** model, which stands for **Linear Structural Relationships**, whose name comes from the software developed by the same name (Jöreskog and van Thillo, 1972). The LISREL model, using most of the same notation as in Jöreskog and van Thillo (1972), is given by the following equations:

$$\eta = \mathbf{B}\eta + \Gamma\xi + \zeta$$
$$\mathbf{Y} = \Lambda_{\mathbf{Y}}\eta + \epsilon$$
$$\mathbf{X} = \Lambda_{\mathbf{X}}\xi + \delta,$$

where

- $\boldsymbol{\eta}$ is an $m$-dimensional random vector of latent dependent, or endogenous, variables;

- $\boldsymbol{\xi}$ is an $n$-dimensional random vector of latent independent, or exogenous, variables;

- $\mathbf{Y}$ is a $p$-dimensional vector of observed indicators of the dependent latent variables $\boldsymbol{\eta}$;

- $\mathbf{X}$ is a $q$-dimensional vector of observed indicators of the independent latent variables $\boldsymbol{\xi}$;

- $\boldsymbol{\epsilon}$ is a $p$-dimensional vector of measurement errors in $\mathbf{Y}$;

- $\boldsymbol{\delta}$ is a $q$-dimensional vector of measurement errors in $\mathbf{X}$;

- $\Lambda_{\mathbf{Y}}$ is a $p \times m$ matrix of coefficients of the regression of $\mathbf{Y}$ on $\boldsymbol{\eta}$;

- $\Lambda_{\mathbf{X}}$ is a $q \times n$ matrix of coefficients of the regression of $\mathbf{X}$ on $\boldsymbol{\xi}$;

- $\Gamma$ is an $m \times n$ matrix of coefficients of the $\boldsymbol{\xi}$-variables in the structural relationship;

- $\mathbf{B}$ is an $m \times m$ matrix of coefficients of the $\boldsymbol{\eta}$-variables in the structural relationship such that $\mathbf{B}$ has zeros on the diagonal, and $(\mathbf{I}_m - \mathbf{B})$ is required to be non-singular;

- $\boldsymbol{\zeta}$ is an $m$-dimensional vector of equation errors (random disturbances) in the structural relationship between $\boldsymbol{\eta}$ and $\boldsymbol{\xi}$.

Assumptions of this model include:

- $E(\boldsymbol{\zeta}) = \mathbf{0}$, $E(\boldsymbol{\epsilon}) = \mathbf{0}$, $E(\boldsymbol{\delta}) = \mathbf{0}$, $E(\boldsymbol{\xi}) = \mathbf{0}$, and $E(\boldsymbol{\eta}) = \mathbf{0}$;

- $\mathrm{Cov}(\boldsymbol{\zeta}) = \Psi$, $\mathrm{Cov}(\boldsymbol{\epsilon}) = \Sigma_\epsilon$, $\mathrm{Cov}(\boldsymbol{\delta}) = \Sigma_\delta$, and $\mathrm{Cov}(\boldsymbol{\xi}) = \Phi$;

- $\boldsymbol{\zeta}$, $\boldsymbol{\epsilon}$, and $\boldsymbol{\delta}$ are mutually uncorrelated;

- $\boldsymbol{\zeta}$ is uncorrelated with $\boldsymbol{\xi}$;

- $\boldsymbol{\epsilon}$ is uncorrelated with $\boldsymbol{\eta}$;

- $\boldsymbol{\delta}$ is uncorrelated with $\boldsymbol{\xi}$.

The quantities $\boldsymbol{\eta}$ and $\boldsymbol{\xi}$ are the latent variables for this model. $\mathbf{Y}$ and $\mathbf{X}$ are the observed values (i.e., the data) and are linearly related to $\boldsymbol{\eta}$ and $\boldsymbol{\xi}$ through the coefficient matrices $\Lambda_{\mathbf{Y}}$ and $\Lambda_{\mathbf{X}}$, respectively. Because $\mathbf{Y}$ and $\mathbf{X}$ are observed, their respective equations in the LISREL model are sometimes called the **measurement equations**.

The LISREL model often cannot be solved explicitly and so other routines must be used. Depending on the complexity of the model being investigated, estimation methods such as 2SLS can be used or other methods which use a combination of a least squares criterion and a maximum likelihood criterion. We will not explore the details of the estimation procedures for SEMs, but numerous texts are devoted to SEMs; e.g., Bollen (1989), Kline (2016), and Schumacker and Lomax (2016).

## 11.5   Dilution

Consider fitting a hyperplane for the multiple linear regression model

$$\mathbf{Y} = \mathbf{X}^{\mathrm{T}}\boldsymbol{\beta} + \boldsymbol{\epsilon}$$

and then estimating the gradients (slopes) of the hyperplane for each predictor. Variation in the response causes imprecision in the estimated slopes, but they are unbiased since the procedure calculates the right gradients. However, measurement error in the predictors causes bias in the estimated gradients as well as imprecision. Basically, the more variability there is in a given predictor's measurement, the closer the estimated gradient gets to 0 instead of the true gradients. This "dilution" of the gradient toward 0 for a given predictor is referred to as **regression dilution**, **attenuation**, or **attenuation bias**. In fact, the reliability ratio $\lambda_j$ given in Equation (11.3) is an attenuating factor for the corresponding $\beta_j$.

The random noise in predictor variables induce a bias, but the random noise in the response variable does not. Recall that multiple regression is not symmetric as the hyperplane of best fit for predicting $\mathbf{Y}$ from $\mathbf{X}$ (the usual multiple linear regression) is not the same as the hyperplane of best fit for predicting any one of the predictors from the remaining predictors and $\mathbf{Y}$. This was indicated in the discussion of orthogonal regression and will be illustrated in one of the following examples.

Regression dilution bias can be handled a number of ways, such as through the measurement error models we discussed in this chapter. However, many more methods are available in the literature for other relevant measurement error models. One can refer to the texts by Fuller (1987) and Carroll et al. (2006) for a deeper treatment.

## 11.6   Examples

**Example 11.6.1.** *(Blood Alcohol Concentration Data)*
Krishnamoorthy et al. (2001) analyzed data based on a study to compare the

blood alcohol concentration (BAC) of $n = 15$ subjects using two different methods. The variables are:

- $Y$: BAC based on a breath estimate in a laboratory

- $X$: BAC obtained using the Breathalyzer Model 5000

Krishnamoorthy et al. (2001) analyzed these data in the context of a calibration problem. There is a certain amount of measurement error in obtaining these measurements, so it might not be entirely clear which measurement could possibly be used to explain the variation in the other measurement. So we proceed with orthogonal regression.

First we present the summary of the coefficients for the ordinary least squares fit when laboratory estimate $(Y)$ is regressed on the breathalyzer estimate $(X)$:

```
###########################
Coefficients:
            Estimate Std. Error t value Pr(>|t|)
(Intercept) 0.001347   0.008920   0.151    0.882
breath      0.958008   0.069548  13.775 3.93e-09 ***
---
Signif. codes:  0 *** 0.001 ** 0.01 * 0.05 . 0.1   1

Residual standard error: 0.01366 on 13 degrees of freedom
Multiple R-squared:  0.9359,Adjusted R-squared:  0.9309
F-statistic: 189.7 on 1 and 13 DF,  p-value: 3.931e-09
###########################
```

Next we present the summary of the coefficients for the ordinary least squares fit when the breathalyzer estimate $(X)$ is regressed on the laboratory estimate $(Y)$:

```
###########################
Coefficients:
            Estimate Std. Error t value Pr(>|t|)
(Intercept) 0.006238   0.008848   0.705    0.493
labtest     0.976901   0.070920  13.775 3.93e-09 ***
---
Signif. codes:  0 *** 0.001 ** 0.01 * 0.05 . 0.1   1

Residual standard error: 0.01379 on 13 degrees of freedom
Multiple R-squared:  0.9359,Adjusted R-squared:  0.9309
F-statistic: 189.7 on 1 and 13 DF,  p-value: 3.931e-09
###########################
```

Using the above estimates, we then find that the total least squares solution

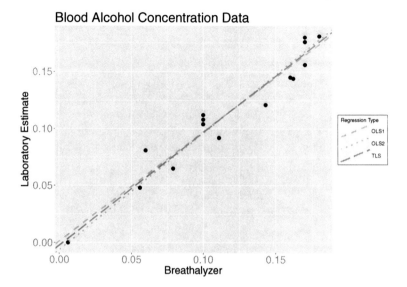

**FIGURE 11.1**
Scatterplot of the blood alcohol concentration dataset with the ordinary least squares fits from regressing $Y$ on $X$, $X$ on $Y$, and the total least squares fit.

is $\boldsymbol{\beta}^{\perp} = (-0.002,\ 0.990)^{\mathrm{T}}$. Since these estimates are close to 0 and 1, respectively, this indicates fairly close agreement between the values of the two variables.

Figure 11.1 provides a scatterplot of laboratory estimate $(Y)$ versus the breathalyzer estimate. The three lines shown are the ordinary least squares line for regressing $Y$ on $X$, the ordinary least squares line for regressing $X$ on $Y$, and the orthogonal regression line. The three regression lines are presented to show how three different estimates for the same dataset can look, albeit the orthogonal regression line is close to the ordinary least squares line when regressing $Y$ on $X$. Normally, you only repeat the orthogonal regression line if you are proceeding under the assumption that there are measurement errors in both variables and it is unclear as to which role each variable should play.

**Example 11.6.2.** *(U.S. Economy Data)*
Suppose we are interested in modeling a simultaneous set of equations in the context of instrumental variables regression. We consider a U.S. economy dataset (taken over $n = 20$ years) as presented in Kmenta (1986). The variables are:

- $Y$: food consumption per capita

- $X_1$: ratio of food prices to general consumer prices

- $X_2$: disposable income in constant U.S. dollars

- $X_3$: ratio of preceding year's prices received by farmers to general consumer prices

- $X_4$: time in years

We first wish to estimate (using 2SLS) the supply equation where $Y$ is the endogenous variable, $X_1$, $X_3$, and $X_4$ are the exogenous variables, and $X_2$, $X_3$, and $X_4$ are the instrumental variables (a variable can appear as both an exogenous and instrumental variable). The 2SLS results are as follows:

```
############################
2SLS Estimates

Model Formula: Y ~ X1 + X3 + X4

Instruments: ~X2 + X3 + X4

Residuals:
   Min. 1st Qu.  Median    Mean 3rd Qu.    Max.
-4.8720 -1.2590  0.6415  0.0000  1.4750  3.4860

               Estimate  Std. Error t value   Pr(>|t|)
(Intercept) 49.53244170 12.01052641 4.12409 0.00079536 ***
X1           0.24007578  0.09993385 2.40235 0.02878451 *
X3           0.25560572  0.04725007 5.40964 5.7854e-05 ***
X4           0.25292417  0.09965509 2.53800 0.02192877 *
---
Signif. codes:  0 *** 0.001 ** 0.01 * 0.05 . 0.1   1

Residual standard error: 2.4575552 on 16 degrees of freedom
############################
```

Notice that all of the exogenous variables are statistically significant at the 0.05 significance level when accounting for the effect of the instrumental variables.

We next wish to estimate (using 2SLS) the supply equation where $Y$ is the endogenous variable, $X_1$ and $X_2$ are the exogenous variables, and $X_2$, $X_3$, and $X_4$ are the instrumental variables. The 2SLS results are as follows:

```
############################
2SLS Estimates

Model Formula: Y ~ X1 + X2

Instruments: ~X2 + X3 + X4

Residuals:
   Min. 1st Qu.  Median    Mean 3rd Qu.    Max.
```

```
-3.4300 -1.2430 -0.1895   0.0000   1.5760   2.4920
```

```
             Estimate  Std. Error  t value   Pr(>|t|)
(Intercept) 94.63330387 7.92083831 11.94738 1.0762e-09 ***
X1          -0.24355654 0.09648429 -2.52431    0.021832 *
X2           0.31399179 0.04694366  6.68869 3.8109e-06 ***
---
Signif. codes:  0 *** 0.001 ** 0.01 * 0.05 . 0.1   1

Residual standard error: 1.9663207 on 17 degrees of freedom
##############################
```

Notice that both of the exogenous variables are statistically significant at the 0.05 significance level when accounting for the effect of the instrumental variables.

**Example 11.6.3.** *(Arsenate Assay Data)*
Ripley and Thompson (1987) analyzed data on arsenate ion measured in natural river waters, which was determined by two assay methods. The sample size is $n = 30$. The variables are:

- $Y$: measurements taken by non-selective reduction, cold trapping, and atomic emission spectroscopy (micrograms/liter)

- $X$: measurements taken by continuous selective reduction and atomic absorption spectrometry (micrograms/liter)

We are interested in fitting a Deming regression line to these data. From previous data sources, a ratio of the variances $\eta = 0.79$ is assumed. Thus, the estimates for the Deming regression model are

```
##############################
Intercept     Slope sigma.aas sigma.aes
0.4133107 0.8803372 0.8414230 0.7478731
##############################
```

Figure 11.2 shows a plot of these data with the Deming regression fit overlaid as well as the ordinary least squares fit. Clearly, these are comparable fits, but the differences in the two fits becomes more noticeable at the seemingly large influential observation with an atomic absorption spectrometry measurement of 20 micrograms/liter.

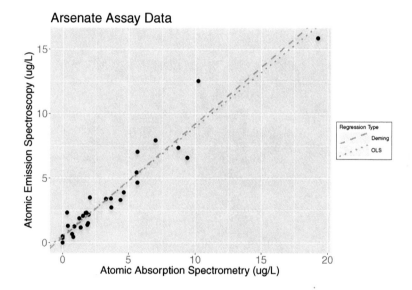

**FIGURE 11.2**

Scatterplot of the arsenate assay dataset with the ordinary least squares fit and the Deming regression fit.

# 12

# Weighted Least Squares and Robust Regression Procedures

So far we have utilized ordinary least squares for estimating the linear regression model. However, aspects of the fit, such as nonconstant error variance or outliers, may require a different method for estimating the regression model. This chapter provides an introduction to some of the other available methods for performing such estimation. To help with the discussions in this chapter, recall that the ordinary least squares estimate is

$$\hat{\beta}_{OLS} = \underset{\beta}{\mathrm{argmin}} \sum_{i=1}^{n} \epsilon_i^2$$
$$= (\mathbf{X}^T\mathbf{X})^{-1}\mathbf{X}^T\mathbf{Y},$$

where we have used the subscript OLS to differentiate it from some of the other estimates of $\beta$ that we will introduce.

## 12.1 Weighted Least Squares

The method of ordinary least squares assumes that there is constant variance in the errors (which is called **homoskedasticity**). The method of **weighted least squares** can be used when the ordinary least squares assumption of constant variance in the errors is violated (which is called **heteroskedasticity**). In some cases, weighted least squares may also help correct correlation between observations. The model under consideration is

$$\mathbf{Y} = \mathbf{X}\beta + \epsilon,$$

where now $\epsilon$ is assumed to be (multivariate) normally distributed with mean vector $\mathbf{0}$ and nonconstant variance–covariance matrix $\sigma^2\mathbf{W}$, where $\mathbf{W}$ is a diagonal matrix. Thus, the residuals will reflect this nonconstant dispersion. In the ordinary least squares setting, the assumption is that the errors have a constant variance–covariance matrix $\sigma^2\mathbf{I}_n$. It can be shown that there exists a unique, nonsingular symmetric matrix $\mathbf{P}$ such that $\mathbf{P}^T\mathbf{P} = \mathbf{W}$. Therefore, (by skipping over much of the matrix algebra) we can write $\epsilon^* = \mathbf{P}^{-1}\epsilon$, which now has constant variance–covariance matrix $\sigma^2 \mathbf{I}_n$. Notice that

$$\epsilon^* = \mathbf{P}^{-1}\epsilon$$
$$= \mathbf{P}^{-1}(\mathbf{Y} - \mathbf{X}\beta),$$

which in turn yields the weighted least squares estimate

$$\hat{\beta}_{\text{WLS}} = \underset{\beta}{\text{argmin}} \sum_{i=1}^{n} \epsilon_i^{*2}$$
$$= (\mathbf{X}^T \mathbf{W}^{-1} \mathbf{X})^{-1} \mathbf{X}^T \mathbf{W}^{-1} \mathbf{Y},$$

where $\mathbf{W}^{-1}$ is the matrix of weights. Clearly the use of weights will impact the widths of any calculated statistical intervals.

The most common application of weighted least squares occurs when the observations are independent, but have different variances. So

$$\sigma^2 \mathbf{W} = \sigma^2 \begin{pmatrix} 1/w_1 & 0 & \cdots & 0 \\ 0 & 1/w_2 & \cdots & 0 \\ \vdots & \vdots & \ddots & \vdots \\ 0 & 0 & \cdots & 1/w_n \end{pmatrix} = \sigma^2 \begin{pmatrix} \sigma_1^2 & 0 & \cdots & 0 \\ 0 & \sigma_2^2 & \cdots & 0 \\ \vdots & \vdots & \ddots & \vdots \\ 0 & 0 & \cdots & \sigma_n^2 \end{pmatrix}.$$

With this setting, we can make a few observations:

- In theory, the weights for this purpose should be $w_i = \frac{1}{\sigma_i^2}$, where $\sigma_i^2$ is the variance of $\epsilon_i$, the $i^{\text{th}}$ error.

- Often times, we set $\sigma^2 = 1$ in this setting, so the theoretical weighted SSE is $\sum_{i=1}^{n} w_i \epsilon_i^2 = \sum_{i=1}^{n} \frac{1}{\sigma_i^2} \epsilon_i^2 = \sum_{i=1}^{n} (\frac{\epsilon_i}{\sigma_i})^2$. Thus, we are minimizing the sum of squared standardized errors; i.e., where the errors have mean $\mathbf{0}$ and variance–covariance matrix $\mathbf{I}_n$.

In practice, it may be difficult to obtain the structure of $\mathbf{W}$ at first (especially if dependency is an issue), so sometimes it is necessary to make the assumption that $\mathbf{W} = \mathbf{I}_n$ in an attempt to discover something about the form of $\mathbf{W}$. This can be done via residual diagnostics.

Here are some possible variance and standard deviation estimates that have been suggested for the weights:

- If a residual plot of a predictor (or the response) exhibits a megaphone shape, then regress the absolute values of the residuals against that predictor (or the response). The resulting fitted values of this regression can be used as estimates of $\sigma_i$.

- If a residual plot of the squared residuals against a predictor (or the response) exhibits an upward trend, then regress the squared residuals against that predictor (or the response). The resulting fitted values of this regression can be used as estimates of $\sigma_i^2$.

In practice, the difficulty with weighted least squares is determining estimates of the variances.

The values of the weights may be based on theory or prior research, but all of the theory underlying weighted least squares assumes that the weights are known. Typically, the weighted least squares estimates of the coefficients will be nearly the same as the the unweighted estimates. To see the effect of weighting in residual diagnostic plots, it is necessary to construct all such plots using standardized residuals, because ordinary residuals will look the same whether weights are used or not. Deeper treatment of weighting in regression is discussed in Carroll and Ruppert (1988).

## 12.2   Robust Regression Methods

The ordinary least squares estimates for linear regression is optimal when all of the regression assumptions are valid. But as we have seen, if these assumptions are not met, then least squares estimates can be misleading. Residual diagnostics can help identify where the breakdown in assumptions occur, but can be time consuming and sometimes reflect a great deal of subjectivity. **Robust regression** methods have less restrictive assumptions than those for ordinary least squares regression. These methods attempt to dampen the influence of outlying cases in order to provide a better fit to the majority of the data.

Outliers have a tendency to pull the least squares fit too far in their direction by receiving much more "weight" than they deserve. Typically, you would expect that each observation's weight would be about $1/n$ in a dataset with $n$ observations. However, outliers may receive considerably more weight, leading to distorted estimates of the regression coefficients. It can then become difficult to identify the outliers since their residuals are much smaller than they should be. As we have seen, scatterplots may be used to assess outliers when a small number of predictors are present. However, the complexity added by additional independent variables hides the outliers from view in these scatterplots. Robust regression downweights the influence of outliers, which makes their residuals larger and easier to identify. Being able to adequately assess such weights allows one to use weighted least squares. However, this may be difficult in practice and so other techniques are often used.

For our first robust regression method, suppose we have a dataset of size $n$ such that

$$y_i = \mathbf{x}_i^{\mathrm{T}} \boldsymbol{\beta} + \epsilon_i$$
$$\Rightarrow \epsilon_i(\boldsymbol{\beta}) = y_i - \mathbf{x}_i^{\mathrm{T}} \boldsymbol{\beta},$$

where $i = 1, \ldots, n$. Here we have rewritten the error term as $\epsilon_i(\boldsymbol{\beta})$ to reflect the error terms dependency on the regression coefficients. Ordinary least squares is sometimes known as $L_2$-**norm regression** since it is minimizing the $L_2$-norm of the residuals; i.e., the square of the residuals. Thus, observations with high residuals (and high squared residuals) will pull the least squares fit more in that direction. An alternative is to use what is sometimes known as **least absolute deviation** (or $L_1$-**norm regression**), which minimizes the $L_1$-norm of the residuals; i.e., the absolute value of the residuals. Formally defined, the least absolute deviation estimator is

$$\hat{\beta}_{\mathrm{LAD}} = \underset{\beta}{\operatorname{argmin}} \sum_{i=1}^{n} |\epsilon_i(\beta)|,$$

which in turn minimizes the absolute value of the residuals (i.e., $|r_i|$).

Another common robust regression method falls into a class of estimators called **M-estimators**. There are also other related classes such as $R$-estimators and $S$-estimators, whose properties we will not explore, but one can refer to Huber and Ronchetti (2009). $M$-estimators attempt to minimize the sum of a chosen function $\rho(\cdot)$ which is acting on the residuals. Formally defined, $M$-estimators are given by

$$\hat{\beta}_{\mathrm{M}} = \underset{\beta}{\operatorname{argmin}} \sum_{i=1}^{n} \rho(\epsilon_i(\beta)).$$

The $M$ stands for "maximum likelihood" since $\rho(\cdot)$ is related to the likelihood function for a suitable assumption of a distribution on the residuals. Notice that, if assuming normality, then $\rho(z) = \frac{1}{2}z^2$ results in the ordinary least squares estimate.

Some $M$-estimators are influenced by the scale of the residuals, so a scale-invariant version of the $M$-estimator is used:

$$\hat{\beta}_{\mathrm{M}} = \underset{\beta}{\operatorname{argmin}} \sum_{i=1}^{n} \rho\left(\frac{\epsilon_i(\beta)}{\tau}\right),$$

where $\tau$ is a measure of the scale. An estimate of $\tau$ is given by

$$\hat{\tau} = \frac{\operatorname{med}_i|r_i - \tilde{r}|}{0.6745},$$

where $\tilde{r}$ is the median of the residuals. Minimization of the above is accomplished primarily in two steps:

1. Set $\frac{\partial \rho}{\partial \beta_j} = 0$ for each $j = 0, 1, \ldots, p - 1$, resulting in a set of $p$ nonlinear equations

$$\sum_{i=1}^{n} x_{i,j} \psi\left(\frac{\epsilon_i}{\tau}\right) = 0,$$

where $\psi(z) = \rho'(z)$. $\psi(\cdot)$ is called the **influence function**.

2. A numerical method called **iteratively reweighted least squares (IRLS)** is used to iteratively estimate the weighted least squares estimate until a stopping criterion is met. Specifically, for iterations $t = 0, 1, \ldots$

$$\hat{\beta}^{(t+1)} = (\mathbf{X}^{\mathrm{T}}[\mathbf{W}^{-1}]^{(t)}\mathbf{X})^{-1}\mathbf{X}^{\mathrm{T}}[\mathbf{W}^{-1}]^{(t)}\mathbf{y},$$

where $[\mathbf{W}^{-1}]^{(t)} = \operatorname{diag}(w_1^{(t)}, \ldots, w_n^{(t)})$ such that

$$w_i^{(t)} = \begin{cases} \dfrac{\psi[(y_i - \mathbf{x}_i^t\beta^{(t)})/\hat{\tau}^{(t)}]}{(y_i\mathbf{x}_i^t\beta^{(t)})/\hat{\tau}^{(t)}}, & \text{if } y_i \neq \mathbf{x}_i^{\mathrm{T}}\beta^{(t)}; \\ 1, & \text{if } y_i = \mathbf{x}_i^{\mathrm{T}}\beta^{(t)}. \end{cases}$$

More generally, $w(z) = \psi(\cdot)/z$, where $w(\cdot)$ is called the **weighting function**.

Four common functions chosen in $M$-estimation (and there are *many* others) are given below:

1. **Andrews's Sine** (see Andrew, 1974):

$$\rho(t) = \begin{cases} k[1 - \cos(t/k)], & \text{if } |t| < \pi k; \\ 2k, & \text{if } |t| \geq \pi k \end{cases}$$

$$\psi(t) = \begin{cases} \sin(t/k), & \text{if } |t| < \pi k; \\ 0, & \text{if } |t| \geq \pi k \end{cases}$$

$$w(t) = \begin{cases} \dfrac{\sin(t/k)}{t/k}, & \text{if } |t| < \pi k; \\ 0, & \text{if } |t| \geq \pi k, \end{cases}$$

where $k \approx 1.339$.

2. **Hampel's Method** (see Hampel, 1996):

$$\rho(t) = \begin{cases} t^2/2, & \text{if } |t| \leq a; \\ a^2/2 + a(|t| - a), & \text{if } a < |t| \leq b; \\ (a/2)\left(2b - a + (|t| - b)\left(1 + \frac{c - |t|}{c - b}\right)\right), & \text{if } b < |t| < c; \\ (a/2)(b - a + c), & \text{if } c \leq |t| \end{cases}$$

$$\psi(t) = \begin{cases} t, & \text{if } |t| \leq a; \\ a[\operatorname{sgn}(t)], & \text{if } a < |t| \leq b; \\ \frac{a(c - |t|)}{c - b}, & \text{if } b < |t| < c; \\ 0, & \text{if } c \leq |t| \end{cases}$$

$$w(t) = \begin{cases} 1, & \text{if } |t| \leq a; \\ a/|t|, & \text{if } a < |t| \leq b; \\ \frac{a(c - |t|)}{|t|(c - b)}, & \text{if } b < |t| < c; \\ 0, & \text{if } c \leq |t|, \end{cases}$$

where $a = 1.5k$, $b = 3.5k$, $c = 8k$, and $k \approx 0.902$.

3. **Huber's Method** (see Huber, 1964):

$$\rho(t) = \begin{cases} t^2/2, & \text{if } |t| < k; \\ k(|t| - k/2), & \text{if } |t| \geq k \end{cases}$$

$$\psi(t) = \begin{cases} t, & \text{if } |t| < k; \\ c[\text{sgn}(t)], & \text{if } |t| \geq k \end{cases}$$

$$w(t) = \begin{cases} 1, & \text{if } |t| < k; \\ k/|t|. & \text{if } |t| \geq k, \end{cases}$$

where $k \approx 1.345$.

4. **Tukey's Biweight** (see Huber and Ronchetti, 2009):

$$\rho(t) = \begin{cases} \frac{k^2}{6}\left\{1 - [1 - (\frac{t}{k})^2]^3\right\}, & \text{if } |t| < k; \\ k^2/6, & \text{if } |t| \geq k \end{cases}$$

$$\psi(t) = \begin{cases} t[1 - (\frac{t}{k})^2]^2, & \text{if } |t| < k; \\ 0, & \text{if } |t| \geq k \end{cases}$$

$$w(t) = \begin{cases} [1 - (\frac{t}{k})^2]^2, & \text{if } |t| < k; \\ 0, & \text{if } |t| \geq k, \end{cases}$$

where $k \approx 4.685$.

Notice in the above that there is a suggested value of $k$ provided. These values of $k$ are chosen so that the corresponding $M$-estimator gives 95% efficiency at the normal distribution.

Finally, one lesser used robust regression procedure involves **winsorizing** the residuals. Traditional winsorization of univariate data involves the calculation of the mean where a specified percentage of the smallest and largest values are set equal to the most extreme remaining values. For example, suppose you have the $n = 5$ (ordered) data values 3, 12, 15, 17, 42. The mean of this data is 17.8. The 20% winsorized mean would involve taking the lowest 20% of the data (i.e., the value 3) and setting it to the most extreme remaining value of 12, while doing the same for the largest values would result in setting the value of 42 to 17. Thus, the winsorized dataset is 12, 12, 15, 17, 17, and the winsorized mean is 14.6.

Let $\alpha$ be the level of winsorization performed; e.g., 10% in the above example. In **winsorized regression**, we find

$$\hat{\beta}_{\text{WIN}} = \underset{\beta}{\text{argmin}} \sum_{i=1}^{n} \left[\tilde{\epsilon}_i^{(\alpha)}(\beta)\right]^2,$$

which in turn minimizes the square of the $\alpha$-level winsorized residuals; i.e., $[\tilde{r}_i^{(\alpha)}]^2$. Note that clearly an algorithm is necessary to find such a solution. For example, we can start with the (ordered) residuals from the ordinary least squares fit, winsorize the residuals, adjust the predicted values based on the winsorized residuals, and then refit the regression line based on these adjusted quantities. There are many ways to potentially apply winsorization in a least squares context. A general discussion is found in Yale and Forsythe (1976).

## 12.3 Theil–Sen and Passing–Bablok Regression

In the simple linear regression setting, there are various regression estimators used that are robust to outliers. We discuss a couple of them in this section.

### Theil–Sen Regression

In ordinary least squares, the $\hat{\beta}_1$ estimate of $\beta$ is also the solution to

$$\rho(\mathbf{x}, \epsilon(\beta_1)) = 0,$$

where $\epsilon(\beta_1)$ is a vector of the $\epsilon_i(\beta_1)$ values, but rewritten to emphasize that we are interested in $\beta_1$. $\rho(\cdot)$ is the Pearson's correlation coefficient. A non-parametric analog is to instead use Kendall's $\tau_b$; i.e., solve

$$\tau_b(\mathbf{x}, \epsilon(\beta_1)) = 0.$$

The solution to the above is called the **Theil–Sen estimator** (Theil, 1950; Sen, 1968) and can be written as

$$\hat{\beta}_{1;\text{TS}} = \text{med}\left\{ \frac{y_i - y_j}{x_i - x_j} : x_i \neq x_j, i < j = 1, \ldots, n \right\}.$$

In other words, we calculate the Theil–Sen estimator by drawing a line segment between each of the $n(n-1)/2$ unique pairs of data points, retain the slopes, and then take the median of these slopes. The Theil–Sen estimator for the intercept is given by $\hat{\beta}_{0;\text{TS}} = \text{med}_i\left\{ y_i - \hat{\beta}_{1;\text{TS}}x_i \right\}$. A $100 \times (1-\alpha)\%$ confidence interval for $\beta_{1;\text{TS}}$ can be obtained by taking the middle $100 \times (1 - \alpha)\%$ of the slopes.

### Passing–Bablok Regression

The Theil–Sen estimator is actually biased towards zero if there is measurement error in both the predictor and the response. Moreover, it is not symmetric in the two variables. One correction is to use a **Passing–Bablok estimator**, of which there are multiple such estimators proposed in the papers by Passing and Bablok (1983, 1984) and Passing et al. (1988). One of the estimators is found as follows.

For $\beta_1$, proceed in the same manner as in the Theil–Sen setup, but disregard all slopes between two data points for which the slopes are 0 or $-1$. Take the shifted median of all of these slopes where the shift is calculated as the number of slopes that are less than $-1$. This is our estimated Passing–Bablok slope, say, $\hat{\beta}_{1;\text{PB}}$. Like the Theil–Sen estimator for the intercept, the estimate for the Passing–Bablok intercept is then $\hat{\beta}_{0;\text{PB}} = \text{med}_i \left\{ y_i - \hat{\beta}_{1;\text{PB}} x_i \right\}$.

Overall, the Passing–Bablok regression procedure is typically used when there is no more than a few outliers in the data.

## 12.4   Resistant Regression Methods

The next methods we discuss are often used interchangeably with robust regression methods. However, there is a subtle difference between the two methods that is not always clear in the literature. Whereas robust regression methods attempt to only dampen the influence of outlying cases, **resistant regression methods** use estimates that are *not* influenced by any outliers. This comes from the definition of **resistant statistics**, which are measures of the data that are not influenced by outliers, such as the median. This is best accomplished by **trimming** the data, which "trims" extreme values from either end (or both ends) of the range of data values.

There is also one other relevant term when discussing resistant regression methods. Suppose we have a dataset $x_1, x_2, \ldots, x_n$. The **order statistics** are simply defined to be the data values arranged in increasing order and are written as $x_{(1)}, x_{(2)}, \ldots, x_{(n)}$. Therefore, the minimum and maximum of this dataset are $x_{(1)}$ and $x_{(n)}$, respectively. As we will see, the resistant regression estimators provided here are all based on the ordered residuals.

We present four commonly used resistant regression methods:

1. The **least quantile of squares** method minimizes the squared order residual (presumably selected as it is most representative of where the data is expected to lie) and is formally defined by

$$\hat{\beta}_{\text{LQS}} = \underset{\beta}{\text{argmin}} \left[ \epsilon_{(\nu)}(\beta) \right]^2,$$

where $\nu = nP$ is the $P^{\text{th}}$ percentile (i.e., $0 < P \leq 1$) of the empirical data. If $\nu$ is not an integer, then specify $\nu$ to be either the next greatest or lowest integer value. A specific case of the least quantile of squares method is where $P = 0.50$ (i.e., the median) and is called the **least median of squares** method. The estimate is written as $\hat{\beta}_{\text{LMS}}$.

2. The **least trimmed sum of squares** method minimizes the sum of the $h$ smallest squared residuals and is formally defined by

$$\hat{\beta}_{\text{LTS}} = \underset{\beta}{\text{argmin}} \sum_{i=1}^{h} \left[ \epsilon_{(i)}(\beta) \right]^2,$$

where $h \leq n$. If $h = n$, then you just obtain $\hat{\beta}_{OLS}$.

3. The **least trimmed sum of absolute deviations** method minimizes the sum of the $h$ smallest absolute residuals and is formally defined by

$$\hat{\beta}_{LTA} = \underset{\beta}{argmin} \sum_{i=1}^{h} |\epsilon(\beta)|_{(i)},$$

where again $h \leq n$. If $h = n$, then you just obtain $\hat{\beta}_{LAD}$. Note that while it may seem like obtaining $\hat{\beta}_{LAD}$ would be similar to other optimization routines, such as those for the other resistant regression methods discussed, it is actually computationally quite complex and not all numerical routines provide reliable parameter estimates; see Hawkins and Olive (1999) for a discussion of relevant algorithms.

So which method from robust or resistant regressions do we use? In order to guide you in the decision-making process, you will want to consider both the theoretical benefits of a certain method as well as the type of data you have. The theoretical aspects of these methods that are often cited include their breakdown values and overall efficiency. **Breakdown values** are a measure of the proportion of contamination (due to outlying observations) that an estimation method can withstand and still maintain being robust against the outliers. **Efficiency** is a measure of an estimator's variance relative to another estimator. When the efficiency is as small as it can possibly be, then the estimator is said to be "best." For example, the least quantile of squares method and least trimmed sum of squares method both have the same maximal breakdown value for certain $P$, the least median of squares method is of low efficiency, and the least trimmed sum of squares method has the same efficiency (asymptotically) as certain $M$-estimators. As for your data, if there appear to be many outliers, then a method with a high breakdown value should be used. A preferred solution is to calculate many of these estimates for your data and compare their overall fits, but this will likely be computationally expensive.

## 12.5 Resampling Techniques for $\hat{\beta}$

In this section, we discuss a few commonly used **resampling** techniques, which are used for estimating characteristics of the sampling distribution of $\hat{\beta}$. While we discuss these techniques for the regression parameter $\beta$, it should be noted that they can be generalized and applied to any parameter of interest. These procedures are inherently nonparametric and some of them can be used for constructing nonparametric confidence intervals for the parameter(s) of interest. More details on the resampling methods discussed below are given in the classic text by Efron and Tibshirani (1993).

**Permutation Tests**

**Permutation tests** (also known as **randomization tests** or **exact tests**) are a class of statistical tests where the distribution of the test statistic under the null hypothesis is obtained by calculating all possible values of the test statistic under rearrangements (or permutations) of the labels of the observed data. These are computationally-intensive techniques that predate computers and were first introduced in Fisher (1935). If the sample size $n$ is even moderately large, then calculation of all $n!$ permutations quickly becomes prohibitive. Instead, we can randomly sample without replacement $M < n!$ of the possible permutations.

For a regression setting, the permutation test is calculated as follows:

---

`Permutation Test for Regression`

1. Randomly permute the values in $\mathbf{Y}$ among the values in $\mathbf{X}$ $M$ times. This gives the permutation sample $(\mathbf{Y}_1^\#, \mathbf{X}), \ldots, (\mathbf{Y}_M^\#, \mathbf{X})$.

2. For $m = 1, \ldots, M$, calculate the permutation replications $\beta_m^\# = \beta(\mathbf{Y}_m^\#, \mathbf{X})$, which are the estimates of $\boldsymbol{\beta}$ based on the $m^{\text{th}}$ permutation sample.

3. To conduct a test of

$$H_0 : \beta_j = 0$$
$$H_A : \beta_j \neq 0$$

for each $\beta_j$ in the vector $\boldsymbol{\beta}$, let $t_{obs}(\beta_j) = \hat{\beta}_j / \text{s.e.}(\hat{\beta}_j)$ be the test statistic based on the observed data and $t_m^\#(\beta_j) = \hat{\beta}_{j;m}^\# / \text{s.e.}(\hat{\beta}_{j;m}^\#)$ be the test statistic based on the $m^{\text{th}}$ permutation sample. Then,

$$\widehat{ASL}_{\text{perm}} = M^{-1} \sum_{m=1}^{M} \mathrm{I}\{|t_m^\#(\beta_j)| \geq |t_{obs}(\beta_j)|\}$$

is the approximate **achieved significance level (ASL)** for the test.

---

The ASL is virtually the same type of quantity as a $p$-value, but calculated in the context of a resampling scheme. We then compare $\widehat{ASL}_{\text{perm}}$ to a significance level, like $\alpha = 0.05$, to determine the strength of the evidence for or against the relationship specified in $H_0$ above. If $n!$ is not computationally prohibitive, the above can give the exact permutation distribution by replacing the $M$ random permutations with all $n!$ unique permutations. This testing paradigm can also be easily modified for one-sided tests.

## The Bootstrap

**Bootstrapping** is a method where you resample your data with replacement in order to approximate the distribution of the observed data.[1] Bootstrapping allows us to estimate properties of our estimator and to conduct inference for the parameter(s) of interest. Bootstrapping procedures are very appealing and possess a number of asymptotic properties; however, they can be computationally intensive. In nonstandard situations, such as some of the robust and resistant regression models discussed in this chapter, bootstrapping provides a viable alternative for providing standard errors and confidence intervals for the regression coefficients and predicted values. When in the regression setting, there are two primary bootstrapping methods that may be employed. Before we make a distinction between these approaches, we present some basics about the bootstrap.

In bootstrapping, we treat the sample as the population of interest. We then draw $B$ samples ($B$ is usually at least 1000) of size $n$ from the original sample with replacement. These are called the **bootstrap samples**. For each bootstrap sample, we then calculate the statistic(s) of interest. These are called the **bootstrap replications**. We then use the bootstrap replications to characterize the sampling distribution for the parameter(s) of interest. One assumption that bootstrapping heavily relies upon is that the sample approximates the population fairly well. Thus, bootstrapping does not usually work well for small samples as they are likely not representative of the underlying population. Bootstrapping methods should be relegated to medium sample sizes or larger, but what constitutes a "medium sample size" is subjective.

Now suppose that it is desired to obtain standard errors and confidence intervals for the regression coefficients. The standard deviation of the $B$ estimates provided by the bootstrapping scheme is the bootstrap estimate of the standard error for the respective regression coefficient. Furthermore, bootstrap confidence intervals are traditionally found by sorting the $B$ estimates of a regression coefficient and selecting the appropriate percentiles from the sorted list. For example, a 95% bootstrap confidence interval would be given by the 2.5[th] and 97.5[th] percentiles from the sorted list. Other statistics may be computed in a similar manner.

Now we can turn our attention to the two bootstrapping techniques available in the regression setting. Assume for both methods that the sample consists of $n$ responses (which we denote by the $n$-dimensional vector $\mathbf{Y}$), each with $p$ predictors (which we represent using the $n \times p$ matrix $\mathbf{X}$). The parame-

---

[1] Traditional bootstrapping as we present here is considered a nonparametric procedure. However, there are also parametric bootstraps available in the literature. See Chapter 6 of Efron and Tibshirani (1993) for a discussion about the parametric bootstrap.

ter of interest is the vector of regression coefficients $\boldsymbol{\beta}$, which may or may not have dimension $p$ depending on the type of regression model being estimated.

We can first bootstrap the observations. In this setting, the bootstrap samples are selected from the original pairs of data. So the pairing of a response with its measured predictor is maintained. This method is appropriate for observational data; i.e., the predictor levels were not predetermined. Specifically, the algorithm is as follows:

---

**Bootstrap the Observations**

1. Draw a sample of size $n$ with replacement from the observed data $(\mathbf{Y}, \mathbf{X})$. Repeat this a total of $B$ times to get the bootstrap sample $(\mathbf{Y}_1^*, \mathbf{X}_1^*), \ldots, (\mathbf{Y}_B^*, \mathbf{X}_B^*)$.

2. For $b = 1, \ldots, B$, calculate the bootstrap replications $\boldsymbol{\beta}_b^* = \boldsymbol{\beta}(\mathbf{Y}_b^*, \mathbf{X}_b^*)$, which are the estimates of $\boldsymbol{\beta}$ based on the $b^{\text{th}}$ bootstrap sample.

3. Estimate the sample variance–covariance using the bootstrap replications as follows:

$$\widehat{\text{Var}}_{\text{boot}} = \frac{1}{B-1} \sum_{b=1}^{B} (\hat{\boldsymbol{\beta}}_b^* - \bar{\boldsymbol{\beta}}^*)(\hat{\boldsymbol{\beta}}_b^* - \bar{\boldsymbol{\beta}}^*)^{\text{T}},$$

where $\bar{\boldsymbol{\beta}}^* = B^{-1} \sum_{b=1}^{B} \hat{\boldsymbol{\beta}}_b^*$. The square root of the diagonals of the variance–covariance matrix give the respective bootstrap estimates of the standard errors.

---

Alternatively, we can bootstrap the residuals. The bootstrap samples in this setting are selected from what are called the **Davison–Hinkley modified residuals** (Davison and Hinkley, 1997), given by

$$e_i^* = \frac{e_i}{\sqrt{1 - h_{i,i}}} - \frac{1}{n} \sum_{j=1}^{n} \frac{e_j}{\sqrt{1 - h_{j,j}}},$$

where the $e_1, \ldots, e_n$ are the original observed regression residuals. We do not simply just use the original residuals because these will lead to biased results. In each bootstrap sample, the randomly sampled modified residuals are added to the original fitted values forming new values of the response. Thus, the original structure of the predictors will remain the same while only the response will be changed.

## Bootstrap the Residuals

Let $e^*$ be the $n$-dimensional vector of the Davison–Hinkley modified residuals. We then have the following bootstrap algorithm:

1.  Draw a sample of size $n$ with replacement from the Davison–Hinkley modified residuals. Repeat this a total of $B$ times to get the bootstrap sample $(\mathbf{Y}_1^{**}, \mathbf{X}), \ldots, (\mathbf{Y}_B^{**}, \mathbf{X})$, where $\mathbf{Y}_b^{**} = \mathbf{Y} + \mathbf{e}_b^{**}$ such that $\mathbf{e}_b^{**}$ is the $b^{\text{th}}$ bootstrap sample of the Davison–Hinkley modified residuals, $b = 1, \ldots, B$.

2.  For $b = 1, \ldots, B$, calculate the bootstrap replications $\hat{\boldsymbol{\beta}}_b^{**} = \boldsymbol{\beta}(\mathbf{Y}_b^{**}, \mathbf{X})$, which are the estimates of $\boldsymbol{\beta}$ based on the $b^{\text{th}}$ bootstrap sample.

3.  Estimate the sample variance–covariance using the bootstrap replications as follows:

$$\widehat{\text{Var}}_{\text{boot}} = \frac{1}{B-1} \sum_{b=1}^{B} (\hat{\boldsymbol{\beta}}_b^{**} - \bar{\boldsymbol{\beta}}^{**})(\hat{\boldsymbol{\beta}}_b^{**} - \bar{\boldsymbol{\beta}}^{**})^{\mathrm{T}},$$

where $\bar{\boldsymbol{\beta}}^{**} = B^{-1} \sum_{b=1}^{B} \hat{\boldsymbol{\beta}}_b^{**}$.

The above method is appropriate for designed experiments where the levels of the predictor are predetermined. Also, since the residuals are sampled and added back at random, we must assume the variance of the residuals is constant. If not, this method should not be used.

Finally, inference based on either of the above bootstrap procedures can also be carried out. For illustrative purposes, assume that bootstrapping was performed on the pairs of observations. For a single parameter from $\boldsymbol{\beta}$, say, $\beta_j$, a $100 \times (1 - \alpha)\%$ bootstrap confidence interval based on normality is given by

$$\hat{\beta}_j^* \pm z_{1-\alpha/2} \times \text{s.e.}_{\text{boot}}(\hat{\beta}_j^*).$$

A fully nonparametric $100 \times (1 - \alpha)\%$ bootstrap confidence interval[2] for each regression coefficient $\beta_j$ in $\boldsymbol{\beta}$ is given by

$$\left( \beta_{j, \lfloor \frac{\alpha}{2} \times n \rfloor}^*, \beta_{j, \lceil (1 - \frac{\alpha}{2}) \times n \rceil}^* \right),$$

where the lower and upper bounds are the $100 \times (\alpha/2)^{\text{th}}$ and $100 \times (1 - \alpha/2)^{\text{th}}$ percentiles, respectively, from the bootstrap replications of $\beta_j$. Here, $\lfloor \cdot \rfloor$ is the **floor function**, which returns the largest integer less than or equal to its

---

[2]This is also called the **percentile interval**. There are many possible ways to construct confidence intervals for $\beta_j$ based on bootstrap samples. See, for example, those discussed in Efron and Tibshirani (1993).

argument, and $\lceil \cdot \rceil$ is the **ceiling function**, which returns the smallest integer greater than or equal to its argument. We can also construct joint percentile (bootstrap) confidence intervals using a Bonferroni correction on the above formula for each $j$.

We can also use the bootstrap sample to conduct a test of

$$H_0 : \beta_j = 0$$
$$H_A : \beta_j \neq 0$$

for each $\beta_j$. Let $t_{obs}(\beta_j) = \hat{\beta}_j / \text{s.e.}(\hat{\beta}_j)$ be the test statistic based on the observed data and $t_b^*(\beta_j) = \hat{\beta}_{j;b}^* / \text{s.e.}(\hat{\beta}_{j;b}^*)$ be the test statistic based on the $b^{\text{th}}$ bootstrap sample. Then,

$$\widehat{ASL}_{\text{boot}} = B^{-1} \sum_{b=1}^{B} \mathrm{I}\{|t_b^*(\beta_j)| \geq |t_{obs}(\beta_j)|\}$$

is the approximate ASL for the test. We then compare $\widehat{ASL}_{\text{boot}}$ to a significance level, like $\alpha = 0.05$, to determine the strength of the evidence for or against the relationship specified in $H_0$ above. This testing paradigm can also be easily modified for one-sided tests.

**The Jackknife**

The **jackknife** was first introduced by Quenoille (1949) and is similar to the bootstrap. It is used in statistical inference to estimate the bias and standard error (variance) of a statistic when a random sample of observations is used for the calculations. The basic idea behind the jackknife variance estimator lies in systematically recomputing the estimator of interest by leaving out one or more observations at a time from the original sample. From this new set of replicates of the statistic, an estimate for the bias and variance of the statistic can be calculated, which can then be used to calculate jackknife confidence intervals.

We present one of the more common jackknifing procedures, which is called the **delete-$d$ jackknife**. This procedure is based on the **delete-$d$ subsamples**, which are subsamples of our original data constructed by omitting $d$ of the observations, where $n = rd$ for some integer $r$. Assuming the same initial set-up as in the bootstrap discussion, below are the steps for the delete-$d$ jackknife:

---

Delete-$d$ Jackknife

---

1. Let $s_1, \ldots, s_K$ be an index that denotes all unique subsets of size $n - d$, chosen without replacement, such that $K = \binom{n}{d}$. This gives the delete-$d$ jackknife samples $(\mathbf{Y}_{(s_1)}, \mathbf{X}_{(s_1)}), \ldots, (\mathbf{Y}_{(s_K)}, \mathbf{X}_{(s_K)})$.

2. For $k = 1, \ldots, K$, calculate the jackknife replications $\hat{\boldsymbol{\beta}}^{\dagger}_{s_k} = \boldsymbol{\beta}(\mathbf{Y}_{(s_k)}, \mathbf{X}_{(s_k)})$, which are based on the delete-$d$ jackknife sample determined by the subset $s_k$.

3. Estimate the sample variance–covariance using the bootstrap replications as follows:

$$\widehat{\mathrm{Var}}_{\mathrm{jack}} = \frac{r}{\binom{n}{d}} \sum_{k=1}^{K} (\hat{\boldsymbol{\beta}}^{\dagger}_{s_k} - \bar{\boldsymbol{\beta}}^{\dagger})(\hat{\boldsymbol{\beta}}^{\dagger}_{s_k} - \bar{\boldsymbol{\beta}}^{\dagger})^{\mathrm{T}},$$

where $\bar{\boldsymbol{\beta}}^{\dagger} = \sum_{k=1}^{K} \hat{\boldsymbol{\beta}}^{\dagger}_{s_k} / \binom{n}{d}$. The square root of the diagonals of the variance–covariance matrix give the respective delete-$d$ jackknife estimates of the standard errors.

---

When $d = 1$, the above reduces to what is usually considered the standard jackknife.

For a single parameter from $\boldsymbol{\beta}$, say, $\beta_j$, a $100 \times (1 - \alpha)\%$ jackknife confidence interval based on normality is given by

$$\hat{\beta}^{\dagger}_j \pm z_{1-\alpha/2} \times \mathrm{s.e.}_{\mathrm{jack}}(\hat{\beta}^{\dagger}_j).$$

A fully nonparametric $100 \times (1 - \alpha)\%$ jackknife confidence interval for the regression coefficient $\beta_j$ is given by

$$\left( \beta^{\dagger}_{j, \lfloor \frac{\alpha}{2} \times n \rfloor}, \beta^{\dagger}_{j, \lceil (1 - \frac{\alpha}{2}) \times n \rceil} \right).$$

Notice that these confidence intervals are constructed in a similar manner as the bootstrap version. There are, again, various different jackknife confidence intervals that one can construct. In particular, we can construct joint jackknife confidence intervals using a Bonferroni correction on the above formula for each $j$.

While for moderately sized data the jackknife requires less computation, there are some drawbacks to using the jackknife. Since the jackknife is using fewer samples, it is only using limited information about $\hat{\boldsymbol{\beta}}$. In fact, the jackknife can be viewed as an approximation to the bootstrap. Specifically, it is a linear approximation to the bootstrap in that the two are roughly equal for linear estimators. Moreover, the jackknife can perform quite poorly if the estimator of interest is not sufficiently smooth, like the median or when your sample size is too small. See Chapter 11 of Efron and Tibshirani (1993) for more on the jackknife.

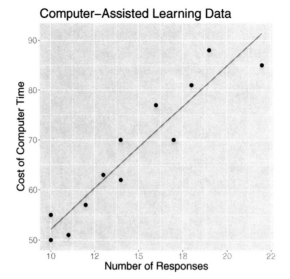

**FIGURE 12.1**
Scatterplot of the computer-assisted learning dataset.

## 12.6   Examples

**Example 12.6.1.** *(Computer-Assisted Learning Data)*
Kutner et al. (2005) presented data collected from a study of computer-assisted learning by $n = 12$ students in an effort to assess the cost of computer time. The variables are:

- $Y$: cost of the computer time (in cents)

- $X$: total number of responses in completing a lesson

A scatterplot of the data is given in Figure 12.1 with the ordinary least squares line overlaid. From this scatterplot, a simple linear regression seems appropriate for explaining this relationship.

Below is the summary of the simple linear regression fit for these data:

```
###############################
Coefficients:
                Estimate Std. Error t value Pr(>|t|)
(Intercept)      19.4727     5.5162   3.530  0.00545 **
num.responses     3.2689     0.3651   8.955 4.33e-06 ***
---
Signif. codes:  0 '***' 0.001 '**' 0.01 '*' 0.05 '.' 0.1 ' ' 1
```

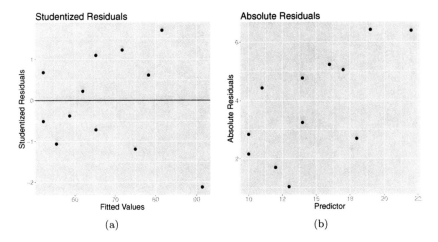

**FIGURE 12.2**
(a) Scatterplot of the Studentized residuals versus the fitted values and (b) scatterplot of the absolute residuals versus the predictor.

```
Residual standard error: 4.598 on 10 degrees of freedom
Multiple R-Squared: 0.8891,    Adjusted R-squared: 0.878
F-statistic: 80.19 on 1 and 10 DF,  p-value: 4.33e-06
##############################
```

Both the plots of the Studentized residuals versus the fitted values and the absolute residuals versus the predictor values are given in Figure 12.2. Nonconstant variance is possible here as there is a slight megaphone pattern emerging for the Studentized residuals (Figure 12.2(a)) and there is a slight increasing trend in the absolute residuals versus the predictor values (Figure 12.2(b)). Thus, we will turn to weighted least squares.

The weights we will use will be based on regressing the absolute residuals against the predictor. Specifically, we will take 1 over the fitted values from this regression. Furthermore, $x_3 = 22$ yields the largest fitted value since it is the largest predictor and the relationship between the absolute residuals versus the predictors is positive, thus it would get the smallest weight. Just to penalize that point a little more, we will take 1/2 the value from its fitted value as its final weight.

The summary of this weighted least squares fit is as follows:

```
##############################
Coefficients:
            Estimate Std. Error t value Pr(>|t|)
(Intercept)   16.6009    5.2833   3.142   0.0105 *
num.responses  3.4880    0.3708   9.406 2.78e-06 ***
```

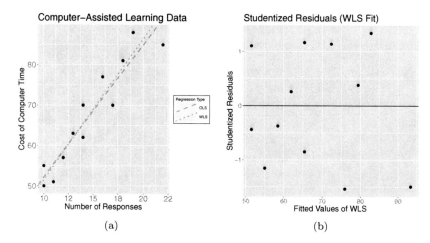

(a)                                            (b)

**FIGURE 12.3**
(a) A plot of the computer-assisted learning data with the ordinary least squares line (dashed line) and the weighted least squares line (dotted line). (b) A plot of the Studentized residuals versus the fitted values when using the weighted least squares method.

```
---

Signif. codes:   0 '***' 0.001 '**' 0.01 '*' 0.05 '.' 0.1 ' ' 1

Residual standard error: 3.023 on 10 degrees of freedom
Multiple R-Squared: 0.8985,      Adjusted R-squared: 0.8883
F-statistic: 88.48 on 1 and 10 DF,  p-value: 2.777e-06
#############################
```

Notice that the regression estimates have not changed much from the ordinary least squares method. Again, usually the estimates will be fairly close, although sometimes there is a noticeably larger change in the intercept estimate. Figure 12.3(a) shows both the ordinary least squares and weighted least squares lines overlaid on the same scatterplot. Furthermore, Figure 12.3(b) shows the Studentized residuals. Notice how we have slightly improved the megaphone shape and that the Studentized residuals appear to be more randomly scattered about 0.

**Example 12.6.2.** *(Arsenate Assay Data, cont'd)*
Let us return to the arsenate assay data from Example 11.6.3. Figure 12.4(a) gives a plot of the Studentized residuals versus the fitted values and Figure 12.4(b) plot of the Cook's distance values. Clearly, there are some values that are a little more extreme with respect to the other observations. While these are not very detrimental looking for estimation purposes, we fit some of the robust and resistant regression procedures discussed in this chapter.

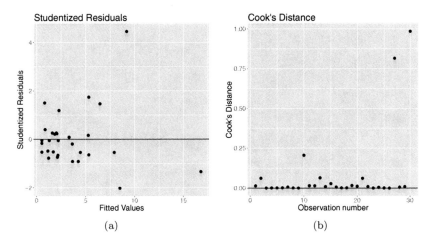

**FIGURE 12.4**
(a) A plot of the Studentized residuals versus the fitted values for the ordinary least squares fit of the arsenate assay data. (b) A plot of the corresponding Cook's distance values.

Table 12.1 gives the ordinary least squares estimates, Theil–Sen regression estimates, and Passing–Bablok regression estimates as well as their 95% confidence intervals. Generally, the estimates for the slope as well as the 95% confidence intervals are similar across the three methods. However, the intercepts show more of difference. This is not uncommon as the intercept terms tends to be affected more by observations that are a bit more extreme with respect to the data. Figure 12.5 shows the data with these three estimators overlaid. Notice that they all produce similar estimates.

**TABLE 12.1**
A comparison of the ordinary least squares, Theil–Sen, and Passing–Bablok regression estimates with 95% confidence intervals

| Method | Intercept (95% CI) | Slope (95% CI) |
|---|---|---|
| Ordinary Least Squares | 0.5442 (0.0178, 1.0705) | 0.8446 (0.7481, 0.9412) |
| Theil–Sen | 0.4581 (0.3015, 0.7400) | 0.8207 (0.7287, 0.9231) |
| Passing–Bablok | 0.4296 (0, 0.6203) | 0.8439 (0.7712, 1.0366) |

Table 12.2 gives the robust regression and scale estimates for the various procedures discussed in this chapter. Generally, the estimates for the slope

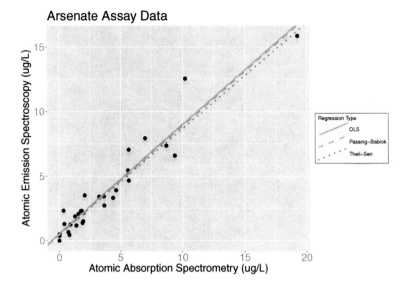

**FIGURE 12.5**
Scatterplot of the arsenate assay dataset with the ordinary least squares fit,
Passing–Bablok fit, and Theil–Sen fit.

are all similar. We again have some variability with the intercept estimates as
well as with the scale estimates using these different procedures. The constant
values used for the $M$-estimators are those specified earlier in the chapter such
that it yields 95% efficiency at the normal distribution.

Table 12.3 gives the regression estimates obtained using the resistant procedures discussed in this chapter. Recall with these data that $n = 30$. We
calculated each criterion based on including $27/30 = 0.90$ (90%) of the data.
Compared with the previous methods, we observe a bit more variability depending on the criterion employed. In particular, the least trimmed sum of
squares has an intercept that is now greater than 0 while the slope is noticeably different than the approximate values we have been calculating. This is
to be expected since, again, each procedure is performing the optimization
with respect to a different criterion.

In general, the choice of which method to employ is subjective. One can
use various arguments for their choice, such as ease of calculation, general
features of the data (e.g., breakdown values or high leverage values observed),
or efficiency between competing estimators.

Let us focus on the least absolute deviation estimates for these data. For
less standard estimators, we typically use resampling schemes to estimate the
standard errors or obtain confidence intervals. For the least absolute deviation
estimates, we will employ the nonparametric bootstrap and delete-1 jackknife.

**TABLE 12.2**

Robust regression and scale estimates for the various procedures discussed in this chapter

| Robust Method | Intercept Estimate | Slope Estimate | Scale Estimate |
|---|---|---|---|
| Least Absolute Deviation | −0.4093 | 1.1061 | 1.1857 |
| 10% Winsorized Regression | −0.8461 | 1.2993 | 2.2109 |
| *M*-Estimators | | | |
| Andrews's Sine | −0.2987 | 1.0890 | 0.9227 |
| Hampel's Method | −0.3454 | 1.1320 | 0.8585 |
| Huber's Method | −0.4428 | 1.1874 | 0.7754 |
| Tukey's Biweight | −0.4857 | 1.2131 | 0.7563 |

**TABLE 12.3**

Resistant regression estimates for the various procedures discussed in this chapter

| Robust Method | Intercept Estimate | Slope Estimate |
|---|---|---|
| Least Quantile of Squares | −0.9865 | 1.1911 |
| Least Trimmed Sum of Squares | 0.1452 | 0.8650 |
| Least Trimmed Sum of Absolute Deviations | −0.5720 | 1.2547 |

For the bootstrap, we draw $B = 1000$ bootstrap samples. The estimated bias and standard error for each estimated regression parameter is as follows:

```
#############################
        original      bias     std. error
t1* -0.4092737  0.14426915   0.3813678
t2*  1.1061453 -0.02392077   0.1712346
#############################
```

For the delete-1 jackknife, this gives us a total of $\binom{30}{1} = 30$ samples. The estimated bias and standard error for each estimated regression parameter is as follows:

```
#############################
        original        bias std. error
j1* -0.4092737 -0.2617489   0.5302596
j2*  1.1061453  0.4668158   0.3295412
#############################
```

Notice that the absolute bias and standard error for each estimate in the jackknife scheme are larger than the corresponding estimates in the bootstrap scheme. This is not unexpected given that we only have a sample size of $n = 30$.

**TABLE 12.4**
Bootstrap and delete-1 jackknife 90% confidence intervals for the least absolute deviation regression estimates

| Method | Intercept (90% CI) | Slope (90% CI) |
|---|---|---|
| Bootstrap | $(-0.7429, 0.4041)$ | $(0.7987, 1.2661)$ |
| Jackknife | $(-0.6100, -0.4039)$ | $(1.1037, 1.2449)$ |

Using the bootstrap and delete-1 jackknife samples, we also obtain 90% confidence intervals using the respective samples. These are given in Table 12.4 and can be interpreted similarly as the traditional confidence intervals. Notice that the bootstrap confidence interval includes 0 while the jackknife confidence interval is completely below 0. If using these results to perform a test about using a regression through the origin, different conclusions would be reached depending on the procedure used. Moreover, remember that if you were to perform another bootstrap resulting in a different sample of $B = 1000$, then the estimated intervals given above will be slightly different due to the randomness of the resampling process.

**Example 12.6.3.** *(Pulp Property Data, cont'd)*
Let us return to the pulp property data from Example 3.6.3 and run the permutation test. The $t$-tests for the intercept and slope yielded test statistics of 15.774 and $-3.178$, respectively. These are the test statistics based on the observed data. Since $n = 7$ for these data, obtaining the exact permutation is not prohibitive. All test statistics for $\beta_0$ and $\beta_1$ based on the $7! = 5040$ permutation samples are given in Figure 12.6. The approximate ASL for the test about the intercept is 0.016, while the approximate ASL for the test about the slope is 0.026. Each of these would indicate that there is reasonably strong evidence against the respective null hypothesis.

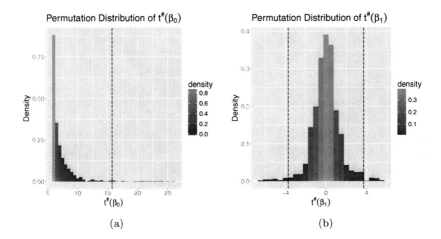

**FIGURE 12.6**
Histograms of the permutation distributions of the test statistics for (a) $\beta_0$ and (b) $\beta_1$ from the simple linear regression estimates for the pulp property data. The vertical lines indicate the observed value of the test statistics used in the calculation of the ASL.

# 13

---

## Correlated Errors and Autoregressive Structures

---

Recall that one of the assumptions when building a linear regression model is that the errors are independent. This section discusses methods for handling dependent errors. In particular, the dependency usually appears because of a temporal component. Error terms correlated over time are said to be **autocorrelated** or **serially correlated**. When error terms are autocorrelated, some issues arise when using ordinary least squares, such as:

- estimated regression coefficients are still unbiased, but they no longer have the minimum variance property;

- the MSE may seriously underestimate the true variance of the errors;

- the standard error of the regression coefficients may seriously underestimate the true standard deviation of the estimated regression coefficients; and

- statistical intervals and inference procedures are no longer strictly applicable.

We also consider the setting where a dataset has a temporal component that affects the analysis.

The development of statistical models for time series involves a lot of mathematical details. We only provide some details at a very superficial level. Fortunately, there are many texts devoted to the topic of time series, ranging from the applied to very mathematical. A selection of texts on the topic includes Brockwell and Davis (1991), Hamilton (1994), Enders (2003), Shumway and Stoffer (2011), and Montgomery et al. (2015).

---

## 13.1 Overview of Time Series and Autoregressive Structures

A **time series** is a sequence of measurements of the same variable(s) made over time. A common notation for a time series $X$ is

$$X = \{X_t : t \in T\},$$

where $T$ is referred to as the **index set**. Usually the index set is the set of natural numbers (i.e., $\{0, 1, 2, \ldots\}$) and are representing evenly-spaced times, for example, daily, monthly, or yearly. This section provides a general overview of time series and relevant terminology. While some of the concepts are presented in a more general setting, it is helpful to have this terminology established for presenting regression analyses with autocorrelated errors.

Let us consider the problem in which we have the variables $X_1, \ldots, X_{p-1}$ and $Y$, all measured as time series. For example, $Y$ might be the monthly highway accidents on a stretch of interstate highway and $X_1, \ldots, X_{p-1}$ might be other variables like the monthly amount of travel on the interstate and the monthly amount of precipitation received in that area. Here, the index set corresponds to monthly measurements. A **(multiple) time series regression** model can be written as:

$$Y_t = \mathbf{X}_t^{\mathrm{T}} \boldsymbol{\beta} + \epsilon_t. \tag{13.1}$$

To emphasize that we have measured values over time, we use the subscript $t$ from the index set representation rather than the usual $i$. The difficulty in time series regression is that the errors ($\epsilon_t$) are often correlated with each other. In other words, we have a dependency *between* the errors.

One may consider situations in which the error at one specific time is linearly related to the error at the previous time. That is, the errors themselves have a simple linear regression model that can be written as

$$\epsilon_t = \rho \epsilon_{t-1} + \omega_t. \tag{13.2}$$

Here, $|\rho| < 1$ is called the **autocorrelation parameter** and the $\omega_t$ term is a new error term which will have the usual regression assumptions that we make about errors. So this model says that the error at time $t$ is predictable from a fraction of the error at time $t - 1$ plus some new perturbation $\omega_t$.

Our model for the $\epsilon_t$ errors of the original $Y$ versus $\mathbf{X}$ regression is called an **autoregressive model for the errors**. One reason the errors might have an autoregressive structure is that the $Y$ and $\mathbf{X}$ variables at time $t$ may be (and most likely are) related to the $Y$ and $\mathbf{X}$ measurements at time $t - 1$. These relationships are being absorbed into the error term of our simple regression model that only relates $Y$ and $\mathbf{X}$ measurements made at concurrent times. Notice that the autoregressive model for the errors is a violation of the assumption that we have independent errors, which creates theoretical difficulties for ordinary least squares estimates of the regression coefficients. There are several different methods for estimating the regression parameters of the $Y$ versus $\mathbf{X}$ relationship when we have errors with an autoregressive structure and we will introduce a few of these in this chapter.

One additional note concerns the difference in terminology here. A time series regression is when we regress one value from a time series on values from another time series (e.g., $Y_t$ on $\mathbf{X}_t$). An autoregressive model is when a value from a time series is regressed on previous values from that same time series (e.g., $Y_t$ on $Y_{t-1}$). Both models will likely (but it is not necessary in the strict sense of these models) have an autoregressive structure on the errors.

The most fundamental time series problem involves creating a model for a single variable and to examine how an observation of that variable at time $t$ relates to measurements at previous times. For example, we may be able to predict a measure of global temperature this year $(Y_t)$ using measurements of global temperature in the previous two years $(Y_{t-1}, Y_{t-2})$. The autoregressive model for doing so would then be

$$Y_t = \beta_0 + \beta_1 Y_{t-1} + \beta_2 Y_{t-2} + \epsilon_t. \tag{13.3}$$

In this regression model, the two previous time periods have become the predictors and the errors might either have our usual assumptions about error terms or could possibly also have an autoregressive structure.

The **order** of an autoregression is the number of immediately preceding values in the series that are used to predict the value at the present time. A first-order autoregression (written as AR(1)) is a simple linear regression in which the value of the series at any time $t$ is a straight-line function of the value at time $t - 1$. The model in (13.3) is a second-order autoregression (written as AR(2)) since the value at time $t$ is predicted from the values at times $t - 1$ and $t - 2$. More generally, a $k^{\text{th}}$-order autoregression (written as AR($k$)) is a multiple linear regression in which the value of the series at any time $t$ is a (linear) function of the values at times $t - 1, t - 2, \ldots, t - k$. For most of this chapter, the focus is on the setting where only a first-order autoregressive structure is assumed on the errors.

## 13.2    Properties of the Error Terms

Suppose we have a (multiple) time series regression where the error terms follow an AR(1) structure:

$$Y_t = \mathbf{X}_t^{\mathrm{T}} \boldsymbol{\beta} + \epsilon_t$$
$$\epsilon_t = \rho \epsilon_{t-1} + w_t,$$

where $|\rho| < 1$ and the $w_t$ are *iid* $\mathcal{N}(0, \sigma^2)$. The error terms $\epsilon_t$ still have mean 0 and constant variance:

$$\mathrm{E}(\epsilon_t) = 0$$
$$\mathrm{Var}(\epsilon_t) = \frac{\sigma^2}{1 - \rho^2}.$$

However, the covariance between adjacent error terms is

$$\mathrm{Cov}(\epsilon_t, \epsilon_{t-1}) = \rho \left( \frac{\sigma^2}{1 - \rho^2} \right),$$

which implies the coefficient of correlation between adjacent error terms is

$$\text{Corr}(\epsilon_t, \epsilon_{t-1}) = \frac{\text{Cov}(\epsilon_t, \epsilon_{t-1})}{\sqrt{\text{Var}(\epsilon_t)\text{Var}(\epsilon_{t-1})}} = \rho,$$

which is the autocorrelation parameter we introduced earlier.

The assumption of independent error terms in previous regression models results in the errors having variance–covariance matrix $\sigma^2 \mathbf{I}_n$. In time series regression models, all of the off-diagonals are no longer 0. But, with the formulas we established above, we can construct the off-diagonals for the variance–covariance matrix of the error terms for the AR(1) model. The covariance between error terms that are $k$ time periods apart is called the **autocovariance function** and is given by:

$$\text{Cov}(\epsilon_t, \epsilon_{t-k}) = \rho^k \left( \frac{\sigma^2}{1 - \rho^2} \right).$$

The coefficient of correlation between these two error terms is called the **autocorrelation function (ACF)** and is given by:

$$\text{Corr}(\epsilon_t, \epsilon_{t-k}) = \rho^k.$$

This value of $k$ is the time gap being considered and is called the **lag**. A **lag 1 autocorrelation** (i.e., $k = 1$ in the above) is the correlation between values that are one time period apart. More generally, a **lag $k$ autocorrelation** is the correlation between values that are $k$ time periods apart.

Now, we may formally write out the variance–covariance matrix for the AR(1) model:

$$\Sigma(1) = \begin{pmatrix} \gamma & \gamma\rho & \gamma\rho^2 & \cdots & \gamma\rho^{n-1} \\ \gamma\rho & \gamma & \gamma\rho & \cdots & \gamma\rho^{n-2} \\ \gamma\rho^2 & \gamma\rho & \gamma & \cdots & \gamma\rho^{n-3} \\ \vdots & \vdots & \vdots & \ddots & \vdots \\ \gamma\rho^{n-1} & \gamma\rho^{n-2} & \gamma\rho^{n-3} & \cdots & \gamma \end{pmatrix},$$

where

$$\gamma = \frac{\sigma^2}{1 - \rho^2}.$$

For the sake of completion, assume now that the error terms follow an AR($k$) structure:

$$Y_t = \mathbf{X}_W^{\text{T}} \boldsymbol{\beta} + \epsilon_t$$

$$\epsilon_t = \sum_{i=1}^{k} \rho_i \epsilon_{t-i} + \omega_t,$$

where $|\rho_i| < 1$ for all $i = 1, \ldots, k$ and the $w_t$ are *iid* $\mathcal{N}(0, \sigma^2)$. The autocovariance function and ACF are found by solving the system of **Yule–Walker equations**:

$$\sigma(1) = \rho_1\sigma^2 + \rho_2\sigma(1) + \rho_3\sigma(2) + \ldots + \rho_k\sigma(k-1)$$
$$\sigma(2) = \rho_1\sigma(1) + \rho_2\sigma^2 + \rho_3\sigma(1) + \ldots + \rho_k\sigma(k-2)$$
$$\vdots$$
$$\sigma(k) = \rho_1\sigma(k-1) + \rho_2\sigma(k-2) + \rho_3\sigma(k-3) + \ldots + \rho_k\sigma^2,$$

where

$$\sigma(k) = \text{Cov}(\epsilon_t, \epsilon_{t-k})$$

is the lag $k$ autocovariance and

$$\rho_k = \frac{\sigma(k)}{\sigma^2}$$

is the lag $k$ autocorrelation. Upon solving these, the variance–covariance matrix for the errors, $\Sigma(k)$, can be constructed.

The ACF is a way to measure the linear relationship between an error at time $t$ and the errors at previous times. If we assume an AR($k$) model, then we may wish to only measure the connection between $\epsilon_t$ and $\epsilon_{t-k}$ and filter out the linear influence of the random variables that lie in between (i.e., $\epsilon_{t-1}, \epsilon_{t-2}, \ldots, \epsilon_{t-(k-1)}$), which requires a transformation on the errors. One way to measure this is by calculating the correlation of the transformed errors, which is called the **partial autocorrelation function (PACF)**. The PACF values can be found by using another set of the Yule–Walker equations for an AR($k$) model, which we will not present here. After skipping the mathematical details, the $k^{\text{th}}$ PACF (denoted by $\phi_k$) is given by:

$$\phi_k = \frac{|P_k^*|}{|P_k|},$$

where

$$P_k = \begin{pmatrix} 1 & \rho_1 & \cdots & \rho_{k-1} \\ \rho_1 & 1 & \cdots & \rho_{k-2} \\ \vdots & \vdots & \ddots & \vdots \\ \rho_{k-1} & \rho_{k-2} & \cdots & 1 \end{pmatrix}$$

and $P_k^*$ is the matrix $P_k$ with the $k^{\text{th}}$ column replaced by the vector of correlations $(\rho_1, \rho_2, \ldots, \rho_k)^{\text{T}}$.

Notice that the difference between the ACF and the PACF is that the latter controls for the influence of the other lags. The PACF is most useful for identifying the order of an autoregressive model. Specifically, sample partial autocorrelations that are significantly different from 0 indicate lagged terms of $\epsilon$ that are predictors of $\epsilon_t$. A helpful way to differentiate between ACF and

PACF is to think of them as rough analogues to $R^2$ and partial $R^2$ values. For example, the PACF for an AR(1) structure is

$$\phi_1 = \frac{\rho}{1} = \rho$$

and the PACF for an AR(2) structure is

$$\phi_2 = \frac{\begin{vmatrix} 1 & \rho_1 \\ \rho_1 & \rho_2 \end{vmatrix}}{\begin{vmatrix} 1 & \rho_1 \\ \rho_1 & 1 \end{vmatrix}} = \frac{\rho_2 - \rho_1^2}{1 - \rho_1^2}.$$

## 13.3   Testing and Remedial Measures for Autocorrelation

In this section we present some formal tests and remedial measures for handling autocorrelation. Further details on these tests can be found in, for example, Brockwell and Davis (1991) and Shumway and Stoffer (2011).

**Durbin–Watson Test**

We usually assume that the error terms are independent unless there is reason to think that this assumption has been violated. Most often the violation occurs because there is a known temporal component to how the observations were drawn. There are various ways to diagnose this violation, but one of the easiest ways is to produce a scatterplot of the residuals versus their time ordering. If the data are independent, then the residuals should look randomly scattered about 0. However, if a noticeable pattern emerges — particularly one that is cyclical — then dependency is likely an issue.

Recall that for a first-order autocorrelation with the errors is modeled as:

$$\epsilon_t = \rho \epsilon_{t-1} + w_t,$$

where $|\rho| < 1$ and the $w_t$ are *iid* $\mathcal{N}(0, \sigma^2)$. If we suspect first-order autocorrelation with the errors, then a formal test exists regarding the parameter $\rho$. This is called the **Durbin–Watson test** (Durbin and Watson, 1950, 1951, 1971) and is constructed as:

$$H_0 : \rho = 0$$
$$H_A : \rho \neq 0.$$

So the null hypothesis of $\rho = 0$ means that $\epsilon_t = w_t$ (i.e., that the error terms are not correlated with the previous error term), while the alternative hypothesis of $\rho \neq 0$ means the error terms are either positively or negatively correlated with the previous error term. Often a researcher will already have an indication of whether the errors are positively or negatively correlated based on the science of the study. For example, a regression of oil prices (in dollars per barrel) versus the gas price index will surely have positively correlated errors. When the researcher has an indication of the direction of the correlation, then the Durbin-Watson test also accommodates the one-sided alternatives $H_A : \rho < 0$ for negative correlations or $H_A : \rho > 0$ for positive correlations, for which the latter would be used in the oil example.

The test statistic for the Durbin–Watson test based on a dataset of size $n$ is given by:

$$D = \frac{\sum_{t=2}^{n}(e_t - e_{t-1})^2}{\sum_{t=1}^{n} e_t^2},$$

where $e_t = y_t$ are the residuals from the ordinary least squares fit. Exact critical values are difficult to obtain, but tables for certain significance values can be used to make a decision. The tables provide a lower and upper bound, called $d_L$ and $d_U$, respectively, which depend on the matrix of predictors. Thus, it is not possible to present a comprehensive table of bounds. Decision rules based on the critical values are as follows:

- If $D < d_L$, then reject $H_0$.

- If $D > d_U$, then fail to reject $H_0$.

- If $d_L \leq D \leq d_U$, then the test is inconclusive.

While the prospect of having an inconclusive test result is less than desirable, there are some programs that use exact and approximate procedures for calculating a $p$-value. These procedures require certain assumptions on the data, which we will not discuss. One "exact" method is based on the beta distribution for obtaining $p$-values.

**Breusch–Godfrey Test**
Consider the (multiple) time series regression where the error terms follow an AR($k$) structure:

$$Y_t = \mathbf{X}_t^\mathsf{T}\boldsymbol{\beta} + \epsilon_t$$

$$\epsilon_t = \sum_{j=1}^{k} \rho_j \epsilon_{t-j} + \omega_t,$$

where $|\rho_j| < 1$, $j = 1, \ldots, k$, and the $\omega_t$ are *iid* $\mathcal{N}(0, \sigma^2)$. The **Breusch–Godfrey test** (Breusch, 1978; Godfrey, 1978) allows us to test for these higher-order correlations; i.e.,

$$H_0 : \rho_j = 0 \text{ for all } j$$
$$H_A : \rho_j \neq 0 \text{ for some } j,$$

which cannot be tested using the Durbin–Watson test. Let $\hat{\epsilon}_t, \hat{\epsilon}_{t-1}, \ldots, \hat{\epsilon}_{t-k}$ be the ordinary least square estimates of the lagged residuals. First, fit the following auxiliary regression model:

$$\hat{\epsilon}_t = \mathbf{X}_t^{\mathsf{T}}\boldsymbol{\beta} + \sum_{j=1}^{k} \rho_j \hat{\epsilon}_{t-j} + \omega_t.$$

and calculate the $R^2$ value. Then, $mR^2 \sim \chi_k^2$, where $m = n - k$ is the number of observations available for estimating the AR($k$) regression equation. Note that this test is based on the idea of Lagrange multiplier testing, just like White's test in Chapter 10.

### ACF and PACF Plots

Graphical approaches to assessing the lag of a time series include looking at the ACF and PACF values versus the lag. In a plot of ACF versus the lag, if large ACF values and a non-random pattern are indicated, then likely the values are serially correlated. In a plot of PACF versus the lag, the pattern will usually appear random, but large PACF values at a given lag indicate this value as a possible choice for the order of an autoregressive model. It is important that the choice of the order makes sense. For example, suppose you have your blood pressure readings for every day over the past two years. You may find that an AR(1) or AR(2) model is appropriate for modeling your blood pressure. However, the PACF may indicate a large partial autocorrelation value at a lag of 17, but such a large order for an autoregressive model likely does not make much sense.

Approximate bounds can also be constructed for each of these plots to aid in determining large values. For the ACF, approximate $(1 - \alpha) \times 100\%$ significance bounds for the $i^{\text{th}}$ autocorrelation are given by:

$$\pm z_{1-\alpha/2} \sqrt{\frac{1 + 2\sum_{j=1}^{i-1} r_j^2}{n}},$$

where $r_j$ is the $j^{\text{th}}$ autocorrelation value. Notice that these bounds widen as a function of lag. For the PACF, approximate $(1-\alpha) \times 100\%$ significance bounds are given by $\pm z_{1-\alpha/2}/\sqrt{n}$, which is the same for all lag values. Values lying outside of either of these bounds are indicative of an autoregressive process.

### Ljung–Box Q Test

The **Ljung–Box Q test** or **Portmanteau test** (Ljung and Box, 1978) is used to test whether or not observations over time are random and independent. In particular, for a given $k$, it tests the following:

$H_0$ : the autocorrelations up to lag $k$ are all 0

$H_A$ : the autocorrelations of one or more lags differ from 0.

The test statistic is calculated as:

$$Q_k = n(n+2) \sum_{i=1}^{k} \frac{r_j^2}{n-j},$$

which is approximately $\chi_k^2$-distributed.

**Box–Pierce Test**
The **Box–Pierce test** (Box and Pierce, 1978) is nearly identical to the Ljung–Box Q test, except that the test statistic is calculated as:

$$Q_{BP;k} = n \sum_{i=1}^{k} r_j^2,$$

which is also approximately $\chi_k^2$-distributed.

**Cochrane–Orcutt Procedure**
When autocorrelated error terms are found to be present, then one of the first remedial measures should be to investigate the omission of a key predictor variable. If such a predictor does not aid in reducing/eliminating autocorrelation of the error terms, then certain transformations on the variables can be performed. We discuss four transformations which are designed for AR(1) errors. Methods for dealing with errors from an AR($k$) process exist in the literature, but are much more technical in nature. However, we present one possible solution for the AR($k$) setting at the end of this section.

The first of the three transformation methods we discuss is called the **Cochrane–Orcutt procedure**, which involves an iterative process after identifying the need for an AR(1) process:

1. Estimate $\rho$ for
   $$\epsilon_t = \rho \epsilon_{t-1} + \omega_t$$
   by performing a regression through the origin. Call this estimate $r$.

2. Transform the variables from the multiple regression model
   $$y_t = \beta_0 + \beta_1 x_{t,1} + \ldots + \beta_{p-1} x_{t,p-1} + \epsilon_t$$
   by setting $y_t^* = y_t - r y_{t-1}$ and $x_{t,j}^* = x_{t,j} - r x_{t-1,j}$ for $j = 1, \ldots, p-1$. Call these estimates $\hat{\beta}_0^*, \ldots, \hat{\beta}_{p-1}^*$. Note that we will not have transformed values for $t = 1$.

3. Regress $y_t^*$ on the transformed predictors using ordinary least squares. Look at the error terms for this fit and determine if autocorrelation is still present (such as using the Durbin–Watson test). If autocorrelation is still present, then iterate this procedure. If it appears to be corrected, then transform the estimates back to

their original scale by setting $\hat{\beta}_0 = \hat{\beta}_0^*/(1 - r)$ and $\hat{\beta}_j = \hat{\beta}_j^*$ for $j = 1, \ldots, p - 1$. Notice that only the intercept parameter requires a transformation. Furthermore, the standard errors of the regression estimates for the original scale can also be obtained by setting s.e.$(\hat{\beta}_0)$ = s.e.$(\hat{\beta}_0^*)/(1-r)$ and s.e.$(\hat{\beta}_j)$ = s.e.$(\hat{\beta}_j^*)$ for $j = 1, \ldots, p-1$.

There are a few things to note about the Cochrane–Orcutt approach. First is that it does not always work properly. This primarily occurs because if the errors are positively autocorrelated, then $r$ tends to underestimate $\rho$. When this bias is serious, then it can seriously reduce the effectiveness of the Cochrane–Orcutt procedure. Also, the correct standard errors from the Cochrane–Orcutt procedure are larger than the incorrect values from the simple linear regression on the original data. If ordinary least squares estimation is used when the errors are autocorrelated, the standard errors often are underestimated. It is also important to note that this does not always happen. Overestimation of the standard errors is an "on average" tendency over all problems.

## Prais–Winsten Procedure
The **Prais–Winsten procedure** (Prais and Winsten, 1954) is identical to the Cochrane–Orcutt procedure, except that it does not drop the first observation. The Prais–Winsten procedure simply includes the following transformation for $t = 1$ in Step 2 of the Cochrane–Orcutt procedure:

$$y_1^* = \sqrt{1 - r^2} y_1.$$

Then, $y_1^*$ is also included in Step 3 where ordinary least squares is performed.

## Hildreth–Lu Procedure
The **Hildreth–Lu procedure** (Hildreth and Lu, 1960) is a more direct method for estimating $\rho$. After establishing that the errors have an AR(1) structure, follow these steps:

1. Select a series of candidate values for $\rho$ (presumably values that would make sense after you assessed the pattern of the errors).

2. For each candidate value, regress $y_t^*$ on the transformed predictors using the transformations established in the Cochrane–Orcutt procedure. Retain the SSEs for each of these regressions.

3. Select the value which minimizes the SSE as an estimate of $\rho$.

Notice that this procedure is similar to the Box–Cox transformation discussed in Chapter 4 and that it is not iterative like the Cochrane–Orcutt procedure.

## First Differences Procedure
Since $\rho$ is frequently large for AR(1) errors, especially in economics data, one could simply set $\rho = 1$ in the transformed model of the previous three

procedures. This procedure is called the **first differences procedure** and simply regresses $y_t^* = y_t - y_{t-1}$ on the $x_{t,j}^* = x_{t,j} - x_{t-1,j}$ for $j = 1, \ldots, p-1$. The estimates from this regression are then transformed back by setting $\hat{\beta}_j = \hat{\beta}_j^*$ for $j = 1, \ldots, p-1$ and $\hat{\beta}_0 = \bar{y} - (\hat{\beta}_1 \bar{x}_1 + \ldots + \hat{\beta}_{p-1} \bar{x}_{p-1})$.

**Generalized Least Squares**

Finally, as mentioned in Chapter 11, weighted least squares can also be used to reduce autocorrelation by choosing an appropriate weighting matrix. In fact, the method used is more general than weighted least squares. The method of weighted least squares uses a diagonal matrix to help correct for non-constant variance. However, in a model with correlated errors, the errors have a more complicated variance–covariance structure, such as $\Sigma(1)$ given earlier for the AR(1) model. Thus, the weighting matrix for the more complicated variance–covariance structure is non-diagonal and utilizes the method of **generalized least squares**, of which weighted least squares is a special case. In particular, when $\mathrm{Var}(\mathbf{Y}) = \mathrm{Var}(\boldsymbol{\epsilon}) = \Omega$, the objective is to find a matrix $\Lambda$ such that:

$$\mathrm{Var}(\Lambda \mathbf{Y}) = \Lambda \Omega \Lambda^{\mathrm{T}} = \sigma^2 \mathbf{I}_n,$$

The generalized least squares estimator (sometimes called the **Aitken estimator**) takes $\Lambda = \sigma \Omega^{1/2}$ and is given by

$$\hat{\boldsymbol{\beta}}_{\mathrm{GLS}} = \underset{\boldsymbol{\beta}}{\mathrm{argmin}} \| \Lambda(\mathbf{Y} - \mathbf{X}\boldsymbol{\beta}) \|^2$$
$$= (\mathbf{X}^{\mathrm{T}} \Omega^{-1} \mathbf{X})^{-1} \mathbf{X}^{\mathrm{T}} \Omega \mathbf{Y}.$$

There is no optimal way of choosing such a weighting matrix $\Omega$. $\Omega$ contains $n(n+1)/2$ parameters, which is often too many parameters to estimate, so restrictions must be imposed. The estimate will heavily depend on these restrictions and thus makes the method of generalized least squares difficult to use unless you are savvy (and lucky) enough to choose helpful restrictions. For more on the theory of generalized least squares, see Chapter 6 of Amemiya (1985).

**Feasible Generalized Least Squares**

If one does not assume $\Omega$ is known, a strategy to obtain an estimate, say, $\hat{\Omega}$, is **feasible generalized least squares**. Feasible generalized least squares is a two-step approach:

1. Estimate the multiple linear regression model using ordinary least squares or using another consistent (but inefficient) estimator. Then, take the raw residuals and use these to estimate the variance–covariance matrix. At this point, the residuals need to be checked to see if they, say, follow a time series. If so, incorporate such additional constraints to obtain the estimate of $\hat{\Omega}$.

2. Calculate the feasible generalized least squares estimate as follows:

$$\hat{\beta}_{\text{FGLS}} = (\mathbf{X}^{\text{T}}\hat{\Omega}^{-1}\mathbf{X})^{-1}\mathbf{X}^{\text{T}}\hat{\Omega}\mathbf{Y}.$$

In the first step, the residuals are often used in estimating a robust version of the variance–covariance matrix. Such estimators include the **Newey–West estimator** (Newey and West, 1987), **heteroskedasticity and autocorrelation (HAC) estimator** (Andrews, 1991), and the **Eicker–White estimator** (Eicker, 1963; White, 1980). These and other estimators are discussed in Chapter 4 of Kleiber and Zeileis (2008). An iterative approach to feasible generalized least squares can also be implemented.

**Prediction Issues**

When calculating predicted values, it is important to utilize $\epsilon_t = \rho\epsilon_{t-1} + \omega_t$ as part of the process. Our estimated regression equation with AR(1) errors is

$$\hat{y}_t = \hat{\beta}_0 + \hat{\beta}_1 x_t + e_t = \hat{\beta}_0 + \hat{\beta}_1 x_t + re_{t-1}.$$

Values of $\hat{y}_t$ are computed iteratively.

- Assume $e_0 = 0$ (error before $t = 1$ is 0), compute $\hat{y}_1$ and $e_1 = y_1 - \hat{y}_1$.

- Use the value of $e_1 = y_1 - \hat{y}_1$ when computing $\hat{y}_2 = \hat{\beta}_0 + \hat{\beta}_1 x_1 + re_1$.

- Determine $e_2 = y_2 - \hat{y}_2$, and use that value when computing $\hat{y}_3 = \hat{\beta}_0 + \hat{\beta}_1 x_3 + re_2$.

- Iterate.

## 13.4 Advanced Methods

This section provides a few more advanced techniques used in time series analysis. While they are more peripheral to the autoregressive error structures that we have discussed, they are germane to this text since these models are constructed in a regression framework.

### 13.4.1 ARIMA Models

Most time series models have a regression structure, so it is beneficial to introduce a general class of time series models called **autoregressive integrated moving average** or **ARIMA models**. They are also referred to as **Box–Jenkins models**, due to the systematic methodology of identifying, fitting, checking, and utilizing ARIMA models; see Box et al. (2008). Before we present a general ARIMA model, we need to introduce a few additional concepts.

Suppose we have the time series $Y = \{Y_t : t \in T\}$. If the value for each $Y$ is determined *exactly* by a mathematical formula, then the series is said to be **deterministic**. If the future values of $Y$ can only be described through their probability distribution, then the series is said to be a **stochastic process**. A special class of stochastic processes is a **stationary stochastic process**, which occurs when the probability distribution for the process is the same for all starting values of $t$. Stationary stochastic processes are completely defined by their mean, variance, and autocorrelation functions. When a time series exhibits nonstationary behavior, then part of our objective will be to transform it into a stationary process. More details can be found, for example, in Chapter 1 of Shumway and Stoffer (2011).

When stationarity is not an issue, then we can define an **autoregressive moving average** or **ARMA** model as follows:

$$Y_t = \sum_{i=1}^{p} \phi_i Y_{t-i} + a_t - \sum_{j=1}^{q} \theta_j a_{t-j},$$

where $\phi_1, \ldots, \phi_p$ are the autoregressive parameters to be estimated, $\theta_1, \ldots, \theta_q$ are the moving average parameters to be estimated, and $a_1, \ldots, a_t$ are a series of unknown random errors (or residuals) that are assumed to follow a normal distribution. This is also referred to as an ARMA$(p, q)$ model. The model can be simplified by introducing the **Box–Jenkins backshift operator**, which is defined by the following relationship:

$$B^p X_t = X_{t-p},$$

such that $X = \{X_t : t \in T\}$ is any time series and $p < t$. Using the backshift notation yields the following:

$$\left(1 - \sum_{i=1}^{p} \phi_i B^i\right) Y_t = \left(1 - \sum_{j=1}^{q} \theta_j B^j\right) a_t,$$

which is often reduced further to

$$\phi_p(B) Y_t = \theta_q(B) a_t,$$

where $\phi_p(B) = (1 - \sum_{i=1}^{p} \phi_i B^i)$ and $\theta_q(B) = (1 - \sum_{j=1}^{q} \theta_j B^j)$.

When time series exhibit nonstationary behavior, which commonly occurs in practice, then the ARMA model presented above can be extended and written using differences:

$$W_t = Y_t - Y_{t-1}$$
$$= (1 - B)Y_t$$
$$W_t - W_{t-1} = Y_t - 2Y_{t-1} + Y_{t-2}$$
$$= (1 - B)^2 Y_t$$

$$\vdots$$

$$W_t - \sum_{k=1}^{d} W_{t-k} = (1 - B)^d Y_t,$$

where $d$ is the order of **differencing**. Replacing $Y_t$ in the ARMA model with the differences defined above yields the formal ARIMA$(p, d, q)$ model:

$$\phi_p(B)(1 - B)^d Y_t = \theta_q(B)a_t.$$

An alternative way to deal with nonstationary behavior is to simply fit a linear trend to the time series and then fit a Box–Jenkins model to the residuals from the linear fit. This will provide a different fit and a different interpretation, but is still a valid way to approach a process exhibiting non-stationarity.

Seasonal time series can also be incorporated into a Box–Jenkins framework. The following general model is usually recommended:

$$\phi_p(B)\Phi(P)(B^s)(1 - B)^d(1 - B^s)^D Y_t = \theta_q(B)\Theta_Q(B^s)a_t,$$

where $p$, $d$, and $q$ are as defined earlier, $s$ is a (known) number of seasons per timeframe (e.g., months or years), $D$ is the order of the seasonal differencing, and $P$ and $Q$ are the autoregressive and moving average orders, respectively, when accounting for the seasonal shift. Moreover, the operator polynomials $\phi_p(B)$ and $\theta_q(B)$ are as defined earlier, while $\Phi_P(B) = (1 - \sum_{i=1}^{P} \Phi_i B^{s \times i})$ and $\Theta_Q(B) = (1 - \sum_{j=1}^{Q} \Theta_j B^{s \times j})$. Luckily, the maximum value of $p$, $d$, $q$, $P$, $D$, and $Q$ is usually 2, so the resulting expression is relatively simple.

With all of the technical details provided above, we can now provide a brief overview of how to implement the Box–Jenkins methodology in practice, which (fortunately) most statistical software packages can perform:

1. **Model Identification:** Using plots of the data, autocorrelations, partial autocorrelations, and other information, a class of simple ARIMA models is selected. This amounts to approximating an appropriate value for $d$ followed by estimates for $p$ and $q$ as well as $P$, $D$, and $Q$ in the seasonal time series setting.

2. **Model Estimation:** The autoregressive and moving average parameters are found via an optimization method, like maximum likelihood.

3. **Diagnostic Checking:** The fitted model is checked for inadequacies by studying the autocorrelations of the residual series; i.e., the time-ordered residuals.

The above is then iterated until there appears to be minimal to no improvement in the fitted model.

## 13.4.2   Exponential Smoothing

The techniques of the previous section can all be used in the context of **forecasting**, which is the art of modeling patterns in the data that are usually visible in time series plots and then extrapolated into the future. In this section, we discuss exponential smoothing methods that rely on **smoothing parameters**, which are parameters that determine how fast the weights of the series decay. For each of the three methods we discuss below, the smoothing constants are found objectively by selecting those values which minimize one of the three error-size criterion below:

$$\text{MSE} = \frac{1}{n} \sum_{t=1}^{n} e_t^2$$

$$\text{MAE} = \frac{1}{n} \sum_{t=1}^{n} |e_t|$$

$$\text{MAPE} = \frac{100}{n} \sum_{t=1}^{n} \left| \frac{e_t}{Y_t} \right|.$$

The error in the above expressions is the difference between the actual value of the time series at time $t$ and the fitted value at time $t$, which is determined by the smoothing method employed. Moreover, MSE is (as usual) the mean square error, MAE is the mean absolute error, and MAPE is the mean absolute percent error.

Exponential smoothing methods also require initialization since the forecast for period one requires the forecast at period zero, which by definition we do not have. Several methods have been proposed for generating starting values. Most commonly used is the **backcasting method**, where the series is reversed so that we forecast into the past instead of into the future. Once we have done this, we then switch the series back and apply the exponential smoothing algorithm in the regular manor. A lot of the original work on the following smoothing methods is due to Holt (2004) and Winters (1960).

**Single exponential smoothing** smooths the data when no trend or seasonal components are present. The equation for this method is:

$$\hat{Y}_t = \alpha \left( Y_t + \sum_{i=1}^{r} (1 - \alpha)^i Y_{t-i} \right),$$

where $\hat{Y}_t$ is the forecasted value of the series at time $t$ and $\alpha$ is the smoothing constant. Note that $r < t$, but $r$ does not have to equal $t-1$. From the previous equation, we see that the method constructs a weighted average of the observations. The weight of each observation decreases exponentially as we move back in time. Hence, since the weights decrease exponentially and averaging is a form of smoothing, the technique was named exponential smoothing. An equivalent ARIMA(0,1,1) model can be constructed to represent the single exponential smoother.

**Double exponential smoothing** (also called **Holt's method**) smooths the data when a trend is present. The double exponential smoothing equations are:

$$L_t = \alpha Y_t + (1 - \alpha)(L_{t-1} + T_{t-1})$$
$$T_t = \beta(L_t - L_{t-1}) + (1 - \beta)T_{t-1}$$
$$\hat{Y}_t = L_{t-1} + T_{t-1},$$

where $L_t$ is the level at time $t$, $\alpha$ is the weight (or smoothing constant) for the level, $T_t$ is the trend at time $t$, $\beta$ is the weight (or smoothing constant) for the trend, and all other quantities are defined as earlier. An equivalent ARIMA(0,2,2) model can be constructed to represent the single exponential smoother.

Finally, **Holt–Winters exponential smoothing** smooths the data when trend and seasonality are present; however, these two components can be either additive or multiplicative. For the additive model, the equations are:

$$L_t = \alpha(Y_t - S_{t-p}) + (1 - \alpha)(L_{t-1} + T_{t-1})$$
$$T_t = \beta(L_t - L_{t-1}) + (1 - \beta)T_{t-1}$$
$$S_t = \delta(Y_t - L_t) + (1 - \delta)S_{t-p}$$
$$\hat{Y}_t = L_{t-1} + T_{t-1} + S_{t-p}.$$

For the multiplicative model, the equations are:

$$L_t = \alpha(Y_t/S_{t-p}) + (1 - \alpha)(L_{t-1} + T_{t-1})$$
$$T_t = \beta(L_t - L_{t-1}) + (1 - \beta)T_{t-1}$$
$$S_t = \delta(Y_t/L_t) + (1 - \delta)S_{t-p}$$
$$\hat{Y}_t = (L_{t-1} + T_{t-1})S_{t-p}.$$

For both sets of equations, all quantities are the same as they were defined in the previous models, except now we also have that $S_t$ is the seasonal component at time $t$, $\delta$ is the weight (or smoothing constant) for the seasonal component, and $p$ is the seasonal period.

### 13.4.3   Spectral Analysis

Suppose we believe that a time series $Y = \{Y_t : t \in T\}$ contains a periodic (cyclic) component. **Spectral analysis** takes the approach of specifying a time series as a function of trigonometric components. A model for the periodic component is

$$Y_t = R\cos(ft + d) + e_t,$$

where $R$ is the **amplitude** of the variation, $f$ is the **frequency**[1] of periodic variation, and $d$ is the **phase**. Using the trigonometric identity $\cos(A + B) = \cos(A)\cos(B) - \sin(A)\sin(B)$, we can rewrite the above model as

$$Y_t = a\cos(ft) + b\sin(ft) + e_t,$$

where $a = R\cos(d)$ and $b = -R\sin(d)$. Thus, the above is a multiple regression through the origin model with two predictors.

Variation in time series may be modeled as the sum of several different individual waves occurring at different frequencies. Generalizing the above model as a sum of $k$ frequencies yields

$$Y_t = \sum_{j=1}^{k} R_j \cos(f_j t + d_j) + e_t,$$

which can be rewritten as

$$Y_t = \sum_{j=1}^{k} a_j \cos(f_j t) + \sum_{j=1}^{k} b_j \sin(f_j t) + e_t.$$

If the $f_j$ are known constants and we let $X_{t,r} = \cos(f_r t)$ and $Z_{t,r} = \sin(f_r t)$, then the above can be rewritten as the multiple regression model

$$Y_t = \sum_{j=1}^{k} a_j X_{t,j} + \sum_{j=1}^{k} b_j Z_{t,j} + e_t,$$

which is an example of a **harmonic regression**.

**Fourier analysis** provides a framework for approximating or representing functions using the sum of trigonometric functions, which is called the Fourier series representation of the function. Spectral analysis is a type of Fourier analysis such that the sum of sine and cosine terms approximate a time series that includes a random component. Thus, the coefficients in the harmonic regression (i.e., the $a_j$s and $b_j$s, $j = 1, \ldots, k$) may be estimated using, say, ordinary least squares.

Figures can also be constructed to help in the spectral analysis of a time series. While we do not develop the details here, the basic methodology consists of partitioning the total sum of squares into quantities associated with each frequency, like an ANOVA. From these quantities, histograms of the frequency (or wavelength) can be constructed, which are called **periodograms**. A smoothed version of the periodogram, called a **spectral density**, can also be constructed and is generally preferred to the periodogram. More details on spectral analysis can be found in Chapter 4 of Shumway and Stoffer (2011).

---

[1]If we take $2\pi/f$, then this is called the **wavelength**.

## 13.5    Examples

**Example 13.5.1.** *(Google Stock Data)*
The first dataset we analyze is for identifying the order of an autoregressive model. The dataset consists of $n = 105$ values which are the closing stock price of a share of Google stock during February 7th and July 7th of 2005. Thus, the measured variable is:

- $Y$: closing price of one share of Google stock (measured in U.S. dollars)

A plot of the stock prices versus time is presented in Figure 13.1. There appears to be a moderate linear pattern, suggesting that the first-order autoregressive model

$$Y_t = \beta_0 + \beta_1 Y_{t-1} + \epsilon_t$$

could be useful.

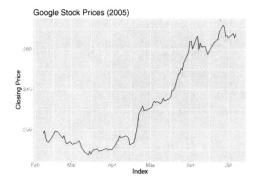

**FIGURE 13.1**
A plot of the price of one share of Google stock between February 7th and July 7th of 2005.

Figure 13.2 gives plots of the autocorrelations and partial autocorrelations for the data. The plot of the autocorrelations (Figure 13.2(a)) definitely exhibits a non-random pattern and the high values at earlier lags suggest an autoregressive model. Since the values are so high for so many lower values of lag, we also examine the plot of partial autocorrelations (Figure 13.2(b)). Here we notice that there is a significant spike at a lag of 1 and much lower spikes for the subsequent lags. This is assessed using the approximate 95% confidence bounds (dashed blue lines). Thus, an AR(1) model would likely be feasible for this dataset. This can also noted by simply looking at the numerical values used to construct these plots:

##############################

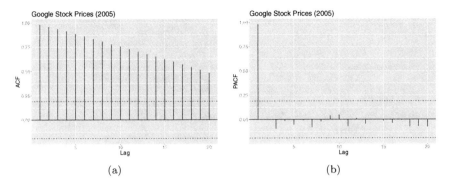

(a)                               (b)

**FIGURE 13.2**
The (a) autocorrelations and (b) partial autocorrelations versus lag for the
Google stock data.

```
Autocorrelations of series 'stock', by lag

    0     1     2     3     4     5     6
1.000 0.979 0.959 0.935 0.912 0.886 0.861
    7     8     9    10    11    12    13
0.833 0.804 0.778 0.754 0.727 0.702 0.675
   14    15    16    17    18    19    20
0.650 0.624 0.598 0.572 0.543 0.513 0.480


Partial autocorrelations of series 'stock', by lag

     1      2      3      4      5      6
 0.979  0.003 -0.095 -0.013 -0.053 -0.008
     7      8      9     10     11     12
-0.079 -0.021  0.038  0.045 -0.072  0.012
    13     14     15     16     17     18
-0.044  0.000 -0.014 -0.037 -0.007 -0.074
    19     20
-0.067 -0.076
#############################
```

It is obvious from the trend of the autocorrelations and the relatively high
value of the lag 1 partial autocorrelation that an AR(1) model would likely be
feasible for this dataset. This is further justified by the small $p$-values obtained
for the Ljung–Box Q test and Box–Pierce test:

```
#############################
Box-Ljung test
```

```
data:   stock
X-squared = 103.59, df = 1, p-value < 2.2e-16

Box-Pierce test

data:   stock
X-squared = 100.69, df = 1, p-value < 2.2e-16
##############################
```

The parameter estimates for the AR(1) model are $\hat{\beta}_0 = -0.3585$ and $\hat{\beta}_1 = 1.0059$, which yields the fitted values given in Figure 13.3.

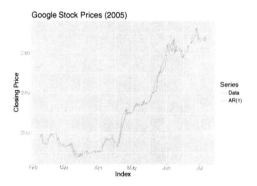

**FIGURE 13.3**
The Google stock data with the fitted values from the AR(1) model.

**Example 13.5.2.** *(Cardiovascular Data)*
Shumway et al. (1988) studied the possible effects of temperature and pollution particulates on daily mortality in Los Angeles County. The data were collected weekly between the years 1970 and 1979. While many variables were measured, we will focus on building a model with the following variables:

- $Y$: cardiovascular mortality

- $X_1$: average weekly temperature

- $X_2$: measured particulates

Time series plots of each of these variables are given in Figures 13.4(a) - 13.4(c). Clearly there is some seasonal trend in these data. We then fit the multiple linear regression model

$$Y_t = \beta_0 + \beta_1 X_{t,1} + \beta_2 X_{t,2} + \epsilon_t,$$

where $t = 1, \ldots, 508$. The residuals to this fit are given in Figure 13.4(d). This plot indicates some possible autocorrelation in the residuals.

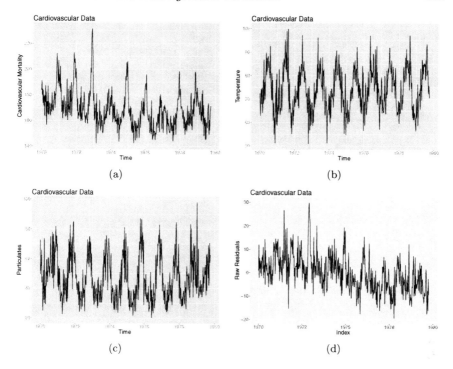

**FIGURE 13.4**
Time series plots of (a) cardiovascular mortality, (b) temperatures, (c) particulates, and (d) raw residuals for the multiple linear regression model fit of the cardiovascular data.

Figure 13.4 gives plots of the autocorrelations and partial autocorrelations for the data. The plot of the autocorrelations (Figure 13.5(a)) exhibits a nonrandom pattern and the high values at earlier lags suggest an autoregressive model. We also examine the plot of partial autocorrelations (Figure 13.5(b)). Here we notice that there is a significant spike at a lag of 1 and much lower spikes for the subsequent lags. This is assessed using the approximate 95% confidence bounds (dashed blue lines). Thus, an AR(1) model would likely be feasible for this dataset. The results from the Durbin–Watson test confirm this assessment:

```
#############################
 lag Autocorrelation D-W Statistic p-value
   1       0.5379932       0.9211727       0
 Alternative hypothesis: rho != 0
#############################
```

Moreover, the same results are obtained using the Breusch–Godfrey test:

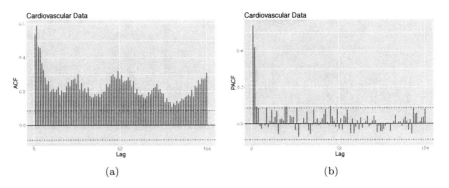

(a)                                          (b)

**FIGURE 13.5**
The (a) autocorrelations and (b) partial autocorrelations versus lag for the cardiovascular data.

```
#############################
Breusch-Godfrey test for serial correlation of order up to 1

data:  out
LM test = 151.46, df = 1, p-value < 2.2e-16
#############################
```

We also fit an AR(2) model for comparison. A likelihood ratio test between these two models indicates that the AR(1) model is appropriate:

```
#############################
      Model df    AIC      BIC    logLik    Test L.Ratio p-value
out1      1  5 3277.000 3298.152 -1633.500
out2      2  6 3173.895 3199.277 -1580.947 1 vs 2 105.105  <.0001
#############################
```

In the above, out1 and out2 are for the AR(1) and AR(2) fits, respectively.

Table 13.1 provides the maximum likelihood estimates of the regression coefficients and correlation parameters under both of these models. We also provide the generalized least squares estimates of these parameters using what is called the power variance function as the weight function; see Chapter 5 of Pinheiro and Bates (2000) for a discussion of this function. When comparing the maximum likelihood estimates versus the generalized least squares estimates, the weighting function did not change the estimates much. Other weighting functions could, however, result in more substantial changes in the estimates.

**Example 13.5.3.** *(Natural Gas Prices Data)*
Wei (2005) presented and analyzed monthly observations of spot prices for natural gas from January 1988 to October 1991 for the states of Louisiana

**TABLE 13.1**
Maximum likelihood and generalized least
squares estimates for the cardiovascular data

| Parameter | AR(1) | | AR(2) | |
|---|---|---|---|---|
| | MLE | GLS | MLE | GLS |
| $\beta_0$ | 76.134 | 73.956 | 79.613 | 80.018 |
| $\beta_1$ | 0.097 | 0.131 | 0.039 | 0.032 |
| $\beta_2$ | 0.113 | 0.105 | 0.133 | 0.134 |
| $\rho_1$ | 0.788 | 0.799 | 0.433 | 0.430 |
| $\rho_2$ | — | — | 0.435 | 0.437 |

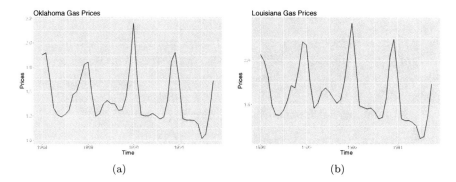

(a)          (b)

**FIGURE 13.6**
Spot prices of natural gas for (a) Oklahoma and (b) Louisiana.

and Oklahoma. We are interested in building a model of the Oklahoma spot
prices as a function of the Louisiana spot prices. The measured variables are:

- $Y$: Louisiana spot prices for natural gas (dollars per million British thermal
  units)

- $X$: Oklahoma spot prices for natural gas (dollars per million British thermal
  units)

Figure 13.6 provides time series plots of these natural gas prices, which clearly
indicate correlations in the trends.

Figure 13.7 gives plots of the autocorrelations and partial autocorrelations
for the data. An AR(1) model seems appropriate, as confirmed by the Durbin-
Watson test:

```
#############################
 lag Autocorrelation D-W Statistic p-value
  1        0.4735974     0.9323707   0.002
 Alternative hypothesis: rho != 0
#############################
```

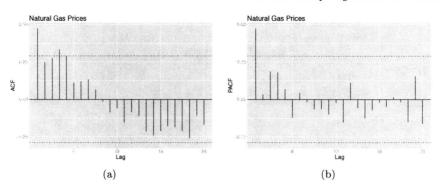

(a)                                                                 (b)

**FIGURE 13.7**
The (a) autocorrelations and (b) partial autocorrelations versus lag for the natural gas prices data.

Below is the output for the Cochrane-Orcutt procedure:

```
#############################
Coefficients:
                Estimate Std. Error t value Pr(>|t|)
XB(Intercept)   0.08111    0.06021   1.347    0.185
XBLA            0.82567    0.03596  22.963   <2e-16 ***
---
Signif. codes:  0 *** 0.001 ** 0.01 * 0.05 . 0.1   1

Residual standard error: 0.05176 on 43 degrees of freedom
Multiple R-squared:  0.9946,Adjusted R-squared:  0.9943
F-statistic:  3951 on 2 and 43 DF,  p-value: < 2.2e-16
```

```
$rho
[1] 0.5304893
#############################
```

Using the above fit, the Durbin–Watson test statistic for the first two lags are 2.102 and 1.861, respectively. These are both small test statistics and, thus, yield small $p$-values. This indicates that the Cochrane–Orcutt procedure with an AR(1) model has accurately characterized the correlation.

We can also use the Prais–Winsten procedure in an analogous manner. Below is the output:

```
#############################
[[1]]
```

```
Call:
lm(formula = fo)

Residuals:
     Min        1Q     Median        3Q       Max
-0.164697 -0.025339 -0.003329  0.024111  0.150474

Coefficients:
          Estimate Std. Error t value Pr(>|t|)
Intercept  0.06442    0.06060   1.063    0.294
LA         0.83973    0.03544  23.693   <2e-16 ***
---
Signif. codes:  0 *** 0.001 ** 0.01 * 0.05 . 0.1   1

Residual standard error: 0.05221 on 44 degrees of freedom
Multiple R-squared:  0.9945,Adjusted R-squared:  0.9942
F-statistic:  3974 on 2 and 44 DF,  p-value: < 2.2e-16


[[2]]
       Rho Rho.t.statistic Iterations
 0.5537178        4.353201          9
##############################
```

Using the above fit, the Durbin–Watson test statistic for the first two lags are 2.077 and 1.891, respectively. Just like the results from the Cochrane–Orcutt procedure, these are both small test statistics and, thus, have small $p$-values. This indicates that the Prais–Winsten procedure with an AR(1) model has accurately characterized the correlation.

Figure 13.8 shows the SSE values for different values of $\rho$ using the Hildreth–Lu procedure. The smallest SSE value is found at 0.530. Below is the output for the Hildreth–Lu procedure using $\hat{\rho} = 0.530$:

```
##############################
Coefficients:
            Estimate Std. Error t value Pr(>|t|)
(Intercept)  0.03816    0.02829   1.349    0.185
x            0.82562    0.03595  22.966   <2e-16 ***
---
Signif. codes:  0 *** 0.001 ** 0.01 * 0.05 . 0.1   1

Residual standard error: 0.05176 on 43 degrees of freedom
Multiple R-squared:  0.9246,Adjusted R-squared:  0.9229
F-statistic: 527.4 on 1 and 43 DF,  p-value: < 2.2e-16
##############################
```

Using the above fit, the Durbin–Watson test output is as follows:

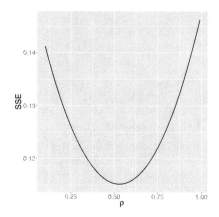

**FIGURE 13.8**
The SSE values for the Hildreth–Lu procedure.

```
##############################
 lag Autocorrelation D-W Statistic p-value
  1      -0.13827187       2.100808   0.868
  2      -0.03461041       1.860599   0.726
 Alternative hypothesis: rho[lag] != 0
##############################
```

The test for lags 1 and 2 are not significant. Thus, the Hildreth–Lu procedure for an AR(1) model has accurately characterized the correlation.

Finally, below is the output for the first differences procedure:

```
##############################
Coefficients:
          Estimate Std. Error t value Pr(>|t|)
diff(LA)  0.85601    0.03845   22.26   <2e-16 ***
---
Signif. codes:  0 *** 0.001 ** 0.01 * 0.05 . 0.1   1

Residual standard error: 0.05793 on 44 degrees of freedom
Multiple R-squared:  0.9185,Adjusted R-squared:  0.9166
F-statistic: 495.7 on 1 and 44 DF,  p-value: < 2.2e-16
##############################
```

Using the above fit, the Durbin–Watson test output is as follows:

```
##############################
 lag Autocorrelation D-W Statistic p-value
  1      -0.3576411        2.660344   0.016
  2      -0.1695991        2.199113   0.314
```

## TABLE 13.2
Smoothing parameter estimates for the four exponential smoothers

| Smoothing | Single | Double | Holt–Winters | |
|:---:|:---:|:---:|:---:|:---:|
| Parameter | Exponential | Exponential | Additive | Multiplicative |
| $\alpha$ | 1.0000 | 0.9999 | 0.2480 | 0.2756 |
| $\beta$ | 0.0032 | — | 0.0345 | 0.0327 |
| $\delta$ | — | — | 1.0000 | 0.8707 |

```
Alternative hypothesis: rho[lag] != 0
#############################
```

For this fit, it appears that there is still significant autocorrelation present. Thus, the results from this procedure have not characterized the autocorrelation as well as the other procedures for these data.

**Example 13.5.4.** *(Air Passengers Data)*
We next look at a classic dataset from Box et al. (2008). This dataset contains the monthly totals of airline passengers from 1949 to 1960. Thus, the measured variable is:

- $Y$: monthly total of airline passengers

We fit the four types of exponential smoothing models using the MSE as the minimization criterion. Each smoother is plotted with the raw data in Figure 13.9 while the smoothing parameter estimates are reported in Table 13.2. Visually, there are not many differences in the fits, but obviously different parameterizations are used depending on the smoother. Since seasonality is clearly present, we would use one of the Holt–Winters exponential smoothers. But fits for all four smoothers are provided for illustration.

Next, we fit an ARIMA(1, 1, 1) model. Below are the estimates and standard errors for the autoregressive and moving average regression coefficients:

```
#############################
Coefficients:
          ar1      ma1
      -0.4741   0.8634
s.e.   0.1159   0.0720

sigma^2 estimated as 962.2:  log likelihood = -694.34,
aic = 1394.68
#############################
```

One could proceed to try other ARIMA fits and compare their relative fits using the loglikelihood or model selection criteria. Proceeding with this model, we then forecast the next six months of international airline passengers:

```
#############################
```

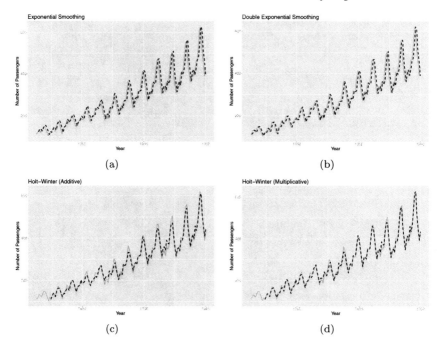

**FIGURE 13.9**
Time series plots of the air passengers data (solid blue lines) with fits (dashed black lines) for (a) single exponential smoother, (b) double exponential smoother, (c) additive Holt–Winters smoother, and (d) multiplicative Holt–Winters smoother.

```
$pred
          Jan      Feb      Mar      Apr      May      Jun
1961 475.7282 454.9963 464.8255 460.1654 462.3748 461.3273

$se
          Jan      Feb      Mar      Apr      May      Jun
1961 31.01914 53.09757 64.92951 76.30576 85.60247 94.23753
#############################
```

Obviously the standard errors increase the further out the forecasts are made. This is due to the greater uncertainty when compared to performing forecasts at a more immediate time increment.

**Example 13.5.5.** *(Mouse Liver Data)*
Menet et al. (2012) presented and analyzed nascent RNA-sequence data as a way to quantify circadian transcriptional activity in the livers of mice. While much data is available, we look at a particular sequence consisting at $t = 1, \ldots, 12$ times. Thus, the measured variable is:

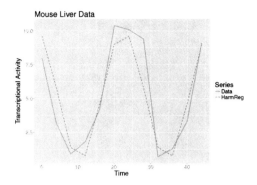

**FIGURE 13.10**
The mouse liver data (solid line) with the fitted harmonic regression model
(dashed line).

- $Y$: circadian transcriptional activity measurement

We fit a harmonic regression model to these data where the frequency $f$ is set
to 24; i.e., $f$ is not estimated. Estimates of the amplitude $R$ and phase $d$ are
given below:

```
#############################
$pars
        amp       phi
1 0.9206755 5.886998
#############################
```

The data and the harmonic regression fit are plotted in Figure 13.10. Based
solely on a visual assessment, this appears to be an appropriate model.

# 14

## *Crossvalidation and Model Selection Methods*

One problem often faced when building a regression model is that there is a large set of candidate predictor variables that could be used for predicting the response variable. The goal is to be parsimonious in that the best subset of the predictors are to be identified and unnecessary predictors do not end up in the final model. The procedures described here attempt to accomplish this, but not without a few caveats. First is that it is important to understand what these procedures are trying to tell us about the data. An informed researcher is crucial for proper utilization of these techniques. Some of these criteria will be based on how large (or small) the value is for a particular model relative to the criterion's value for other models. It is not necessarily the case that the largest (or smallest) is the best. Also, it is important to note that the procedures described here will not identify the need for transformations, so there is still the need to construct the usual diagnostic plots to assess if any regression assumptions appear to be violated.

## 14.1 Crossvalidation

As noted in Hastie et al. (2009), we might be interested in two separate goals:

1. **Model selection**: choosing the "best" model based on some performance criterion.

2. **Model assessment**: upon choosing a model, estimating its prediction error on new data.

The remainder of this chapter deals with techniques for accomplishing these goals. If we have enough data, a general strategy is to partition the data into three subsets (also called **data splitting**):

1. The **training set**, which we use to develop and estimate the model.

2. The **validation set**, which we use to estimate prediction error for model selection.

3. The **test set**, which we use to evaluate the reasonableness and predictive ability of our selected model; i.e., model assessment.

We also refer to the three sets above as the **training sample, validation sample**, and **test sample**, respectively.

However, we are often faced with data problems where we simply do not have enough sample to partition the data as described above. Instead, the step where we use the validation sample to select the "best" model is often approximated by comparing analytical measures calculated for each model (e.g., $\sqrt{\text{MSE}}$, $R^2$, or information criteria) or resampling routines.

One popular class of techniques is **crossvalidation**, which is a model evaluation technique for assessing how the results from a regression analysis (or any statistical analysis) generalize to independent datasets. Three primary variations exist for crossvalidation:

- The first is **$K$-fold crossvalidation**, which partitions the data into $K$ parts which are (roughly) equal in size. For each part, we use the remaining $K-1$ parts to estimate the model of interest (i.e., the training samples) and test the predictability of the model with the remaining part (i.e., the testing sample). We then evaluate the criterion of interest, which is done for each of the $K$ parts and then the criterion results are averaged. The disadvantage is that the computation time grows as $K$ increases.

- The second is the **holdout method of crossvalidation**, which is the case when $K = 2$ for the $K$-fold crossvalidation. The advantage of this method is that it is usually preferable to residual diagnostic methods and takes no longer to compute. However, its evaluation can have high variance since evaluation may depend on which data points end up in the training sample and which end up in the test sample.

- The third is the **leave-one-out crossvalidation**, which is the case when $K = n$ for the $K$-fold crossvalidation. That means that $n$ separate datasets are trained on all of the data (except one point) and then prediction is made for that one point. The evaluation of this method is very good, but often computationally expensive.

More generally, let $\kappa : \{1, \dots, n\} \to \{1, \dots, K\}$ be an indexing function that indicates to which of the $K$ subsets each observation $i$ is allocated. For example, if observation $i$ were assigned to the second subset, then $\kappa(i) = 2$. Let $\boldsymbol{\beta}_{(-K_h)}$ be the estimated regression coefficients when the data subset $K_h$ is excluded. The **crossvalidation estimate of prediction error** is calculated as

$$\widehat{CV} = \frac{1}{n} \sum_{i=1}^{n} (y_i - \mathbf{X}\boldsymbol{\beta}_{(-\kappa(i))})^2. \tag{14.1}$$

Intuitively, the formula for $\widehat{CV}$ says we calculate the MSE where each observation's predicted value is calculated under the fitted model from which it was excluded. Then, the lower the crossvalidation estimate of prediction error, the better we can claim that the model performs at predicting.

An important point to reiterate is that as $K \to n$, crossvalidation becomes a computationally expensive procedure, especially if the number of predictors is large and you are comparing many models. Unless you specify how to partition the data, the results for different numerical runs of $\widehat{CV}$ will differ if the computer does the partition randomly. If doing a random partition, you should set a seed for random number generation so that you can reproduce your results. If possible, try doing multiple random partitions of the data and compare the $\widehat{CV}$ to assess the sensitivity the results. See Chapter 7 of Hastie et al. (2009) for an accessible presentation of the details of crossvalidation.

## 14.2 PRESS

The **prediction sum of squares** (or **PRESS**) is a model validation method used to assess a model's predictive ability; see Allen (1974). For a dataset of size $n$, PRESS is calculated by omitting each observation individually and then the remaining $n-1$ observations are used to estimate a regression model. The estimated model is then used to estimate the value of the omitted observation, which, recall, we denote by $\hat{y}_{i(i)}$. We then calculate the $i^{\text{th}}$ PRESS residual as the difference $y_i - \hat{y}_{i(i)}$, which then gives us the following formula for PRESS:

$$\text{PRESS} = \sum_{i=1}^{n} (y_i - \hat{y}_{i(i)})^2.$$

In general, the smaller the PRESS value, the better the model's predictive ability.

PRESS can also be used to calculate the **predicted** $R^2$ (denoted by $R^2_{pred}$) which is generally more intuitive to interpret than PRESS itself. It is defined as

$$R^2_{pred} = 1 - \frac{\text{PRESS}}{\text{SSTO}}$$

and is a helpful way to validate the predictive ability of your model without selecting another sample or splitting the data into training and test datasets in order to assess the predictive ability. Together, PRESS and $R^2_{pred}$ can help prevent overfitting because both are calculated using observations not included in the model estimation. Overfitting refers to models that appear to provide a good fit for the dataset at hand, but fail to provide valid predictions for new observations.

You may notice that $R^2$ and $R^2_{pred}$ are similar in form. While they will not be equal to each other, it is possible to have $R^2$ quite high relative to $R^2_{pred}$, which implies that the fitted model is data-dependent; i.e., the model is overfitting. However, unlike $R^2$, $R^2_{pred}$ ranges from values below 0 to 1. $R^2_{pred} < 0$ occurs when the underlying PRESS gets inflated beyond the level of the correct SSTO. In such a case, we simply truncate $R^2_{pred}$ at 0.

Finally, if the PRESS value appears to be large due to a few outliers, then a variation on PRESS (using the absolute value as a measure of distance) may also be calculated:

$$\text{PRESS}^* = \sum_{i=1}^{n} |y_i - \hat{y}_{i(i)}|,$$

which also leads to

$$R^{2*}_{pred} = 1 - \frac{\text{PRESS}^*}{\text{SSTO}}.$$

---

## 14.3   Best Subset Procedures

The **best subset procedure** (sometimes referred to as **all possible regressions**) usually reports various statistics to identify the "best" few models of each possible size. By the "size" of the model, we mean the number of variables in the model. Some common statistics used and their interpretations are as follows:

$R^2$: We would like this to be as large as possible, but the difficulty is that as the size of the model increases, so does the value of $R^2$. Thus, we cannot simply choose a model based on the highest $R^2$. Instead, we look to see where $R^2$ more or less levels off after a particular size of the model, thus indicating that the added variables are unnecessary.

$R^2$-**adjusted:** The interpretation is about the same as for $R^2$. The difference is that $R^2$-adjusted can actually start decreasing when more predictors are in the model since we penalize the $R^2$ value when adding unnecessary variables. The formula for $R^2$-adjusted, which we denote by $R^2_{adj}$, is the following:

$$R^2_{adj} = \frac{\frac{\text{SSTO}}{n-1} - \frac{\text{SSE}}{n-p}}{\frac{\text{SSTO}}{n-1}}$$

$$= 1 - (n-1)\frac{\text{MSE}}{\text{SSTO}},$$

where $p$ is the number of regression parameters in the model being considered and $n$ is the sample size.

**Mallow's $C_p$:** Developed by Mallow (1973), this statistic compares the MSE for the model being considered, MSE(reduced), to the MSE for a model that includes all predictor variables, MSE(full). For a particular model, the formula is:

$$C_p = \frac{\text{SSE(reduced)}}{\text{MSE(full)}} - (n - 2p)$$

$$= (n - p)\left[\frac{\text{MSE(reduced)}}{\text{MSE(full)}} - 1\right] + p,$$

where, again, $p$ is the number of regression parameters in the model being considered and $n$ is the sample size. While $C_p$ has an expected value of $p$, a low value is desirable and models with the lowest $C_p$ values are the "best." To be a potentially good $C_p$, we should have $C_p \leq p$. Moreover, if there is a significant lack of fit for the candidate model (or bias), then there is nothing to prevent $C_p < 0$, which happens in extreme cases.

**S $= \sqrt{\text{MSE}}$:** As we saw before, this is the standard deviation of the residuals. Here, a low value is good. Usually the value will level off or possibly begin increasing when unnecessary variables are added to the model. Note that it is possible for the MSE to increase when a variable is added to a model. While the SSE cannot increase, it can happen that it may only decrease slightly and the effect of dividing by $n - p$ with $p$ increased by 1 can result in an increased MSE.

**CV:** The **coefficient of variation** (or **CV**) is a relative measure of dispersion, It is computed by taking the root mean square error and dividing by the mean of the response variable. The meaning of CV has little value by itself, but can be helpful when comparing various models for the same dataset. Mathematically, we have

$$\text{CV} = \frac{\text{S}}{\bar{y}}.$$

Ideally, we would search for models with smaller CV values relative to the other fits.

**AAPE:** The **average of the absolute percent errors** (or **AAPE**) is another measure of goodness-of-fit of the regression model to the data. Like the CV, it is usually not useful by itself, but rather when comparing various models to the same dataset. Mathematically, we have

$$\text{AAPE} = \frac{1}{n}\sum_{i=1}^{n}\left|\frac{y_i - \hat{y}_i}{y_i^*}\right|,$$

where

$$y_i^* = \begin{cases} y_i, & \text{if } y_i \neq 0; \\ \hat{y}_i, & \text{if } y_i = 0; \\ 1, & \text{if } y_i = \hat{y}_i = 0. \end{cases}$$

Ideally, we would search for models with smaller AAPE values relative to the other fits.

The method for comparing each of the above's relative performance is to plot the value of the statistic versus the number of predictors it was calculated for.

One algorithm to identify the best subset of predictors is the **McHenry's best subset algorithm** (McHenry, 1978), which seeks a subset that provides a maximum or minimum value of the specified criterion.[1] It is extremely fast and typically finds the best (or very near best) subset in most situations. Assuming we use $R^2$ as our statistic, the algorithm runs as follows:

---

McHenry's Best Subset Algorithm

1. Find the best single-variable model according to the largest $R^2$ value.

2. Find the best two-variable model by including each of the remaining variables one at a time and selects the one that adds the most to the $R^2$.

3. Omit the first variable and sequentially try each of the remaining variables in its place to determine if they improve the $R^2$. If a better variable is found, it is kept instead of the originally chosen first variable.

4. For the "better" variable, switch it with the other candidate variables. Continue this until no switching results in a better subset according to the $R^2$.

---

## 14.4  Statistics from Information Criteria

Some statistical software may also give values of statistics referred to as *information criteria* or *Bayesian information statistics*. For regression models, these statistics combine information about the SSE, number of parameters in the model, and the sample size. A low value, compared to values for other possible models, is good. Some data analysts feel that these statistics give a more realistic comparison of models than the $C_p$ statistic because $C_p$ tends to make models seem more different than they actually are.

Three information criterion that we present are called **Akaike's Information Criterion (AIC)**, the **Bayesian Information Criterion (BIC)** (which is sometimes called **Schwartz's Bayesian Criterion (SBC)**), and

---

[1] In the case of multiple responses (i.e., multivariate regression), a minimum Wilks' lambda is found. The Wilks' lambda is discussed in Chapter 22.

**Amemiya's Prediction Criterion (APC).** The respective formulas are as follows:

$$\text{AIC}_p = n \log(\text{SSE}) - n \log(n) + 2p \tag{14.2}$$

$$\text{BIC}_p = n \log(\text{SSE}) - n \log(n) + p \log(n) \tag{14.3}$$

$$\text{APC}_p = \frac{(n+p)}{n(n-p)} \text{SSE}, \tag{14.4}$$

where, again, $p$ is the number of regression parameters in the model being considered and $n$ is the sample size. Notice that the only difference between AIC and BIC is the multiplier of $p$, the number of parameters. The BIC places a higher penalty on the number of parameters in the model so will tend to reward more parsimonious (smaller) models. This leads to one criticism of AIC in that it tends to overfit models, especially in asymptotic situations; i.e., when the sample size tends to $\infty$. Refer to the text by Konishi and Kitagawa (2008) for a more detailed discussion about these and other information criteria.

## 14.5 Stepwise Procedures for Identifying Models

Stepwise methods for model selection are used in several areas of applied statistics, so it is worth knowing about them. However, in regression, the best subsets procedure will give more reliable results and should be used whenever possible. The procedures in this section should not be used as the sole decision in selecting your final model. They should either be used as an initial screening of your data or to provide additional justification for the model you plan to select.

There are a few fundamental stepwise methods used in practice. First is **forward selection**, where we begin with no predictor variables in the model and add variables, one at a time, in some optimal way. This procedure starts with no candidate variables in the model and then adds a variable to the model that increases $R^2$ the most and is still statistically significant; i.e., the $p$-value for that added predictor is less than some significance level $\alpha$. Sometimes, the criterion used is just looking at the most significant predictor and seeing if its $p$-value is less than $\alpha$. Then the procedure stops when no more predictors meet the specified criterion. This helps to provide an initial screening when you have a large number of candidate variables and the best subsets procedure would be too computationally expensive.

Second is **backward elimination**, where we begin with all potential predictor variables in a model and then remove "weak" variables, one at a time, until a desirable stopping point is reached. This procedure starts with all candidate variables in the model and then removes a variable from the model that decreases $R^2$ the most and is not statistically significant; i.e., the $p$-value for that added predictor is greater than some significance level $\alpha$. However, since the procedure works its way down in this manner, a large $R^2$ is almost always retained. We know the more variables in a model the higher the $R^2$, which does not always imply we should retain all of those variables. So backward elimination is not usually preferred. Sometimes, the criterion used is just looking at the least significant predictor and seeing if its $p$-value is greater than $\alpha$, which appears more sensible than using the $R^2$ criterion. Then the procedure stops when no more predictors are removed based on the specified criterion. A critical concept to note with both this procedure and the forward selection procedure is that each step (adding a variable or removing one) is conditional on the previous step. For instance, in forward selection we are adding a variable to those already selected.

A third method is **stepwise regression**, which combines forward selection and backward elimination.[2] The basic direction of the steps is forward (adding variables), but if a variable becomes nonsignificant, it is removed from the equation (a backward elimination). It is also important to discuss the stopping rules employed. Stopping points for stepwise regression are almost always defined in terms of $p$-values for testing $\beta_k = 0$, where $\beta_k$ is the coefficient that multiplies variable $X_k$ in a multiple regression.

- In a forward step, the best available variable is only added if its associated $p$-value is less than a specified $\alpha$-level. The procedure stops when no available variables meet this criterion.

- In a backward step, the weakest variable in the present equation is only eliminated if its associated $p$-value is greater than a specified $\alpha$-level. The procedure stops when all $p$-values are smaller than the specified level.

A fourth method is **sequential replacement regression** (Miller, 1984), which is similar to stepwise regression. In this method, you start with a certain number of predictors $q < p$. You then sequentially replace predictors such that those predictors improving performance are retained. In contrast to stepwise regression, you do not systematically remove all weak predictors at each step of the process.

## 14.6   Example

**Example 14.6.1.** *(Punting Data, cont'd)*
Let us continue with our analysis of the punting data discussed in Example 10.5.1. Recall that there are five predictors of interest. For this part of the analysis, we will run all possible regressions, forward selection, backward

---

[2]Sometimes the methods of forward selection and backward elimination are directly referred to as stepwise regression.

**TABLE 14.1**

Various statistics for the all possible regressions procedure

| Statistic | Number of Predictors | | | | |
|:---:|:---:|:---:|:---:|:---:|:---:|
| | 1 | 2 | 3 | 4 | 5 |
| S | 0.259 | 0.192 | 0.194 | 0.206 | 0.220 |
| $R^2$ | 0.743 | 0.871 | 0.882 | 0.882 | 0.882 |
| $R^2_{adj}$ | 0.720 | 0.846 | 0.842 | 0.823 | 0.798 |
| $C_p$ | 6.234 | 0.641 | 2.026 | 4.000 | 6.000 |
| $BIC_p$ | −12.553 | −18.959 | −17.484 | −14.967 | −12.403 |

elimination, and sequential replacement regression. We will retain the best fitting models consisting of 1, 2, 3, 4, and 5 predictors for comparison.

When running these procedures, we find that the all possible regressions, forward selection, and backward elimination all yield the same results. Specifically, for the best models selected for each number of predictors in the model, the predictors selected are as follows:

```
#############################
  RStr LStr RFlex LFlex  OStr
1 FALSE TRUE FALSE FALSE FALSE
2 FALSE TRUE FALSE FALSE  TRUE
3 FALSE TRUE FALSE  TRUE  TRUE
4  TRUE TRUE FALSE  TRUE  TRUE
5  TRUE TRUE  TRUE  TRUE  TRUE
#############################
```

The sequential replacement regression differs in the model selected with four predictors:

```
#############################
  RStr LStr RFlex LFlex  OStr
1 FALSE TRUE FALSE FALSE FALSE
2 FALSE TRUE FALSE FALSE  TRUE
3 FALSE TRUE FALSE  TRUE  TRUE
4  TRUE TRUE  TRUE  TRUE FALSE
5  TRUE TRUE  TRUE  TRUE  TRUE
#############################
```

Note that the intercept is included in each of the models above. These different results are also shown in Figure 14.1. Given that the all subsets regression does an exhaustive search, which will not always be feasible for big data problems, we proceed to interpret the results obtained from that method.

Table 14.1 gives the various statistics for the different models. Clearly, the model with two predictors is the best model. Specifically, it has the smallest values of S, $C_p$, and $BIC_p$ and the largest value of $R^2_{adj}$ when compared to the other models. While $R^2$ will always increase as more predictors are added, it is

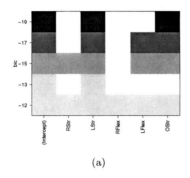
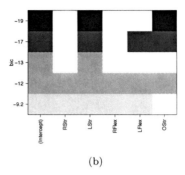

(a)                                              (b)

**FIGURE 14.1**
The BIC values for the different models fit using. (a) These are the results
for all possible regressions, forward selection, and backward elimination. (b)
These are the results for the sequential replacement regression.

clear that there are only minor increases in $R^2$ when there are more than two
predictors in the model. The output for the final model with left leg strength
and overall leg strength as predictors is given below:

```
##############################
Coefficients:
            Estimate Std. Error t value Pr(>|t|)
(Intercept) 1.10765    0.35605     3.11  0.01104 *
LStr        0.01370    0.00284     4.83  0.00069 ***
OStr        0.00429    0.00136     3.15  0.01029 *
---
Signif. codes:  0 *** 0.001 ** 0.01 * 0.05 . 0.1   1

Residual standard error: 0.192 on 10 degrees of freedom
Multiple R-squared:  0.871,Adjusted R-squared:  0.846
F-statistic: 33.9 on 2 and 10 DF,  p-value: 3.53e-05
##############################
```

We note that performing a stepwise regression also converges to the above
solution.

For the purpose of illustrating validation procedures, we will consider the
three models identified by the all possible regression procedure for the one-,
two-, and three-predictor settings. Call these M1, M2, and M3, respectively.
The PRESS and $R^2_{pred}$ values for these three models are given in Table 14.2.
Clearly, M2 has the smallest PRESS value and the largest $R^2_{pred}$ value. Thus, it
would have the better predictive ability relative to the models being compared.
The PRESS residuals are also provided in Figure 14.2.

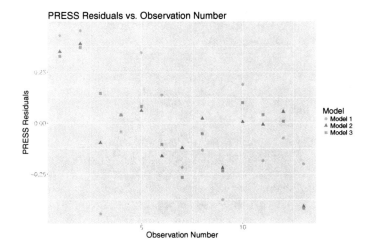

**FIGURE 14.2**
Scatterplot of PRESS residuals for the three different models.

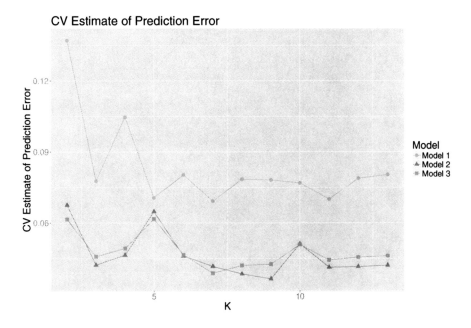

**FIGURE 14.3**
Prediction error of $K$-fold crossvalidation for the three models.

**TABLE 14.2**
PRESS and $R^2_{pred}$ values for the
three models

|           | M1    | M2    | M3    |
|-----------|-------|-------|-------|
| PRESS     | 1.046 | 0.547 | 0.599 |
| $R^2_{pred}$ | 0.635 | 0.809 | 0.791 |

Figure 14.3 gives the results for $K$-fold crossvalidation for $K = 2, \ldots, 13$. For most values of $K$, M2 tends to have the same or smaller estimate of prediction error than M3, while it always has a smaller estimate relative to M1. These results all indicate the good predictive ability of M2 relative to M1 and M3. Note that for the crossvalidation, partitioning of the data was done at random, so different partitionings will yield different, though likely comparable, numerical results.

# Part IV

# Advanced Regression Models

# 15

## Mixed Models and Some Regression Models for Designed Experiments

Designing experiments allows one to properly characterize variation of the conditions under consideration. Experimental design is a broad area in Statistics and provides researchers with tools, strategies, and knowledge to avoid shortcomings that could arise in the analysis stage. Most models in experimental design are also referred to as **ANOVA models** because the objective is to characterize significant sources of variation. The majority of these models are linear models and can also be written using a linear regression model. Experimental design is a very extensive topic, so this chapter discusses only some specific models where regression plays a more prominent role. Some classical applied texts on experimental design include Kuehl (2000), Box et al. (2005), and Montgomery (2013). A couple of theoretical texts on the topic are Pukelsheim (1993) and Bailey (2008).

## 15.1 Mixed Effects Models

Thus far, we have illustrated regression analyses on data where the predictors are observed with the response. But when considering regression in the context of ANOVA models and experimental design, the independent variables are controlled at specific levels of interest. Moreover, the independent variables are typically called **treatments** or **effects**. The treatments often represent the unique combination of different **factor levels**. These are the terms we will use throughout the remainder of this chapter. In particular, all of the regression models presented thus far have included predictors where the levels are assumed fixed. Thus, they are referred to as **fixed effects**.

Sometimes the factor levels that comprise the treatments are chosen at random from a larger population of all possible levels. The researcher's goals are then focused on drawing conclusions about the *entire* population of levels and not just those that would have been chosen if it were treated as a fixed effect. Typically, they are assumed to be drawn from a normal distribution, although more complicated distributions can be assumed. Since the levels are assumed to be drawn at random from a larger population, they will be referred to as **random effects**.

Models that include both fixed effects and random effects are called **mixed effects models** or simply **mixed models**. The standard mixed effects regression model is as follows:

$$\mathbf{Y} = \mathbf{Z}\boldsymbol{\alpha} + \mathbf{X}\boldsymbol{\beta} + \boldsymbol{\epsilon}, \tag{15.1}$$

where

- $\mathbf{Y}$ is an $n$-dimensional response vector such that $\text{E}[\mathbf{Y}] = \mathbf{X}\boldsymbol{\beta}$.

- $\boldsymbol{\alpha}$ is an unknown $q$-dimensional vector of random effects coefficients, which we take to be multivariate normally distributed with mean $\mathbf{0}$ and variance–covariance matrix $\boldsymbol{\Omega}$.

- $\mathbf{Z}$ is a known $n \times q$ design matrix relating $\mathbf{Y}$ to $\boldsymbol{\alpha}$.

- $\boldsymbol{\beta}$ is an unknown $p$-dimensional vector of fixed effects coefficients.

- $\mathbf{X}$ is a known $n \times p$ design matrix relating $\mathbf{Y}$ to $\boldsymbol{\beta}$.

- $\boldsymbol{\epsilon}$ is an $n$-dimensional random error term that is distributed independently of the $\boldsymbol{\alpha}$ (i.e., $\text{Cov}(\boldsymbol{\epsilon}, \boldsymbol{\alpha}) = \mathbf{0}$) and assumed to be multivariate normally distributed with mean $\mathbf{0}$ and variance–covariance matrix $\boldsymbol{\Sigma}$.[1]

Using the assumptions above, we then maximize the joint density of $\mathbf{Y}$ and $\mathbf{Z}$ for $\boldsymbol{\beta}$ and $\boldsymbol{\alpha}$ using the following mixed model equation:

$$\begin{pmatrix} \mathbf{X}^T\boldsymbol{\Sigma}^{-1}\mathbf{X} & \mathbf{X}^T\boldsymbol{\Sigma}^{-1}\mathbf{Z} \\ \mathbf{Z}^T\boldsymbol{\Sigma}^{-1}\mathbf{X} & \mathbf{Z}^T\boldsymbol{\Sigma}^{-1}\mathbf{Z} + \boldsymbol{\Omega}^{-1} \end{pmatrix} \begin{pmatrix} \hat{\boldsymbol{\beta}} \\ \hat{\boldsymbol{\alpha}} \end{pmatrix} = \begin{pmatrix} \mathbf{X}^T\boldsymbol{\Sigma}^{-1}\mathbf{Y} \\ \mathbf{Z}^T\boldsymbol{\Sigma}^{-1}\mathbf{Y} \end{pmatrix}. \tag{15.2}$$

Due to the Gauss–Markov Theorem (Theorem 1), we know that $\hat{\boldsymbol{\beta}}$ and $\hat{\boldsymbol{\alpha}}$ are best unbiased linear estimators and predictors for $\boldsymbol{\beta}$ and $\boldsymbol{\alpha}$, respectively. Clearly (15.2) does not yield closed-form solutions. Two primary strategies are often used. The first is to use some numerical optimization routine, like the **Expectation-Maximization (EM) algorithm** of Dempster et al. (1977). Another approach, which ensures that you achieve unbiased estimates of variance components, is to use the method of **restricted maximum likelihood** or **REML**.[2] REML uses a modified form of the likelihood function that is based on contrasts calculated from the data. More details on REML for mixed effects can be found in Pinheiro and Bates (2000). Since REML is usually unbiased, it is typically the preferred strategy for small sample sizes.

---

[1] This variance–covariance matrix is often assumed to be $\sigma^2 \mathbf{I}_n$.

[2] The "R" in REML is sometimes taken to mean "residual" or "reduced."

## 15.2 ANCOVA

Various experimental layouts use ANOVA models to analyze components of variation in the experiment. The corresponding ANOVA tables are constructed to compare the means of several levels of one or more factors. For example, a **one-way ANOVA** can be used to compare six different dosages of blood pressure pills and the mean blood pressure of individuals who are taking one of those six dosages. In this case, there is one factor (which would also be called the treatment) with six different levels. Suppose further that there are four different exercise regimens of interest in this study. Then a **two-way ANOVA** can be used since we have two factors — the dosage of the pill and the exercise regimen of the individual taking the pill. Each of the 24 unique combinations of medicine dosage and exercise regimen constitute the treatment levels. Furthermore, an interaction term can be included if we suspect that the dosage a person is taking and their exercise regimen have a combined effect on the response. As you can see, you can extend to the more general $k$-**way ANOVA** (with or without interactions) for the setting with $k$ factors. However, dealing with $k > 2$ can sometimes be challenging when interpreting the results.

Another important facet of ANOVA models is that, while they use least squares for estimation, they differ from the way categorical variables are handled in a regression model. In an ANOVA model, there is a parameter estimated for the factor level means and these are used for the linear model of the ANOVA. This differs slightly from a regression model which estimates a regression coefficient for, say, $m - 1$ indicator variables when there are $m$ levels of the categorical variable; i.e., the leave-one-out method. Also, ANOVA models utilize ANOVA tables, which are broken down by each factor. ANOVA tables for regression models simply test if the regression model has *at least* one variable which is a significant predictor of the response. More details on these differences are discussed in Kuehl (2000), Box et al. (2005), and Montgomery (2013).

When there is also a continuous variable measured with each response, which is called a **covariate** (or sometimes a **concomitant variable**), then the ANOVA model should include this continuous variable. This model is called an **analysis of covariance** (or **ANCOVA**) model. One difference in how an ANCOVA model is specified is that an interaction between the covariate and *each* factor is always tested first. The reason is that an ANCOVA is conducted to investigate the overall relationship between the response and the covariate while assuming this relationship is true for *all* groups; i.e., for *all* levels of a given factor. If, however, this relationship does differ across the groups, then the overall regression model is inaccurate. This assumption is called the assumption of **homogeneity of slopes**. It is assessed by testing for parallel slopes, which involves testing the interaction term between the covariate and each factor in the ANCOVA table. If the interaction is not statistically significant, then you can claim parallel slopes and proceed to build the ANCOVA model. If the interaction is statistically significant, then the regression model used is not appropriate and an ANCOVA model should not be used.

There are various ways to define an ANCOVA model. We present a couple of such models for a single-factor setting that could, potentially, be unbalanced. Since we are in a single-factor setting, this means that the treatment levels are simply the levels of that factor. Thus, treatment and factor can essentially be interchanged in most of the following discussion. Let the number of subjects for the $i^{\text{th}}$ treatment level be denoted by $n_i$ and the total number of cases be $N = \sum_{i=1}^{t} n_i$. The **one-way ANOVA model** with a fixed effect is given by

$$Y_{i,j} = \mu + \tau_i + \epsilon_{i,j}, \tag{15.3}$$

where

- $i = 1, \ldots, t$ is the index over the treatment levels;

- $j = 1, \ldots, n_i$ is the index over the experimental units from treatment $i$;

- $Y_{i,j}$ is the measured response of observation $j$ in treatment $i$;

- $\mu$ is the grand (or overall) mean;

- $\tau_i$ is the deviation from the grand mean due to the $i^{\text{th}}$ treatment such that $\sum_{i=1}^{t} \tau_i = 0$; and

- $\epsilon_{i,j}$ is the experimental error.

Let the response and covariate level for the $j^{\text{th}}$ case of the $i^{\text{th}}$ treatment level be given by $Y_{i,j}$ and $X_{i,j}$, respectively. We start with the ANOVA model in (15.3) and add another term reflecting the relationship between the response and the covariate. At first, one might consider approximating the following linear relationship:

$$Y_{i,j} = \mu + \tau_i + \beta X_{i,j} + \epsilon_{ij}.$$

In the above, $\beta$ is a regression coefficient for the relation between $Y_{i,j}$ and $X_{i,j}$, however, the constant $\mu$ is no longer an overall mean as in (15.3). We can make the constant an overall mean and simplify calculations by centering the covariate about the overall mean $(\bar{X}_{..})$ as in the **one-way ANCOVA model**:

$$Y_{i,j} = \mu + \tau_i + \beta(X_{i,j} - \bar{X}_{..}) + \epsilon_{ij}, \tag{15.4}$$

where all of the notation is as before, but now we also have $\beta$ as a regression coefficient for the relationship between the response and covariate and $X_{i,j}$ are observed constants.

We can also rewrite model (15.3) as a multiple linear regression model using the leave-one-out approach; i.e., one that is linear in the parameters $\mu, \tau_1, \ldots, \tau_{t-1}$. The model is as follows:

$$Y_{i,j} = \mu + \tau_1 I_{ij,1} + \tau_2 I_{ij,2} + \cdots + \tau_{t-1} I_{ij,t-1} + \epsilon_{ij}, \qquad (15.5)$$

where for each $i^* = 1, \ldots, t - 1$

$$I_{ij,i^*} = \begin{cases} 1 & \text{if case from treatment level } i^*; \\ -1 & \text{if case from treatment level } t; \\ 0 & \text{otherwise.} \end{cases}$$

Then, we can directly obtain the ordinary least squares estimates of $\mu, \tau_1, \ldots, \tau_{t-1}$. When a covariate is present, we then have the following multiple linear regression formulation:

$$Y_{i,j} = \mu + \tau_1 I_{ij,1} + \tau_2 I_{ij,2} + \cdots + \tau_{t-1} I_{ij,t-1} + \beta(X_{i,j} - \bar{X}..) + \epsilon_{ij}, \quad (15.6)$$

where $\beta$ is a regression coefficient for the relationship between the response and covariate $X_{i,j}$.

The key assumptions in ANCOVA are:

1. normality of error terms;

2. equality of error variances for different treatment levels;

3. uncorrelatedness of error terms;

4. linearity of regression relation with covariates (i.e., appropriate model specification); and

5. equality of slopes of the different treatment regression lines.

The first four assumptions are checked and corrected using the standard residual diagnostics and remedial measures. The last assumption requires us to test for parallel slopes; i.e., that all treatment levels have regression lines with the same slope $\beta$. The regression model in (15.6) can be generalized to allow for different slopes for the treatment levels by introducing cross-product interaction terms. Letting $\gamma_1, \ldots, \gamma_{r-1}$ be the regression coefficients for the interaction terms, we have our generalized model as

$$\begin{aligned} Y_{i,j} &= \mu + \tau_1 I_{ij,1} + \cdots + \tau_{t-1} I_{ij,t-1} + \beta(X_{i,j} - \bar{X}..) \\ &+ \gamma_1 I_{ij,1}(X_{i,j} - \bar{X}..) + \cdots + \gamma_{r-1} I_{ij,t-1}(X_{i,j} - \bar{X}..) + \epsilon_{ij}. \end{aligned}$$

The test for parallel slopes is then:

$$H_0 : \gamma_1 = \cdots = \gamma_{t-1} = 0$$
$$H_A : \text{at least one } \gamma_j \text{ is not } 0, \text{ for } j = 1, \ldots, t - 1.$$

We can then apply the general linear $F$-test where the model above is the full model and the model in (15.6) without the interaction terms is the reduced model. A high $p$-value indicates that we have parallel slopes (or homogeneity of slopes) and can therefore use an ANCOVA model.

We would then proceed to test for reduction in variance due to the covariate using

$$H_0 : \beta = 0$$
$$H_A : \beta \neq 0,$$

with test statistic $F^* \sim F_{1,N-t-1}$. Finally, we would test for significance of the (adjusted) treatment effects using

$$H_0 : \tau_1 = \cdots = \tau_{t-1} = 0$$
$$H_A : \text{not all } \tau_j \text{ equal } 0,$$

with test statistic $F^* \sim F_{t-1,N-t-1}$.

## 15.3    Response Surface Regression

A **response surface model (RSM)** is a method for determining a surface predictive model based on one or more variables. A response surface regression basically involves fitting a polynomial regression where the predictors have a certain structure. The response surface regression is intended to be a reasonable approximation to the true model, which is usually a complicated nonlinear surface. Response surface regression and RSMs in general are commonly used in industrial experiments. The models are used to identify values of the given factor variables to optimize a response. If each factor is measured at three or more values, then a quadratic response surface can be estimated using ordinary least squares regression. The predicted optimal value can be found from the estimated surface if the surface is shaped like a hill or valley. If the estimated surface is more complicated or if the optimum is far from the region of the experiment, then the shape of the surface can be analyzed to indicate the directions in which future experiments should be performed. Hence, RSMs should be viewed as an iterative process in order to identify regions of optimal operating conditions.

In polynomial regression, the predictors are often continuous with a large number of different values. In response surface regression, the factors (say, $k$ factors) typically represent a quantitative measure where their factor levels (say, $p$ levels) are equally spaced and established at the design stage of the experiment. This is called a $p^k$ **factorial design** because the analysis involves all of the $p^k$ different treatment combinations. Our goal is to find a polynomial approximation that works well in a specified region (ideally, a region of optimal operating conditions) of the predictor space. As an example, we may be performing an experiment with $k = 2$ factors where one of the factors is a certain chemical concentration in a mixture. The factor levels for the chemical concentration are 10%, 20%, and 30%; i.e., $p = 3$. The factors are then coded in the following way:

$$X_{i,j}^* = \frac{X_{i,j} - [\max_i(X_{i,j}) + \min_i(X_{i,j})]/2}{[\max_i(X_{i,j}) - \min_i(X_{i,j})]/2},$$

where $i = 1, \ldots, n$ indexes the sample and $j = 1, \ldots, k$ indexes the factor. For our example, assuming the chemical concentration factor's label is $j = 1$, we have

$$X_{i,1}^* = \begin{cases} \frac{10-[30+10]/2}{[30-10]/2} = -1, & \text{if } X_{i,1} = 10\%; \\[2ex] \frac{20-[30+10]/2}{[30-10]/2} = 0, & \text{if } X_{i,1} = 20\%; \\[2ex] \frac{30-[30+10]/2}{[30-10]/2} = +1, & \text{if } X_{i,1} = 30\%. \end{cases}$$

Some aspects which differentiate a response surface regression model from the general context of a polynomial regression model include:

- In a response surface regression model, $p$ is usually 2 or 3, while $k$ is usually the same value for each factor. More complex models can be developed outside of these constraints, but are typically less common.

- The factors are treated as categorical variables. Therefore, the design matrix **X** will have a noticeable pattern reflecting the way the experiment was designed.

- The number of factor levels must be at least as large as the number of factors; $p \geq k$.

- If examining a response surface with interaction terms, then the model *must* obey the hierarchy principle. This is not required of general polynomial models, although it is usually recommended.

- The number of factor levels must be greater than the order of the model; i.e., $p > h$.

- The number of observations $n$ must be greater than the number of terms in the model, including all higher-order terms and interactions. It is desirable to have a larger $n$. A rule of thumb is to have at least 5 observations per term in the model.

- Typically response surface regression models only have two-way interactions while polynomial regression models can (in theory) have $k$-way interactions.

As we change the factor levels in an RSM, it is implicitly assumed that the response changes in a regular manner that can be adequately represented by a smooth response surface. We assume that the mean of the response variable is a function of the quantitative factor levels represented by the variables

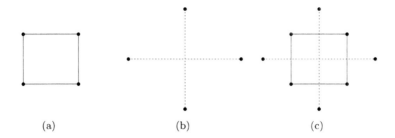

(a)                                    (b)                                    (c)

**FIGURE 15.1**
(a) The points of a square portion of a design with factor levels coded at $\pm 1$. This is how a $2^2$ factorial design is coded. (b) Illustration of the axial (or star) points of a design at $(+a,0)$, $(-a,0)$, $(0,-a)$, and $(0,+a)$. (c) A diagram which shows the combination of the previous two diagrams with the design center at $(0,0)$. This final diagram is how a composite design is coded.

$X_1, X_2, \ldots, X_k$. The true underlying function is, of course, unknown, but polynomial functions tend to provide good approximations. Below are the mean functions for the response models of interest, written in terms of $k = 2$ factors:

- **Steepest Ascent Model** (first-order model):

$$E[Y|X_1, X_2] = \beta_0 + \beta_1 x X_1 + \beta_2 X_2$$

- **Screening Response Model** (first-order model with interaction):

$$E[Y|X_1, X_2] = \beta_0 + \beta_1 X_1 + \beta_2 X_2 + \beta_{12} X_1 X_2$$

- **Optimization Model** (second-order model with interaction):

$$E[Y|X_1, X_2] = \beta_0 + \beta_1 X_1 + \beta_2 X_2 + \beta_{11} X_1^2 + \beta_{22} X_2^2 + \beta_{12} X_1 X_2$$

The response surface regression models we discussed above are for a factorial design. Figure 15.1 shows how a factorial design can be diagrammed as a square using factorial points. More elaborate designs can be constructed, such as a **central composite design**, which takes into consideration axial (or star) points, which are also illustrated in Figure 15.1. Figure 15.1 is a design with two factors while Figure 15.2 is a design with three factors.

We mentioned that response surface regression follows the hierarchy principle. However, some texts and software report ANOVA tables that do not quite follow the hierarchy principle. While fundamentally there is nothing wrong with these tables, it really boils down to a matter of terminology. If the hierarchy principle is not in place, then technically you are just performing a polynomial regression.

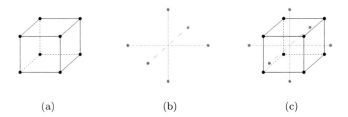

(a)    (b)    (c)

**FIGURE 15.2**
(a) The points of a cube portion of a design with factor levels coded at the corners of the cube. This is how a $2^3$ factorial design is coded. (b) Illustration of the axial (or star) points of this design. (c) A diagram which shows the combination of the previous two diagrams with the design center at (0,0). This final diagram is how a composite design is coded.

Table 15.1 gives a list of all possible terms when assuming an $h^{\text{th}}$-order response surface model with $k$ factors. For any interaction that appears in the model (e.g., $X_i^{h_1} X_j^{h_2}$ such that $h_2 \leq h_1$), then the hierarchy principle says that at least the main factor effects for $1, \ldots, h_1$ must appear in the model, that all $h_1$-order interactions with the factor powers of $1, \ldots, h_2$ must appear in the model, and all order interactions less than $h_1$ must appear in the model. Fortunately, response surface regression models (and polynomial models for that matter) rarely go beyond $h = 3$.

For the next step, an ANOVA table is usually constructed to assess the significance of the model. Since the factor levels are all essentially treated as categorical variables, the designed experiment will usually include replicates for certain factor level combinations. This is unlike multiple regression in an observational study where the predictors are usually assumed to be continuous and no predictor level combinations are assumed to be replicated. Thus, a formal lack-of-fit test is also usually performed. Furthermore, the SSR is broken down into the components making up the full model, so you can formally test the contribution of those components to the fit of your model.

An example of a response surface regression ANOVA is given in Table 15.2. Since it is not possible to compactly show a generic ANOVA table nor to compactly express the formulas, this example is for a quadratic model with linear interaction terms. The formulas will be similar to their respective quantities defined earlier. For this example, assume that there are $k$ factors, $n$ observations, $m$ unique levels of the factor level combinations, and $q$ total regression parameters are needed for the full model. In Table 15.2, the SSR can be decomposed into the following partial sums of squares values:

**TABLE 15.1**

All terms that could be included in a response surface regression
model, where the indices for the factors are $i, j = 1, \ldots, k$

| Effect | Relevant Terms |
|---|---|
| Main Factor | $X_i, X_i^2, X_i^3, \ldots, X_i^h$ for all $i$ |
| Linear Interaction | $X_i X_j$ for all $i < j$ |
| Quadratic Interaction | $X_i^2 X_j$ for $i \neq j$ <br> and $X_i^2 X_j^2$ for all $i < j$ |
| Cubic Interaction | $X_i^3 X_j, X_i^3 X_j^2$ for $i \neq j$ <br> $X_i^3 X_j^3$ for all $i < j$ |
| $\vdots$ | $\vdots$ |
| $h^{\text{th}}$-order Interaction | $X_i^h X_j, X_i^h X_j^2, X_i^h X_j^3, \ldots, X_i^h X_j^{h-1}$ for $i \neq j$ <br> $X_i^h X_j^h$ for all $i < j$ |

**TABLE 15.2**

ANOVA table for a response surface regression model with linear,
quadratic, and linear interaction terms

| Source | df | SS | MS | F |
|---|---|---|---|---|
| **Regression** | $q - 1$ | SSR | MSR | MSR/MSE |
| Linear | $k$ | SSLIN | MSLIN | MSLIN/MSE |
| Quadratic | $k$ | SSQUAD | MSQUAD | MSQUAD/MSE |
| Interaction | $q - 2k - 1$ | SSINT | MSINT | MSINT/MSE |
| **Error** | $n - q$ | SSE | MSE | |
| Lack of Fit | $m - q$ | SSLOF | MSLOF | MSLOF/MSPE |
| Pure Error | $n - m$ | SSPE | MSPE | |
| **Total** | $n - 1$ | SSTO | | |

- The sum of squares due to the linear component:

$$\text{SSLIN} = \text{SSR}(X_1, X_2, \ldots, X_k).$$

- The sum of squares due to the quadratic component:

$$\text{SSQUAD} = \text{SSR}(X_1^2, X_2^2, \ldots, X_k^2 | X_1, X_2 \ldots, X_k).$$

- The sum of squares due to the linear interaction component:

$$\text{SSINT} = \text{SSR}(X_1 X_2, \ldots, X_1 X_k, X_2 X_k, \ldots, X_{k-1} X_k | X_1, X_2 \ldots, X_k,$$
$$X_1^2, X_2^2, \ldots, X_k^2).$$

Other analysis techniques are commonly employed in response surface regression. For example, **canonical analysis**, a multivariate analysis technique,
uses the eigenvalues and eigenvectors in the matrix of second-order parameters to characterize the shape of the response surface. For example, it can help

determine if the surface is flat or have some noticeable shape like a hill or a valley. There is also **ridge analysis**, which computes the estimated ridge of optimum response for increasing radii from the center of the original design. More details on these techniques and RSMs is found in Myers et al. (2009).

## 15.4    Mixture Experiments

A **mixture experiment** is a specific response surface experiment with two or more factors that are ingredients in a mixture in which their proportions sum to 1. Therefore, levels of one factor are not independent of the levels of the other factor(s) that comprise the mixture. Typically, the ingredients in a mixture are measured in terms of weight or volume. Some examples of mixtures include:

- blend ratio of different motor fuels for a type of gasoline;

- mixture of nitrogen, phosphorus, and potassium for a particular fertilizer formulation; and

- mixture of different plasticizers for producing a durable vinyl product.

The factors in a mixture experiment are referred to as the **components** of the overall mixture. Let $k$ denote the number of components in a mixture. Then, the $k$ components are given by

$$\left\{ X_i : 0 \leq X_i \leq 1, \sum_{i=1}^{k} X_i, i = 1, \ldots, k \right\}. \tag{15.7}$$

There are two primary designs used in mixture experiments. The first is the **simplex lattice design**, where we use the designation $\{k, h\}$ to indicate that we have $k$ components and we wish to estimate a polynomial response surface model of degree $h$. The proportions assumed for each component in the mixture take the values

$$X_i = 0, \frac{1}{h}, \frac{2}{h}, \ldots, 1, \quad i = 1, \ldots, k,$$

where are all possible mixtures of the above proportions are used. In general, the number of point in a $\{k, h\}$ simplex lattice design is $N = \binom{k+h-1}{h}$. The first column of Table 15.3 gives the design points for a $\{3, 3\}$ simplex lattice design while Figure 15.3(a) shows these points on a plot using the simplex coordinate system.

The second design used is a **simplex centroid design**, which consists of mixtures containing $1, 2, \ldots, k$ components each in equal proportions, where we use the designation $\{k\}$ to indicate that we have a mixture design with

**TABLE 15.3**

Design points for a $\{3,3\}$ simplex lattice design (SLD) and a $\{3\}$ simplex centroid design (SCD)

| $\{3,3\}$ SLD | | | $\{3\}$ SCD | | |
|---|---|---|---|---|---|
| $X_1$ | $X_2$ | $X_3$ | $X_1$ | $X_2$ | $X_3$ |
| 1 | 0 | 0 | 1 | 0 | 0 |
| 0 | 1 | 0 | 0 | 1 | 0 |
| 0 | 0 | 1 | 0 | 0 | 1 |
| $\frac{2}{3}$ | $\frac{1}{3}$ | 0 | $\frac{1}{2}$ | $\frac{1}{2}$ | 0 |
| $\frac{2}{3}$ | 0 | $\frac{1}{3}$ | $\frac{1}{2}$ | 0 | $\frac{1}{2}$ |
| 0 | $\frac{2}{3}$ | $\frac{1}{3}$ | 0 | $\frac{1}{2}$ | $\frac{1}{2}$ |
| $\frac{1}{3}$ | $\frac{2}{3}$ | 0 | $\frac{1}{3}$ | $\frac{1}{3}$ | $\frac{1}{3}$ |
| $\frac{1}{3}$ | 0 | $\frac{2}{3}$ | | | |
| 0 | $\frac{1}{3}$ | $\frac{2}{3}$ | | | |
| $\frac{1}{3}$ | $\frac{1}{3}$ | $\frac{1}{3}$ | | | |

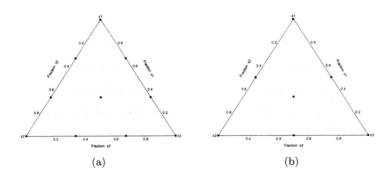

(a)                                         (b)

**FIGURE 15.3**

Use of simplex coordinate axes to illustrate a (a) $\{3,3\}$ simplex lattice design and (b) $\{3\}$ simplex centroid design.

$k$ components. Thus, for a $\{k\}$ simplex centroid design, there are total of $2^{k-1}$ points, consisting of all $k$ single-component mixtures, all possible two-component mixtures with proportions of $\frac{1}{2}$, all possible three-component mixtures with proportions of $\frac{1}{3}$, etc. The second column of Table 15.3 gives the design points for a $\{3\}$ simplex centroid design while Figure 15.3(a) shows these points on a plot using the simplex coordinate system. One drawback of such a design is that nearly all of the design points lie on the boundary of the experimental region and it is usually desirable to have more design points interior to this region. This can be accomplished by augmenting the design with additional points.

Models for mixture experiments differ from traditional RSMs due to the constraint in (15.7). To reflect this constraint, the models will not include an intercept term. Below are the mean functions for some commonly-used models in mixture experiments, where $\underline{X} = (X_1, \ldots, X_k)^{\mathrm{T}}$:

- **Linear**:

$$E[Y|\underline{X}] = \sum_{i=1}^{k} \beta_i X_i$$

- **Quadratic**:

$$E[Y|\underline{X}] = \sum_{i=1}^{k} \beta_i X_i + \sum_{i<j}^{k} \beta_{i,j} X_i X_j$$

- **Full Cubic**:

$$E[Y|\underline{X}] = \sum_{i=1}^{k} \beta_i X_i + \sum_{i<j}^{k} \beta_{i,j} X_i X_j + \sum_{i<j}^{k} \alpha_{i,j} X_i X_j (X_i - X_j)$$

$$+ \sum_{i<j<l}^{k} \beta_{i,j,l} X_i X_j X_l$$

- **Special Cubic**:

$$E[Y|\underline{X}] = \sum_{i=1}^{k} \beta_i X_i + \sum_{i<j}^{k} \beta_{i,j} X_i X_j + \sum_{i<j<l}^{k} \beta_{i,j,l} X_i X_j X_l$$

In the full cubic model, the $\alpha_{i,j}$ are still regression parameters, but use of a $\beta$ for the notation becomes cumbersome. Note that you should not confuse the terminology of the models above with the terminology used for polynomial regression models discussed earlier, especially for quadratic and cubic models. Estimation of the models above can easily be carried out using ordinary least squares. More on RSMs for mixture experiments can be found in Cornell (2002) and Myers et al. (2009).

## 15.5   Examples

**Example 15.5.1.** *(Sleep Study Data)*
Belenky et al. (2003) presented a dataset on the average reaction time per
day for 18 subjects in a sleep deprivation study. On day 0, the subjects were
allowed to have their normal amount of sleep. They were then restricted to 3
hours of sleep per night over a period of an additional 9 days. The variables
are:

- $Y$: average reaction time (in milliseconds)

- $X$: day of study

- $Z$: subject

Plots of the average reaction time versus the day for each subject are provided
in the panel plots of Figure 15.4.

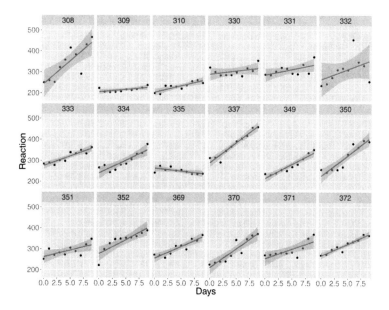

**FIGURE 15.4**
Average reaction time (milliseconds) versus days of sleep deprivation by subject. Each subject has an individual ordinary least squares line overlaid with
a 95% confidence band.

These data were analyzed by Bates et al. (2015) using a linear mixed
effects regression framework. As seen in Figure 15.4, each subject's reaction
time increases approximately linearly with the number of sleep-deprived days.

However, as noted by Bates et al. (2015), the subjects also appear to vary in the slopes and intercepts of these relationships. Thus, we can use a model with random slopes and intercepts. Fitting this model using REML, we have the following output:

```
###############################
REML criterion at convergence: 1743.6

Scaled residuals:
    Min      1Q  Median      3Q     Max
-3.9536 -0.4634  0.0231  0.4634  5.1793

Random effects:
 Groups    Name         Variance Std.Dev. Corr
 Subject  (Intercept)   612.09   24.740
          Days           35.07    5.922   0.07
 Residual               654.94   25.592
Number of obs: 180, groups:  Subject, 18

Fixed effects:
            Estimate Std. Error t value
(Intercept)  251.405      6.825   36.84
Days          10.467      1.546    6.77

Correlation of Fixed Effects:
     (Intr)
Days -0.138
###############################
```

Thus, the estimated of the standard deviations of the random effects for the intercept and the slope are about 24.74 and 5.92, respectively. The fixed-effects intercept and slope coefficients are 251.41 and 10.47, respectively, which can be interpreted as their respective estimated population mean values. Note that we can also obtain (biased) estimates using the loglikelihood as a minimization criterion:

```
###############################
     AIC     BIC  logLik deviance df.resid
  1763.9  1783.1  -876.0   1751.9      174

Scaled residuals:
    Min      1Q  Median      3Q     Max
-3.9416 -0.4656  0.0289  0.4636  5.1793

Random effects:
 Groups    Name         Variance Std.Dev. Corr
```

```
Subject  (Intercept) 565.52    23.781
         Days             32.68     5.717    0.08
  Residual                654.94    25.592
Number of obs: 180, groups:  Subject, 18

Fixed effects:
              Estimate Std. Error t value
(Intercept)   251.405      6.632   37.91
Days           10.467      1.502    6.97

Correlation of Fixed Effects:
     (Intr)
Days -0.138
###############################
```

For this dataset, the results are not much different from using the REML criterion.

**Example 15.5.2.** *(Cracker Promotion Data)*
Kutner et al. (2005) presented marketing data on the sales of crackers for a particular company. The treatment is the type of promotional strategy used, of which three were considered. $N = 15$ stores were selected for the study and each store was randomly assigned one of the promotion types; i.e., $n_i = n = 5$ stores assigned to each type of promotion. Other relevant conditions under the control of the company (e.g., price and advertising) were held constant for all stores in the study. In particular, the variables in this study are:

- $Y$: the number of cases of the crackers sold during the promotional period

- $I$: an indicator for which of the three marketing strategies was employed (treatment)

- $X$: cracker sales during the preceding sales period (covariate)

Figure 15.5 is a scatterplot of the cracker promotion data. It looks like there is an effect due to the different treatments. Moreover, the change in the response as the covariate changes appears to be similar for each treatment; i.e., parallel slopes. Thus, following the same analysis as in Kutner et al. (2005), we use a one-way ANCOVA.

Below is the ANCOVA table which includes the treatment by covariate interaction term; i.e., the interaction between promotional treatment and sales from the previous period:

```
###############################
Anova Table (Type III tests)

Response: y
            Sum Sq Df F value    Pr(>F)
```

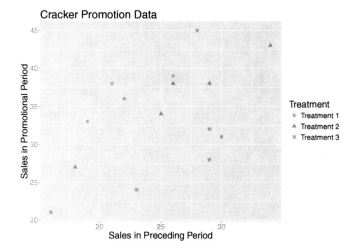

**FIGURE 15.5**
Scatterplot of the cracker promotion data.

```
(Intercept)  48.742  1 13.9172  0.004693 **
treat         1.263  2  0.1803  0.837923
x           243.141  1 69.4230 1.597e-05 ***
treat:x       7.050  2  1.0065  0.403181
Residuals    31.521  9
---
Signif. codes:  0 *** 0.001 ** 0.01 * 0.05 . 0.1   1
##############################
```

The row for the interaction gives us the results for the test of $H_0 : \beta_1 = \beta_2 = \beta_3$. This test has $F^* = 1.007 \sim F_{2,9}$, which has a $p$-value of 0.403. Therefore, this test is not statistically significant and we can claim that the assumption of equal slopes is appropriate.

Proceeding with the assumption of homogeneity of slopes, we construct the ANCOVA table for the sales during the treatment period using sales from the preceding period as a covariate:

```
##############################
Anova Table (Type III tests)
```

Response: y

| | Sum Sq | Df | F value | Pr(>F) | |
|---|---|---|---|---|---|
| (Intercept) | 66.16 | 1 | 18.867 | 0.001168 | ** |
| x | 269.03 | 1 | 76.723 | 2.731e-06 | *** |
| treat | 417.15 | 2 | 59.483 | 1.264e-06 | *** |
| Residuals | 38.57 | 11 | | | |

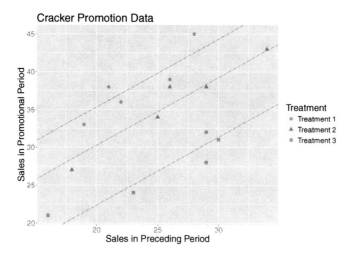

**FIGURE 15.6**
Scatterplot of the cracker promotion data with the estimated regression equations for the three promotional treatments.

```
---
Signif. codes:  0 *** 0.001 ** 0.01 * 0.05 . 0.1   1
#############################
```

For testing $H_0 : \beta = 0$, which determines the significance of the reduction in error variance due to sales from the preceding period, we have $F^* = 76.723 \sim F_{1,11}$ with a $p$-value of $2.731 \times 10^{-6}$. This is a highly significant result, therefore, the addition of the covariate has significantly reduced experimental error variability. For testing $H_0 : \tau_1 = \tau_2 = 0$, which determines equality of the adjusted treatment means, we have $F^* = 59.483 \sim F_{2,11}$ with a $p$-value of $1.264 \times 10^{-6}$. This result is also highly significant. Therefore, the treatment means adjusted for sales from the previous period are significantly different.

Below is the regression output for this ANCOVA model:

```
#############################
Coefficients:
             Estimate Std. Error t value Pr(>|t|)
(Intercept)    4.3766     2.7369   1.599    0.138
x              0.8986     0.1026   8.759 2.73e-06 ***
treat1        12.9768     1.2056  10.764 3.53e-07 ***
treat2         7.9014     1.1887   6.647 3.63e-05 ***
---
Signif. codes:  0 *** 0.001 ** 0.01 * 0.05 . 0.1   1
```

```
Residual standard error: 1.873 on 11 degrees of freedom
Multiple R-squared:  0.9403,Adjusted R-squared:  0.9241
F-statistic: 57.78 on 3 and 11 DF,  p-value: 5.082e-07
###############################
```

Finally, the estimated regression equations for the three promotional treatments are:

$$\hat{y}_{1,j} = 38.2 + 0.899(x_{1,j} - 23.2)$$
$$\hat{y}_{2,j} = 36.0 + 0.899(x_{2,j} - 26.4)$$
$$\hat{y}_{3,j} = 27.2 + 0.899(x_{3,j} - 25.4)$$

In Figure 15.6, the three estimated regression equations are overlaid on the scatterplot of the data.

**Example 15.5.3.** *(Odor Data)*
John (1971) analyzed data from an experiment that was designed to determine the effects of three factors in reducing the unpleasant odor in a chemical product being sold for household use. The variables in this experiment are:

- $Y$: measure of chemical's odor

- $X_1$: temperature at time of measurement (40, 80, and 120)

- $X_2$: gas-liquid ratio (0.3, 0.5, and 0.7)

- $X_3$: packing height (2, 4, and 6)

Each factor has three levels, but the design was not constructed as a full factorial design; i.e., it is not a $3^3$ design. Nonetheless, we can still analyze the data using a response surface regression. The factors are coded as follows:

$$X_1^* = \frac{X_1 - 80}{40}, \quad X_2^* = \frac{X_2 - 0.5}{0.2}, \quad X_3^* = \frac{X_3 - 4}{2}.$$

First we will fit a response surface regression model consisting of all of the first-order and second-order terms. The summary of this fit is given below:

```
###############################
            Estimate Std. Error t value Pr(>|t|)
(Intercept) -30.6667    10.8396 -2.8291  0.02218 *
temp.c      -12.1250     6.6379 -1.8266  0.10518
ratio.c     -17.0000     6.6379 -2.5611  0.03359 *
height.c    -21.3750     6.6379 -3.2202  0.01224 *
temp.c^2     32.0833     9.7707  3.2836  0.01113 *
ratio.c^2    47.8333     9.7707  4.8956  0.00120 **
height.c^2    6.0833     9.7707  0.6226  0.55087
---
Signif. codes:  0 *** 0.001 ** 0.01 * 0.05 . 0.1   1
```

```
Multiple R-squared:  0.8683,Adjusted R-squared:  0.7695
F-statistic: 8.789 on 6 and 8 DF,  p-value: 0.003616

Stationary point of response surface:
   temp.c   ratio.c  height.c
0.1889610 0.1777003 1.7568493

Stationary point in original units:
      temp       ratio      height
87.5584416  0.5355401  7.5136986
############################
```

In the above output, there is a stationary point reported, both in the coded and original units. This is a point that can be a minimum, maximum, or saddle point on the response surface. Regarding the rest of the output, the square of height is the least statistically significant, so we will drop that term and rerun the analysis. The summary of this new fit is given below:

```
############################
            Estimate Std. Error t value  Pr(>|t|)
(Intercept)  -26.9231      8.7068 -3.0922 0.0128842 *
temp.c       -12.1250      6.4081 -1.8921 0.0910235 .
ratio.c      -17.0000      6.4081 -2.6529 0.0263504 *
height.c     -21.3750      6.4081 -3.3356 0.0087205 **
I(temp.c^2)   31.6154      9.4045  3.3617 0.0083660 **
I(ratio.c^2)  47.3654      9.4045  5.0365 0.0007031 ***
---
Signif. codes:  0 *** 0.001 ** 0.01 * 0.05 . 0.1   1

Multiple R-squared:  0.8619,Adjusted R-squared:  0.7852
F-statistic: 11.23 on 5 and 9 DF,  p-value: 0.001169
############################
```

By omitting the square of height, the temperature main effect has now become marginally significant. Note that the square of temperature is statistically significant. Since we are building a response surface regression model, we must obey the hierarchy principle. Therefore temperature will be retained in the model.

Finally, contour and surface plots can also be generated for the estimated response surface regression model. Figures 15.7(a) and 15.7(b) are the contour and perspective plots of odor, respectively, at the middle level of height (4). Figures 15.7(c) and 15.7(d) are the contour and perspective plots of odor, respectively, at the middle level of ratio (0.5). Figures 15.7(e) and 15.7(f) are the contour and perspective plots of odor, respectively, at the middle level of temperature (80). In the first two figures, you can see approximately where a minimum occurs when height is fixed. For the other four figures, you can see a stationary ridge when ratio and temperature, respectively, are fixed.

**Example 15.5.4.** *(Yarn Fiber Data)*
Cornell (2002) discussed a mixture experiment regarding the fiber blend that will be spun into yarn to make draperies. For the experiment, the researchers are interested in if a quadratic model is necessary when considering the following variables:

- $Y$: yarn elongation

- $X_1$: polyethylene component

- $X_2$: polystyrene component

- $X_3$: polypropylene component

The design points for this mixture experiment are given in Figure 15.8, which is a $\{3, 2\}$ simplex lattice design.

We next perform a general linear $F$-test for comparing the quadratic and linear mixture experiment models. The results of this test are given below:

```
#############################
Analysis of Variance Table

Model 1: y ~ x1 + x2 + x3 - 1
Model 2: y ~ x1 + x2 + x3 + x1:x2 + x1:x3 + x2:x3 - 1
  Res.Df    RSS Df Sum of Sq      F    Pr(>F)
1     12 77.227
2      9  6.560  3    70.667 32.317 3.786e-05 ***
---
Signif. codes:  0 *** 0.001 ** 0.01 * 0.05 . 0.1   1
#############################
```

The quadratic model is found to be highly significant. The estimates for the model are given below:

```
#############################
Coefficients:
        Estimate Std. Error t value Pr(>|t|)
x1       11.7000     0.6037  19.381 1.20e-08 ***
x2        9.4000     0.6037  15.571 8.15e-08 ***
x3       16.4000     0.6037  27.166 6.01e-10 ***
x1:x2    19.0000     2.6082   7.285 4.64e-05 ***
x1:x3    11.4000     2.6082   4.371  0.00180 **
x2:x3    -9.6000     2.6082  -3.681  0.00507 **
---
Signif. codes:  0 *** 0.001 ** 0.01 * 0.05 . 0.1   1
```

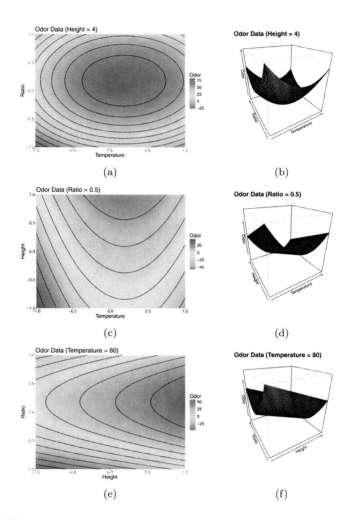

**FIGURE 15.7**
(a) Contour and (b) perspective plots of odor at the middle level of height.
(c) Contour and (d) perspective plots of odor at the middle level of ratio. (e)
Contour and (f) perspective plots of odor at the middle level of temperature.

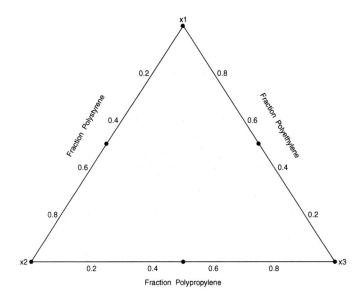

**FIGURE 15.8**
Design points for the yarn fiber mixture experiment.

```
Residual standard error: 0.8537 on 9 degrees of freedom
Multiple R-squared:  0.9977,Adjusted R-squared:  0.9962
F-statistic: 658.1 on 6 and 9 DF,  p-value: 2.271e-11
#############################
```

Note that $\hat{\beta}_3 > \hat{\beta}_1 > \hat{\beta}_2$, which tells us that mixtures with larger polypropylene components will produce higher yarn elongations.

The overall relationships can also be illustrated using a contour plot. Figure 15.9(a) is the contour plot for the linear mixture experiment model and Figure 15.9(b) is the contour plot for the quadratic mixture experiment model. If the quadratic model had not been significant, then the contour plot would have looked similar to that for the linear model. But, as shown with the test results above, the quadratic model does improve the approximation of the mixture response surface.

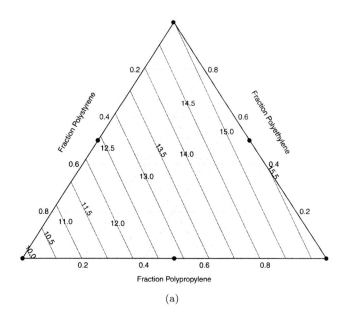

(a)

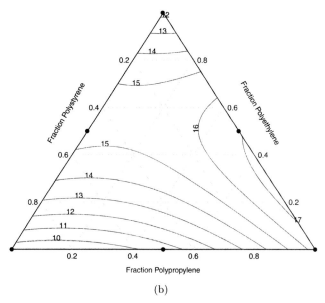

(b)

**FIGURE 15.9**
(a) Contours of the estimated yarn elongation using the linear mixture experiment model. (b) Contours of the estimated yarn elongation using the quadratic mixture experiment model.

# 16

---

## Biased Regression, Regression Shrinkage, and Dimension Reduction

---

In this chapter, we discuss various methods for reducing multicollinearity and more generally for handling a large number of predictors. We discuss some popular methods for producing biased regression estimates when faced with a high degree of multicollinearity. The assumptions made for these methods are mostly the same as in the multiple linear regression model; linearity, constant variance, and independence. Any apparent violation of these assumptions must be dealt with first. For problems with a large number of predictors, we discuss some modern dimension reduction techniques that provide a broader framework than the discrete processes presented in Chapter 14.

---

### 16.1    Regression Shrinkage and Penalized Regression

Suppose we want to find the least squares estimate of the model $\mathbf{Y} = \mathbf{X}\boldsymbol{\beta} + \boldsymbol{\epsilon}$, but subject to a set of equality constraints $\mathbf{A}\boldsymbol{\beta} = \mathbf{a}$. It can be shown by using Lagrange multipliers that

$$\hat{\boldsymbol{\beta}}_{\mathrm{CLS}} = \hat{\boldsymbol{\beta}}_{\mathrm{OLS}} - (\mathbf{X}^{\mathrm{T}}\mathbf{X})^{1}\mathbf{A}^{\mathrm{T}}[\mathbf{A}(\mathbf{X}^{\mathrm{T}}\mathbf{X})^{-1}\mathbf{A}^{\mathrm{T}}]^{-1}[\mathbf{A}\hat{\boldsymbol{\beta}}_{\mathrm{OLS}} - \mathbf{a}],$$

which is called the **constrained least squares** estimator. This is helpful when you wish to restrict $\boldsymbol{\beta}$ from being estimated in various regions of the $p$-dimensional predictor space.

However, you can also have more complicated constraints (e.g., inequality constraints, quadratic constraints, etc.) in which case more sophisticated optimization techniques need to be utilized. The constraints are imposed to restrict the range of $\boldsymbol{\beta}$ and so any corresponding estimate can be thought of as a **shrinkage estimate** as they are covering a smaller range than the ordinary least squares estimates. Oftentimes we hope to shrink our estimates to 0 by imposing certain constraints, but this may not always be possible.

Through the use of Lagrange multipliers, we can formulate the constrained minimization problems for shrinkage estimators as a **penalized sum of squares** problem:

$$\hat{\boldsymbol{\beta}}_{\mathrm{S}} = \underset{\boldsymbol{\beta}}{\operatorname{argmin}} \|\mathbf{Y} - \mathbf{X}\boldsymbol{\beta}\|^2 + P(\boldsymbol{\lambda}, \boldsymbol{\beta}), \tag{16.1}$$

where $P(\cdot)$ is a penalization function that depends on the (vector) of complexity parameters $\boldsymbol{\lambda}$ and the vector of regression coefficients $\boldsymbol{\beta}$. As we will see in the methods discussed below and for some of the models we discuss in Chapter 17, $\boldsymbol{\lambda}$ is usually a single nonnegative value.

**Ridge Regression**

In Chapter 7, we discussed multicollinearity and various ways to reduce such linear dependencies between the predictors. When multicollinearity occurs, the ordinary least squares estimates are still unbiased, but the variances are very large. However, we can add a degree of bias to the estimation process, thus reducing the variance and standard errors. This is known as the **bias-variance tradeoff** due to the functional relationship between the two values.

Perhaps one of the more popular biased regression technique to deal with multicollinearity is **ridge regression** (Hoerl and Kennard, 1970). Ridge regression is a method providing shrinkage estimators although they are biased. The estimators are obtained by using an $L_2$-penalty in the penalized sum of squares formulation in (16.1):

$$\hat{\beta}_{\text{RIDGE}} = \underset{\boldsymbol{\beta}}{\operatorname{argmin}} \|\mathbf{Y} - \mathbf{X}\boldsymbol{\beta}\|^2 + \lambda \sum_{j=1}^{p-1} \beta_j^2.$$

In other words, the constraint is $\sum_{j=1}^{p} \beta_j^2 \leq t$. Note that the intercept term $\beta_0$ is not included in this constraint and, hence, not included in the penalty term. The exclusion of $\beta_0$ becomes clear through the use of standardizing the variables to obtain the ridge regression estimates.

Recall from Chapter 7 that $\mathbf{X}^*$ and $\mathbf{Y}^*$ are the standardized versions of the $\mathbf{X}$ matrix and and $\mathbf{Y}$ vector, respectively. Remember that we have removed the column of 1s in forming $\mathbf{X}^*$. Because of this, we no longer estimate an intercept term, $\beta_0$. When using the standardized variables, the regression model of interest becomes:

$$\mathbf{Y}^* = \mathbf{X}^*\boldsymbol{\beta}^* + \boldsymbol{\epsilon}^*,$$

where $\boldsymbol{\beta}^*$ is now a $(p-1)$-dimensional vector of standardized regression coefficients and $\boldsymbol{\epsilon}^*$ is an $n$-dimensional vector of errors pertaining to this standardized model. Thus, the least squares estimates are

$$\hat{\boldsymbol{\beta}}^* = (\mathbf{X}^{*\text{T}}\mathbf{X}^*)^{-1}\mathbf{X}^{*\text{T}}\mathbf{Y}^*$$
$$= \mathbf{r}_{XX}^{-1}\mathbf{r}_{XY},$$

where $\mathbf{r}_{XX}$ is the $(p-1) \times (p-1)$ correlation matrix of the predictors and $\mathbf{r}_{XY}$ is the $(p-1)$-dimensional matrix of correlation coefficients between the predictors and the response. Moreover, $\hat{\boldsymbol{\beta}}^*$ is a function of correlations,

$$\text{E}(\hat{\boldsymbol{\beta}}^*) = \hat{\boldsymbol{\beta}}^*,$$

and

$$\text{Var}(\hat{\boldsymbol{\beta}}^*) = \sigma^2 \mathbf{r}_{XX}^{-1} = \mathbf{r}_{XX}^{-1}.$$

For the variance–covariance matrix, $\sigma^2 = 1$ because we have standardized all of the variables.

In ridge regression, the complexity parameter $\lambda$ is also called a **biasing constant**, which is added to the diagonal elements of the correlation matrix. Recall that a correlation matrix has 1s down the diagonal, so it can sort of be thought of as a *ridge*. Mathematically, we have

$$\tilde{\boldsymbol{\beta}} = (\mathbf{r}_{XX} + \lambda \mathbf{I}_{(p-1)})^{-1} \mathbf{r}_{XY},$$

where $0 < \lambda < 1$, but is usually less than 0.3. The amount of bias in this estimator is given by

$$\text{E}(\tilde{\boldsymbol{\beta}} - \boldsymbol{\beta}^*) = \left[(\mathbf{r}_{XX} + \lambda \mathbf{I}_{(p-1)})^{-1} \mathbf{r}_{XX} - \mathbf{I}_{(p-1)}\right] \boldsymbol{\beta}^*,$$

and the variance–covariance matrix is given by

$$\text{Var}(\tilde{\boldsymbol{\beta}}) = (\mathbf{r}_{XX} + \lambda \mathbf{I}_{(p-1)})^{-1} \mathbf{r}_{XX} (\mathbf{r}_{XX} + \lambda \mathbf{I}_{(p-1)}^{-1}).$$

Since $\tilde{\boldsymbol{\beta}}$ is calculated on the standardized variables, these are called the **standardized ridge regression estimates**. We can transform back to the original scale, which are called the **ridge regression estimates**, by

$$\tilde{\beta}_j^\dagger = \left(\frac{s_Y}{s_{X_j}}\right) \tilde{\beta}_j$$

$$\tilde{\beta}_0^\dagger = \bar{Y} - \sum_{j=1}^{p-1} \tilde{\beta}_j^\dagger \bar{X}_j,$$

where $j = 1, 2, \ldots, p - 1$.

How do we choose $\lambda$? Many methods exist, but there is no agreement on which to use, mainly due to instability in the estimates asymptotically. Two analytical methods and a graphical method are primarily used. The first analytical method is called the **fixed point method** (Hoerl et al., 1975) and uses the estimates provided by fitting the correlation transformation via ordinary least squares. This method suggests using

$$\lambda = \frac{(p-1)\text{MSE}^*}{\hat{\boldsymbol{\beta}}^{*\text{T}} \hat{\boldsymbol{\beta}}^*},$$

where MSE$^*$ is the mean square error obtained from the respective fit.

The second analytical method is the **Hoerl–Kennard iterative method** (Hoerl and Kennard, 1975). This method calculates

$$\lambda^{(t)} = \frac{(p-1)\text{MSE}^*}{\tilde{\boldsymbol{\beta}}_{\lambda^{(t-1)}}^{\text{T}} \tilde{\boldsymbol{\beta}}_{\lambda^{(t-1)}}},$$

where $t = 1, 2, \ldots$. Here, $\tilde{\boldsymbol{\beta}}_{\lambda^{(t-1)}}$ pertains to the ridge regression estimates obtained when the biasing constant is $\lambda^{(t-1)}$. This process is repeated until the difference between two successive estimates of $\lambda$ is negligible. The starting value for this method ($\lambda^{(0)}$) is chosen to be the value of $\lambda$ calculated using the fixed point method.

Perhaps the most common method is a graphical method. The **ridge trace** is a plot of the estimated ridge regression coefficients versus $\lambda$. The value of $\lambda$ is picked where the regression coefficients appear to have stabilized. The smallest value of $\lambda$ is chosen as it introduces the smallest amount of bias.

There are criticisms regarding ridge regression. One major criticism is that ordinary inference procedures are not available since exact distributional properties of the ridge estimator are not known. Another criticism is in the subjective choice of $\lambda$. While we mentioned a few of the methods here, there are numerous methods found in the literature, each with its own limitations. On the flip-side of these arguments lie some potential benefits of ridge regression. For example, it can accomplish what it sets out to do, and that is to reduce multicollinearity. Also, occasionally ridge regression can provide an estimate of the mean response which is good for new values that lie outside the range of our observations; i.e., extrapolation. The mean response found by ordinary least squares is known not to be good for extrapolation.

## LASSO

Another common regression shrinkage procedure is **least absolute shrinkage and selection operator** or **LASSO**. LASSO is also concerned with the setting of finding the least squares estimate subject to the set of inequality constraints $\sum_{j=1}^{p} |\beta_j| \leq t$. This is an $L_1$-penalty since we are looking at an $L_1$-norm.[1] Here, $t \geq 0$ is a tuning parameter which the user sets to control the amount of shrinkage. If we let $\hat{\boldsymbol{\beta}}$ be the ordinary least squares estimate and let $t_0 = \sum_{j=1}^{p} |\beta_j|$, then values of $t < t_0$ will cause shrinkage of the solution toward 0 and some coefficients may be exactly 0. Because of this, LASSO is also accomplishing model (or subset) selection, as we can omit those predictors from the model whose coefficients become exactly 0.

The penalized sum of squares formulation for the LASSO using (16.1) is

$$\hat{\beta}_{\text{LASSO}} = \underset{\boldsymbol{\beta}}{\text{argmin}} \|\mathbf{Y} - \mathbf{X}\boldsymbol{\beta}\|^2 + \lambda \sum_{j=1}^{p-1} |\beta_j|.$$

Note the similarity to the ridge regression penalized sum of squares problem. The only difference is that we use the $L_1$-penalty $\sum_{j=1}^{p-1} |\beta_j|$ for the LASSO and the $L_2$-penalty $\sum_{j=1}^{p-1} (\beta_j)^2$ for ridge regression. As noted in Hastie et al. (2009), the constraint for the LASSO makes the solutions nonlinear in $\mathbf{Y}$, so there is no closed-form expression as in ridge regression.

---

[1] Ordinary least squares minimizes with respect to the equality constraint $\sum_{i=1}^{n} e_i^2$, which is an $L_2$-penalty.

It is important to restate the purposes of LASSO. Not only does it shrink the regression estimates, but it also provides a way to accomplish subset selection. Furthermore, ridge regression also serves this dual purpose, although we introduced ridge regression as a way to deal with multicollinearity and not as a first line effort for shrinkage. The way subset selection is performed using ridge regression is by imposing the inequality constraint $\sum_{j=1}^{p} \beta_j^2 \leq t$, which is an $L_2$-penalty. Many competitors to LASSO are available in the literature (such as regularized least absolute deviation and Dantzig selectors), but LASSO is one of the more commonly used methods. Computing the LASSO solution is a quadratic programming problem. It should be noted that there are numerous efficient algorithms available for estimating the LASSO and its competitors. We refer to the text by Hastie et al. (2015) for a thorough treatment of such algorithms.

**Elastic Net Regression**

Zou and Hastie (2005) note that the penalty function used for the LASSO has several limitation, such as if a group of predictors are highly correlated, then the LASSO will tend to select only one variable from that group and ignore the others. To overcome such issues, they proposed **elastic net regression** (also known as **Tikhonov regularization**), which combines the ridge regression and LASSO penalties into the following penalization sums of squares problem:

$$\hat{\beta}_{\text{ENET}} = \underset{\beta}{\text{argmin}} \|\mathbf{Y} - \mathbf{X}\beta\|^2 + \lambda_1 \sum_{j=1}^{p-1} |\beta_j| + \lambda_2 \sum_{j=1}^{p-1} \beta_j^2.$$

The above problem has a unique minimum since the inclusion of the quadratic penalty term makes it a convex optimization problem. Moreover, note that ridge regression and LASSO are both special cases of the elastic net regression problem. When $\lambda_1 = 0$, we have ridge regression. When $\lambda_2 = 0$, we have the LASSO.

**LARS**

Another approach used in variable selection, which is commonly discussed in the context of data mining, is **least angle regression** or **LARS** (Efron et al., 2004). LARS is a stagewise procedure that uses a simple mathematical formula to accelerate the computations relative to other variable selection procedures that we have discussed. Only $p$ steps are required for the full set of solutions, where $p$ is the number of predictors. Following the general description in Efron et al. (2004), the LARS procedure starts with all coefficients equal to zero and then finds the predictor most correlated with the response, say $\mathbf{X}_{j_1}$. Next, take the largest step possible in the direction of this predictor until another predictor, say $\mathbf{X}_{j_2}$, has as much correlation with the current residual. LARS then proceeds in a direction equiangular between the two predictors until a third variable, say $\mathbf{X}_{j_3}$, is identified which has as much correlation with the current residual. LARS then proceeds equiangularly between $\mathbf{X}_{j_1}$, $\mathbf{X}_{j_2}$, and

$\mathbf{X}_{j_3}$ — along the **least angle direction** — until a fourth variable enters. This continues until all $p$ predictors have entered the model and then the analyst studies these $p$ models to determine which yields an appropriate level of parsimony.

**Forward Stagewise Regression**

A related methodology to LARS is **forward stagewise regression** (Efron et al., 2004), which has some parallels with the forward selection procedure discussed in Chapter 14. Forward stagewise regression starts by taking the residuals between the response values and their mean; i.e., all of the regression slopes are set to 0. Call this vector $\mathbf{r}$. Then, find the predictor most correlated with $\mathbf{r}$, say $\mathbf{X}_{j_1}$. Update the regression coefficient $\beta_{j_1}$ by setting $\beta_{j_1} = \beta_{j_1}^* + \delta_{j_i}$, where $\delta_{j_i} = \epsilon \times \text{corr}(\mathbf{r}, \mathbf{X}_{j_1})$ for some $\epsilon > 0$ and $\beta_{j_1}^*$ is the old value of $\beta_{j_1}$. Finally, update $\mathbf{r}$ by setting it equal to $(\mathbf{r}^* - \delta_{j_1} \mathbf{X}_{j_1})$, such that $\mathbf{r}^*$ is the old value of $\mathbf{r}$. Repeat this process until no predictor has a correlation with $\mathbf{r}$.

LARS and forward stagewise regression are very computationally efficient. In fact, a slight modification to the LARS algorithm can calculate all possible LASSO estimates for a given problem. Moreover, a different modification to LARS efficiently implements forward stagewise regression. In fact, the acronym for LARS includes an "S" at the end to reflect its connection to LASSO and forward stagewise regression.

## 16.2    Principal Components Regression

The method of **principal components regression** transforms the predictor variables to their principal components and then uses those transformed variables to help data that suffer from multicollinearity. Principal components of $\mathbf{X}^{*T}\mathbf{X}^*$ are extracted using the **singular value decomposition** (**SVD**) method, which says there exist orthogonal matrices $\mathbf{U}_{n \times (p-1)}$ and $\mathbf{P}_{(p-1) \times (p-1)}$ (i.e., $\mathbf{U}^T\mathbf{U} = \mathbf{P}^T\mathbf{P} = \mathbf{I}_{(p-1)}$) such that

$$\mathbf{X}^* = \mathbf{UDP}^T.$$

$\mathbf{P}$ is called the **(factor) loadings matrix** while the **(principal component) scores matrix** is defined as

$$\mathbf{Z} = \mathbf{UD},$$

such that

$$\mathbf{Z}^T\mathbf{Z} = \Lambda.$$

Here, $\Lambda$ is a $(p-1) \times (p-1)$ diagonal matrix consisting of the nonzero eigenvalues of $\mathbf{X}^{*T}\mathbf{X}^*$ on the diagonal. Without loss of generality, we assume that the eigenvalues are in decreasing order down the diagonal: $\lambda_1 \geq \lambda_2 \geq \ldots \geq$

$\lambda_{p-1} > 0$. Notice that $\mathbf{Z} = \mathbf{X}^*\mathbf{P}$, which implies that each entry of the $\mathbf{Z}$ matrix is a linear combination of the entries of the corresponding column of the $\mathbf{X}^*$ matrix. This is because the goal of principal components is to only keep those linear combinations which help explain a larger amount of the variation as determined by using the eigenvalues described below.

Next, we regress $\mathbf{Y}^*$ on $\mathbf{Z}$. The model is

$$\mathbf{Y}^* = \mathbf{Z}\boldsymbol{\beta} + \boldsymbol{\epsilon}^*,$$

which has the least squares solution

$$\hat{\boldsymbol{\beta}}_{\mathbf{Z}} = (\mathbf{Z}^T\mathbf{Z})^{-1}\mathbf{Z}^T\mathbf{Y}^*.$$

Severe multicollinearity is identified by very small eigenvalues and, thus, can be corrected by omitting those components which have small eigenvalues. This guides us in keeping only those linear combinations which are (hopefully) meaningful. Since the $i^{\text{th}}$ entry of $\hat{\boldsymbol{\beta}}_{\mathbf{Z}}$ corresponds to the $i^{\text{th}}$ component, simply set those entries of $\hat{\boldsymbol{\beta}}_{\mathbf{Z}}$ to 0 which have correspondingly small eigenvalues. For example, suppose you have 10 predictors and, hence, 10 principal components. You find that the last three eigenvalues are relatively small and decide to omit these three components. Therefore, you set the last three entries of $\hat{\boldsymbol{\beta}}_{\mathbf{Z}}$ equal to 0.

With this value $\hat{\boldsymbol{\beta}}_{\mathbf{Z}}$, we can transform back to get the coefficients on the $\mathbf{X}^*$ scale by

$$\hat{\boldsymbol{\beta}}_{\text{PC}} = \mathbf{P}\hat{\boldsymbol{\beta}}_{\mathbf{Z}}.$$

This is a solution to

$$\mathbf{Y}^* = \mathbf{X}^*\boldsymbol{\beta}^* + \boldsymbol{\epsilon}^*.$$

Notice that we have not reduced the dimension of $\hat{\boldsymbol{\beta}}_{\mathbf{Z}}$ from the original calculation, but we have only set certain values equal to 0. Furthermore, as in ridge regression, we can transform back to the original scale by

$$\hat{\beta}^{\dagger}_{\text{PC},j} = \left(\frac{s_Y}{s_{X_j}}\right)\hat{\beta}_{\text{PC},j}$$

$$\hat{\beta}^{\dagger}_{\text{PC},0} = \bar{y} - \sum_{j=1}^{p-1}\hat{\beta}^{\dagger}_{\text{PC},j}\bar{X}_j,$$

where $j = 1, 2, \ldots, p-1$.

How do you choose the number of eigenvalues to omit? This can be accomplished by looking at the cumulative percent variation explained by each of the $(p-1)$ components. For the $j^{\text{th}}$ component, this percentage is

$$\frac{\sum_{i=1}^{j}\lambda_i}{\lambda_1 + \lambda_2 + \ldots + \lambda_{p-1}} \times 100\%,$$

where $j = 1, 2, \ldots, p-1$. Remember, the eigenvalues are in decreasing order.

A common rule of thumb is that once you reach a component that explains roughly $80\% - 90\%$ of the variation, then you can omit the remaining components.

Principal components only depends on the covariance or correlation matrix of $\mathbf{X}$, the matrix of predictors. We do not make any distributional assumption, such as multivariate normality.[2] There are a couple of different structures we might assume:

- If we explicitly assume that the data do not follow a multivariate normal distribution and we want a linear transformation that provides statistical independence between its components, then we can use **independent component analysis** (Jutten and Herault, 1991; Comon, 1994; Hyvärinen and Oja, 2000). The procedure for independent component analysis is similar to principal component analysis, but with different decompositions applied to give the independent scores matrix.

- Another approach is to use latent variables, as discussed with SEMs in Chapter 11, to represent the variables in $\mathbf{X}$. This inclusion of additional randomness results in a different factor loadings matrix and a factor scores matrix.

After obtaining the transformations under either of these multivariate procedures, we can then use those transformed variables as predictors in an analogous manner to the principal components regression procedure discussed above. The resulting procedures are called **independent component regression** and **factor analysis regression**. For more details on these regressions, see Westad (2005) and Basilevsky (1981), respectively.

## 16.3   Partial Least Squares

We next look at a procedure that is very similar to principal components regression. Here, we will attempt to construct the $\mathbf{Z}$ matrix from the last section in a different manner such that we are still interested in models of the form

$$\mathbf{Y}^* = \mathbf{Z}\boldsymbol{\beta} + \boldsymbol{\epsilon}^*.$$

Notice that in principal components regression that the construction of the linear combinations in $\mathbf{Z}$ do not rely whatsoever on the response $\mathbf{Y}^*$. Yet, we use the estimate $\hat{\boldsymbol{\beta}}_{\mathbf{Z}}$ (from regressing $\mathbf{Y}^*$ on $\mathbf{Z}$) to help us build our final estimate. The method of **partial least squares** allows us to choose the linear combinations in $\mathbf{Z}$ such that they predict $\mathbf{Y}^*$ as best as possible. We proceed to describe a common way to estimate with partial least squares.

---

[2]It is worth noting that principal components derived for multivariate normal populations provide interpretations of density ellipsoids.

First, define

$$\mathbf{SS}^T = \mathbf{X}^{*T}\mathbf{Y}^*\mathbf{Y}^{*T}\mathbf{X}^*.$$

We construct the score vectors (i.e., the rows of $\mathbf{Z}$) as

$$\mathbf{z}_i = \mathbf{X}^*\mathbf{r}_i,$$

for $i = 1, \ldots, p-1$. The challenge becomes to find the $\mathbf{r}_i$ values. $\mathbf{r}_1$ is just the first eigenvector of $\mathbf{SS}^T$. For $i = 2, \ldots, p-1$, $\mathbf{r}_i$ maximizes

$$\mathbf{r}_{i-1}^T\mathbf{SS}^T\mathbf{r}_i,$$

subject to the constraint

$$\mathbf{r}_{i-1}^T\mathbf{X}^{*T}\mathbf{X}^*\mathbf{r}_i = \mathbf{z}_{i-1}^T\mathbf{z}_i = 0.$$

Next, we regress $\mathbf{Y}^*$ on $\mathbf{Z}$, which has the least squares solution

$$\hat{\beta}_{\mathbf{Z}} = (\mathbf{Z}^T\mathbf{Z})^{-1}\mathbf{Z}^T\mathbf{Y}^*.$$

As in principal components regression, we can transform back to get the coefficients on the $\mathbf{X}^*$ scale by

$$\hat{\beta}_{\text{PLS}} = \mathbf{R}\hat{\beta}_{\mathbf{Z}},$$

which is a solution to

$$\mathbf{Y}^* = \mathbf{X}^*\beta^* + \epsilon^*.$$

In the above, $\mathbf{R}$ is the matrix where the $i^{\text{th}}$ column is $\mathbf{r}_i$. Furthermore, as in both ridge regression and principal components regression, we can transform back to the original scale by

$$\hat{\beta}_{\text{PLS},j}^\dagger = \left(\frac{s_Y}{s_{X_j}}\right)\hat{\beta}_{\text{PLS},j}$$

$$\hat{\beta}_{\text{PLS},0}^\dagger = \bar{y} - \sum_{j=1}^{p-1}\hat{\beta}_{\text{PLS},j}^\dagger\bar{X}_j,$$

where $j = 1, 2, \ldots, p-1$.

The method described above is a version of the **SIMPLS** algorithm (de Jong, 1993). Another commonly used algorithm is **nonlinear iterative partial least squares (NIPALS)**. NIPALS is also more commonly used when you have a multivariate responses for each observation. While we do not discuss the differences between these algorithms any further, we do discuss regression models when we have multivariate responses in Chapter 22.

One final note concerns the connection between some of the methods we have presented. Strictly for the purpose of prediction, Stone and Brooks (1990) formulated a general method that synthesizes the methods of ordinary least squares, partial least squares, and principal components regression. The method is an integrated procedure that treats ordinary least squares and principal components regression at opposite ends of a continuous spectrum, with partial least squares lying in between. The procedure involve selecting the values of two control parameters via crossvalidation:

1. $\alpha \in [0,1]$, where the values of 0, $\frac{1}{2}$, and 1 correspond to ordinary least squares, partial least squares, and principal components regression, respectively, and

2. $\omega$, which is the total number of predictors accepted.

Thus, we can identify an optimal regression procedure to use based on the values obtained by crossvalidation of the above control parameters. Stone and Brooks (1990) refer to this procedure as **continuum regression**.

## 16.4    Other Dimension Reduction Methods and Sufficiency

Thus far, we have discussed methods that, essentially, involve transformations of a large number of predictors to reduce the number of predictors, but such that they are still informative about the response. Generally speaking, these methods can all be considered a form of dimension reduction. This section deals with what are, perhaps, more modern methods for dimension reduction. Significant development of these methods has occurred since around 1990. We will use a lot of the common terminology and common notation used for these methods; cf. Duan and Li (1991), Li (1991), Cook and Weisberg (1991), Cook (1998a), Cook and Li (2004), and Li and Wang (2007). Some of the details for these methods require more linear algebra than the methods we have discussed thus far. Refer to Appendix B for a refresher on some of the basic vector space concepts used in the discussions below.

One reason for dimension reduction is to alleviate problems caused by the **curse of dimensionality** (Bellman, 1961), which means that estimation and inference on the same number of data points in higher-dimensional spaces becomes difficult due to the sparsity of the data in the volume of the higher-dimensional spaces compared to the volume of the lower-dimensional space. For example, consider 100 points on the unit interval [0,1], then imagine 100 points on the unit square $[0,1] \times [0,1]$, then imagine 100 points on the unit cube $[0,1] \times [0,1] \times [0,1]$, and so on. As the dimension increases, the sparsity of the data makes it more difficult to make any relevant inferences about the data. The curse of dimensionality can also result in some methods having a computational cost that increases exponentially with the dimension.

For this section, let $\underline{X}$ be a $p$-dimensional vector where we *do not* assume that the first entry is a 1 for an intercept, but we do assume that $p$ is quite large with respect to $n$; i.e., $p >> n$. Just like our previous regression problems, we are interested in the conditional distribution of $Y|\underline{X}$. Mathematically, we define a **dimension reduction** as a function $\mathcal{R}(\underline{X})$ that maps $\underline{X}$ to a $k$-dimensional subset of the reals such that $k < p$. Furthermore, we say that a dimension reduction is **sufficient** if the distribution of $Y|\mathcal{R}(\underline{X})$ is the same as that for $Y|\underline{X}$. For $Y|\underline{X}$, the **structural dimension**, $d$, is the smallest dimension reduction that is still sufficient and maps $\underline{X}$ to a $d$-dimensional subspace. Many of the methods we discuss below combine these notions of dimension reduction and sufficiency, which are called **sufficient dimension reduction** methods.

Let $\mathbf{B} = (\boldsymbol{\beta}_1, \ldots, \boldsymbol{\beta}_k)$ be a $p \times k$ matrix with rank $k \le p$, where the $\boldsymbol{\beta}_h$, $h = 1, \ldots, k$ are unknown projection vectors for the model

$$Y = f(\mathbf{B}^{\mathrm{T}}\underline{X}, \epsilon).$$

In the above, $\mathrm{E}(\epsilon|\underline{X})$ is random error and $Y$ only depends on $\underline{X}$ through a $k$-dimensional subspace. The unknown $\boldsymbol{\beta}_h$ which span this space are called the **effective dimension reduction directions** (or **EDR-directions**) while the corresponding space they span is called the **effective dimension reduction space** (or **EDR-space**). We use $\mathcal{S}(\mathbf{B})$ to denote the subspace spanned by $\mathbf{B}$.

Now suppose that $Y$ is conditionally independent of $\underline{X}$ given $X|\mathbf{B}^{\mathrm{T}}X$, written as

$$Y \perp\!\!\!\perp \underline{X}|\mathbf{B}^{\mathrm{T}}\underline{X}.$$

Then, the corresponding subspace $\mathcal{S}(\mathbf{B})$ is called the **dimension reduction subspace** or **DRS**. If we are interested in the mean $E(Y|\underline{X})$, then the subspace for

$$Y \perp\!\!\!\perp E(Y|\underline{X})|\mathbf{B}^{\mathrm{T}}\underline{X}$$

is called the **mean DRS**. The intersection of all DRSs is called a **central subspace** (**CS**) and is written as $\mathcal{S}_{Y|\underline{X}}$. The intersection of all mean DRSs is called a **central mean subspace** (**CMS**) and is written as $\mathcal{S}_{\mathrm{E}(Y|\underline{X})}$. For estimating the CS or the CMS, the following two conditions due to Cook and Li (2002) are assumed for many of the dimension reduction methods:

1. **Linearity condition**: $\mathrm{E}(\underline{X}|\mathbf{P}_\mathcal{S}\underline{X})$ is a linear function of $\underline{X}$,

2. **Constant variance–covariance matrix condition**: $\mathrm{Var}(\underline{X}|\mathbf{P}_\mathcal{S}\underline{X})$ is a non-random matrix,

where $\mathbf{P}_\mathcal{S}$ is a projection matrix onto the subspace $\mathcal{S}$, which is either $\mathcal{S}_{Y|\underline{X}}$ or $\mathcal{S}_{\mathrm{E}(Y|\underline{X})}$ for estimating the CS or CMS, respectively. For more details on the CS and CMS, see Cook (1998a) and Cook and Li (2002), respectively.

We do not present details of the algorithms typically employed for the respective method, but cite the primary references for those algorithms. We note, however, that algorithms for most of the dimension reduction methods have a similar flavor as the steps listed for the methods we already discussed in this chapter. For each method below, we give a brief overview that includes limitations, benefits, and if the method estimates the CS or CMS.

**(Sliced) Inverse Regression**
In our preliminary discussion, we emphasized our interest in $E(Y|\underline{X})$, which is a $p$-dimensional surface. **Inverse regression** works with the curve computed by $E(\underline{X}|Y)$, which consists of $p$ one-dimensional regressions. In fact, it is more convenient to use the centered inverse regression curve

$$E(\underline{X}|Y) - E(\underline{X}),\qquad\qquad(16.2)$$

which, under certain conditions (Li, 1991), lies in a $k$-dimensional subspace spanned by the vectors $\Sigma_{\underline{X}}\beta_h$, $h = 1,\dots,k$, where $\Sigma_{\underline{X}}$ is the variance–covariance matrix of $\underline{X}$. **Sliced inverse regression**, or **SIR** (Duan and Li, 1991; Li, 1991), provides a crude estimate of the centered inverse regression curve in (16.2) by dividing the range of $Y$ into $H$ non-overlapping intervals called **slices**, which are then used to compute the sample means of each slice. The algorithm for SIR is used to find the EDR-directions using the standardized version of $\underline{X}$:

$$\underline{Z} = \Sigma_{\underline{X}}^{-1/2}(\underline{X} - E(\underline{X})),\qquad\qquad(16.3)$$

which implies that $\mathcal{S}_{Y|\underline{Z}} = \Sigma_{\underline{X}}^{1/2}\mathcal{S}_{Y|\underline{X}}$. SIR was designed to estimate the EDR-space. However, under the linearity condition stated earlier, SIR estimates the CS; see discussions in Li and Wang (2007) and Adragni and Cook (2009). One modification to SIR is to fit smooth parametric curves on the $p$ inverse regression curves using multivariate linear models.[3] This approach is called **parametric inverse regression**, or **PIR** (Bura, 1997; Bura and Cook, 2001). Moreover, Cook and Ni (2005) studied collectively the inverse regression family and derived the optimal member of this family, which they call the **inverse regression estimator**.

**SAVE**
One known problem with SIR is that it is known to fail in some situations. For example, when the response surface is symmetric about the origin, SIR does not find the EDR-direction. This difficulty can be overcome by also using the conditional variance–covariance matrix $Var(\underline{X}|Y)$. This more comprehensive approach leads to **sliced average variance estimation**, or **SAVE** (Cook and Weisberg, 1991; Cook and Lee, 1999). Using the standardized quantity $\underline{Z}$, like the SIR algorithm and assuming both the linearity and constant variance–covariance matrix assumptions from earlier, the SAVE algorithm estimates the CS. As noted in Li and Wang (2007), SAVE is not very efficient when estimating monotone trends for small to moderate sample sizes.

**PHD**
SIR and SAVE are first and second moment methods, respectively, each having difficulties with finding the EDR-space. SIR and SAVE are also methods that

---

[3]Multivariate linear regression models are discussed in Chapter 22.

only estimate vectors in the CS. The first method we discuss that estimates vectors in the CMS is the **principal Hessian directions**, or **PHD**, method of Li (1992). First, the **response-based (y-based) Hessian matrix**

$$\mathbf{H_y} = E((Y - E(Y))\underline{X}\underline{X}^{\mathrm{T}}) \tag{16.4}$$

of the regression function $f(\underline{X}) = E(Y|\underline{X})$ at any point, say $\underline{X}_0$, will be degenerate along any directions that are orthogonal to the EDR-space $\mathcal{S}(\mathbf{B})$. Li (1992) defined the PHDs as the eigenvectors of $\hat{\mathbf{H}}_y = E(\mathbf{H_y})$, which are then used to define a new coordinate axis. Assuming normally distributed predictors and applying Stein's lemma (Stein, 1981), PHD can then estimate multiple vectors in the CMS. Cook (1998b) presented an alternative approach to PHD using the **residual-based (r-based) Hessian matrix**

$$\mathbf{H_r} = E((Y - E(Y) - E(Y\underline{X}^{\mathrm{T}})\underline{X})\underline{X}\underline{X}^{\mathrm{T}}), \tag{16.5}$$

which was noted to be generally superior to use compared to the y-based Hessian matrix. Like SAVE, PHD is not very efficient when estimating monotone trends.

## MAVE
Other methods exist that have fewer restrictions than the dimension reduction methods discussed thus far, such as the **minimum average variance estimation**, or **MAVE**, estimation of Xia et al. (2002). Consider the following regression model for dimension reduction:

$$Y = g(\mathbf{B}_0^{\mathrm{T}}\underline{X}) + \epsilon,$$

where $g(\cdot)$ is an unknown smooth link function,[4] $\mathbf{B}_0 = (\boldsymbol{\beta}_1^*, \ldots, \boldsymbol{\beta}_k^*)$ is a $p \times k$ orthogonal matrix, and $E(\epsilon|\underline{X})$. Note with this last condition that $\epsilon$ is allowed to depend on $\underline{X}$. For this model, the $\boldsymbol{\beta}_h^*$, $h = 1, \ldots, k$, are the EDR-directions and the $k$-dimensional subspace $\mathbf{B}_0^{\mathrm{T}}\underline{X}$ is the EDR-space. The direction $\mathbf{B}_0$ is found as the solution to the following:

$$\operatorname*{argmin}_{\mathbf{B}} \left\{ E[Y - E(Y|\mathbf{B}^{\mathrm{T}}\underline{X})]^2 \right\}.$$

Then, at the population level, the MAVE procedure minimizes

$$E[\mathrm{Var}(Y|\mathbf{B}^{\mathrm{T}}\underline{X})] = E[Y - E(Y|\mathbf{B}^{\mathrm{T}}\underline{X})]^2$$

among all orthogonal matrices $\mathbf{B}$. As noted in Cook (2002), MAVE is a method for estimating the CMS.

## Iterative Hessian Transformation
As noted in Cook and Li (2004), if $\underline{X}$ has an elliptical distribution, then

---

[4]Link functions are presented in Chapters 20 and 23.

the linearity condition stated earlier holds for all $p$-dimensional subspaces. If $\underline{X}$ is multivariate normal, then the more stringent constant variance–covariance matrix condition holds. However, if there is significant heteroskedasicity among the predictors, another dimension reduction method based on the **iterative Hessian transformation**, or **IHT**, can be used (Cook and Li, 2004, 2002). Let $\beta_{\text{OLS}}$ denote the ordinary least squares vector for our regression problem. Using the r-based Hessian matrix $\mathbf{H}_r$ in (16.5) and under the linearity conditions stated earlier, the subspace spanned by $(\beta_{\text{OLS}}, \mathbf{H}_r\beta_{\text{OLS}}, \mathbf{H}_r^2\beta_{\text{OLS}}, \ldots, \mathbf{H}_r^{p-1}\beta_{\text{OLS}})$ is the same as that spanned by $(\beta_{\text{OLS}}, \mathbf{H}_r\beta_{\text{OLS}}, \mathbf{H}_r^2\beta_{\text{OLS}}, \ldots, \mathbf{H}_r^{k-1}\beta_{\text{OLS}})$, which is contained in the CMS. This is the IHT subspace. Cook and Li (2002) then use sequential testing of hypotheses to determine the dimension $k$ of the IHT subspace. Those same authors also note that the IHT method can be used with the y-based Hessian matrix $\mathbf{H}_y$ in (16.4), but abandon its usage due to the superiority of the r-based Hessian matrix.

### Graphical Regression

One can also take a graphical-based approach to dimension reduction for estimating the CS. The approach of **graphical regression** can be implemented when $p \geq 2$. For $p = 2$, we construct a 3D plot with $Y$ on the vertical axis and $X_1$ and $X_2$ on the other axes. Rotating the plot about the vertical axis gives 2D plots of all linear combinations of $\mathbf{X}^{\text{T}}\beta$. If no dependence is visually detected, then one can infer that $Y \perp\!\!\!\perp \underline{X}$. Otherwise, the structural dimension is $d = 1$ or $d = 2$; see Chapter 7 of Cook (1998a) for making this determination. When $p > 2$, let $X_1$ and $X_2$ denote any two of the predictors and then $\underline{X}_3$ be a vector of the remaining $p - 2$ predictors. The graphical regression approach seeks to determine if there are constants $c_1$ and $c_2$ such that $Y \perp\!\!\!\perp \underline{X}|(c_1X_1 + c_2X_2, \underline{X}_3)$. The structural dimension can be assessed using a sequence of 3D plots employed in a somewhat analogous manner as for the $p = 2$ setting described above. One would essentially iterate through the different predictors retaining those where relationships are visually detected in the 3D plots. For more details on graphical regression, see Chapter 7 of Cook (1998a) and the references therein.

### Contour Regression

Most of the previous dimension reduction procedures are limited by the assumptions they make, especially about the shape of the distribution of $\underline{X}$. Li et al. (2005) proposed a method that targets **contour directions**, which are directions along which the response surface is flat. Contour directions span the orthogonal complement of the CS, so these are equivalently estimated. This method is called **contour regression**, or **CR**, and is capable of capturing all directions in the CS without relying on the constraint of special patterns, such as monotone or U-shaped trends. The intuition of CR depends on the notion of **empirical directions**, which, for a sample of size $n$, are the $\binom{n}{2}$ directions $(\mathbf{x}_i - \mathbf{x}_j)$, $i \neq j$, $i, j = 1, \ldots, n$. Then, CR extracts a subset of the empirical directions characterized by having a small variation in $Y$ and performs a principal component analysis on those extracted directions. This allows us to, thus, characterize a specific set of directions for which the conditional distribution $Y|\underline{X}$ genuinely depends on $\underline{X}$. CR has a number of benefits. Namely, it provides exhaustive estimation of the CS under ellipticity of $\underline{X}$. Moreover, under some additional mild conditions, it provides robustness against non-ellipticity of $\underline{X}$, there are various appealing asymptotic properties achieved regardless of $p$ and the structural dimension $d$, and it is a computationally efficient procedure. See Li et al. (2005) for a full description of this procedure.

**Directional Regression**
Li and Wang (2007) introduced **directional regression**, or **DR**, which works under a similar setup as CR. It also works with the empirical directions, but regresses them against $Y$ in the $L_2$-sense and then performs estimation and testing under this paradigm. Recall that we discussed $L_1$- and $L_2$-penalties earlier in our discussion on penalized regression. DR is a method that estimates the CS. Li and Wang (2007) outline specify various conditions for the subspace of DR to be equal to $\mathcal{S}_{Y|\underline{X}}$ as well as where the subspace of DR is equal to the subspace of SAVE.

## 16.5    Examples

**Example 16.5.1.** *(Prostate Cancer Data)*
Stamey et al. (1989) examined the correlation between the level of prostate-specific antigen and various clinical measures in men who were about to receive a prostatectomy. The data consist of $n = 93$ men. The variables in this study are as follows:

- $Y$: cancer volume (on the log scale)

- $X_1$: prostate weight (on the log scale)

- $X_2$: age

- $X_3$: amount of benign prostatic hyperplasia (on the log scale)

- $X_4$: an indicator where 1 indicates the presence of seminal vesicle invasion and 0 otherwise

- $X_5$: capsular penetration (on the log scale)

- $X_6$: Gleason score, which is a value assigned based on a cancer biopsy

- $X_7$: percentage of Gleason scores 4 or 5

- $X_8$: prostate specific antigen (on the log scale)

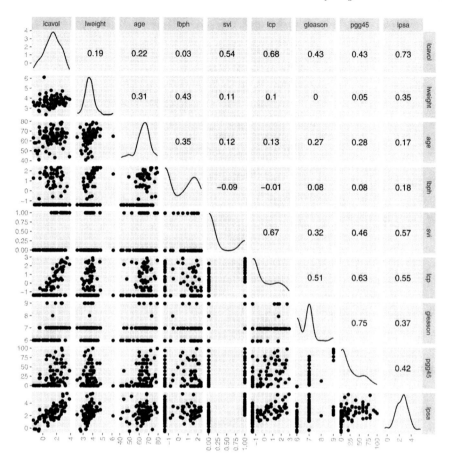

**FIGURE 16.1**
Matrix of pairwise scatterplots of the full prostate cancer data.

The prostate cancer data were also analyzed in Hastie et al. (2009) using ridge regression. In their analysis, they dropped the intercept term during estimation. We will retain the intercept in order to compare results from various shrinkage methods. In particular, we randomly divide the data into a training set of size 67 and test set of size 30. We then compare the residual sum of squares (RSS) and PRESS values for various shrinkage methods. A matrix of scatterplots of the full dataset is given in Figure 16.1. The first column gives the pairwise scatterplots between the response (log of cancer volume) versus each of the eight predictors. The remaining pairwise scatterplots are between the predictor variables. While these scatterplots indicate moderately high correlations between predictor variables, our focus is just on a comparison of the RSS and PRESS values for the different shrinkage methods.

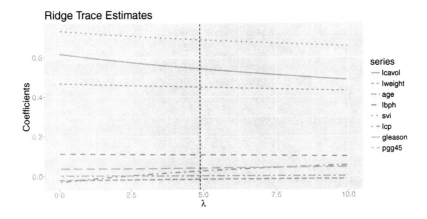

**FIGURE 16.2**
Ridge regression trace plot with the ridge regression coefficient on the *x*-axis.

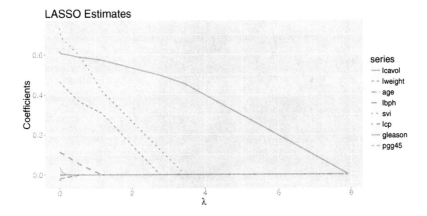

**FIGURE 16.3**
Trace plot of penalty terms for the LASSO fit.

First we fit a ridge regression to the data. The ridge trace plot is given in Figure 16.2. A vertical line is drawn at 4.89, the value of $\lambda$ selected using the Hoerl-Kennard iterative method. We then fit the LASSO to these data. A trace plot of the penalty terms is given in Figure 16.3. This plot shows that the majority of coefficients shrink to 0 once the penalty term reaches about 3.

We also obtain the fits from LARS, forward stagewise regression, principal components regression, and partial least squares. For comparison, we also obtain the ordinary least squares estimates for the linear model. The RSS and PRESS values for all of these methods are as follows:

```
############################
        OLS   Ridge  LASSO   LARS    F-S     PCR      PLS
RSS    0.6195 0.4028 0.4203 0.4083 0.4089  0.4187   0.5422
PRESS  9.2149 9.1260 9.2494 9.2149 9.4999 12.4655  10.7653
############################
```

With the exception of the ordinary least squares and partial least squares approach, all of the methods have comparable RSS values. However, principal components regression and partial least squares yield noticeably larger PRESS values compared to the other methods.

**Example 16.5.2.** *(Automobile Features Data)*
Quinlan (1993b) analyzed a modified dataset on city-cycle fuel consumption (in miles per gallon) to be predicted in various features pertaining to the individual automobiles. The modified dataset (after omitting observations with missing values) has $n = 392$ automobiles. We will focus on the following variables in this dataset:

- $Y$: miles per gallon achieved by the vehicle

- $X_1$: number of cylinders, between 4 and 8

- $X_2$: engine displacement (in cubic inches)

- $X_3$: engine horsepower

- $X_4$: vehicle weight (in pounds)

- $X_5$: time to acceleration from 0 to 60 miles per hour (in seconds)

Notice that the number of cylinders ($X_1$) is actually a discrete predictor. However, for the purpose of this example, we will treat it as a continuous variable.

First we fit a multiple linear regression model with miles per gallon as the response and the five other variables listed above as predictors. The $VIF$ values for these predictors are as follows:

```
############################
cylinders displacement   horsepower      weight acceleration
   10.631       19.535        8.916      10.430        2.609
############################
```

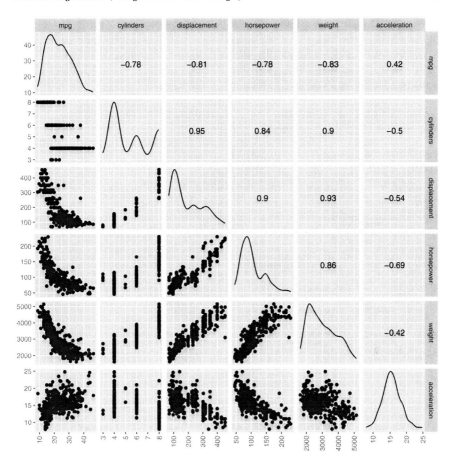

**FIGURE 16.4**
Matrix of pairwise scatterplots of the automobile features data.

Clearly there is an indication of some moderate to strong multicollinearity.
This is also indicated in the matrix of scatterplots of these data (Figure 16.4),
which shows some large correlations between the predictor values. We will run
various dimension reduction regression routines on these data. For each of the
routines below, the estimated basis vectors for the central (mean) subspace
are also able to be obtained, but for brevity, we simply report on the test
results for the number of dimensions.

First we run the SIR method:

```
##############################
Large-sample Marginal Dimension Tests:
                Stat df   p.value
```

```
0D vs >= 1D 389.132 35 0.0000000
1D vs >= 2D  62.354 24 0.0000295
2D vs >= 3D  15.854 15 0.3918438
3D vs >= 4D   4.206  8 0.8381084
##############################
```

The above output suggest that $k = 2$ dimensions is appropriate for these data. Next we run the SAVE method:

```
##############################
Large-sample Marginal Dimension Tests:
              Stat df(Nor) p.value(Nor) p.value(Gen)
0D vs >= 1D 594.89      105    0.000e+00    1.660e-13
1D vs >= 2D 324.78       70    0.000e+00    0.000e+00
2D vs >= 3D 176.25       42    0.000e+00    3.212e-08
3D vs >= 4D  73.57       21    9.295e-08    1.813e-05
##############################
```

The above output suggest that $k = 3$ dimensions is appropriate for these data. Next we run the PHD method:

```
##############################
Large-sample Marginal Dimension Tests:
              Stat df    p.value
0D vs >= 1D 266.78 15 0.000e+00
1D vs >= 2D 119.42 10 0.000e+00
2D vs >= 3D  53.53  6 9.158e-10
3D vs >= 4D  28.97  3 2.277e-06
##############################
```

The above agrees with the SAVE output in selecting $k = 3$. Finally, we run the IRE method:

```
##############################
Large-sample Marginal Dimension Tests:
              Test df      p.value iter
0D vs > 0D 795.786 35 1.511e-144    0
1D vs > 1D  95.362 24  1.835e-10    3
2D vs > 2D  23.093 15  8.219e-02    6
3D vs > 3D   6.669  8  5.728e-01    7
4D vs > 4D   1.125  3  7.710e-01    5
##############################
```

The above agrees with the SAVE and PHD output in selecting $k = 3$. While the result from SIR indicated $k = 2$ dimensions, there is agreement between the other three methods that $k = 3$ dimensions is the appropriate dimension size. Hence, we could justifiably proceed with using $d = 3$ dimensions.

# 17

# Piecewise, Nonparametric, and Local Regression Methods

This chapter focuses on regression models with greater flexibility for modeling complex relationships. We first discuss models where different regressions are fit depending on which area of the predictor space is being modeled. We then discuss nonparametric models which, as the name suggests, are models free of distributional assumptions and subsequently do not have regression coefficients readily available for estimation. However, because we assume that the underlying relationships are some sort of smooth curves, we desire our estimates to be smooth as well. Hence, we refer to such estimators as **smoothers**.

## 17.1 Piecewise Linear Regression

A model that proposes a different linear relationship for different intervals or regions of the predictor(s) is called a **piecewise linear regression model**.[1] The predictor values at which the slope changes are called **knots** or **changepoints**. Such models are helpful when you expect the linear trend of your data to change once you hit some threshold. Usually the knot values are already predetermined due to previous studies or standards that are in place. However, there are various methods for estimating the knot values. We will not explore such methods here, but refer to the methods reviewed by Chen et al. (2011).

For simplicity, we construct the piecewise linear regression model for the setting with one predictor and also briefly discuss how this can be extended to the setting with multiple predictors. The piecewise linear regression model with one predictor and one knot value ($\xi_1$) is

$$Y = \beta_0 + \beta_1 X_1 + \beta_2 (X_1 - \xi_1) \mathrm{I}\{X_1 > \xi_1\} + \epsilon, \qquad (17.1)$$

---

[1] Piecewise linear regression goes by many different names in the literature, including **breakpoint regression**, **broken-stick regression**, **changepoint regression**, and **segmented regression**.

where $I\{\cdot\}$ is the indicator function such that

$$I\{X_1 > \xi_1\} = \begin{cases} 1, & \text{if } X_1 > \xi_1; \\ 0, & \text{otherwise.} \end{cases}$$

More compactly, we can rewrite the model in (17.1) as

$$Y = \beta_0 + \beta_1 X_1 + \beta_2 (X_1 - \xi_1)_+ + \epsilon, \qquad (17.2)$$

where the notation $Z_+$ refers to the **truncated power function** (occasionally called the **hinge function**) given by

$$Z_+ = \begin{cases} Z, & \text{if } Z > 0; \\ 0, & \text{if } Z \le 0. \end{cases}$$

So, when $X_1 \le \xi_1$, the simple linear regression line is

$$E(Y|X_1) = \beta_0 + \beta_1 X_1$$

and when $X_1 > \xi_1$, the simple linear regression line is

$$E(Y|X_1) = (\beta_0 - \beta_2 \xi_1) + (\beta_1 + \beta_2) X_1.$$

An example of such a regression model is given in Figure 17.1(a).

For more than one knot value, we can extend the above regression model to incorporate other indicator values. Suppose we have $c$ knot values, $\xi_1, \xi_2, \ldots, \xi_c$, and we have $n$ observations. Then the piecewise linear regression model is written as

$$y_i = \beta_0 + \beta_1 x_{i,1} + \beta_2 (x_{i,1} - \xi_1)_+ + \ldots + \beta_{c+1} (x_{i,1} - \xi_c)_+ + \epsilon_i.$$

We can write the above model more compactly using matrix notation as

$$\mathbf{Y} = \mathbf{X}\boldsymbol{\beta} + \boldsymbol{\epsilon},$$

where $\boldsymbol{\beta}$ is a $(c+2)$-dimensional vector and

$$\mathbf{X} = \begin{pmatrix} 1 & x_{1,1} & (x_{1,1} - \xi_1)_+ & \cdots & (x_{1,1} - \xi_c)_+ \\ 1 & x_{2,1} & (x_{2,1} - \xi_1)_+ & \cdots & (x_{2,1} - \xi_c)_+ \\ \vdots & \vdots & \vdots & \ddots & \vdots \\ 1 & x_{n,1} & (x_{n,1} - \xi_1)_+ & \cdots & (x_{n,1} - \xi_c)_+ \end{pmatrix}.$$

An example of a piecewise linear regression model with one predictor and two knot values is given in Figure 17.1(c). The above $\mathbf{X}$ matrix can be extended to include more than one predictor with an arbitrary number of knots, however, the notation does become a bit cumbersome.

Sometimes there may also be a discontinuity at the knots. This is easily reflected in the piecewise linear model we constructed above by adding one

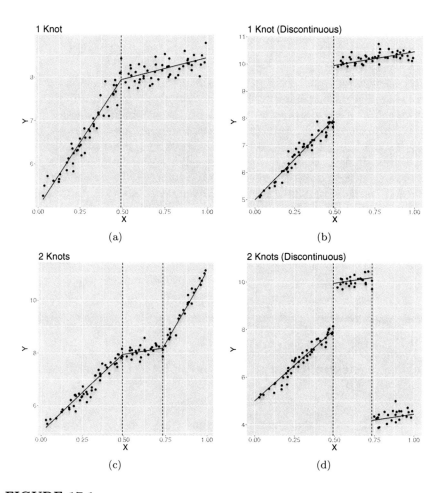

**FIGURE 17.1**

(a) Piecewise linear regression with 1 knot. (b) Discontinuous piecewise linear regression with 1 knot. (c) Piecewise linear regression with 2 knots. (d) Discontinuous piecewise linear regression with 2 knots.

more term to the model. For each $\xi_j$ where there is a discontinuity, include the corresponding indicator random variable $\mathrm{I}\{X_1 > \xi_j\}$ as a regressor in the piecewise linear regression model. Thus, the $\mathbf{X}$ matrix would have the column vector

$$
\begin{pmatrix}
\mathrm{I}\{x_{1,1} > \xi_j\} \\
\mathrm{I}\{x_{2,1} > \xi_j\} \\
\vdots \\
\mathrm{I}\{x_{n,1} > \xi_j\}
\end{pmatrix}
$$

appended to it for each $\xi_j$ where there is a discontinuity. Examples of piecewise linear regression models with one predictor and a discontinuity at one knot value and discontinuities at two knot values are given in Figures 17.1(b) and 17.1(d), respectively. Sometimes, such a discontinuity structure is purposely used for an experimental pretest-posttest design called a **regression continuity design**. The purpose is to assign a cutoff or threshold above or below where a treatment is assigned. This design was first used by Thistlewaite and Campbell (1960) to study how public recognition for a student's scholastic achievement affects the likelihood that they will receive merit-based scholarships as well as their attitudes and career plans. Extending discontinuities to the case of more than one predictor is analogous to the piecewise linear regression case without discontinuities.

## 17.2   Local Regression Methods

**Nonparametric regression** attempts to find a functional relationship between $Y$ and $\underline{X}$:

$$
Y = m(\underline{X}) + \epsilon, \tag{17.3}
$$

where $m$ is the regression function to estimate[2] and $\mathrm{E}(\epsilon) = 0$. It is not necessary to assume constant variance and, in fact, one typically assumes that $\mathrm{Var}(\epsilon) = \sigma^2(\underline{X})$, where $\sigma^2(\cdot)$ is a continuous, bounded function.

When taking the nonparametric approach to regression, we make as few assumptions about the regression function $m$ as possible. We use the observed data to learn as much as possible about the potential shape of $m$, thus allowing $m$ to be very flexible, yet smooth. We take two primary approaches to obtaining a nonparametric estimate of $m$.

First we use kernel-based methods, which take local averages of the response value. Essentially, at a given value of $\underline{X}$, we consider a window of nearby predictor values and obtain an average of the corresponding response

---

[2]In other words, we are still interested in the regression problem for modeling how the expected value of the response changes with respect to the predictor variables; i.e., $\mathrm{E}(Y|\underline{X}) = m(\underline{X})$.

values. Kernel methods for regression are a fairly straightforward extension of kernel density estimation for univariate data.

Second, we consider what are formally called **local regression** methods, which make no global assumptions about the function $m$. Global assumptions are made in standard linear regression about the response curve through the regression coefficient vector $\beta$, which we assume properly models all of our data. However, local regression assumes that $m$ can be well-approximated locally by a member from a simple class of parametric functions, such as a constant, straight-line, quadratic curve, etc. What drives local regression is Taylor's theorem from calculus, which says that any continuous function — in our case, $m$ — can be approximated with a polynomial.

For simplicity, we present some of these nonparametric regression models assuming a single predictor $X$. However, we briefly mention how estimation is extended to the multiple predictors setting.

**Linear Smoother**

One aspect to note about the nonparametric estimators we present in this section is that they are **linear smoothers**. For a sample of size $n$, we are interested in the nonparametric regression equation

$$y_i = m(x_i) + \epsilon_i,$$

where the $E(\epsilon_i) = 0$ and $Var(\epsilon_i) = \sigma^2(x_i)$ such that $\sigma^2(\cdot)$ is a continuous, bounded function. An estimator $\hat{m}_n$ of $m$ is a linear smoother if, for each $x_0$, there exists a weighting vector $l(x_0) = (l_1(x_0), \ldots, l_n(x_0))^T$ such that

$$\hat{m}_n(x_0) = \sum_{i=1}^{n} l_i(x_0) y_i, \tag{17.4}$$

where $\sum_{i=1}^{n} l_i(x_0) = 1$. Letting $\mathbf{Y}$ be our vector of response values, $\hat{\mathbf{m}} = (\hat{m}_n(x_1), \ldots, \hat{m}_n(x_n))^T$, and $\mathbf{L}$ be an $n \times n$ matrix whose $i^{\text{th}}$ row is $l(x_i)^T$, we have

$$\hat{\mathbf{m}} = \mathbf{LY}. \tag{17.5}$$

We call the vector $\hat{\mathbf{m}}$ the fitted values for the linear smoother and $\mathbf{L}$ the **smoothing matrix**, which is analogous to the hat matrix for multiple linear regression. Moreover, we call $\nu = tr(\mathbf{L})$ the **effective degrees of freedom** for the corresponding smoother.

For the linear smoother $\hat{m}_n(x_0)$ in (17.4), we are actually estimating $\bar{m}_n(x_0)$, not $m(x_0)$; i.e.,

$$\bar{m}_n(x_0) = E(\hat{m}_n(x_0)) = \sum_{i=1}^{n} l_i(x_0) m(x_i).$$

For simplicity, assume that $Var(\epsilon_i) = \sigma^2 < \infty$, a constant. Then,

$$Var(\hat{m}_n(x_0)) = \sigma^2 \|l(x_0)\|^2.$$

We can then construct a $100 \times (1 - \alpha)\%$ confidence band for $\bar{m}_n(x_0)$ by

$$(\hat{m}_n(x_0) - c_{1-\alpha}\hat{\sigma}\|\mathbf{l}(x_0)\|, \hat{m}_n(x_0) + c_{1-\alpha}\hat{\sigma}\|\mathbf{l}(x_0)\|),$$

for some $c_{1-\alpha} > 0$ and $a \leq x \leq b$. The quantity $c_{1-\alpha}$ is found using an expression called the **tube formula**. Details of this formula and the construction of such confidence bands are found in Sun and Loader (1994) and Faraway and Sun (1995).

### Kernel Regression

One way of estimating $m(\cdot)$ is to use **density estimation**, which approximates the probability density function of a random variable. Let us consider the non-regression univariate setting. Assuming we have $n$ independent observations $x_1, \ldots, x_n$ from the random variable $X$ with probability density function $f$, the **kernel density estimator** $\hat{f}_h(x_0)$ for estimating the density at $x_0$ (i.e., $f(x_0)$) is defined as

$$\hat{f}_h(x_0) = \frac{1}{nh} \sum_{i=1}^{n} K\left(\frac{x_i - x_0}{h}\right).$$

Here, $K(\cdot)$ is called the **kernel function** and $h$ is a smoothing parameter called the **bandwidth**. $K(\cdot)$ is a function often resembling a probability density function, but with no parameters to be estimated. Some common kernel functions are provided in Table 17.1. $h$ controls the window width around $x_0$ which we perform the density estimation. Thus, a kernel density estimator is essentially a weighting scheme, dictated by the choice of kernel, which takes into consideration the proximity of a point in the dataset near $x$ when given a bandwidth $h$. Furthermore, more weight is given to points near $x_0$ and less weight is given to points farther from $x_0$.[3]

Clearly, kernel density estimation requires a kernel and a bandwidth. Theoretically, we can choose these using the **mean integrated squared error**, or **MISE**, as an optimality criterion:

$$\text{MISE} = \mathrm{E}\left[\int (\hat{f}_h(x) - f(x))^2 dx\right].$$

Based on some asymptotic results, the following **asymptotic MISE**, or **AMISE** can also be used:

$$\text{AMISE} = \frac{\int K(u)^2 du}{nh} + \frac{h^4}{4}\left(\int K(u)^2 du\right)^2 \int f''(u)^2 du,$$

---

[3]The beta kernel is actually a general class of kernels characterized by the shape parameter $g \geq 0$. Common values of $g$ are 0, 1, 2, and 3, which are called the uniform, Epanechnikov, biweight, and triweight kernels, respectively. Despite calling $g$ a "parameter," it is not estimated, but rather set to the value based on which kernel we intend on using. $\mathcal{B}(\cdot, \cdot)$ is the beta function, which is defined later in Chapter 18.

**TABLE 17.1**

A table of common kernel functions

| Kernel | $K(u)$ |
|---|---|
| Uniform | $\frac{1}{2}\mathrm{I}\{|u| \leq 1\}$ |
| Triangular | $(1 - |u|)\mathrm{I}\{|u| \leq 1\}$ |
| Tricube | $(1 - |u|^3)\mathrm{I}\{|u| \leq 1\}$ |
| Beta | $\frac{(1-u^2)^g}{\mathcal{B}(0.5,g+1)}\mathrm{I}\{|u| \leq 1\}$ |
| Gaussian | $\frac{1}{\sqrt{2\pi}}e^{-\frac{1}{2}u^2}$ |
| Cosinus | $\frac{1}{2}(1 + \cos(\pi u))\mathrm{I}\{|u| \leq 1\}$ |
| Optcosinus | $\frac{\pi}{4}\left(\cos(\frac{\pi u}{2})\right)\mathrm{I}\{|u| \leq 1\}$ |

where $f''$ is the second derivative of the unknown density function $f$. Of course, the fact that $f$ is unknown means neither the MISE nor the AMISE can be directly calculated. So various data-based methods can be employed to implement the above. Based on the AMISE and some deeper asymptotic arguments, it can be shown that there are expressions for an optimal kernel and an optimal bandwidth. These require a greater development of theory than what we cover here, so we defer to, for example, Chapter 3 of Härdle et al. (2004b) for a discussion of such details.

Typically, the choice of kernel does not affect the kernel density estimate very much, whereas the choice of bandwidth is more important. While we stated above that a theoretically optimal value of $h$ exists, there are many rule-of-thumb quantities available in the literature for the bandwidth. Perhaps one of most common is the **Silverman's rule of thumb** (Silverman, 1998):

$$h^* = \left(\frac{4\hat{\sigma}^5}{3n}\right)^{1/5} \approx 1.06\hat{\sigma}n^{1/5},$$

where $\hat{\sigma}^2$ is the sample variance of the data. One can also choose the bandwidth using crossvalidation.

With some basics of kernel density estimation established, we can now turn to the regression setting. To perform a **kernel regression** of $Y$ on $X$, a common estimate of $m_h(\cdot)$ is the **Nadaraya–Watson** estimator (Nadaraya, 1964; Watson, 1964):

$$\hat{m}_h(x) = \frac{\sum_{i=1}^{n} K\left(\frac{x_i - x}{h}\right) y_i}{\sum_{i=1}^{n} K\left(\frac{x_i - x}{h}\right)}, \tag{17.6}$$

where $m$ has been subscripted to note its dependency on the bandwidth. As you can see, this kernel regression estimator is just a weighted sum of the observed responses, where the kernel is the weighting function.

It is also possible to construct approximate confidence intervals and confidence bands using the Nadaraya–Watson estimator, but under some restrictive assumptions. The notation used in the formulas below follows that used in Chapter 4 of Härdle et al. (2004b). An approximate $100 \times (1 - \alpha)\%$ pointwise confidence interval is given by

$$\hat{m}_h(x) \pm z_{1-\frac{\alpha}{2}} \sqrt{\frac{\hat{\sigma}_h^2(x) \|K\|_2^2}{nh\hat{f}_h(x)}},$$

where $h = cn^{-1/5}$ for some constant $c > 0$, $\|K\|_2^2 = \int K(u)^2 du$, and

$$\hat{\sigma}_h^2(x) = \frac{\frac{1}{n} \sum_{i=1}^{n} K\left(\frac{x_i - x}{h}\right) \{y_i - \hat{m}_h(x)\}^2}{\sum_{i=1}^{n} K\left(\frac{x_i - x}{h}\right)}.$$

Next, let $h = n^{-\delta}$ for $\delta \in (\frac{1}{5}, \frac{1}{2})$. Then, under certain regularity conditions, an approximate $100 \times (1 - \alpha)\%$ confidence band is given by

$$\hat{m}_h(x) \pm z_{n,\alpha} \sqrt{\frac{\hat{\sigma}_h^2(x) \|K\|_2^2}{nh\hat{f}_h(x)}},$$

where

$$z_{n,\alpha} = \left\{ \frac{-\log\{-\frac{1}{2}\log(1 - \alpha)\}}{(2\delta \log(n))^{1/2}} + d_n \right\}^{1/2}$$

and

$$d_n = (2\delta \log(n))^{1/2} + (2\delta \log(n))^{-1/2} \log\left( \frac{1}{2\pi} \sqrt{\frac{\|K'\|_2^2}{\|K\|_2^2}} \right)^{1/2},$$

where $K'$ is the first derivative of the kernel function $K$.

What we developed in this section is only for the case of one predictor. If you have $p$ predictors, then one needs to use a **multivariate kernel density estimator (MVKDE)** at a point $\mathbf{x}_0 = (x_{0,1}, \ldots, x_{0,p})^{\mathrm{T}}$. A general form for an MVKDE is

$$\hat{m}(\mathbf{x}_0) = \frac{1}{n} \sum_{i=1}^{n} K_{\mathbf{H}}(\mathbf{x}_0 - \mathbf{x}_i),$$

where $\mathbf{x}_i = (x_{i,1}, x_{i,2}, \ldots, x_{i,p})^{\mathrm{T}}$, $i = 1, \ldots, n$ are our $p$-dimensional data vectors, $\mathbf{H}$ is the symmetric, positive definite $p \times p$ bandwidth (or smoothing) matrix, $K$ is a symmetric multivariate kernel density function, and $K_{\mathbf{H}}(\mathbf{x}) = |\mathbf{H}|^{-1/2} K(\mathbf{H}^{-1/2}\mathbf{x})$. In practice, it is common to work with a more tractable MVKDE, such as the **product kernel**:

$$\hat{m}(\mathbf{x}_0) = \frac{1}{n} \sum_{i=1}^{n} \prod_{j=1}^{p} \frac{1}{h_j} K\left(\frac{x_{i,j} - x_{0,j}}{h_j}\right).$$

Multivariate kernels require more advanced methods and are difficult to use as datasets with more predictors will often suffer from the curse of dimensionality. Additional details are found in the text by Scott (2015).

**Local Polynomial Regression**

Local polynomial modeling is similar to kernel regression estimation, but the fitted values are now produced by a locally weighted regression rather than by a locally weighted average. The theoretical basis for this approach is to do a Taylor series expansion around a value $x_0$:

$$m(x) \approx m(x_0) + m^{(1)}(x_0)(x - x_0) + \frac{m^{(2)}(x_0)}{2}(x - x_0)^2$$

$$+ \ldots + \frac{m^{(q)}(x_0)}{q!}(x - x_0)^q,$$

for $x$ in a neighborhood of $x_0$, where $m^{(j)}$ denotes the $j$th derivative of $m$. The above is a polynomial of degree $q$, which we parameterize in a way such that:

$$m(x_0) \approx \beta_0 + \beta_1(x - x_0) + \beta_2(x - x_0)^2 + \ldots + \beta_q(x - x_0)^q, \qquad |x_i - x| \le h,$$

where $\beta_j = m^{(j)}(x_0)/j!$, $j = 0, 1, \ldots, q$. Note that the $\beta$ parameters are considered functions of $x$ and, hence, are local.

Local polynomial fitting minimizes

$$\sum_{i=1}^{n} K\left(\frac{x_i - x_0}{h}\right) \left[ y_i - \sum_{j=0}^{q} \beta_j(x_i - x_0)^j \right]^2$$

with respect to the $\beta_j$ terms. This is just a weighted least squares problem with the weights given by the $K\left(\frac{x_i - x_0}{h}\right)$ quantities. Then, letting

$$\mathbf{X} = \begin{pmatrix} 1 & (x_1 - x_0) & \cdots & (x_1 - x_0)^q \\ 1 & (x_2 - x_0) & \cdots & (x_2 - x_0)^q \\ \vdots & \vdots & \ddots & \vdots \\ 1 & (x_n - x_0) & \cdots & (x_n - x_0)^q \end{pmatrix}$$

and $\mathbf{W} = \operatorname{diag}\left(K\left(\frac{x_1-x_0}{h}\right), \ldots, K\left(\frac{x_n-x_0}{h}\right)\right)$, the **local least squares** estimate can be written as

$$\hat{\boldsymbol{\beta}} = \underset{\boldsymbol{\beta}}{\operatorname{argmin}}(\mathbf{Y} - \mathbf{X}\boldsymbol{\beta})^{\mathrm{T}}\mathbf{W}(\mathbf{Y} - \mathbf{X}\boldsymbol{\beta})$$
$$= (\mathbf{X}^{\mathrm{T}}\mathbf{W}^{-1}\mathbf{X})^{-1}\mathbf{X}^{\mathrm{T}}\mathbf{W}^{-1}\mathbf{Y}. \tag{17.7}$$

Thus, we can estimate the $j^{\mathrm{th}}$ derivative of $m(x_0)$ by

$$\hat{m}^{(j)}(x_0) = j!\hat{\beta}_j.$$

When $q = 0$, we have a local constant estimator, which is just the Nadaraya–Watson estimator. When $q = 1$, we have a local linear estimator, when $q = 2$, we have a local quadratic estimator, and so on.

For any $x_0$, we can perform inference on the $\beta_j$ (or the $m^{(j)}(x_0)$) terms in a manner similar to weighted least squares. We can also construct approximate pointwise confidence intervals and confidence bands as was done in the kernel regression setting. Details on such inference considerations for local polynomial regression models can be found in the text by Fan and Gijbels (1996).

## LOESS

The next local regression method we discuss is the **locally estimated scatterplot smoother**, or **LOESS**.[4] LOESS fits a low-degree polynomial, usually no greater than degree 2, using those points closest to each $X$ value. The closest points are given the greatest weight in the smoothing process, thus limiting the influence of outliers.

Suppose we have a sample of size $n$ with the pairs of observations $(x_1, y_1), \ldots, (x_n, y_n)$. LOESS follows a basic algorithm as follows:

---

**LOESS**

1. Select a set of values partitioning $[x_{(1)}, x_{(n)}]$. Let $x_0$ be an individual value in this set.

2. For each observation, calculate the distance

$$d_i(x_0) = |x_i - x_0|.$$

Let $s$ be the number of observations in the neighborhood of $x_0$. The **neighborhood** is formally defined as the $s$ smallest values of $d_i$ where $s = \lceil \gamma n \rceil$. $\gamma$ the **span**, which is the proportion of points to be selected.

---

[4]There is also another version of LOESS called LOWESS, which stands for locally weighted scatterplot smoother. While in the literature the acronyms are often used interchangeably, sometimes LOESS is attached to the setting of smoothing with a weighted second degree (quadratic) polynomial while LOWESS is attached to the setting of smoothing a first degree (linear) polynomial.

3. Perform a weighted regression using only the points in the neighborhood. The weights are given by the tricube weight function

$$
w_i(x_0) = \begin{cases} \left(1 - \left|\frac{x_i - x_0}{d_s(x_0)}\right|^3\right)^3, & \text{if } |x_i - x_0| < d_s(x_0); \\ 0, & \text{if } |x_i - x_0| \geq d_s(x_0), \end{cases}
$$

where $d_s(x_0) = \max_i\{d_i(x_0)\}$ is the the largest distance in the neighborhood of observations closest to $x_0$. Thus, the data points closest to $x_0$ have the most influence on the fit, while those outside the span have no influence on the fit.

4. Calculate the estimated parameter vector $\hat{\beta}_{\text{LOESS}}$ using the same local least squares estimates from (17.7), but where $\mathbf{W} = \text{diag}(w_1(x_0), \ldots, w_n(x_0))$. For LOESS, usually $q = 2$ is sufficient.

5. Iterate the above procedure for another value of $x_0$ and calculate all of the fitted values as follows:

$$
\hat{\mathbf{Y}}_{\text{LOESS}} = \mathbf{X}\hat{\beta}_{\text{LOESS}}.
$$

---

Since outliers can have a large impact on the local least squares estimates, a robust weighted regression procedure may also be used to lessen the influence of outliers on the LOESS curve. This is done by calculating a different set of robust weights. First, calculate the $q$ LOESS residuals:

$$
\mathbf{r}^* = \mathbf{Y} - \hat{\mathbf{Y}}_{\text{LOESS}}.
$$

Then, calculate the robust weights using the bisquare function:

$$
w_i^* = \begin{cases} \left(1 - \left(\frac{r_i^*}{6\text{MAD}}\right)^2\right)^2, & \text{if } |r_i^*| < 6\text{MAD}; \\ 0, & \text{if } |r_i^*| \geq 6\text{MAD}, \end{cases}
$$

where $\text{MAD} = \text{med}_i\{|r_i^* - \text{med}_i(r_i^*)|\}$ is the **median absolute deviation** of the LOESS residuals. The above is sometimes referred to as **robust LOESS**.

Finally, note that kernel regression is actually a special case of local regression. As with kernel regression, there is also an extension of local regression regarding multiple predictors. It requires use of a multivariate version of Taylor's theorem around the $p$-dimensional point $\mathbf{x}_0$. The model can include main effects, pairwise combinations, and higher-order interactions of the predictors. As noted in Chapter 6 of Hastie et al. (2009), weights can then be defined, such as

$$
w_i(\mathbf{x}_0) = K\left(\frac{\|\mathbf{x}_i - \mathbf{x}_0\|}{\gamma}\right),
$$

where $K(\cdot)$ is typically a kernel that is a radial function and $\gamma$ is again the span. The values of $\mathbf{x}_i$ can also be scaled so that the smoothness occurs the same way in all directions. However, note that this estimation is often difficult due to the curse of dimensionality.

### Local Likelihood Regression

In Appendix C, we briefly define the likelihood and loglikelihood functions for a sample of random variables that are independent and identically distributed according to some parametric distribution. While we have only superficially mentioned aspects of likelihood estimation in regression settings, we can make any parametric regression model local using a local likelihood approach. Assume that each response variable $y_i$ is drawn from a distribution with density $f(y; \theta_i)$, where $\theta_i = \theta(\mathbf{x}_i) = \mathbf{x}_i^T \boldsymbol{\beta}$ is a function linear in the predictors. Let $\ell(y; \theta) = \log(f(y; \theta_i))$. Then, the **loglikelihood regression model** is based on the global loglikelihood of the parameter vector $\boldsymbol{\theta} = (\theta(\mathbf{x}_1), \dots, \theta(\mathbf{x}_n))^T$:

$$\ell(\boldsymbol{\theta}) = \sum_{i=1}^{n} \ell(y_i; \theta_i).$$

We can achieve greater flexibility by using the **local loglikelihood regression model** – local to $\mathbf{x}_0$ – for inference on $\theta(\mathbf{x}_0) = \mathbf{x}_0^T \boldsymbol{\beta}(\mathbf{x}_0)$:

$$\ell(\boldsymbol{\beta}(\mathbf{x}_0)) = \sum_{i=1}^{n} w_i(\mathbf{x}_0) \ell(y_i; \mathbf{x}_i^T \boldsymbol{\beta}(\mathbf{x}_0)),$$

where $w_i(\mathbf{x}_0)$ is a weighting function, typically defined using a kernel function. Then, we can optimize the above loglikelihood to obtain the estimates $\hat{\boldsymbol{\beta}}(\mathbf{x}_0)$ are $\mathbf{x}_0$. More details on local likelihood methods are given in the text by Loader (1999).

## 17.3   Splines

**Splines** are piecewise polynomial functions that have locally simple forms, but possess a high degree of flexibility and smoothness. Splines are a broad class of methods and have an immense amount of research applications at the intersection of applied mathematics and statistics. For the earlier concepts we discuss, we will consider a single predictor, $X$, for simplicity. More details on spline methods can be found in the texts by Wahba (1990) and de Boor (2001).

### Basis Functions

The foundation of splines is based on how we extend the linear model to augment the linear component of $X$ with additional, known functions of $X$:

$$m(x_0) = \sum_{j=1}^{B} \beta_j g_j(x_0), \tag{17.8}$$

where the $g_1, \ldots, g_B$ are known functions called **basis functions**. The basis functions are prespecified and the model is linear in these new variables. Thus, ordinary least squares approaches for model fitting and inference can be employed. We have already seen this basic paradigm earlier in Chapter 6 when we discussed polynomial regression. For example, a quadratic regression model has the following basis functions:

$$g_1(x) = 1, \quad g_2(x) = x, \quad g_3(x) = x^2.$$

However, such a regression model is employing global basis functions, meaning regardless of the region in the predictor space, observations in that region will affect the fit of the entire estimated model. Thus, the use of local basis functions ensure that a given observation only affects the nearby fit and not the fit over the entire predictor space. In other words, we are interested in the same type of local flexibility that we sought out earlier in this chapter with local regression methods.

### $M^{\text{th}}$-Order Splines

An $M^{\text{th}}$**-order spline** is a function $m$ with knots $\xi_1 < \xi_2 < \cdots < \xi_K$ contained in some interval $(a, b)^5$ if

1. $m$ is an $M$-degree polynomial on each of the intervals $(a, \xi_1], [\xi_1, \xi_2], \ldots, [\xi_K, b)$, and

2. $m^{(j)}$, the $j^{\text{th}}$ derivative of $m$, is continuous at $\xi_1, \ldots, \xi_K$ for each $j = 0, 1, \ldots, M - 1$.

One of the most natural ways to parameterize the set of splines with these $K$ knots is to use the **truncated power basis**:

$$h_j(x) = x^{j-1}, \quad j = 1, \ldots, M$$
$$h_{M+k}(x) = (x - \xi_k)_+^{M-1}, \quad k = 1, \ldots, K.$$

Note that a spline thus has $M + K$ degrees of freedom.

Using the truncated power basis functions above, below are a few common $M^{\text{th}}$-order splines used in practice:

- When $M = 1$, this yields a **piecewise constant spline**. All this means is that the data are partitioned into $K + 1$ regions and $\hat{\beta}_j$ is estimated by the mean of the response values in that region.

---

[5] Setting $a = -\infty$ and/or $b = \infty$ is allowed.

- When $M = 2$, this yields a **linear spline**. This is exactly the same model as the (continuous) piecewise linear regression model we discussed at the beginning of this chapter.

- When $M = 3$, this yields a **quadratic spline**. These provide more flexibility over simply fitting straight lines with a linear spline.

- When $M = 4$, this yields a **cubic spline**. Cubic splines are, perhaps, some of the most popular splines.

There are a few useful observations made in Chapter 5 of Hastie et al. (2009) about the $M^{\text{th}}$-order splines. First is that they claim that cubic splines are the lowest-ordered splines such that the discontinuity at the knots is not visible to the human eye. Second, they note that there is no practical reason to go beyond cubic splines, unless one is interested in smooth derivatives. Finally, they also note that, in practice, the most widely-used values are $M \in \{1, 2, 4\}$. Quadratic splines are not as common because greater flexibility is achieved with cubic splines. Diewart and Wales (1992) presented an effective application of quadratic splines for establishing supply-and-demand functions.

## Natural Splines
Splines tend to be even more erratic near the boundaries than regular polynomial fits, thus making extrapolation very treacherous. This gets worse as the order of the spline grows. For $M^{\text{th}}$-order splines where $M$ is odd and greater than 3, we can force the piecewise polynomial to have a lower degree beyond the extreme knots $\xi_1$ and $\xi_K$. This results in what we call a **natural spline** with knots $\xi_1 < \xi_2 < \cdots < \xi_K$, which is a piecewise polynomial function $m$ that is

- a polynomial of degree $M$ on $[\xi_1, \xi_2], [\xi_2, \xi_3], \ldots, [\xi_{K-1}, \xi_K]$,

- a polynomial of degree $(M - 1)/2$ on the half-open intervals $(-\infty, \xi_1]$ and $[\xi_K, \infty)$, and

- continuous with continuous derivatives of orders $1, \ldots, M - 1$ at each of its knot values.

By far, the most common natural splines are **natural cubic splines** $(M = 4)$, which are, again, linear beyond the boundary knots.

## Regression Splines
The knots in the splines introduced above could be placed at each data point, potentially needed to be estimated, or at fixed locations. If we consider estimating $m(x_0) = \mathrm{E}(Y|X = x_0)$ by fitting an $M^{\text{th}}$-order spline at $K$ fixed-knot locations, the choice of which can be tricky, then such splines are also called **regression splines**. Thus, we consider functions of the form $\sum_{j=1}^{M+K+1} \beta_j h_j(x_0)$, where for $j = 1, \ldots, M + K + 1$, the $\beta_j$ are the regression coefficients and $h_j(\cdot)$

are the truncated power basis functions[6] for the $M^{\text{th}}$-order splines over the $K$ fixed-knot locations.

For a sample of size $n$, let $\mathbf{H}$ be the $n \times (M + K + 1)$ **basis matrix** whose $j^{\text{th}}$ column gives the evaluation of $h_j$ over $x_1, \ldots, x_n$. Thus, we are interested in the least squares criterion

$$\hat{\boldsymbol{\beta}}_{\text{RS}} = \underset{\boldsymbol{\beta}}{\operatorname{argmin}} \|\mathbf{Y} - \mathbf{H}\boldsymbol{\beta}\|^2,$$

which gives us the fitted values of the estimator as

$$\hat{\mathbf{m}} = \mathbf{H}\hat{\boldsymbol{\beta}}_{\text{RS}} = \mathbf{LY},$$

where $\mathbf{L} = \mathbf{H}(\mathbf{H}^{\text{T}}\mathbf{H})^{-1}\mathbf{H}^{\text{T}}$. The fitted values above interpolate the observed data. Note that the regression spline is just a linear smoother and, thus, the above matches (17.5). More on regression splines and their usage is discussed in Ruppert et al. (2003).

### Smoothing Splines

As we stated above, the choice of knots for constructing a spline can be tricky. We can avoid this problem by placing knots at all of the predictor values and then control for overfitting using a procedure called **regularization**, which is essentially a penalization problem like we saw with regression shrinkage in Chapter 16. Hence, these types of spline problems are also called **penalized splines**.

First we consider cubic smoothing splines, which are one of the more commonly used smoothing splines. Let $N_1, \ldots, N_n$ denote the collection of natural cubic spline basis functions and $\mathbf{N}$ denote the $n \times n$ design matrix consisting of the basis functions evaluated at the observed values such that the $(i, j)^{\text{th}}$ entry of $\mathbf{N}$ is $N_j(x_i)$ and $m(x_0) = \sum_{j=1}^{n} \beta_j N_j(x_0)$. The objective function to minimize is

$$\hat{\boldsymbol{\beta}}_{\text{CS}} = \|\mathbf{Y} - \mathbf{N}\boldsymbol{\beta}\|^2 + \lambda \boldsymbol{\beta}^{\text{T}} \boldsymbol{\Omega} \boldsymbol{\beta},$$

where $\lambda \boldsymbol{\beta}^{\text{T}} \boldsymbol{\Omega} \boldsymbol{\beta}$ is a **regularization term** such that the $(i, j)^{\text{th}}$ entry of the penalty matrix $\boldsymbol{\Omega}$ is $\int N_i''(t) N_j''(t) dt$. Since smoothing splines are linear smoothers, the fitted values of the estimator are again $\hat{\mathbf{m}} = \mathbf{LY}$, where $\mathbf{L} = \mathbf{N}(\mathbf{N}^{\text{T}}\mathbf{N})^{-1}\mathbf{N}^{\text{T}}$.

More generally, in Chapter 5 of Hastie et al. (2009), they discuss smoothing splines as a functional minimization problem. Consider finding, among all functions $f$ with two continuous derivatives, the one that minimizes the penalized sum of squares problem

$$\hat{\boldsymbol{\beta}}_{\text{SS}} = \underset{\boldsymbol{\beta}}{\operatorname{argmin}} \|\mathbf{Y} - f(\mathbf{X})\|^2 + \lambda \int \{f''(t)\}^2 dt. \tag{17.9}$$

The fixed smoothing parameter $\lambda$ controls the tradeoff between the closeness

---

[6]Other basis functions can be used.

of the data (the first term in the above) and how much the curvature in the function is penalized (the second term in the above). $\lambda$ ranges from very rough (e.g., $\lambda = 0$ means $f$ is any function that interpolates the data) to very smooth (e.g., $\lambda = \infty$ yields the simple linear regression straight-line fit). In fact, the unique minimizer to (17.9) is the natural cubic spline fit $\hat{m}$ from above.

Regularization as we described above seeks a solution through the use of penalized least squares given a value of the smoothing parameter $\lambda$, which can be selected using **generalized crossvalidation** (or **GCV**). Begin by writing

$$\hat{m} = \mathbf{L}_\lambda \mathbf{Y},$$

where we have attached a subscript of $\lambda$ to the smoother matrix $\mathbf{L}$. Then the GCV function is defined as

$$V(\lambda) = \frac{\|(\mathbf{I}_n - \mathbf{L}_\lambda)\mathbf{Y}\|^2 / n}{[\text{tr}(\mathbf{I}_n - \mathbf{L}_\lambda)/n]^2}$$

and the GCV estimate of $\lambda$ is

$$\hat{\lambda} = \underset{\lambda}{\text{argmin}} V(\lambda).$$

### *B*-Splines

Another basis for the set of natural splines that has more stable numeric properties is called the *B*-spline basis. Assuming $K$ knots as we did for most of the previous splines, let $\xi_0 = a$ and $\xi_{K+1} = b$. The *B*-spline basis functions are given by the following recursion formula:

$$B_{i,1}(x_0) = \begin{cases} 1, & \text{if } \xi_i \leq x < \xi_{i+1}; \\ 0, & \text{otherwise}, \end{cases}$$

$$B_{i,t}(x_0) = \frac{x - \xi_i}{\xi_{i+t-1} - \xi_i} B_{i,t-1}(x_0) + \frac{\xi_{i+t} - x}{\xi_{i+t} - \xi_{i+1}} B_{i+1,t-1}(x_0).$$

We can then construct a spline estimator using the above *B*-spline basis functions. In Chapter 3 of Ruppert et al. (2003), the authors note that in the context of regression splines, there is an equivalence relationship between a certain order of the truncated power basis and the *B*-spline basis, resulting in a familiar form for the penalized spline fit.

### Multidimensional Splines

Thus far, we have discussed (regression) splines simply for the setting of a single predictor variable. However, there are various extensions for the multiple predictor setting. The details are quite mathematical for these methods, but some of the details can be found in Chapter 7 of Green and Silverman (1993).

Suppose we have a $p$-dimensional predictor vector measured with each observation and let $\boldsymbol{\xi}_1, \ldots, \boldsymbol{\xi}_K$ be $p$-dimensional knot vectors. The first extension is one that also utilizes $M^{\text{th}}$-ordered polynomials like when we use truncated power or $B$-spline basis functions. When $M$ is odd, we can use functions of the form

$$\phi(\|\mathbf{x_0} - \boldsymbol{\xi_k}\|) = \|\mathbf{x_0} - \boldsymbol{\xi_k}\|^M, \quad k = 1, \ldots, K,$$

which are called **radial basis functions**. Radial basis functions can also be used when $p = 1$, which gives the following set of basis functions:

$$\phi_1(x_0) = 1, \quad \phi_2(x_0) = x_0, \quad \ldots, \quad \phi_{M+1}(x_0) = x_0^M,$$
$$\phi_{M+2}(x_0) = |x_0 - \xi_1|^M, \quad \ldots, \quad \phi_{M+K+1}(x_0) = |x_0 - \xi_K|^M.$$

Another basis sort of mimics the idea of a product kernel for local regression in the multiple predictor setting. Suppose each dimension of the $p$-dimensional predictor vector has its own basis of functions. Namely, $h_{j,r_j}(x_{0,j})$, $r_j = 1, \ldots, R_j$ for representing functions of coordinate $x_{0,j}$, $j = 1, \ldots, p$. Then, the $(R_1 \times R_2 \times \cdots R_p)$-dimensional **tensor product basis** is defined by the functions

$$t_{r_1, r_2, \ldots, r_p}(\mathbf{x_0}) = \prod_{j=1}^{p} h_{j, r_j}(x_{0,j}),$$

which can then be used to represent a $p$-dimensional function

$$m(\mathbf{x_0}) = \sum_{r_p=1}^{R_p} \cdots \sum_{r_2=1}^{R_2} \sum_{r_1=1}^{R_1} \beta_{r_1, r_2, \ldots, r_p} t_{r_1, r_2, \ldots, r_p}(\mathbf{x_0}).$$

The multidimensional analog of smoothing splines is a minimization problem that also uses regularization:

$$\hat{\beta}_{\text{TS}} = \underset{\beta}{\operatorname{argmin}} \|\mathbf{Y} - f(\mathbf{X})\|^2 + \lambda P(f),$$

where $P$ is an appropriate penalty function for stabilizing a function $f$ acting upon a $p$-dimensional input. One of the most common penalty functions when $p = 2$ is

$$P(f) = \int \int \left[ \frac{\partial^2 f(\mathbf{x})}{\partial x_1^2} \right]^2 + 2 \left[ \frac{\partial^2 f(\mathbf{x})}{\partial x_1 \partial x_2} \right]^2 + \left[ \frac{\partial^2 f(\mathbf{x})}{\partial x_2^2} \right]^2 dx_1 dx_2,$$

which is what is known as a **thin-plate spline**. These can be extended to the case of $p > 2$, but the formulas for the basis functions are quite complex. Thin-plate splines share many properties with cubic smoothing splines when $p = 1$, such as if $\lambda \to 0$ or $\lambda \to \infty$, the solution approaches an interpolating function or least squares plane, respectively. In fact, cubic smoothing splines are a member of the family of thin-plate splines. An important feature of thin-plate splines is that they are **isotropic**, which means that curvature in all directions is penalized equally (Wood, 2003).

Finally, we mention two more methods. The first is multivariate adaptive regression splines, which is a method that is also used for decision tree methods for multivariate data. As such, we provide a broader discussion of this topic with the material on tree-based methods in Chapter 24. The second method we mention focuses on a subclass of regularization problems in a space of functions called a **reproducing kernel Hilbert space**. The details for this approach are quite technical, but can be found in Kimeldorf and Wahba (1971) and Wahba (1990).

## 17.4 Other Nonparametric Regression Procedures

Next, we provide a brief discussion of some other nonparametric regression procedures.

### Regressogram

One simple nonparametric smoother is called the **regressogram**, which is a portmanteau of "regression" and "histogram." Conceptually, the regressogram is a very simple extension of a histogram for regression data. First, we divide the horizontal axis (i.e., the $x$-axis) into bins of equal width, which is the smoothing parameter $h$ for this smoother. We then take the sample average of the response values in each bin, resulting in a step function. Specifically, suppose we have a sample of size $n$ such that $a \leq x_i \leq b$ for $i = 1, \ldots, n$. We partition $(a, b)$ into $k$ equally-spaced bins, $B_1, \ldots, B_k$. Each bin has a width of $h$. Define $\hat{m}_h(x_0)$ for $x_0 \in B_j$ as

$$\hat{m}_h(x_0) = \frac{1}{n_j} \sum_{i=1}^{n} y_i \mathrm{I}\{x_i \in B_j\},$$

where $n_j$ is the number of data points in bin $B_j$. Note that as the binwidth $h$ decreases, the resulting estimator becomes less smooth.

### Nearest-Neighbor Regression

Assume now that instead of using a fixed bandwidth to locally estimate a regression relationship, we use a fixed number of data points, say $k$, which we again call the span. This approach is called $k$-**nearest neighbor regression**. To develop estimators for this set-up, suppose we have a sample of size $n$ with one response and $p$ predictors. For a fixed $p$-dimensional point $\mathbf{x}_0$, let

$$d_i(\mathbf{x}_0) = \|\mathbf{x}_0 - \mathbf{x}_i\|, \tag{17.10}$$

where $0 \leq d_{(1)}(\mathbf{x}_0) \leq d_{(2)}(\mathbf{x}_0) \leq \ldots \leq d_{(n)}(\mathbf{x}_0)$ are the corresponding order statistics. Let $R_{\mathbf{x}_0} \equiv d_{(k)}(\mathbf{x}_0)$ denote the $k^{\text{th}}$ order statistic of the Euclidean distance values at the fixed point $\mathbf{x}_0$. Then, the $k$-nearest neighbor estimator is given by

$$\hat{m}(\mathbf{x}_0) = \frac{1}{k} \sum_{i=1}^{n} \mathrm{I}\{\|\mathbf{x}_i - \mathbf{x}_0\| \leq R_{\mathbf{x}_0}\} y_i,$$

which can also be viewed as a kernel estimator with a uniform kernel:

$$\hat{m}(\mathbf{x}_0) = \frac{\sum_{i=1}^{n} K\left(\frac{\|\mathbf{x}_i - \mathbf{x}_0\|}{R_{\mathbf{x}_0}}\right) y_i}{\sum_{i=1}^{n} K\left(\frac{\|\mathbf{x}_i - \mathbf{x}_0\|}{R_{\mathbf{x}_0}}\right)}.$$

The above estimator can be generalized to using other kernel functions, like those listed in Table 17.1. Again, notice that unlike the local regression methods we discussed earlier in this chapter, the width of the neighborhood in nearest-neighbor regression will typically change.[7] If we estimate $m(\cdot)$ at a point where the data are sparse, then we may, in fact, have the $k$-nearest neighbors to that point being quite spread out, resulting in a wide neighborhood. As noted in Altman (1992), if the population variance is constant and there are no replicates, then the variability of the estimator will be the same in every neighborhood. However, there is no clear way to use nearest-neighbor regression in the presence of replicates. See Altman (1992) for a comparison between kernel regression and nearest-neighbor regression.

### Median Smoothing Regression

Our objective thus far has been to estimate the conditional expectation $m(\mathbf{x}_0) = \mathrm{E}(Y|\underline{X} = \mathbf{x}_0)$. An approach that is more robust to outliers is to estimate the **conditional mean function**:

$$m(\mathbf{x}_0) = \mathrm{med}(Y|\underline{X} = \mathbf{x}_0).$$

**Median smoothing regression** provides a framework for estimating the conditional median function using the $k$-nearest neighbors approach. Let $y_1(\mathbf{x}_0), y_2(\mathbf{x}_0), \ldots, y_k(\mathbf{x}_0)$ be the response values corresponding to the $k$-nearest neighbors of $\mathbf{x}_0$. In other words, these are the response values corresponding to the predictor values that are the $k$-nearest neighbors based on the order statistics $0 \leq d_{(1)}(\mathbf{x}_0) \leq d_{(2)}(\mathbf{x}_0) \leq \ldots \leq d_{(k)}(\mathbf{x}_0)$ as defined in (17.10). Then, the median smoother is defined as

$$\hat{m}(\mathbf{x}_0) = \mathrm{med}\{y_1(\mathbf{x}_0), y_2(\mathbf{x}_0), \ldots, y_k(\mathbf{x}_0)\}.$$

### Orthogonal Series Regression

As we noted in our discussion on smoothing splines, under certain regularity conditions we can represent functions as a series of basis functions. One such series of basis functions is a Fourier series, which we briefly highlighted in our discussion in Chapter 13 on performing a spectral analysis for time series data. Suppose that $m(\cdot)$ can be represented by a Fourier series:

$$m(\mathbf{x}_0) = \sum_{j=0}^{\infty} \beta_j \phi_j(\mathbf{x}_0),$$

---

[7] An exception would be if the predictor values were chosen, say, from a properly designed experiment.

where for $j = 0, \ldots, \infty$, $\phi_j$ is a known basis and the $\beta_j$ are unknown Fourier coefficients. As noted in Chapter 4 of Härdle et al. (2004b), we obviously cannot estimate an infinite number of coefficients from a finite number of observations $n$, so we must choose a finite number of terms $h < n$ to include in our Fourier series representation. For estimation, one approach is to simply consider the first $h + 1$ terms in the (Fourier) series representation:

$$m(\mathbf{x}_0) = \sum_{j=0}^{h} \beta_j \phi_j(\mathbf{x}_0).$$

Then, for our sample of size $n$, regress the $y_i$ on the $\phi_0(\mathbf{x}_i), \phi_1(\mathbf{x}_i), \ldots, \phi_h(\mathbf{x}_i)$ and obtain estimates of $\beta_0, \beta_1, \ldots, \beta_h$.

Second, we can choose the $\phi_j$ to be an orthogonal basis; i.e.,

$$\int w(x)\phi_k(x)\phi_l(x)dx = \delta_{kl}c_l,$$

where $w(\cdot)$ is a weighting function, $c_l$ is some quantity that (possibly) depends on $l$, and

$$\delta_{kl} = \begin{cases} 1, & \text{if } k = l; \\ 0, & \text{if } k \neq l; \end{cases}$$

which is the **Kronecker delta**. If $c_l \equiv 1$, then the $\phi_j$ are also orthonormal. Types of orthonormal bases commonly considered are trigonometric (as done in time series), splines, and polynomials. Examples of orthogonal (or orthonormal) polynomials include Hermite polynomials, Jacobi polynomials, and Legendre polynomials. The $j^{\text{th}}$ Fourier coefficient can be written as the expectation of the respective basis function:

$$\beta_j = \sum_{k=0}^{\infty} \beta_k \delta_{jk} = \sum_{k=0}^{\infty} \phi_j(x)\phi_k(x)dx = \int \phi_j(x)m(x)dx = \mathrm{E}[\phi_j(X)].$$

Then, the estimate $\hat{\beta}_j$ can be found using a number of methods. For an overview of such estimation procedures and other practical considerations of orthogonal series, see Efromovich (2010) and the references therein.

One specific orthonormal series is based on the **wavelet basis**, where the functions are calculated using

$$\phi_{jk}(x_0) = 2^{j/2}\phi(2^j x_0 - k),$$

where $2^j$ is a scaling factor, $k$ is a shift factor, and

$$\phi(t) = \begin{cases} -1, & \text{if } 0 \leq t \leq \frac{1}{2}; \\ 1, & \text{if } \frac{1}{2} < t \leq 1; \\ 0, & \text{otherwise} \end{cases}$$

is called the **mother Haar wavelet**. For **wavelet regression** we estimate the coefficients in a somewhat similar manner to the notions referenced above, but we also utilize a notion of nonlinear shrinkage called **thresholding**, which is intended to keep large coefficients and then set the others equal to 0. More on the technical details of wavelet regression, such as thresholding, can be found in Chapter 9 of Wasserman (2006) and the references therein.

## 17.5   Examples

**Example 17.5.1.** *(Gamma-Ray Burst Data)*
Gamma-ray bursts (GRBs) are short, intense flashes of gamma-ray radiation that occur at (seemingly) random times and locations in space. These phenomena are the brightest electromagnetic events known to occur in the universe. There are copious data collected on GRBs and we will analyze one such dataset as reported in Blustin et al. (2006)[8]. This dataset consists of $n = 63$ observations with the following two variables:

- $Y$: X-ray flux (in units of $10^{-11}$ erg/cm$^2$/s, 2-10 keV)

- $X$: time (in seconds)

A plot of the raw GRB data is given in Figure 17.2(a). Blustin et al. (2006) analyzed these data using a broken power law model, which is essentially a log-log piecewise regression. The log-transformed data are given in Figure 17.2(b). First we fit a simple linear regression model to the log-transformed data (Model 1). We then fit two different piecewise models: one with a knot at $\xi_1 = 7.0$ (Model 2) and another with a knot at $\xi_1 = 8.5$ (Model 3). The BIC values for these three fitted models suggest we select Model 2:

```
#############################
    Model.1   Model.2   Model.3
BIC 28.57357 -62.15637 -25.27641
#############################
```

The solid line overlaid on the log-transformed data in Figure 17.2(b) is the estimated Model 2 piecewise linear regression line. The coefficient estimates for both the log of time and the term with the knot are found to be statistically significant:

```
#############################
Coefficients:
            Estimate Std. Error t value Pr(>|t|)
(Intercept)  9.11007    0.18977   48.01   <2e-16 ***
```

---

[8] Each observation is actually binned data to ensure a count of at least 20 counts per bin, which is often done by astronomers to produce error bars for uncertainty quantification.

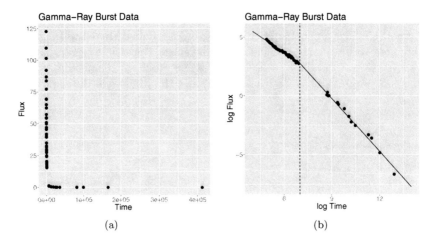

(a)                                          (b)

**FIGURE 17.2**
(a) The raw GRB data. (b) The GRB data on a log-log scale with the estimated piecewise linear regression model with a knot at $\xi_1 = 7.0$ overlaid.

```
lTIME       -0.90550    0.03108  -29.13   <2e-16 ***
lTIME.7     -0.58833    0.04055  -14.51   <2e-16 ***
---
Signif. codes:  0 *** 0.001 ** 0.01 * 0.05 . 0.1   1

Residual standard error: 0.1327 on 60 degrees of freedom
Multiple R-squared:  0.9974,Adjusted R-squared:  0.9973
F-statistic: 1.153e+04 on 2 and 60 DF,  p-value: < 2.2e-16
#############################
```

**Example 17.5.2.** *(Quasar Data)*
Quasars are the brightest objects found in the universe. One way quasars
are identified is through their redshift, which occurs when its electromagnetic
radiation is increased in wavelength. The dataset we will examine is a catalog
of $n = 77,429$ quasars from the $5^{th}$ data release of the Sloan Digital Sky
Survey Schneider et al. (2007). The following two variables are of interest:

- $Y$: difference in magnitudes between the $r$ and $i$ photometric bands

- $X$: measure of redshift

Figure 17.3 provides a scatterplot of these data. Clearly the relationship between these two variables is highly nonlinear. Moreover, we have no theory to
suggest a possible model. Thus, we attempt to characterize a functional relationship between these two variables using nonparametric regression methods,
which is also suggested in Chapter 6 of Feigelson and Babu (2012).

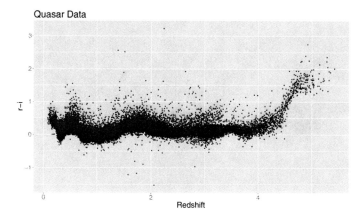

**FIGURE 17.3**
Scatterplot of the quasar dataset.

The scatterplot in Figure 17.3 can be improved upon, especially since it does not provide a good indication of how dense the data are in certain regions. We can do so by converting the scatterplot into a gray-scale image using a two-dimensional **averaged shifted histogram estimator**, which was proposed in Chapter 5 of Scott (2015). We then obtain three different nonparametric regression fits for these data using LOESS, smoothing splines, and a Nadaraya–Watson estimate. Since the data are so dense, we will focus on smaller window of these local techniques so as to highlight possibly interesting nonlinear features. We'll briefly compare estimates of the MSE (or residual standard error) for these three fits.

For the LOESS fit, we used local quadratic fitting (i.e., degree of $q = 2$) and a span of $\gamma = 0.10$. Some of the relevant output for the LOESS fit is as follows:

```
##############################
Number of Observations: 77429
Equivalent Number of Parameters: 29.84
Residual Standard Error: 0.12
Trace of smoother matrix: 33   (exact)
##############################
```

For the smoothing spline fit, we used 100 knot values and estimated the smoothing parameter using leave-one-out crossvalidation. The value identified is $\omega = 0.15067$. Some of the relevant output for the smoothing spline fit is as follows:

```
##############################
Smoothing Parameter: spar=0.15067
```

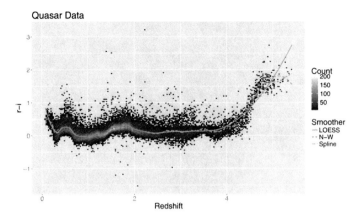

**FIGURE 17.4**
Scatterplot of the quasar dataset using an average shifted histogram. Nonparametric regression fits using LOESS, smoothing splines, and a Nadaraya–Watson estimate are also shown. Colors are in e-book version.

```
Equivalent Degrees of Freedom (Df): 90.44524
Residual standard error: 0.10266
Penalized Criterion: 452.95821
GCV: 0.01392
#############################
```

For the fit using a Nadaraya–Watson estimator, we used a Gaussian kernel and a fixed bandwidth value of $h = 0.10$. Some of the relevant output for this kernel regression fit is as follows:

```
#############################
Kernel Regression Estimator: Local-Constant
Bandwidth Type: Fixed
Residual standard error: 0.1216598
R-squared: 0.5440073
#############################
```

Based on all of the above fits, they have residual standard errors that are all close; however, the smoothing spline fit has the smallest of the three fits.

Figure 17.4 is a plot of the quasars data using the averaged shifted histogram estimator. The three nonparametric regression fits are overlaid. There is generally good visual agreement between the three fits. However, the smoothing spline fit appears to be picking up some additional features relative to the other two fits. Moreover, there tends to be some disagreement toward the boundary of the redshift values. For kernel regression methods — which in this example we use the Nadaraya–Watson estimator — this is a result of the well-known boundary bias problem. For all three methods, general trends near the boundary can be more heavily-influenced based on how "fine" you perform estimation through quantities like the span and smoothness parameters.

**Example 17.5.3.** *(LIDAR data)*
Sigrist (1994) presented data from an experiment involving light detection and ranging, or LIDAR, from two laser sources. This classic dataset is of size $n = 221$ and the variables in this experiment are:

- $Y$: logarithm of the ratio of the light received from the two laser sources

- $X$: distance traveled before the light is reflected back to its laser source

Transforming the original ratio measurement using a logarithm was proposed in Ruppert et al. (2003), who use this dataset throughout their textbook to demonstrate various nonparametric and semiparametric regression procedures. A scatterplot of the data is given in Figure 17.5(a).

We proceed to fit various nonparametric smoothers to these data. We merely illustrate the visual summaries of these fits and comment on which appear to provide (relatively speaking) better fits. To motivate what we expect the general shape of the relationship to look like, we first construct a regressogram of these data with 10 bins. This regressogram is given in Figure 17.5.

We first fit the LIDAR data using $k$-nearest neighbor regression, which is given in Figure 17.6(a). Fits using a rectangular (uniform) kernel and a biweight kernel are provided and we use $k = 49$. Both fits are comparable, but use of the biweight kernel appears to provide a slightly smoother fit. Figure 17.6(b) are Wavelet regression fits using soft thresholding and hard thresholding. The soft thresholding results in a smoother fit than the hard thresholding, but both fits are clearly overfitting with respect to the $k$-nearest neighbor regression fit.

Figure 17.7(a) is the fit obtained using median regression with, again, $k = 49$. This fit is clearly not as smooth as those obtained under $k$-nearest neighbor regression; i.e., it is overfitting. Figure 17.7(b) is the orthogonal regression fit based on Legendre polynomials of order 5. This appears to be a good fit, but in some regions (e.g., the portion of the data where there is a clear decreasing trend starting at a range value of about 550) the polynomial form is a bit restrictive and results in some lack of fit. Overall, the $k$-nearest neighbor regression fit with a biweight kernel appears to provide the best fit.

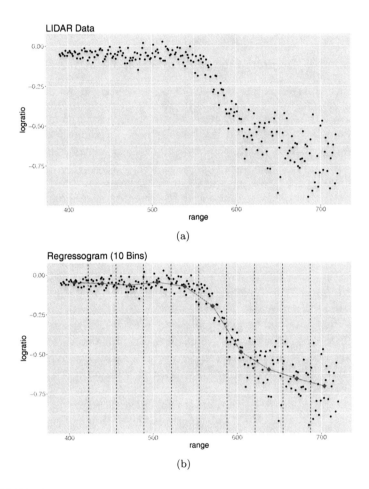

(a)

(b)

**FIGURE 17.5**
(a) Plot of the LIDAR data. (b) Regressogram of the LIDAR data with 10
bins.

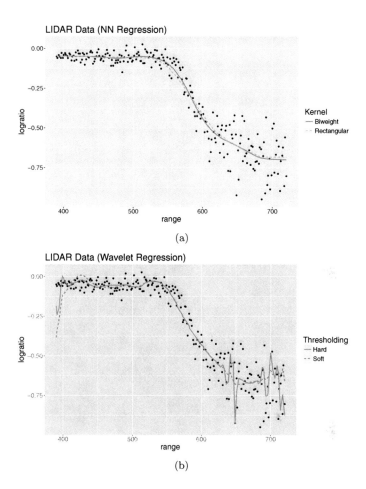

(a)

(b)

**FIGURE 17.6**
Smoother fits for (a) nearest neighbor regression using a rectangular kernel
(red line) and a biweight kernel (green line) and (b) Wavelet regression using
soft thresholding (red line) and hard thresholding (green line).

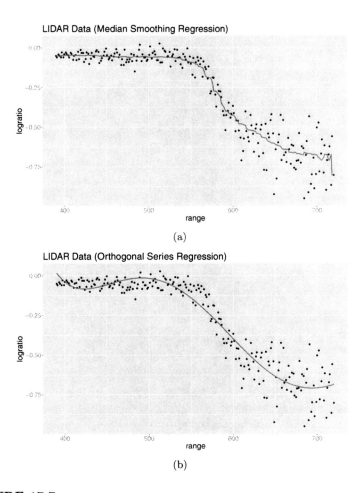

(a)

(b)

**FIGURE 17.7**
Smoother fits for (a) median regression and (b) orthogonal regression based
on Legendre polynomials.

# 18

## Regression Models with Censored Data

Suppose we wish to estimate the parameters of a distribution where only a portion of the data is known. When some of the data has a measurement that exceeds (or falls below) a threshold and only that threshold value is recorded for those observations, then they are said to be **censored**. When some of the data exceeds (or falls below) a threshold, but the data is omitted from the dataset, then those observations are said to be **truncated**. This chapter deals primarily with the analysis of censored data by first introducing the area of reliability (survival) analysis and then presenting some of the basic tools and models from this area as a segue into a regression setting. We also devote a section to discussing truncated regression models.

## 18.1   Overview of Survival and Reliability Analysis

It is helpful to formally define the area of analysis which focuses on estimating models with censored data. **Survival analysis** concerns the analysis of data from biological events associated with the study of animals and humans. **Reliability analysis** concerns the analysis of data from events associated with the study of engineering applications. We will utilize terminology from both areas for the sake of completeness.

Survival (reliability) analysis studies the distribution of lifetimes (failure times). Such studies consist of the elapsed time between an initiating event and a terminal event. For example:

- Consider the amount of time that individuals are in a cancer study. The initiating time could be the diagnosis of cancer or the start of treatment. The terminal event could be death or remission of the disease.

- Consider a study about the lifetime of motors in mechanical punch presses. The initiating time could be the date the presses were first brought online. The terminal event could be complete machine failure or the first time it must be brought off-line for maintenance.

Data collected in studies like the above are a combination of complete and censored values, which means a terminal event has occurred or not occurred, respectively. Formally, let $Y$ be the observed time from the study, $T$ denote the actual event time, which is a latent variable, and $t$ denote some known threshold value where the values are censored. Observations in a study can be censored in the following ways:

- **Right censoring**: This occurs when an observation has dropped out, been removed from a study, or did not reach a terminal event prior to termination of the study. In other words, $Y \leq T$ such that

$$Y = \begin{cases} T, & T < t; \\ t, & T \geq t. \end{cases}$$

- **Left censoring**: This occurs when an observation reaches a terminal event before the first time point in the study. In other words, $Y \geq T$ such that

$$Y = \begin{cases} T, & T > t; \\ t, & T \leq t. \end{cases}$$

- **Interval censoring**: This occurs when a study has discrete time points and an observation reaches a terminal event between two of the time points. In other words, for discrete time increments $0 = t_1 < t_2 < \ldots < t_r < \infty$, we have $Y_1 < T < Y_2$ such that for $j = 1, \ldots, r - 1$,

$$Y_1 = \begin{cases} t_j, & t_j < T < t_{j+1}; \\ 0, & \text{otherwise} \end{cases}$$

and

$$Y_2 = \begin{cases} t_{j+1}, & t_j < T < t_{j+1}; \\ \infty, & \text{otherwise.} \end{cases}$$

- **Double censoring**: This is when all of the above censoring mechanisms can occur in a study.

Moreover, there are two criteria that define the type of censoring in a study. If the experimenter controls the type of censoring, then we have **non-random censoring**, of which there are two types:

- **Type I** or **time-truncated** censoring, which occurs if an observation is still alive (in operation) when a test is terminated after a pre-determined length of time. It is assumed that we know the exact times of failures when failures are present.

- **Type II** or **failure-truncated** censoring, which occurs if an observation is still alive (in operation) when a test is terminated after a pre-determined number of failures is reached. While advantageous in terms of knowing how many failures you will need to test, this approach is typically not used much in practice.

Suppose $T$ has probability density function $f(t)$ with cumulative distribution function $F(t)$. Since we are interested in survival times (lifetimes), the support of $T$ is $(0, +\infty)$. There are three functions usually of interest in a survival (reliability) analysis:

- The **survival function** $S(t)$ (or **reliability function** $R(t)$) is given by:

$$S(t) = R(t) = \int_t^{+\infty} f(x)dx = 1 - F(t).$$

  This is the probability that an individual survives (or something is reliable) beyond time $t$ and is usually the first quantity studied.

- The **hazard rate** $h(t)$ (or **conditional failure rate**) is the probability that an observation at time $t$ will experience a terminal event in the next instant. It is given by:

$$h(t) = \frac{f(t)}{S(t)} = \frac{f(t)}{R(t)}.$$

  The empirical hazard (conditional failure) rate function is useful in identifying which probability distribution to use if it is not already specified.

- The **cumulative hazard function** $H(t)$ (or **cumulative conditional failure function**) is given by:

$$H(t) = \int_0^t h(x)dx.$$

The above terms and quantities are only the basics when it comes to survival (reliability) analysis. However, they provide enough of a foundation for our interests. More details on survival (reliability) analysis can be found in the texts by Klein and Moeschberger (2003) and Tableman and Kim (2004).

## 18.2 Censored Regression Model

We now turn our attention to when a set of predictors (or covariates) is also measured with the observed time. **Censored regression models** relate the unknown variable $T$ — which can be left-censored, right-censored, or both — to a linear combination of the covariates $X_1, \ldots, X_{p-1}$. The model is

$$T = \beta_0 + \beta_1 X_1 + \ldots + \beta_{p-1} X_{p-1} + \epsilon, \tag{18.1}$$

where $\epsilon$ is normally distributed with mean 0 and variance $\sigma^2$ and

$$Y = \begin{cases} t_1, & T \leq t_1; \\ T, & t_1 < T < t_2; \\ t_2, & T \geq t_2. \end{cases}$$

In the above, if $t_1 = -\infty$ or $t_2 = \infty$, then the dependent variable is not left-censored or right-censored, respectively.

Suppose we have a sample of size $n$ and wish to fit the censored regression model in (18.1). The loglikelihood function[1] for estimating the parameters $\beta$ and $\sigma^2$ is

$$\ell(\beta, \sigma) = \sum_{i=1}^{n} \left[ I\{y_i = t_1\} \log \Phi \left( \frac{t_1 - x_i^T \beta}{\sigma} \right) + I\{y_i = t_2\} \log \Phi \left( \frac{x_i^T \beta - t_2}{\sigma} \right) \right.$$
$$\left. + (1 - I\{y_i = t_1\} - I\{y_i = t_2\}) \left( \log \phi \left( \frac{y_i - x_i^T \beta}{\sigma} \right) - \log(\sigma) \right) \right],$$

where $\phi(\cdot)$ and $\Phi(\cdot)$ are the probability density function and cumulative distribution function of a standard normal random variable, respectively. The above likelihood problem can also be modified for the settings with only left-censoring or only right-censoring.

Censored regression models are also generalized versions of what are called **Tobit models** (Tobin, 1958). A **standard Tobit model** uses the same model in (18.1) with the same assumption on the error term, but the dependent variable is subject to left-censoring at a threshold of $t = 0$; i.e.,

$$Y = \begin{cases} T, & T > 0; \\ 0, & T \leq 0. \end{cases}$$

It can be shown for the observed variable $Y$, that

$$E[Y|Y > 0] = \underline{X}^T \beta + \sigma \lambda(\alpha),$$

where $\alpha = (0 - \underline{X}^T \beta)/\sigma$ and

$$\lambda(\alpha) = \frac{\phi(\alpha)}{1 - \Phi(\alpha)},$$

which is called the **inverse Mills ratio** and reappears later in our discussion about truncated regression models. The above standard Tobit regression model can be modified to reflect the more general notions of censoring presented in the previous section as well as more complex notions that introduce additional dependent and latent variables. Amemiya (1984) classified these different type of Tobit models into five categories: Tobit Type I models to Tobit Type V models. The standard Tobit model presented above is classified as a Tobit Type I model.

Finally, we note that censored regression models are part of a broader class of survival (reliability) regression models. However, they are commonly used so that is why it is illustrative to discuss them separately from the broader class of survival (reliability) regression models, which are discussed in the next section.

---

[1]See Appendix C for further details on likelihood functions.

## 18.3 Survival (Reliability) Regression

In our examples from the introduction, some covariates you may also measure include:

- Gender, age, weight, and previous ailments of the cancer patients.

- Manufacturer, metal used for the drive mechanisms, and running temperature of the machines.

Let $\underline{X}^*$ be the a $(p-1)$-dimensional vector of covariates without an additional value of 1 appended as the first entry of the vector. We are interested in the model

$$T^* = \beta_0 + \underline{X}^{*\mathrm{T}}\boldsymbol{\beta}^* + \epsilon, \tag{18.2}$$

where $\boldsymbol{\beta}^*$ is a $(p-1)$-dimensional vector, $T^* = \log(T)$, and $\epsilon$ has a certain distribution. Then

$$T = \exp(T^*) = e^{\beta_0 + \underline{X}^{*\mathrm{T}}\boldsymbol{\beta}^*} e^\epsilon = e^{\beta_0 + \underline{X}^{*\mathrm{T}}\boldsymbol{\beta}^*} T^\sharp,$$

where $T^\sharp = e^{(\epsilon)}$. Thus, the covariate acts multiplicatively on the survival time $T$.

The distribution of $\epsilon$ will allow us to determine the distribution of $T^*$. Each possible probability distribution has a different hazard rate $h(t)$. Furthermore, in a survival *regression* setting, we assume the hazard rate at time $t$ for an individual has the form

$$h(t|\underline{X}^*) = h_0(t)k(\underline{X}^{*\mathrm{T}}\boldsymbol{\beta}^*)$$
$$= h_0(t)e^{\underline{X}^{*\mathrm{T}}\boldsymbol{\beta}^*}.$$

In the above, $h_0(t)$ is called the **baseline hazard** and is the value of the hazard function when $\underline{X}^* = \mathbf{0}$ or when $\boldsymbol{\beta}^* = \mathbf{0}$. Note that in the expression for $T^*$ in (18.2), we separated the intercept term $\beta_0$ as it becomes part of the baseline hazard. Also, $k(\cdot)$ in the equation for $h(t|\underline{X}^*)$ is a specified link function, which for our purposes will be $\exp\{\cdot\}$.[2]

We next outline some of the possible and more common distributions assumed for $\epsilon$. These are summarized in Table 18.1 along with their corresponding density function, survival (reliability) function, and hazard function. These distributions have up to three parameters, which help control various aspects of the density curve: location, scale, and shape. These should be considered with respect to the general shape that the data appear to exhibit or have historically exhibited. Also, there are a few quantities used in Table 18.1 that we define here:

---

[2]More on the role of link functions is discussed in the context of generalized linear models in Chapter 21.

**TABLE 18.1**

Densities, survival functions, and hazard functions of some commonly used distributions in survival regression

| Density Function $f(t)$ | Survival Function $S(t)$ | Hazard Function $h(t)$ |
|---|---|---|
| Normal, $-\infty < \mu < \infty, \sigma^2 > 0, -\infty < t < \infty$ | | |
| $\frac{1}{\sigma}\phi\left(\frac{t-\mu}{\sigma}\right)$ | $1 - \Phi\left(\frac{t-\mu}{\sigma}\right)$ | $\frac{\frac{1}{\sigma}\phi\left(\frac{t-\mu}{\sigma}\right)}{1-\Phi\left(\frac{t-\mu}{\sigma}\right)}$ |
| Truncated Normal, $-\infty < \mu < \infty, \sigma^2 > 0, a < t < b$ | | |
| $\frac{\frac{1}{\sigma}\phi\left(\frac{t-\mu}{\sigma}\right)}{\Phi\left(\frac{b-\mu}{\sigma}\right)-\Phi\left(\frac{a-\mu}{\sigma}\right)}$ | $\frac{\Phi\left(\frac{b-\mu}{\sigma}\right)-\Phi\left(\frac{t-\mu}{\sigma}\right)}{\Phi\left(\frac{b-\mu}{\sigma}\right)-\Phi\left(\frac{a-\mu}{\sigma}\right)}$ | $\frac{\frac{1}{\sigma}\phi\left(\frac{t-\mu}{\sigma}\right)}{\Phi\left(\frac{b-\mu}{\sigma}\right)-\Phi\left(\frac{t-\mu}{\sigma}\right)}$ |
| Lognormal, $\mu > 0, \sigma^2 > 0, t > 0$ | | |
| $\frac{1}{t\sigma}\phi\left(\frac{\log t-\mu}{\sigma}\right)$ | $1 - \Phi\left(\frac{\log t-\mu}{\sigma}\right)$ | $\frac{\phi\left(\frac{\log t-\mu}{\sigma}\right)}{t\sigma\left(1-\Phi\left(\frac{\log t-\mu}{\sigma}\right)\right)}$ |
| Weibull, $\alpha > 0, \beta > 0, 0 < t < \delta$ | | |
| $\alpha\beta(\beta(t-\delta))^{\alpha-1}e^{-(\beta(t-\delta))^{\alpha}}$ | $e^{-(\beta(t-\delta))^{\alpha}}$ | $\alpha\beta(\beta(t-\delta))^{\alpha-1}$ |
| Gumbel, $-\infty < \mu < \infty, \sigma^2 > 0, \infty < t < \infty$ | | |
| $\frac{1}{\sigma^2}\exp\left\{-\left(\frac{t-\mu}{\sigma^2}+e^{-\left(\frac{t-\mu}{\sigma^2}\right)}\right)\right\}$ | $1 - \exp\left\{-e^{-\left(\frac{t-\mu}{\sigma^2}\right)}\right\}$ | $\frac{e^{-\left(\frac{t-\mu}{\sigma^2}\right)}}{\sigma^2\left(\exp\left\{-e^{-\left(\frac{t-\mu}{\sigma^2}\right)}\right\}-1\right)}$ |
| Exponential, $\lambda > 0, t > 0$ | | |
| $\lambda e^{-\lambda t}$ | $e^{-\lambda t}$ | $\lambda$ |
| Logistic, $-\infty < \mu < \infty, \sigma^2 > 0, -\infty < t < \infty$ | | |
| $\frac{e^{-\left(\frac{t-\mu}{\sigma^2}\right)}}{\sigma^2\left(1+e^{-\left(\frac{t-\mu}{\sigma^2}\right)}\right)^2}$ | $\left(1+e^{\left(\frac{t-\mu}{\sigma^2}\right)}\right)^{-1}$ | $\sigma^{-2}\left(1+e^{-\left(\frac{t-\mu}{\sigma^2}\right)}\right)^{-1}$ |
| Log-logistic, $\alpha > 0, \lambda > 0, t > 0$ | | |
| $\frac{\lambda\alpha(\lambda t)^{\alpha-1}}{(1+(\lambda t)^{\alpha})^2}$ | $\frac{1}{1+(\lambda t)^{\alpha}}$ | $\frac{\lambda\alpha(\lambda t)^{\alpha-1}}{1+(\lambda t)^{\alpha}}$ |
| Generalized Gamma, $\alpha > 0, \beta > 0, -\infty < \gamma < \infty, t > \gamma$ | | |
| $\frac{\alpha^{-\beta}(t-\gamma)^{\beta-1}e^{-\left(\frac{t-\gamma}{\alpha}\right)}}{\Gamma(\beta)}$ | $1 - \frac{\mathcal{I}\left(\alpha^{-1}(t-\gamma),\beta\right)}{\Gamma(\beta)}$ | $\frac{\alpha^{-\beta}(t-\gamma)^{\beta-1}e^{-\left(\frac{t-\gamma}{\alpha}\right)}}{\Gamma(\beta)-\mathcal{I}\left(\alpha^{-1}(t-\gamma),\beta\right)}$ |
| Beta, $\alpha > 0, \beta > 0, a \leq t \leq b$ | | |
| $\frac{(t-a)^{\alpha-1}(b-t)^{\beta-1}}{\mathcal{B}(\alpha,\beta)(b-a)^{\alpha+\beta-1}}$ | $\frac{\int_a^t (u-a)^{\alpha-1}(b-u)^{\beta-1}\,du}{\mathcal{B}(\alpha,\beta)(b-a)^{\alpha+\beta-1}}$ | $\frac{(t-a)^{\alpha-1}(b-t)^{\beta-1}}{(b-a)\int_a^t (u-a)^{\alpha-1}(b-u)^{\beta-1}\,du}$ |

- The **gamma function** is defined as

$$\Gamma(z) = \begin{cases} (z-1)! & \text{if } z \text{ is an integer,} \\ \int_0^\infty u^{z-1}e^{-u}du & \text{otherwise.} \end{cases} \qquad (18.3)$$

- The **incomplete gamma function** is defined as

$$\mathcal{I}(a,b) = \int_0^a u^{b-1}e^{-u}du.$$

- The **beta function** is defined as

$$\begin{aligned} \mathcal{B}(a,b) &= \int_0^\infty u^{a-1}(1-u)^{b-1}du \\ &= \frac{\Gamma(a)\Gamma(b)}{\Gamma(a+b)}. \end{aligned}$$

Here is a brief discussion of those distributions that are summarized in Table 18.1:

- The normal distribution with location parameter (mean) $\mu$ and scale parameter (variance) $\sigma^2$. As we have seen, this is one of the more commonly used distributions in statistics, but is infrequently used for lifetime distribution as it allows negative values while lifetimes are always positive. One possibility is to consider a truncated normal or a log transformation.

- The truncated normal distribution is similar to the normal distribution with location parameter (mean) $\mu$ and scale parameter (variance) $\sigma^2$, but the observed values are constrained to be in the range $[a, b]$. Typically, $a$ and $b$ are assumed known. This is more appropriate for a lifetime distribution compared to a regular normal distribution.

- The lognormal distribution with location parameter $\mu$ and scale parameter $\sigma^2$. The parameters in the lognormal and normal distribution are also related. Specifically, if $T$ has a lognormal distribution, then $\log(T)$ has a normal distribution.

- The (three-parameter) Weibull distribution with location parameter $\delta$, scale parameter $\beta$, and shape parameter $\alpha$. The Weibull distribution is probably most commonly used for time to failure data since it is fairly flexible to work with. $\delta$ gives the minimum value of the random variable $T$ and is often set to 0 so that the support of $T$ is positive. Assuming $\delta = 0$ provides the more commonly used two-parameter Weibull distribution.

- The Gumbel distribution (or extreme-value distribution) with location parameter $\mu$ and scale parameter $\sigma^2$ is sometimes used, but more often it is presented due to its relationship to the Weibull distribution. If $T$ has a Weibull distribution, then $\log(T)$ has a Gumbel distribution.

- The exponential distribution with rate parameter $\lambda$. This is, perhaps, one of the most commonly used distributions for lifetime data and it is also one of the easier density functions to work with. The exponential distribution is a model for lifetimes with a constant failure rate. If $T$ has an exponential distribution, then $\log(T)$ has a standard Gumbel distribution; i.e., the scale of the Gumbel distribution is 1. There are other ways to generalize the exponential distribution, which includes incorporating a location (or threshold) parameter, $\delta$. In this case, $\delta$ would be the minimum value of the random variable $T$.

- The logistic distribution with location parameter $\mu$ and scale parameter $\sigma^2$. This distribution is very similar to the normal distribution, but is used in cases where there are heavier tails; i.e., higher probability of the data occurring out in the tails of the distribution.

- The log-logistic distribution with scale parameter $\lambda$ and shape parameter $\alpha$. If $T$ has a log-logistic distribution, then $\log(T)$ has a logistic distribution with location parameter $\mu = -1/\log(\lambda)$ and scale parameter $\sigma = 1/\alpha$. Just like with the exponential distribution, we can generalize the log-logistic distribution to incorporate a location (or threshold) parameter, $\delta$.

- The generalized gamma distribution with location parameter $\gamma$, scale parameter $\alpha$, and shape parameter $\beta$. The gamma distribution is a competitor to the Weibull distribution, but is more mathematically complicated and thus avoided where the Weibull appears to provide a good fit. The gamma distribution also arises because the sum of independent exponential random variables is gamma distributed. $\gamma$ gives the minimum value of the random variable $T$ and is often set to 0 so that the support of $T$ is positive. Setting $\delta = 0$ results in what is usually referred to as the gamma distribution.

- The beta distribution has two shape parameter, $\alpha$ and $\beta$, as well as two location parameters, $a$ and $b$, which denote the minimum and maximum of the data. If the beta distribution is used for lifetime data, then it appears when fitting data which are assumed to have an absolute minimum and absolute maximum. Thus, $a$ and $b$ are almost always assumed known as in the truncated normal.

Note that the above is *not* an exhaustive list, but provides some of the more commonly used distributions in statistical texts and software. Also, there is an abuse of notation in that duplication of certain characters (e.g., $\mu$, $\sigma$, etc.) does *not* imply a mathematical relationship between all of the distributions where those characters appear.

Estimation of the parameters can be accomplished in two primary ways. One approach is to construct a probability of the chosen distribution with your data and then apply least squares regression to this plot. A second, perhaps more pragmatic, approach is to use maximum likelihood estimation as it can be shown to be optimal in most situations. It also provides estimates of standard errors and, thus, confidence limits. Maximum likelihood estimation is commonly performed using a Newton–Raphson algorithm.

---

## 18.4 Cox Proportional Hazards Regression

Recall from the last section that we set $T^* = \log(T)$ where the hazard function is $h(t|\underline{x}^*) = h_0(t)e^{\underline{x}^{*\mathrm{T}}\beta^*}$. The Cox formation of this relationship gives

$$\log(h(t)) = \log(h_0(t)) + \mathbf{x}^{*\mathrm{T}}\beta^*,$$

which yields the following form of the linear regression model:

$$\log\left(\frac{h(t)}{h_0(t)}\right) = \mathbf{x}^{*\mathrm{T}}\beta^*.$$

Exponentiating both sides yields a ratio of the actual hazard rate and baseline hazard rate, which is called the **relative risk**:

$$\frac{h(t)}{h_0(t)} = e^{\mathbf{x}^{*\mathrm{T}}\beta^*}$$

$$= \prod_{j=1}^{p-1} e^{\beta_j x_j}.$$

Thus, the regression coefficients have the interpretation as the relative risk when the value of a covariate is increased by 1 unit. The estimates of the regression coefficients are interpreted as follows:

- A positive coefficient means there is an increase in the risk, which decreases the expected survival (failure) time.

- A negative coefficient means there is a decrease in the risk, which increases the expected survival (failure) time.

- The ratio of the estimated risk functions for two different sets of covariates (i.e., two groups) can be used to examine the likelihood of Group 1's survival (failure) time to Group 2's survival (failure) time. More specifically, let $\mathbf{x}_1^*$ and $\mathbf{x}_2^*$ be the covariates of those two groups. Then,

$$\frac{h(t|\mathbf{x}_1^*)}{h(t|\mathbf{x}_2^*)} = \frac{\exp\{\mathbf{x}_1^{*\mathrm{T}}\beta^*\}}{\exp\{\mathbf{x}_2^{*\mathrm{T}}\beta^*\}} = \exp\{(\mathbf{x}_1^* - \mathbf{x}_2^*)^{\mathrm{T}}\beta^*\}$$

is called the **hazards ratio** and is constant with respect to the time $t$.

Remember, for this model the intercept term has been absorbed by the baseline hazard.

The model we developed above is the **Cox proportional hazards regression** model (Cox, 1972) and does not include $t$ on the right-hand side. Thus, the relative risk and hazards ratio are both constant for all values of $t$. Estimation for this regression model is usually done by maximum likelihood and Newton–Raphson is usually the algorithm used. Usually, the baseline hazard is found nonparametrically, so the estimation procedure for the entire model is said to be semiparametric. Additionally, if there are failure time ties in the data, then the likelihood gets more complex and an approximation to the likelihood is usually used, such as the Breslow Approximation (Breslow, 1974) or the Efron Approximation (Efron, 1977).

## 18.5   Diagnostic Procedures

Depending on the survival regression model being used, the diagnostic measures presented here may have a slightly different formulation. We present general equations for these measures, but the emphasis is on their purpose.

### Cox–Snell Residuals

In the previous regression models we studied, residuals were defined as a difference between observed and fitted values. For survival regression, in order to check the overall fit of a model, the **Cox–Snell residual** for the $i^{\text{th}}$ observation in a dataset is used:

$$r_{C_i} = \hat{H}_0(t_i)e^{\mathbf{x}^{*\mathrm{T}}\hat{\beta}^*}.$$

In the above, $\hat{\beta}^*$ is the maximum likelihood estimate of the regression coefficient vector. $\hat{H}_0(t_i)$ is a maximum likelihood estimate of the **baseline cumulative hazard function** $H_0(t_i)$, defined as:

$$H_0(t) = \int_0^t h_0(x)dx.$$

Notice that $r_{C_i} > 0$ for all $i$. The way we check for goodness-of-fit with the Cox-Snell residuals is to estimate the cumulative hazard rate of the residuals, say $\hat{H}_{r_C}(t_{r_{C_i}})$, from whichever distribution is assumed, and then plot $\hat{H}_{r_C}(t_{r_{C_i}})$ versus $r_{C_i}$. A good fit is indicated if they form roughly a straight line, much like we look for in normal probability plots with residuals from multiple linear regression fits.

## Martingale Residuals

Define a censoring indicator for the $i^{\text{th}}$ observation as

$$\delta_i = \begin{cases} 0, & \text{if observation } i \text{ is censored;} \\ 1, & \text{if observation } i \text{ is uncensored.} \end{cases}$$

In order to identify the best functional form for a covariate given the assumed functional form of the remaining covariates, we use the **Martingale residual** for the $i^{\text{th}}$ observation, which is defined as:

$$\hat{M}_i = \delta_i - r_{C_i}.$$

The $\hat{M}_i$ values fall between the interval $(-\infty, 1]$ and are always negative for censored values. The $\hat{M}_i$ values are plotted against the $x_{i,j}$, where $j$ represents the index of the covariate for which we are trying to identify the best functional form. Plotting a smooth-fitted curve over this dataset will indicate what sort of function (if any) should be applied to $x_{i,j}$. Note that the martingale residuals are not symmetrically distributed about 0, but asymptotically they have mean 0.

## Schoenfeld Residuals

Let $\mathcal{R}(t_i)$ be the **risk set**, which is the set of observations still at risk at time $t_i$; i.e., those subjects that have survived until time $t_i$. For $j = 1, \ldots, p-1$, let

$$\bar{a}_j(t_i) = \frac{\sum_{i \in \mathcal{R}(t_i)} x_{i,j} \exp\{\mathbf{x}_i^{*\text{T}} \hat{\boldsymbol{\beta}}^*\}}{\sum_{i \in \mathcal{R}(t_i)} \exp\{\mathbf{x}_i^{*\text{T}} \hat{\boldsymbol{\beta}}^*\}}.$$

Then, the **Schoenfeld residuals** (Schoenfeld, 1982) are given by

$$C_{i,j} = \delta_i(x_{i,j} - \bar{a}_j(t_i)),$$

which is just the difference between a covariate value for observation $i$ and a weighted average of the values of covariate variables over individuals at risk at time $t$. Thus, we have a set of these residuals for each covariate.

## Score Residuals

The **score residuals** (Therneau et al., 1990) are given by

$$S_{i,j} = C_{i,j} + \sum_{t_h \le t_i} [x_{i,j} - \bar{a}_j(t_h)] \exp\{\mathbf{x}_i^* \hat{\boldsymbol{\beta}}^*\}[\hat{H}_0(t_h) - \hat{H}_0(t_{h-1})].$$

These, again, are calculated for the given subject $i$ with respect to the covariate $j$ and are weighted in a similar manner as the Schoenfeld residuals.

## Deviance Residuals

Outlier detection in a survival regression model can be done using the **deviance residual** for the $i^{\text{th}}$ observation:

$$D_i = \mathbf{sgn}(\hat{M}_i)\sqrt{-2(\ell_i(\hat{\boldsymbol{\theta}}) - \ell_{S_i}(\theta_i))}.$$

For $D_i$, $\ell_i(\hat{\boldsymbol{\theta}})$ is the $i^{\text{th}}$ loglikelihood evaluated at $\hat{\boldsymbol{\theta}}$, which is the maximum likelihood estimate of the model's parameter vector $\boldsymbol{\theta}$. $\ell_{S_i}(\theta_i)$ is the loglikelihood of the saturated model evaluated at the maximum likelihood $\boldsymbol{\theta}$. A **saturated model** is one where $n$ parameters (i.e., $\theta_1, \ldots, \theta_n$) fit the $n$ observations perfectly.

The $D_i$ values should behave like a standard normal sample. A normal probability plot of the $D_i$ values and a plot of $D_i$ versus the fitted $\log(t)_i$ values, will help to determine if any values are fairly far from the bulk of the data. It should be noted that this only applies to cases where light to moderate censoring occur.

### Partial Deviance

We can also consider hierarchical (nested) models. We start by defining the **model deviance**:

$$\Delta = \sum_{i=1}^{n} D_i^2.$$

Suppose we are interested in seeing if adding additional covariates to our model significantly improves the fit from our original model. Suppose we calculate the model deviances under each model. Denote these model deviances as $\Delta_R$ and $\Delta_F$ for the reduced model (our original model) and the full model (our model with all covariates included), respectively. Then, a measure of the fit can be done using the **partial deviance**:

$$\begin{aligned}
\Lambda &= \Delta_R - \Delta_F \\
&= -2(\ell(\hat{\boldsymbol{\theta}}_R) - \ell(\hat{\boldsymbol{\theta}}_F)) \\
&= -2\log\left(\frac{\ell(\hat{\boldsymbol{\theta}}_R)}{\ell(\hat{\boldsymbol{\theta}}_F)}\right),
\end{aligned}$$

where $\ell(\hat{\boldsymbol{\theta}}_R)$ and $\ell(\hat{\boldsymbol{\theta}}_F)$ are the loglikelihood functions evaluated at the maximum likelihood estimates of the reduced and full models, respectively. Luckily, this is a likelihood ratio statistic and has the corresponding asymptotic $\chi^2$ distribution. A large value of $\Lambda$ (large with respect to the corresponding $\chi^2$ distribution) indicates the additional covariates improve the overall fit of the model. A small value of $\Lambda$ means they add nothing significant to the model and you can keep the original set of covariates.

### Inference

Inference is straightforward for censored regression models. Again, let $\boldsymbol{\theta} = (\theta_1, \ldots, \theta_q)^{\text{T}}$ generically represent all of the model parameters, including regression coefficients. Inference and confidence intervals can be constructed based on asymptotic theory. First, for the individual test

$$\begin{aligned}
H_0 &: \theta_j = \theta^* \\
H_A &: \theta_j \neq \theta^*,
\end{aligned}$$

we can construct the test statistic

$$Z^* = \frac{\hat{\theta}_j - \theta^*}{\text{s.e.}(\hat{\theta}_j)},$$

which is based on the maximum likelihood estimator $\hat{\theta}_j$ and is approximately standard normal. In the above, s.e.$(\hat{\theta}_j)$ is the $j^{\text{th}}$ diagonal entry of the inverse of the information matrix $I(\boldsymbol{\theta})$ evaluated at the maximum likelihood estimate $\hat{\boldsymbol{\theta}}$. Moreover, by the duality principle, an approximate $100 \times (1-\alpha)\%$ confidence interval is

$$\hat{\theta}_j \pm z_{1-\alpha/2}\text{s.e.}(\hat{\theta}_j).$$

Joint estimation for the entire parameter vector $\boldsymbol{\theta}$ is typically performed using one of three statistics:

- The **Wald statistic** is calculated as

$$W = (\hat{\boldsymbol{\theta}} - \boldsymbol{\theta}_0)^{\text{T}} I(\boldsymbol{\theta}_0)(\hat{\boldsymbol{\theta}} - \boldsymbol{\theta}_0),$$

  which is approximately $\chi_q^2$ under the null hypothesis.

- The **Rao score statistic** is calculated as

$$S = \frac{\partial}{\partial \boldsymbol{\theta}^{\text{T}}} \ell(\boldsymbol{\theta}_0)^{\text{T}} [I(\boldsymbol{\theta}_0)]^{-1} \frac{\partial}{\partial \boldsymbol{\theta}^{\text{T}}} \ell(\boldsymbol{\theta}_0),$$

  which is also approximately $\chi_q^2$ under the null hypothesis. Here, $\ell(\boldsymbol{\theta}_0)$ is the loglikelihood evaluated at the null value of $\boldsymbol{\theta}_0$.

- The **likelihood ratio test statistic** is calculated as

$$R = -2\left(\ell(\boldsymbol{\theta}_0) - \ell(\hat{\boldsymbol{\theta}})\right),$$

  which is also approximately $\chi_q^2$ under the null hypothesis.

Of course, $p$-values are easily obtained based on any of the above test statistics and their approximate $\chi^2$ distributions.

## 18.6 Truncated Regression Models

**Truncated regression models** (Amemiya, 1973) are used in cases where observations with values for the response variable that are below and/or above certain thresholds are systematically excluded from the sample. Therefore, entire observations are missing so that neither the dependent nor independent variables are known. For example, suppose we had wages and years of schooling for a sample of employees. Some persons for this study are excluded from the sample because their earned wages fall below the minimum wage. So the data would be missing for these individuals.

Truncated regression models are often confused with the censored regression models that we introduced earlier. In censored regression models, only the value of the dependent variable is clustered at a lower and/or upper threshold value, while values of the independent variable(s) are still known. In truncated regression models, entire observations are systematically omitted from the sample based on the lower and/or upper threshold values or potentially from some other mechanism. Regardless, if we know that the data has been truncated, we can adjust our estimation technique to account for the bias introduced by omitting values from the sample. This will allow for more accurate inferences about the entire population. However, if we are solely interested in the population that does not fall outside the threshold value(s), then we can rely on standard techniques that we have already introduced, namely ordinary least squares.

Let us formulate the general framework for truncated distributions. Suppose that $X$ is a random variable with a probability density function $f_X$ and associated cumulative distribution function $F_X$.[3] Consider the two-sided truncation $a < X < b$; i.e., $a$ and $b$ are lower and upper thresholds, respectively. Then the truncated distribution is given by

$$f_X(x|a < X < b) = \frac{g_X(x)}{F_X(b) - F_X(a)},$$

where $g_X(x) = f_X(x)\mathrm{I}\{a < x < b\}$. Similarly, one-sided truncated distributions can be defined by assuming $a$ or $b$ are set at the respective, natural bound of the support for the distribution of $X$; i.e., $F_X(a) = 0$ or $F_X(b) = 1$, respectively. So a bottom-truncated (or left-truncated) distribution is given by

$$f_X(x|a < X) = \frac{g_X(x)}{1 - F_X(a)},$$

while a top-truncated (or right-truncated) distribution is given by

$$f_X(x|X < b) = \frac{g_X(x)}{F_X(b)}.$$

$g_X(x)$ is then defined accordingly for whichever distribution with which you are working.

Consider the multiple linear regression model

$$Y = \underline{X}^\mathrm{T}\beta + \epsilon,$$

where the $\epsilon$ is normally distributed with mean 0 and variance $\sigma^2$. If no truncation (or censoring) is assumed with the data, then normal distribution theory yields

$$Y|\underline{X} \sim \mathcal{N}(\underline{X}^\mathrm{T}\beta, \sigma^2).$$

---

[3]The discrete setting is defined analogously.

When truncating the response, the distribution, and consequently the mean and variance of the truncated distribution, must be adjusted accordingly.

Consider the three possible truncation settings of $a < Y < b$ (two-sided truncation), $a < Y$ (bottom truncation), and $Y < b$ (top truncation). Suppose we have a sample of size $n$ and define the following:

$$\alpha_i = (a - \mathbf{x}_i^{\mathrm{T}}\boldsymbol{\beta})/\sigma,$$
$$\gamma_i = (b - \mathbf{x}_i^{\mathrm{T}}\boldsymbol{\beta})/\sigma, \quad \text{and}$$
$$\psi_i = (y_i - \mathbf{x}_i^{\mathrm{T}}\boldsymbol{\beta})/\sigma.$$

Moreover, recall that $\lambda(z)$ is the inverse Mills ratio applied to the value of $z$ and let

$$\delta(z) = \lambda(z)[(\Phi(z))^{-1} - 1].$$

Then, using the earlier results for the truncated normal distribution, the three different truncated probability density functions are

$$f_{Y|\underline{X}}(y_i|\mathbf{x}_i^{\mathrm{T}},\boldsymbol{\beta},\sigma) = \begin{cases} \frac{\frac{1}{\sigma}\phi(\psi_i)}{1-\Phi(\alpha_i)}, & \text{bottom truncation;} \\[2ex] \frac{\frac{1}{\sigma}\phi(\psi_i)}{\Phi(\gamma_i)-\Phi(\alpha_i)}, & \text{two-sided truncation;} \\[2ex] \frac{\frac{1}{\sigma}\phi(\psi_i)}{\Phi(\gamma_i)}, & \text{top truncation,} \end{cases}$$

while the respective truncated cumulative distribution functions are

$$F_{Y|\underline{X}}(y_i|\mathbf{x}_i^{\mathrm{T}},\boldsymbol{\beta},\sigma) = \begin{cases} \frac{\phi(\psi_i)-\phi(\alpha_i)}{1-\Phi(\alpha_i)}, & \text{bottom truncation;} \\[2ex] \frac{\phi(\psi_i)-\phi(\alpha_i)}{\Phi(\gamma_i)-\Phi(\alpha_i)}, & \text{two-sided truncation;} \\[2ex] \frac{\phi(\psi_i)}{\Phi(\gamma_i)}, & \text{top-truncation.} \end{cases}$$

Furthermore, the means of the three different truncated distributions are

$$\mathrm{E}[Y|\Theta,\underline{X}^{\mathrm{T}}] = \begin{cases} \underline{X}^{\mathrm{T}}\boldsymbol{\beta} + \sigma\lambda(\alpha), & \Theta = \{a < Y\}; \\[2ex] \underline{X}^{\mathrm{T}}\boldsymbol{\beta} + \sigma\left(\frac{\phi(\alpha)-\phi(\gamma)}{\Phi(\gamma)-\Phi(\alpha)}\right), & \Theta = \{a < Y < b\}; \\[2ex] \underline{X}^{\mathrm{T}}\boldsymbol{\beta} - \sigma\delta(\gamma), & \Theta = \{Y < b\}, \end{cases}$$

while the corresponding variances are

$$\mathrm{Var}[Y|\Theta,\underline{X}^{\mathrm{T}}] = \begin{cases} \sigma^2\{1 - \lambda(\alpha)[\lambda(\alpha) - \alpha]\}, & \Theta = \{a < Y\}; \\[2ex] \sigma^2\left\{1 + \frac{\alpha\phi(\alpha)-\gamma\phi(\gamma)}{\Phi(\gamma)-\Phi(\alpha)} - \left(\frac{\phi(\alpha)-\phi(\gamma)}{\Phi(\gamma_i)-\Phi(\alpha)}\right)^2\right\}, & \Theta = \{a < Y < b\}; \\[2ex] \sigma^2\{1 - \delta(\gamma)[\delta(\gamma) + \gamma]\}, & \Theta = \{Y < b\}. \end{cases}$$

Using the distributions defined above, the likelihood function can be found and maximum likelihood procedures can be employed. Note that the likelihood functions will not have a closed-form solution and thus numerical techniques must be employed to find the estimates of $\beta$ and $\sigma$.

It is also important to underscore the type of estimation method used in a truncated regression setting. Maximum likelihood estimation will be used when you are interested in a regression equation that characterizes the entire population, including the observations that were truncated. If you are interested in characterizing just the subpopulation of observations that were not truncated, then ordinary least squares can be used. In the context of the example provided at the beginning of this section, if we regressed wages on years of schooling and were only interested in the employees who made above the minimum wage, then ordinary least squares can be used for estimation. However, if we were interested in all of the employees, including those who happened to be excluded due to not meeting the minimum wage threshold, then maximum likelihood can be employed.

Figure 18.1 shows a simulated dataset of size $n = 50$ to illustrate some of the regression models we introduced in this chapter. In both figures, a threshold of $Y = 50$ is used, of which there are 15 observations that have a response value that fall below this threshold (denoted by the red circles). In Figure 18.1(a), we fit truncated regression models. The maximum likelihood fit and the ordinary least squares fit for the truncated dataset are given by the dotted green line and dashed blue line, respectively. Clearly the maximum likelihood fit provides a better fit than the ordinary least squares fit. For reference, the ordinary least squares fit to the entire dataset is also provided.

In Figure 18.1(b), we fit a Tobit regression line to these data. This line is given by the dotted blue line. Again, the ordinary least squares fit to the entire dataset is also provided. Clearly this shows the impact that censored data can have on the regression estimate.

## 18.7    Examples

**Example 18.7.1.** *(Extramarital Affairs Data)*
Kleiber and Zeileis (2008) analyzed a dataset presented in Fair (1978) about a survey of extramarital affairs conducted in 1969. The survey had a total of $n = 601$ respondents. The measured variables in this dataset are:

- $Y$: number of extramarital affairs during the past year

- $X_1$: age of respondent

- $X_2$: number of years respondent has been married

- $X_3$: whether the respondent has children (yes/no)

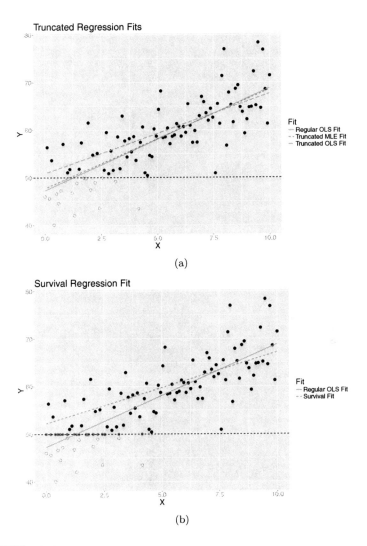

**FIGURE 18.1**

(a) The red circles have been truncated as they fall below 50. The maximum likelihood fit for the truncated regression (dotted green line) and the ordinary least squares fit for the truncated dataset (dashed blue line) are shown. The ordinary least squares line (which includes the truncated values for the estimation) is shown for reference. (b) The same data with a Tobit regression fit provided (dotted blue line). The data has been censored at 50 (i.e., the solid red dots are included in the data). Again, the ordinary least squares line has been provided for reference. Colors are in e-book version.

- $X_4$: self-rating of happiness in the marriage (1 through 5 with a 5 indicating they are "very happy")

Another variable about the respondent's occupation is also in this dataset, but as shown in Kleiber and Zeileis (2008), it is not statistically significant.

Since no fewer than 0 extramarital affairs can be reported, this variable is treated as censored. The number of left-censored values (i.e., reports of 0 affairs) is 451. Moreover, the response is a count variable. Regardless, we proceed to fit a Tobit regression model:

```
##############################
Observations:
          Total  Left-censored      Uncensored Right-censored
            601            451             150              0

Coefficients:
              Estimate Std. Error z value Pr(>|z|)
(Intercept)    9.08289    2.65881   3.416 0.000635 ***
age           -0.16034    0.07772  -2.063 0.039095 *
yearsmarried   0.53890    0.13417   4.016 5.91e-05 ***
religiousness -1.72337    0.40471  -4.258 2.06e-05 ***
rating        -2.26735    0.40813  -5.556 2.77e-08 ***
Log(scale)     2.11310    0.06712  31.482  < 2e-16 ***
---
Signif. codes:  0 *** 0.001 ** 0.01 * 0.05 . 0.1   1


Scale: 8.274


Gaussian distribution
Number of Newton-Raphson Iterations: 4
Log-likelihood: -706.4 on 6 Df
Wald-statistic:  66.4 on 4 Df, p-value: 1.3044e-13
##############################
```

As we can see, the $p$-values for all of the regression coefficients are highly significant. The above output also presents the statistics for a Wald test about the joint significance of the regression coefficients (excluding the intercept) in the model. This test is akin to the general linear $F$-test we discussed for the multiple linear regression setting. The test statistic follows a $\chi^2_{p-1}$ distribution, which in this case $p - 1 = 4$. The $p$-value is very small, thus indicating that the regression coefficients are all significantly different from 0. More details on Wald tests for the single parameter setting are presented in Chapter 20.

**Example 18.7.2.** *(Motorette Data)*
Nelson and Hahn (1972) presented data on the times to failure of $n = 40$ motorettes tested at different operating temperatures. These data are often used to illustrate survival regression models; cf. Kalbfleisch and Prentice (1980),

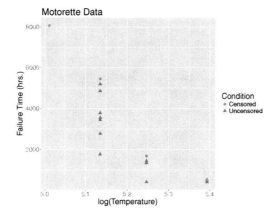

**FIGURE 18.2**
Scatterplot of the motorette failure time data.

Tableman and Kim (2004), and Davison (2003). The measured variables in this dataset are:

- $Y$: failure time (in hours)

- $X$: temperature (in °C)

For the purpose of fitting a survival regression model, we will consider the logarithm of temperature as the predictor. A plot of these data is given in Figure 18.2.

We fit the following survival regression models: Weibull, log-logistic, log-normal, exponential, normal, and logistic. Below are the AIC values and log-likelihood values for each of these fits:

```
#############################
                AIC    LogLike
Weibull      298.0372 -146.0186
LogLogistic  299.5551 -146.7776
LogNormal    302.5186 -148.2593
Exponential  314.4060 -155.2030
Normal       337.8740 -165.9370
Logistic     338.8084 -166.4042
#############################
```

Based on the above output, the Weibull distribution provides the relatively better fit. The survival regression fit assuming a Weibull distribution is given below:

```
#############################
```

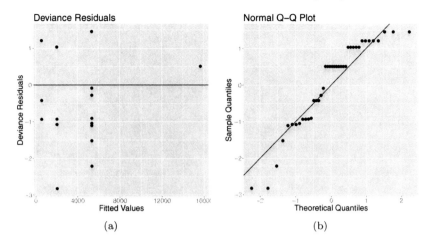

**FIGURE 18.3**
(a) Plot of the deviance residuals. (b) Normal Q-Q plot for the deviance
residuals.

```
                Value Std. Error       z        p
(Intercept)     52.70      3.263   16.15 1.16e-58
log(x)          -8.59      0.621  -13.83 1.61e-43
Log(scale)      -1.13      0.209   -5.38 7.30e-08

Scale= 0.325

Weibull distribution
Loglik(model)= -146    Loglik(intercept only)= -169.5
Chisq= 47.02 on 1 degrees of freedom, p= 7e-12
Number of Newton-Raphson Iterations: 7
n= 40
#############################
```

As we can see, the log of temperature is statistically significant. Furthermore,
the scale, which is estimated at 0.325, is the scale pertaining to the distribution
being fit for this model; i.e., the Weibull. It is also statistically significant.

While we appear to have a decent fitting model, we will turn to looking
at the deviance residuals. Figure 18.3(a) gives a plot of the deviance residuals
versus the fitted values. Overall, this is not a bad fit, though the largest fitted
value could indicate some nonconstant variability. Also, Figure 18.3(b) gives
the NPP plot for these residuals and they do appear to fit along a straight
line. Finally, the partial deviance is $\Lambda = 47.016$, as shown below:

```
#############################
        Df Deviance Resid. Df    -2*LL      Pr(>Chi)
```

```
NULL    NA        NA       38 339.0534              NA
log(x)  1 47.01619          37 292.0372 7.040278e-12
##############################
```

This has a very low *p*-value, thus indicating a highly significant improvement in the fit by including the log of temperature in the model.

**Example 18.7.3.** *(Bone Marrow Transplant Data)*
Avalos et al. (1993) presented data on $n = 43$ bone marrow transplant patients at The Ohio State University Bone Marrow Transplant Unit. Censoring occurs due to the death or relapse of a patient. Klein and Moeschberger (2003) provide an extensive survival analysis of these data, but here, we use a Cox proportional hazards model to model the relationship between time to death or relapse while waiting for a transplant and a number of covariates. The measured variables in this dataset are:

- $Y$: time to death or relapse (in days)

- $X_1$: disease type of patient (either Hodgkin's or non-Hodgkin's lymphoma)

- $X_2$: transplant type (either allogeneic or autogeneic)

- $X_3$: pre-transplant Karnofsky score (i.e., a performance status of a cancer patient's general well-being)

- $X_4$: waiting time to transplant (in months)

We begin by fitting the Cox proportional hazards model with all of the covariates listed above:

```
##############################
                 coef exp(coef) se(coef)     z      p
factor(dtype)2  0.99270   2.69851  0.52320  1.90  0.058
factor(gtype)2 -0.24353   0.78386  0.44300 -0.55  0.583
score          -0.05555   0.94596  0.01215 -4.57 4.8e-06
wtime          -0.00792   0.99211  0.00790 -1.00  0.316

Likelihood ratio test=26.5  on 4 df, p=2.48e-05
n= 43, number of events= 26
##############################
```

It appears that the transplant type and waiting time to transplant are not statistically significant in this model. However, these covariates are still of practical significance for the study undertaken, so we will retain those covariates for the remainder of this analysis.

Figure 18.4 is a plot of the cumulative hazard function based on the Cox–Snell residuals. The red line overlaid on that figure is the $y = x$ line. A good fitting Cox proportional hazards model will have the cumulative hazard function closely following the $y = x$ line. Here, it appears to indicate an overall

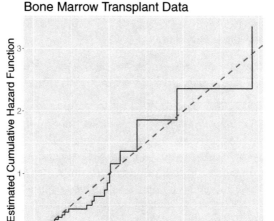

**FIGURE 18.4**
The cumulative hazard function based on the Cox–Snell residuals. The red
line indicates a decent fit for these data. Colors are in e-book version.

good fit, but there does appear to be some room for improvement in the model
given that the larger cumulative hazard function is a bit more variable for the
larger Cox–Snell residuals.

Figure 18.5 is a plot of the Martingale, score, and Schoenfeld residuals
for the two continuous covariates (Karnofsky score and waiting time). The
Martingale residuals (Figures 18.5(a) and 18.5(b)) are assessed to determine
possible nonlinearities in the respective covariate. Based on the LOESS curves
(red curves), it looks like the waiting time might enter the Cox proportional
hazards model nonlinearly. The score residuals (Figures 18.5(c) and 18.5(d))
are used to assess the proportional hazards assumption. If there are any extreme score residuals for a given covariate value, then the proportional hazards
assumption should be used with caution. In this setting, there appears to be
an extreme value at the waiting time near 100, but overall, we would likely
conclude that the proportional hazards assumption is appropriate. Finally, the
Schoenfeld residuals (Figures 18.5(e) and 18.5(f)) are used to detect outlying
covariate values. There are no clear outliers indicated by these plots.

**Example 18.7.4.** *(Durable Goods Data)*
Tobin (1958) presented a dataset from the reinterview portion of 1952 and
1953 Surveys of Consumer Finances. This dataset was used to illustrate when
there are limitations in the dependent variable. The original dataset has 735
observations regarding the number of durable goods purchased, but here we

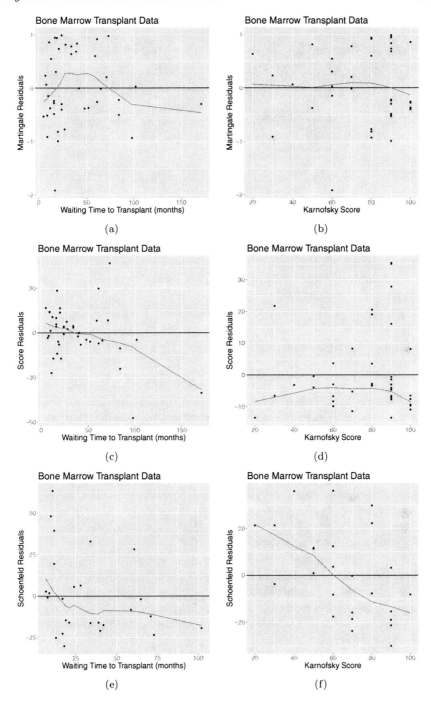

**FIGURE 18.5**
Martingale residuals for (a) the Karnofsky score and (b) waiting time covariates. Score residuals for (a) the Karnofsky score and (b) waiting time
covariates. Schoenfeld residuals for (a) the Karnofsky score and (b) waiting
time covariates.

**TABLE 18.2**
Ordinary least squares fits for the full and truncated datasets and the truncated regression fits; $p$-values given in parentheses

|  | OLS (full) | OLS (reduced) | Truncated |
|---|---|---|---|
| $\hat{\beta}_0$ | 11.074 | 13.778 | 12.598 |
|  | (0.121) | (0.135) | (0.172) |
| $\hat{\beta}_1$ | −0.026 | 0.339 | 0.431 |
|  | (0.754) | (0.094) | (0.056) |
| $\hat{\beta}_2$ | −0.035 | −0.104 | −0.119 |
|  | (0.169) | (0.022) | (0.000) |
| MSE | 2.709 | 1.897 | 1.652 |
| AIC | 101.373 | 32.909 | 31.453 |

analyze a subset of that data of size $n = 20$. The values of 0 are treated as truncated, even though other predictor variables are measured. The measured variables in this dataset are:

- $Y$: durable goods purchased

- $X_1$: age of the head of the spending unit

- $X_2$: ratio of liquid assets holdings ($\times 1000$)

We are interested in the linear relationship between the durable goods purchased and the two predictor variables listed above. The full data consists of these $n = 20$ measurements. The truncated data consists of those observations removed where $Y = 0$. An ordinary least squares fit for the full and truncated data are given in Figures 18.6(a) and 18.6(b), respectively. The truncated regression fit is given in Figure 18.6(c). The fits on the truncated data both appear to be similar, but they are different. However, both of these fits differ substantially from the ordinary least squares fit on the full data. Clearly if one does not properly account for truncated data, then any subsequent estimates and predictions could be misleading.

Table 18.2 provides the regression coefficient estimates, the MSEs, and the AIC values for the three models fit in Figure 18.6. Clearly the slope estimates for the truncated regression fit are more significant than when obtaining the ordinary least squares fit on the full data. Moreover, the truncated regression fit has the smallest AIC, which indicates it is the better fit when comparing these three models.

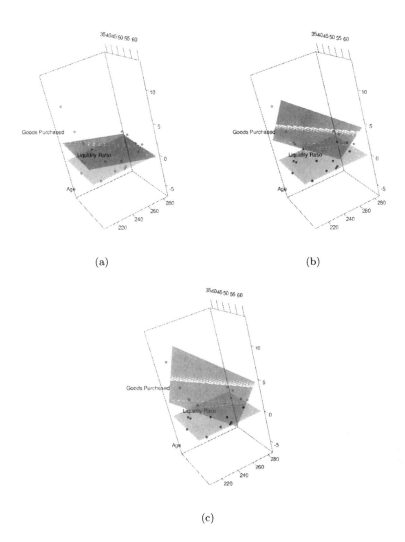

(a)

(b)

(c)

**FIGURE 18.6**
(a) The ordinary least squares fit for the full data. (b) The ordinary least squares fit for the truncated data. (b) The truncated regression fit.

# 19

## Nonlinear Regression

Thus far, we have primarily discussed linear regression models; i.e., models that are linear in terms of the regression coefficients. For example, polynomial regression was used to model curvature in our data by using higher-ordered values of the predictors. However, the final regression model was just a linear combination of higher-ordered predictors. When a regression model is either not linear in its parameters or cannot be transformed to be linear, it is then considered a nonlinear regression model.

### 19.1   Nonlinear Regression Models

The **nonlinear regression model** relating a response $Y$ to $\underline{X}$, a $p$-dimensional vector of predictors, is given by

$$Y = f(\underline{X}, \beta) + \epsilon,$$

where $\beta$ is a vector of $k$ parameters, $f(\cdot)$ is some known regression function, and $\epsilon$ is an error term whose distribution may or may not be normal. Notice that we no longer necessarily have the dimension of the parameter vector tied to the number of predictors. Some examples of nonlinear regression models are:

$$Y = \frac{e^{\beta_0 + \beta_1 X}}{1 + e^{\beta_0 + \beta_1 X}} + \epsilon$$
$$Y = \frac{\beta_0 + \beta_1 X}{1 + \beta_2 e^{\beta_3 X}} + \epsilon$$
$$Y = \beta_0 + (0.4 - \beta_0)e^{-\beta_1(X-5)} + \epsilon.$$

Notice that in each model, $p = 1$, but that the number of parameters is, respectively, 2, 4, and 2.

However, there are some nonlinear models which are actually called **intrinsically linear** because they can be made linear in the parameters by transformation. For example:

$$Y = \frac{\beta_0 X}{\beta_1 + X}$$

361

can be rewritten as

$$\frac{1}{Y} = \frac{1}{\beta_0} + \frac{\beta_1}{\beta_0} X$$

or rather

$$Y' = \theta_0 + \theta_1 X,$$

which is linear in the transformed variables $\theta_0$ and $\theta_1$. In such cases, transforming a model to its linear form often provides better inference procedures and confidence intervals, but one must be cognizant of the effects that the transformation has on the distribution of the errors.

We will discuss some of the basics of fitting and inference with nonlinear regression models. There is a great deal of theory, practice, and computing associated with nonlinear regression and we will only get to scratch the surface of this topic. A couple of texts that provide a deeper treatment of nonlinear regression modeling are Bates and Watts (1988) and Seber and Wild (2003).

We first discuss a handful of common nonlinear regression models. For simplicity, we assume that the error terms in each model are *iid* with mean 0 and constant variance $\sigma^2$.

**Exponential Regression Model**
The **exponential regression model** is

$$Y = \beta_0 + \beta_1 \exp\left\{\beta_2 + \sum_{j=1}^{k} \beta_{k+2} X_k\right\} + \epsilon. \qquad (19.1)$$

Notice that if $\beta_0 = 0$, then the above is intrinsically linear by taking the natural logarithm of both sides. Exponential regression is probably one of the simplest nonlinear regression models. An example is simple interest rate models, which are based on some form of an exponential regression model.

**Michaelis–Menten Models**
In biochemistry, the work of Michaelis and Menten (1913) is regarded as having made some of the most significant contributions to enzyme kinetics. Their work has led to a well-utilized class of nonlinear regression models that relate the reaction rate (response) to the concentration of one or more enzyme substrates (predictors). Three common types of **Michaelis–Menten models** are

- *Classical:*

$$Y = \frac{\beta_1 X_1}{\beta_0 + X_1} + \epsilon.$$

- *Competitive Inhibition:*

$$Y = \frac{\beta_1 X_1}{\beta_0 + X_1 (1 + X_2/\beta_2)} + \epsilon.$$

- *Non-Competitive Inhibition:*

$$Y = \frac{\beta_1 X_1}{(\beta_0 + X_1)(1 + X_2/\beta_2)} + \epsilon.$$

In the above models, $X_1$ and $X_2$ are the concentrations of two different enzyme substrates, $\beta_0$ (to be estimated) is the substrate concentration at which the reaction rate is at half-maximum, and $\beta_1$ and $\beta_2$ are inhibition constants that are also to be estimated.

## Gompertz Growth Curve
The **Gompertz growth curve model** (Gompertz, 1825) is given by

$$Y = \beta_0 + \beta_1 \exp\left\{\beta_2 e^{\beta_3 + \sum_{j=1}^{k} \beta_{k+3} X_k}\right\} + \epsilon, \tag{19.2}$$

which is commonly used to model time series data where growth of a process is slowest at the start and end of the process. Typically, the quantities in the exponents are appropriately constrained so that the lower and upper asymptotes are $\beta_0$ and $\beta_0 + \beta_1$, respectively. An example where Gompertz growth curve models are often used is in assessing the growth of tumors.

## Richards' Curve
The **Richards' curve model** (Richards, 1959), also called the **generalized logistic curve** model, is

$$Y = \beta_0 + \frac{\beta_1}{\left(1 + e^{\beta_3 + \sum_{j=1}^{k} \beta_{k+3} X_k}\right)^{\beta_2}} + \epsilon.$$

When $\beta_0 = 0$ and $\beta_1 = \beta_2 = 1$, we have a standard logistic regression model. In fact, the Gompertz growth curve model, Richards' curve model, and the logistic regression model are all part of a class of models that have sigmoidal curves. Such models are useful in signal detection theory, economics, and engineering.

## Unspecified Functional Form
Nonlinear regression is usually used when there is a known functional form for the response function that is to be fit to the data. However, if such a response function is not assumed, then one can employ a few different strategies:

- Try fitting a general class of nonlinear response functions (e.g., exponential regression models) to the data and compare goodness-of-fit measures and residual diagnostic plots.

- Consider a higher-order polynomial regression, possibly with interactions if more than one predictor is present.

- Explore transformations and see if an intrinsically linear model might provide a reasonable fit.

- Try fitting a more flexible regression model, such as a piecewise linear regression, smoothing spline, or other nonparametric/semiparametric regression models as discussed in Chapters 17 and 23.

## 19.2   Nonlinear Least Squares

Suppose we have a sample of size $n$ where each observation has a measured response and vector of $p$ predictor variables. These variables are related through the nonlinear regression model

$$y_i = f(\mathbf{x}_i, \boldsymbol{\beta}) + \epsilon_i,$$

where the $\epsilon_i$ are *iid* normal with mean 0 and constant variance $\sigma^2$ and $f(\cdot)$ has some specified functional form. For this setting, we can rely on some of the least squares theory we have developed over the course. For other nonnormal error terms, different techniques need to be employed.

First, let

$$Q = \sum_{i=1}^{n} (y_i - f(\mathbf{x}_i, \boldsymbol{\beta}))^2.$$

In order to find

$$\hat{\boldsymbol{\beta}} = \operatorname*{argmin}_{\boldsymbol{\beta}} Q,$$

we first find each of the partial derivatives of $Q$ with respect to $\beta_j$:

$$\frac{\partial Q}{\partial \beta_j} = -2 \sum_{i=1}^{n} (y_i - f(\mathbf{x}_i, \boldsymbol{\beta})) \left( \frac{\partial f(\mathbf{x}_i, \boldsymbol{\beta})}{\partial \beta_j} \right).$$

Then, we set each of the above partial derivatives equal to 0 and the parameters $\beta_j$ are each replaced by $\hat{\beta}_j$:

$$\sum_{i=1}^{n} y_i \left( \frac{\partial f(\mathbf{x}_i, \boldsymbol{\beta})}{\partial \beta_j} \right) \bigg|_{\boldsymbol{\beta} = \hat{\boldsymbol{\beta}}} - \sum_{i=1}^{n} f(\mathbf{x}_i, \hat{\boldsymbol{\beta}}) \left( \frac{\partial f(\mathbf{x}_i, \boldsymbol{\beta})}{\partial \beta_j} \right) \bigg|_{\boldsymbol{\beta} = \hat{\boldsymbol{\beta}}} = 0,$$

for $j = 0, 1, \ldots, k - 1$. The solutions to the critical values of the above partial derivatives are nonlinear in the parameter estimates $\hat{\beta}_j$ and are often difficult to solve, even in the simplest cases. Hence, iterative numerical methods are often employed. However, there are additional numerical issues to consider that can arise when employing such methods.

## 19.2.1 A Few Algorithms

We will discuss a few algorithms used in nonlinear least squares estimation. It should be noted that this is *not* an exhaustive list of algorithms, but rather an introduction to some of the more commonly implemented algorithms.

First let us introduce some notation used in these algorithms:

- Since these numerical algorithms are iterative, let $\hat{\beta}^{(t)}$ be the estimated value of $\beta$ at time $t$. When $t = 0$, this symbolizes a user-specified starting value for the algorithm. When our chosen algorithm has converged, we denote the final estimate obtained by $\hat{\beta}$. This, of course, allows us to define our fitted values and residuals as, respectively,

$$\hat{y}_i = f(\mathbf{x}_i, \hat{\beta})$$

and

$$e_i = y_i - \hat{y}_i.$$

- Let

$$\epsilon = \begin{pmatrix} \epsilon_1 \\ \vdots \\ \epsilon_n \end{pmatrix} = \begin{pmatrix} y_1 - f(\mathbf{x}_1, \beta) \\ \vdots \\ y_n - f(\mathbf{x}_n, \beta) \end{pmatrix}$$

be an $n$-dimensional vector of the error terms and $\mathbf{e}$ is again the residual vector.

- Let

$$\nabla Q(\beta) = \frac{\partial Q(\beta)}{\partial \beta}^{\mathrm{T}} = \begin{pmatrix} \frac{\partial \|\epsilon\|^2}{\partial \beta_0} \\ \vdots \\ \frac{\partial \|\epsilon\|^2}{\partial \beta_{k-1}} \end{pmatrix}$$

be the **gradient** of the sum of squared errors where $Q(\beta) = \|\epsilon\|^2 = \sum_{i=1}^{n} \epsilon_i^2$ is the sum of squared errors.

- Let

$$J(\beta) = \begin{pmatrix} \frac{\partial \epsilon_1}{\partial \beta_0} & \cdots & \frac{\partial \epsilon_1}{\partial \beta_{k-1}} \\ \vdots & \ddots & \vdots \\ \frac{\partial \epsilon_n}{\partial \beta_0} & \cdots & \frac{\partial \epsilon_n}{\partial \beta_{k-1}} \end{pmatrix}$$

be the **Jacobian matrix** of the residuals.

- Let

$$H(\beta) = \frac{\partial^2 Q(\beta)}{\partial \beta^{\mathrm{T}} \partial \beta}$$

$$= \begin{pmatrix} \frac{\partial^2 \|\epsilon\|^2}{\partial \beta_0 \partial \beta_0} & \cdots & \frac{\partial^2 \|\epsilon\|^2}{\partial \beta_0 \partial \beta_{k-1}} \\ \vdots & \ddots & \vdots \\ \frac{\partial^2 \|\epsilon\|^2}{\partial \beta_{k-1} \partial \beta_0} & \cdots & \frac{\partial^2 \|\epsilon\|^2}{\partial \beta_{k-1} \partial \beta_{k-1}} \end{pmatrix}$$

be the **Hessian matrix**; i.e., matrix of mixed partial derivatives and second-order derivatives.

- In the presentation of the numerical algorithms, we use the following notation to simplify the expressions given above:

$$\nabla Q(\hat{\boldsymbol{\beta}}^{(t)}) = \nabla Q(\boldsymbol{\beta})\big|_{\boldsymbol{\beta}=\hat{\boldsymbol{\beta}}^{(t)}}$$

$$H(\hat{\boldsymbol{\beta}}^{(t)}) = H(\boldsymbol{\beta})\big|_{\boldsymbol{\beta}=\hat{\boldsymbol{\beta}}^{(t)}}$$

$$J(\hat{\boldsymbol{\beta}}^{(t)}) = J(\boldsymbol{\beta})\big|_{\boldsymbol{\beta}=\hat{\boldsymbol{\beta}}^{(t)}}$$

### Newton's Method

A common optimization technique is based on **gradient methods**, which attempt to optimize the problem by searching along the gradient of the function at the current point. This is why we defined the gradient function and subsequent derivative matrices. The classical method for nonlinear least squares estimation based on the gradient approach is **Newton's method**, which starts at $\hat{\boldsymbol{\beta}}^{(0)}$ and iteratively calculates

$$\hat{\boldsymbol{\beta}}^{(t+1)} = \hat{\boldsymbol{\beta}}^{(t)} - [H(\hat{\boldsymbol{\beta}}^{(t)})]^{-1}\nabla Q(\hat{\boldsymbol{\beta}}^{(t)})$$

until a convergence criterion is achieved. The difficulty in this approach is that inversion of the Hessian matrix can be computationally difficult. In particular, the Hessian is not always positive definite unless the algorithm is initialized with a good starting value, which may be difficult to find.

### Gauss–Newton Algorithm

A modification to Newton's method is the **Gauss–Newton algorithm**, which, unlike Newton's method, can only be used to minimize a sum of squares function. An advantage with using the Gauss–Newton algorithm is that it no longer requires calculation of the Hessian matrix, but rather approximates it using the Jacobian. The gradient and approximation to the Hessian matrix can be written as

$$\nabla Q(\boldsymbol{\beta}) = 2J(\boldsymbol{\beta})^{\mathrm{T}}\boldsymbol{\epsilon} \quad \text{and} \quad H(\boldsymbol{\beta}) \approx 2J(\boldsymbol{\beta})^{\mathrm{T}}J(\boldsymbol{\beta}).$$

Thus, the iterative approximation based on the Gauss–Newton method yields

$$\hat{\boldsymbol{\beta}}^{(t+1)} = \hat{\boldsymbol{\beta}}^{(t)} - \delta(\hat{\boldsymbol{\beta}}^{(t)})$$
$$= \hat{\boldsymbol{\beta}}^{(t)} - [J(\hat{\boldsymbol{\beta}}^{(t)})^{\mathrm{T}}J(\hat{\boldsymbol{\beta}}^{(t)})]^{-1}J(\hat{\boldsymbol{\beta}}^{(t)})^{\mathrm{T}}\mathbf{e},$$

where we have defined $\delta(\hat{\boldsymbol{\beta}}^{(t)})$ to be everything that is subtracted from $\hat{\boldsymbol{\beta}}^{(t)}$.

**Levenberg–Marquardt Method**

Convergence is not always guaranteed with the Gauss–Newton algorithm because the steps taken by this method may be too large and, thus, lead to divergence. One can incorporate a partial step to circumvent this issue by using

$$\hat{\boldsymbol{\beta}}^{(t+1)} = \hat{\boldsymbol{\beta}}^{(t)} - \alpha\delta(\hat{\boldsymbol{\beta}}^{(t)})$$

such that $0 < \alpha < 1$. However, if $\alpha$ is close to 0, an alternative method is the **Levenberg–Marquardt method**, which calculates

$$\delta(\hat{\boldsymbol{\beta}}^{(t)}) = (J(\hat{\boldsymbol{\beta}}^{(t)})^{\mathrm{T}}J(\hat{\boldsymbol{\beta}}^{(t)}) + \lambda\mathbf{G})^{-1}J(\hat{\boldsymbol{\beta}}^{(t)})^{\mathrm{T}}\mathbf{e},$$

where $\mathbf{G}$ is a positive diagonal matrix (often taken as the identity matrix) and $\lambda$ is the so-called **Marquardt parameter**. The above is optimized for $\lambda$ which limits the length of the step taken at each iteration and improves an ill-conditioned Hessian matrix.

**Additional Numerical Considerations**

For these algorithms, you will want to try the easiest one to calculate for a given nonlinear problem. Ideally, you would like to be able to use the algorithms in the order they were presented. Newton's method will give you an accurate estimate if the Hessian is not ill-conditioned. The Gauss–Newton will give you a good approximation to the solution Newton's method should have arrived at, but convergence is not always guaranteed. Finally, the Levenberg–Marquardt method can take care of computational difficulties arising with the other methods, but searching for $\lambda$ can be tedious. Moreover, just like other optimization algorithms, these can all be sensitive to the starting values. So it may be necessary to try the chosen algorithm from various starting values and assess the appropriateness of the numerical solution.

## 19.3  Approximate Inference Procedures

When we performed estimation and inference on the regression parameters for the multiple linear regression model with normal errors, those procedures were exact and held for any sample size. The theory underlying those procedures is called **normal theory**. Unfortunately, nonlinear regression models with normal errors do not have analogous exact procedures. In particular, the nonlinear least squares estimators (or maximum likelihood estimators) obtained in the previous section are *not* normally distributed, *not* unbiased, and do *not* have the minimum variance property. However, approximate inference procedures can be done by appealing to asymptotic theory. Asymptotic theory in the nonlinear regression setting tells us that, when the sample size is large, nonlinear least squares and maximum likelihood estimates are *approximately*

normally distributed, *almost* unbiased, and have *almost* minimum variance. In
other words, the theoretical properties of the obtained estimators are decent
for all intents and purposes.

Just like inference in multiple linear regression, inferences about nonlinear
regression parameters require an estimate of the error term variance $\sigma^2$. The
estimate is similar to that for multiple linear regression:

$$\text{MSE} = \frac{\text{SSE}}{n-k} = \frac{Q(\hat{\boldsymbol{\beta}})}{n-k} = \frac{\sum_{i=1}^{n}(y_i - \hat{y}_i)^2}{n-k},$$

where $\hat{\boldsymbol{\beta}}$ is the $k$-dimensional vector of final parameter estimates obtained
using whatever optimization algorithm that you employ. Note that the MSE
is not an unbiased estimator of $\sigma^2$, but it has small bias asymptotically.

When the error terms $\epsilon_1, \ldots, \epsilon_n$ are normally distributed with mean 0
and constant variance $\sigma^2$ and the sample size $n$ is reasonably large, then the
sampling distribution of $\hat{\boldsymbol{\beta}}$ is approximately normal with approximate mean

$$\text{E}(\hat{\boldsymbol{\beta}}) \approx \boldsymbol{\beta}$$

and approximate variance–covariance matrix

$$\text{Var}(\hat{\boldsymbol{\beta}}) = \text{MSE}(D(\hat{\boldsymbol{\beta}})^{\text{T}} D(\hat{\boldsymbol{\beta}}))^{-1}.$$

In the above,

$$D(\boldsymbol{\beta}) = \begin{pmatrix} \frac{\partial f(\mathbf{x}_1, \boldsymbol{\beta})}{\partial \beta_0} & \cdots & \frac{\partial f(\mathbf{x}_1, \boldsymbol{\beta})}{\partial \beta_{k-1}} \\ \vdots & \ddots & \vdots \\ \frac{\partial f(\mathbf{x}_n, \boldsymbol{\beta})}{\partial \beta_0} & \cdots & \frac{\partial f(\mathbf{x}_n, \boldsymbol{\beta})}{\partial \beta_{k-1}} \end{pmatrix}$$

is the Jacobian matrix of the nonlinear function $f(\cdot)$ and

$$D(\hat{\boldsymbol{\beta}}) = D(\boldsymbol{\beta})\big|_{\boldsymbol{\beta}=\hat{\boldsymbol{\beta}}}$$

is the evaluation at the final least squares estimate $\hat{\boldsymbol{\beta}}$. The square roots of the
diagonal entries of this matrix are the corresponding **approximate standard
errors**, s.e.$(\hat{\beta}_k)$. Thus, when the sample size is large, the approximate distribution of the estimated regression coefficients is normal with the approximate
means and variances determined above, thus allowing us to carry out inference
in a similar manner as for multiple linear regression.

Sometimes, $D(\hat{\boldsymbol{\beta}})$ may be difficult to calculate. Alternatively, we can use
bootstrapping to obtain estimates of s.e.$(\hat{\beta}_j)$, $j = 0, 1, \ldots, k-1$. When assuming that the error terms have constant variance, we can use the fixed $X$
sampling form of the bootstrap. Suppose we are interested in the nonlinear
regression model

$$y_i = f(\mathbf{x}_i, \boldsymbol{\beta}) + \epsilon_i.$$

The algorithm is as follows:

## Bootstrap for Nonlinear Regression Standard Errors

1. Let $\hat{y}_1, \ldots, \hat{y}_n$ and $e_1, \ldots, e_n$ be the fitted values and residuals from our estimated nonlinear regression model.

2. Draw a random sample with replacement of size $n$ from the residuals. Denote these by $e_1^*, \ldots, e_n^*$, which are called bootstrap sample residuals.

3. Add these residuals to the original fitted values:

$$y_i^* = \hat{y}_i + e_i^*.$$

The values $y_1^*, \ldots, y_n^*$ are our bootstrap responses.

4. Regress the bootstrap responses on the original $\mathbf{x}_1, \ldots, \mathbf{x}_n$ and retain the estimates of the regression coefficients. These bootstrap estimates of the nonlinear regression coefficients are written as $\beta_0^*, \beta_1^*, \ldots, \beta_{k-1}^*$.

5. Repeat the above steps a large number of times, say, $B$ times. The standard deviation of the bootstrap sample of each $\beta_j^*$ estimates s.e.$(\hat{\beta}_j)$.

With the above established, we can now develop approximate inference procedures that closely follow those for the multiple linear regression setting.

### Inference for a Single $\beta_j$

Based on large-sample theory, the following approximate result holds when the sample size is large and the error terms are normally distributed:

$$\frac{\hat{\beta}_j - \beta_j}{\text{s.e.}(\hat{\beta}_j)} \sim t_{n-k},$$

for $j = 0, 1, \ldots, k - 1$. A large-sample test for a single $\beta_j$ is

$$H_0 : \beta_j = \beta^*$$
$$H_A : \beta_j \neq \beta^*.$$

The test statistic used for the above test is then

$$t^* = \frac{\hat{\beta}_j - \beta^*}{\text{s.e.}(\hat{\beta}_j)},$$

which also follows a $t_{n-k}$ distribution. Moreover, an approximate $100 \times (1 - \alpha)\%$ confidence interval for a single $\beta_j$ is

$$\hat{\beta}_j \pm t_{1-\alpha/2, n-k} \text{s.e.}(\hat{\beta}_j).$$

**Inference for Multiple $\beta_j$**

When a large-sample test concerning several $\beta_k$ simultaneously is desired, we use the same approach as for the general linear $F$-test. Let us partition the vector of coefficients as $\boldsymbol{\beta}^{\mathrm{T}} = (\boldsymbol{\beta}_1^{\mathrm{T}}, \boldsymbol{\beta}_2^{\mathrm{T}})$ with dimensions $k_1$ and $k_2$, respectively, such that $k = k_1 + k_2$. Suppose we wish to test

$$H_0 : \boldsymbol{\beta}_2 = \mathbf{0}$$
$$H_A : \boldsymbol{\beta}_2 \neq \mathbf{0}.$$

We first fit the full model (that specified under $H_A$) and obtain the SSE(full) and then fit the reduced model (that specified under $H_0$) and obtain the SSE(reduced). The test statistic is then calculated as

$$F^* = \frac{\frac{\text{SSE(reduced)} - \text{SSE(full)}}{df_E(\text{reduced}) - df_E(\text{full})}}{\frac{\text{SSE(full)}}{df_E(\text{full})}}$$
$$= \frac{\text{SSE(reduced)} - \text{SSE(full)}/k_2}{\text{MSE(full)}},$$

which follows an $F_{k_2, n-k}$ distribution. Moreover, approximate joint confidence intervals for $\beta_2$ can be obtained using the Bonferroni correction. The Bonferroni joint confidence intervals for $\beta_2$ are:

$$\hat{\beta}_{j^*} \pm t_{1-\alpha/2m, n-p} \text{s.e.}(\hat{\beta}_{j^*}),$$

where $j^* = k_1, k_1 + 1, \ldots, k - 1$.

---

## 19.4   Examples

**Example 19.4.1.** *(Puromycin Data)*

Treloar (1974) presented data on an enzymatic reaction involving untreated cells or cells treated with the antibiotic Puromycin. The following two variables were recorded:

- $Y$: reaction velocity (in counts/min/min)

- $X$: concentration of substrate (in ppm)

It has been shown that the classical Michaelis–Menten model is appropriate for these data (Bates and Watts, 1988). We will present an analysis of only those observations treated with Puromycin, which has a sample size of $n = 12$.

We now present some of the results when fitting the Michaelis–Menten model to these data. Note in the output that v, S, Km, and Vmax — which are the traditional quantities used in Michaelis-Menten models — correspond to the values of $Y$, $X$, $\beta_0$, and $\beta_1$ in our setup, respectively.

```
##############################
------
Formula: v ~ S/(S + Km) * Vmax

Parameters:
      Estimate Std. Error t value Pr(>|t|)
Km   6.412e-02  8.281e-03   7.743 1.57e-05 ***
Vmax 2.127e+02  6.947e+00  30.615 3.24e-11 ***
---
Signif. codes:  0 *** 0.001 ** 0.01 * 0.05 . 0.1   1

Residual standard error: 10.93 on 10 degrees of freedom

Number of iterations to convergence: 9
Achieved convergence tolerance: 8.515e-06


------

Residual sum of squares: 1200


------

t-based confidence interval:
              2.5%          97.5%
Km      0.04566999    0.08257208
Vmax 197.20437734 228.16279203


------

Correlation matrix:
            Km        Vmax
Km   1.0000000 0.7650834
Vmax 0.7650834 1.0000000
##############################
```

The above output shows that the estimates for our parameters are $\hat{\beta}_0 = 0.06412$ and $\hat{\beta}_1 = 212.68358$. Both of these terms are significant predictors of the reaction velocity. Approximate 95% confidence intervals for each coefficient are $(0.04567, 0.08257)$ and $(197.20438, 228.16279)$, respectively. A plot of the data with the estimated Michaelis–Menten model is provided in Figure 19.1, which clearly shows a good fit.

**Example 19.4.2.** *(Light Data)*
In Example 9.3.1 of Graybill and Iyer (1994), data on a light experiment were presented. In this experiment, light was transmitted through a chemical solution and an optical reading was recorded. This process helps to determine the concentrations of certain chemicals. The following two variables were recorded:

- $Y$: optical reading

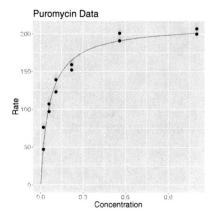

**FIGURE 19.1**
Scatterplot of the Puromycin dataset with the estimated Michaelis–Menten model.

- $X$: concentration of chemical

We will analyze these data using a Gompertz growth curve model with a single predictor, which is the concentration of the chemical. The sample size of this dataset is $n = 12$.

Below is the output for the fitted Gompertz growth curve model:

```
##############################
Parameters:
   Estimate Std. Error t value Pr(>|t|)
b0  0.02875   0.17152   0.168 0.870591
b1  2.72328   0.21054  12.935 4.05e-07 ***
b2 -0.68276   0.14166  -4.820 0.000947 ***
---
Signif. codes:  0 *** 0.001 ** 0.01 * 0.05 . 0.1   1

Residual standard error: 0.2262 on 9 degrees of freedom

Number of iterations to convergence: 4
Achieved convergence tolerance: 9.077e-06
##############################
```

Concentration is a significant predictor of the optical reading in this nonlinear relationship (*p*-value is 0.0009). The MSE is 0.2262 with 9 degrees of freedom. A scatterplot of this data with overlaid estimated Gompertz growth curve is given in Figure 19.2(a). A plot of the raw residuals versus the fitted values is given in Figure 19.2(b), which indicates a decent fit. Based on the residual plot, there may be some nonconstant variance present due to the slight

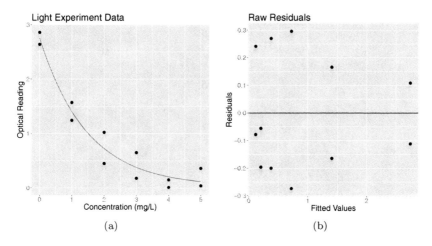

**FIGURE 19.2**
(a) Scatterplot of the light dataset with the estimated Gompertz growth curve model. (b) Scatterplot of the raw residuals versus predicted values from this fit.

decreasing variability that occurs for larger concentration values. One could explore further transformations on these data in an attempt to adjust that feature. However, the appearance of possible nonconstant variance could be due to the small sample size.

**Example 19.4.3.** *(James Bond Data)*
Young (2014) analyzed data on the *James Bond* film franchise. Between 1962 and 2015, $n = 24$ official films have been produced. The following two variables are of interest:

- $Y$: unadjusted U.S. gross (in millions of dollars)

- $X$: year of film's theatrical release

For these data, we are interested in fitting the following Richards' curve model:

$$y_i = \frac{\beta_0}{1 + e^{\beta_1 + \beta_2 x_i}} + \epsilon_i.$$

Below is the output for the fitted Richards' curve model:

```
#############################
Parameters:
        Estimate Std. Error t value Pr(>|t|)
beta0 1415.38421 4828.72387   0.293  0.77231
beta1    4.12521    3.21890   1.282  0.21397
beta2   -0.04893    0.01713  -2.857  0.00944 **
```

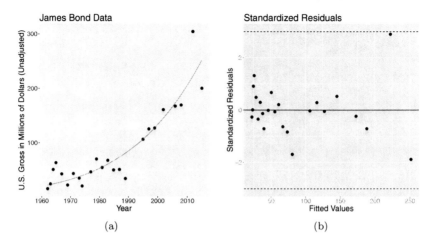

(a)                                              (b)

**FIGURE 19.3**
(a) Scatterplot of the *James Bond* dataset with the estimated Richards' curve
model. (b) Scatterplot of the standardized residuals versus predicted values
from this fit.

```
---
Signif. codes:  0 *** 0.001 ** 0.01 * 0.05 . 0.1   1

Residual standard error: 27.85 on 21 degrees of freedom

Number of iterations to convergence: 5
Achieved convergence tolerance: 5.074e-06
##############################
```

The year is a significant predictor of the unadjusted U.S. gross of the films in
this nonlinear relationship ($p$-value is 0.0094). The MSE is 27.85 with 21 degrees of freedom. Approximate joint 95% confidence intervals for each regression parameter — using a Bonferroni correction to get the 99.167[th] percentile
of a $t_{21}$ distribution — is given as follows:

$$\hat{\theta}_1 : 1415.384 \pm 2.601(4828.724) = (-11145.816, 13976.585)$$

$$\hat{\beta}_0 : 4.125 \pm 2.601(3.219) = (-4.248, 12.499)$$

$$\hat{\beta}_1 : -0.049 \pm 2.601(0.002) = (-0.009, 0.004).$$

A scatterplot of this data with the estimated Richards' curve overlaid is
given in Figure 19.3(a). A plot of the standardized residuals versus the fitted
values is given in Figure 19.3(b), which appears to be a good fit. The most
extreme observation comes close to the threshold of $+3$. This is the film *Skyfall*,
which holds many box office records for the franchise consisting of these 24
films.

# 20

# Regression Models with Discrete Responses

The responses in the models we have presented thus far have either been continuous or integer-valued but are treated as continuous. We now turn to the case where we have responses that are counts. This includes the special case of binary outcomes where the response is to indicate whether or not a particular event occurred.

We focus on the estimation and assessment of three common count regression models: logistic regression, Poisson regression, and negative binomial regression. We then discuss models that characterize different problems that can occur with zero counts in count regression settings.

## 20.1 Logistic Regression

**Logistic regression** models a relationship between predictor variables and a categorical response variable. For example, logistic regression can be used to model the relationship between various measurements of a manufactured specimen (such as dimensions and chemical composition) to predict if a crack greater than 10 mils will occur (a binary variable: either yes or no). Logistic regression estimates a probability of falling into a certain level of the categorical response given a set of predictors. There are three primary types of logistic regression models, depending on the nature of the categorical response variable:

**Binary Logistic Regression:** Used when the response is binary; i.e., it has two possible outcomes. The cracking example given above would utilize binary logistic regression. Other examples of binary responses include a subject passing or failing a test, responding yes or no on a survey, and having high or low blood pressure.

**Nominal Logistic Regression:** Used when there are three or more categories with no natural ordering to the levels. Examples of nominal responses include departments within a firm (e.g., marketing, sales, and human resources), type of search engine used (e.g., Google, Yahoo!, and Bing), and color (e.g., black, red, blue, and orange).

**Ordinal Logistic Regression:** Used when there are three or more categories with a natural ordering to the levels, but the ranking of the levels do not necessarily mean the intervals between them are equal. Examples of ordinal responses could be how you rate the effectiveness of a college course on a scale of 1–5, levels of flavors for hot wings, and medical condition (e.g., good, stable, serious, and critical).

The challenges with logistic regression models is that they include nonnormal error terms, nonconstant error variance, and constraints on the response function; i.e., the response is bounded between 0 and 1. We will discuss these in the logistic regression setting. More details about logistic regression models are found in the text by Hosmer et al. (2013).

### 20.1.1   Binary Logistic Regression

Let $Y$ be a binary response variable. We are interested in modeling the conditional probability $\pi(\underline{X}) \equiv P(Y = 1|\underline{X})$, which is a function of $\underline{X}$.[1] Consider the multiple linear regression model, but where $Y$ can only take on the value of 0 or 1 and $E(\epsilon) = 0$:

$$Y = \underline{X}^{\mathrm{T}}\beta + \epsilon.$$

Moreover, consider $Y|\underline{X}$ to be a Bernoulli random variable, which takes on the value of 0 or 1 with respective probabilities $\pi(\underline{X})$ and $1 - \pi(\underline{X})$. Thus,

$$\pi(\underline{X}) = P(Y = 1|\underline{X}) = E(Y|\underline{X}) = \underline{X}^{\mathrm{T}}\beta.$$

A binary response variable creates a few difficulties compared to the multiple linear regression model where the response is continuous:

1. **Nonnormal Error Terms**: For a binary response variable, the normal assumption on the errors is not appropriate since the error can take on only two values:

$$\epsilon = \begin{cases} 1 - \underline{X}^{\mathrm{T}}\beta, & \text{if } (Y|\underline{X}) = 1; \\ -\underline{X}^{\mathrm{T}}\beta, & \text{if } (Y|\underline{X}) = 0. \end{cases}$$

2. **Nonconstant Error Variance**: The variance of the error term is

$$\begin{aligned} \mathrm{Var}(\epsilon) &= \mathrm{Var}(Y - \pi(\underline{X})|\underline{X}) \\ &= \pi(\underline{X})(1 - \pi(\underline{X})) \\ &= \underline{X}^{\mathrm{T}}\beta(1 - \underline{X}^{\mathrm{T}}\beta). \end{aligned}$$

Note that the above expression depends on $\underline{X}$ and, thus, the error variance will differ at different predictor levels.

---

[1] Note that here $\pi$ is being used to represent a probability function and *not* the irrational number.

3. **Constrained Response Function**: The response function for binary outcomes represents probabilities. Hence, the mean responses are constrained as follows:

$$0 \leq \mathrm{E}(Y|\underline{X}) = \pi(\underline{X}) \leq 1.$$

Logistic regression models give us a framework for incorporating the features discussed above. The multiple binary logistic regression model is as follows:

$$\pi(\underline{X}) = \frac{e^{\beta_0 + \beta_1 X_1 + \ldots + \beta_{p-1} X_{p-1}}}{1 + e^{\beta_0 + \beta_1 X_1 + \ldots + \beta_{p-1} X_{p-1}}}$$
$$= \frac{e^{\underline{X}^\mathrm{T}\beta}}{1 + e^{\underline{X}^\mathrm{T}\beta}}. \tag{20.1}$$

Thus, $\pi(\underline{X})$ is the probability that an observation is in a specified category of the binary $Y$ variable given the predictors $\underline{X}$. For example, it might provide estimates of the probability that an older person has heart disease given other subject-specific variables like health indicators, height, and weight.

With the logistic model, estimates of $\pi(\underline{X})$ from equation (20.1) will always be between 0 and 1. This is because the numerator $e^{\underline{X}^\mathrm{T}\beta}$ is *always* positive being a power of a positive value ($e$) and the denominator of the model is (1+numerator), so it will always be larger than the numerator. With a single predictor ($X$), the theoretical model for $\pi(X)$ has an elongated "S" shape (or sigmoidal shape) with asymptotes at 0 and 1. However, in sample estimates we may not see this sigmoidal shape if the range of the $X$ variable is limited.

There are algebraically equivalent ways to write the logistic regression model in equation (20.1). First is

$$\frac{\pi(\underline{X})}{1 - \pi(\underline{X})} = e^{\underline{X}^\mathrm{T}\beta}, \tag{20.2}$$

which is an equation that describes the odds of being in the current category of interest. By definition, the **odds** for an event is $P/(1 - P)$ such that $P$ is the probability of the event.[2] The second way to write the logistic regression model is

$$\mathrm{logit}(\pi(\underline{X})) \equiv \log\left(\frac{\pi(\underline{X})}{1 - \pi(\underline{X})}\right) = \underline{X}^\mathrm{T}\beta, \tag{20.3}$$

which states that the logarithm of the odds is a linear function of the $\underline{X}$ variables and, thus, is called the **log odds**. This function to calculate the log odds is also called the **logit function**, which is written as logit($\cdot$) on the left-hand side of (20.3). We also note that if $\pi(\underline{X})$ (the probability of success of a Bernoulli trial) is replaced by the $\mu(\underline{X}) = n\pi(\underline{X})$ (the mean of a binomial

---

[2]For example, if you are at the racetrack and there is an 80% chance that a certain horse will win the race, then his odds are $0.80/(1 - 0.80) = 4$, or 4:1.

random variable comprised of $n$ independent Bernoulli trials), then (20.3) can be rewritten as

$$\text{logit}(\mu(\underline{X})) \equiv \log\left(\frac{\mu(\underline{X})}{n - \mu(\underline{X})}\right) = \underline{X}^{\text{T}}\beta,$$

which is called **binomial regression**. For all intents and purposes, subsequent estimates and inferences using the logistic regression model and binomial regression model are virtually the same, with the former being the model most often used in practice.

In order to discuss goodness-of-fit measures and residual diagnostics for binary logistic regression, it is necessary to at least define the likelihood; see Appendix C for a brief discussion about likelihood functions. For a sample of size $n$, $y_1, \ldots, y_n$ are independent and realizations of the independent random variables $Y_1, \ldots, Y_n$, such that $Y_i | \mathbf{X}_i$ is a Bernoulli random variable with probability of success $\pi(\mathbf{X}_i)$. The likelihood for a binary logistic regression is given by:

$$\mathcal{L}(\beta; \mathbf{y}, \mathbf{X}) = \prod_{i=1}^{n} \pi(\mathbf{x}_i)^{y_i} (1 - \pi(\mathbf{x}_i))^{1-y_i}$$

$$= \prod_{i=1}^{n} \left(\frac{e^{\mathbf{x}_i^{\text{T}}\beta}}{1 + e^{\mathbf{x}_i^{\text{T}}\beta}}\right)^{y_i} \left(\frac{1}{1 + e^{\mathbf{x}_i^{\text{T}}\beta}}\right)^{1-y_i}.$$

This yields the loglikelihood:

$$\ell(\beta) = \sum_{i=1}^{n} [y_i e^{\mathbf{x}_i^{\text{T}}\beta} - \log(1 + e^{\mathbf{x}_i^{\text{T}}\beta})].$$

Maximizing the likelihood (or loglikelihood) has no closed-form solution, so a technique like iteratively reweighted least squares (see Chapter 12) is used to find an estimate of the regression coefficients, $\hat{\beta}$. Specifically, for iterations $t = 0, 1, \ldots$

$$\hat{\beta}^{(t+1)} = (\mathbf{X}^{\text{T}}\mathbf{W}^{(t)}\mathbf{X})^{-1}\mathbf{X}^{\text{T}}(\mathbf{W}^{(t)}\mathbf{X}\hat{\beta}^{(t)} + \mathbf{y} - \boldsymbol{\pi}^{(t)}),$$

where $\mathbf{W}^{(t)} = \text{diag}(\pi^{(t)}(\mathbf{x}_i)(1 - \pi^{(t)}(\mathbf{x}_i)))$ is a diagonal weighting matrix, $\pi^{(t)}(\mathbf{x}_i)$ is the estimate of $\pi(\mathbf{x}_i)$ in (20.1) based on the current estimate of $\beta$ (i.e., $\hat{\beta}^{(t)}$), and $\boldsymbol{\pi}^{(t)}$ is the vector of the current estimates of those probabilities.

When the routine has converged, we call that value $\hat{\beta}$, which is the maximum likelihood estimate. We can then obtain fitted values analogous to other regression settings:

$$\hat{y}_i = \hat{\pi}(\mathbf{x}_i) = \left[1 + e^{-\mathbf{x}_i^{\text{T}}\hat{\beta}}\right]^{-1},$$

which are used in calculating various goodness-of-fit measures and residuals.

For the residuals we present, they serve the same purpose as in linear regression. When plotted versus the response, they will help identify suspect data points. It should also be noted that the following is by no way an exhaustive list of diagnostic procedures, but rather some of the more common methods which are used.

## Odds Ratio

The odds ratio, $\theta$ determines the relationship between a predictor and response and is available only when the logit link is used. The odds ratio can be any nonnegative number. An odds ratio of 1 serves as the baseline for comparison and indicates there is no association between the response and predictors. If the odds ratio is greater than 1, then the odds of success are higher for the reference level of the factors (or for higher levels of continuous predictors). If the odds ratio is less than 1, then the odds of success are less for the reference level of the factors (or for higher levels of continuous predictors). Values farther from 1 represent stronger degrees of association. For binary logistic regression, the odds of success are:

$$\frac{\pi(\underline{X})}{1 - \pi(\underline{X})} = e^{\underline{X}^{\mathrm{T}}\boldsymbol{\beta}}.$$

This exponential relationship provides an interpretation for $\boldsymbol{\beta}$. The odds increase multiplicatively by $e^{\beta_j}$ for every one-unit increase in $X_j$. More formally, the odds ratio between two sets of predictors (say, $\underline{X}_{(1)}$ and $\underline{X}_{(2)}$) is given by

$$\theta = \frac{\pi(\underline{X}_{(1)})/\left(1 - \pi(\underline{X}_{(1)})\right)}{\pi(\underline{X}_{(2)})/\left(1 - \pi(\underline{X}_{(2)})\right)}.$$

## Wald Test

The **Wald test** is the test of significance for regression coefficients in logistic regression, which is analogous to how we use $t$-tests in linear regression. For maximum likelihood estimates, the ratio

$$Z = \frac{\hat{\beta}_j}{\text{s.e.}(\hat{\beta}_j)}$$

can be used to test $H_0 : \beta_j = 0$, which follows a standard normal distribution. Furthermore, confidence intervals can be constructed as

$$\hat{\beta}_j \pm z_{1-\alpha/2}\text{s.e.}(\hat{\beta}_j).$$

## Raw Residuals

The raw residual is the difference between the actual response and the estimated probability from the model. The formula for the raw residual is

$$r_i = y_i - \hat{\pi}(\mathbf{x}_i).$$

**Pearson Residuals**

The **Pearson residual** corrects for the unequal variance in the raw residuals by dividing by the standard deviation. The formula for the Pearson residuals is

$$p_i = \frac{r_i}{\sqrt{\hat{\pi}(\mathbf{x}_i)(1 - \hat{\pi}(\mathbf{x}_i))}}.$$

**Deviance Residuals**

Deviance residuals are also popular because the sum of squares of these residuals is the deviance statistic. The formula for the deviance residual is

$$d_i = \mathbf{sgn}(r_i)\sqrt{2\left[y_i \log\left(\frac{y_i}{\hat{\pi}(\mathbf{x}_i)}\right) + (1 - y_i) \log\left(\frac{1 - y_i}{1 - \hat{\pi}(\mathbf{x}_i)}\right)\right]}.$$

**Hat Values**

The hat matrix serves a similar purpose as in the case of linear regression — to measure the influence of each observation on the overall fit of the model. However, the interpretation is not as clear due to its more complicated form. The hat values are given by

$$h_{i,i} = \hat{\pi}(\mathbf{x}_i)(1 - \hat{\pi}(\mathbf{x}_i))\mathbf{x}_i^{\mathrm{T}}(\mathbf{X}^{\mathrm{T}}\mathbf{W}\mathbf{X})\mathbf{x}_i,$$

where as in the iteratively reweighted least squares solution (20.1.1), $\mathbf{W} = \mathrm{diag}(\hat{\pi}(\mathbf{x}_i)(1 - \hat{\pi}(\mathbf{x}_i)))$. As in linear regression, a hat value is large if $h_{i,i} > 2p/n$.

**Studentized Residuals**

We can also report Studentized versions of some of the earlier residuals. The **Studentized Pearson residuals** are given by

$$sp_i = \frac{p_i}{\sqrt{1 - h_{i,i}}}$$

and the **Studentized deviance residuals** are given by

$$sd_i = \frac{d_i}{\sqrt{1 - h_{i,i}}}.$$

**C and C̄**

C and C̄ are extensions of Cook's distance for logistic regression. C̄ measures the overall change in the fits due to deleting the $i^{\mathrm{th}}$ observation for all points excluding the one deleted while C includes the deleted point. They are defined by, respectively,

$$C_i = \frac{p_i^2 h_{i,i}}{(1 - h_{i,i})^2}$$

and

$$\bar{C}_i = \frac{p_i^2 h_{i,i}}{(1 - h_{i,i})}.$$

## Goodness-of-Fit Tests

Overall performance of the fitted model can be measured by two different chi-square tests. There is the **Pearson chi-square statistic**

$$P = \sum_{i=1}^{n} p_i^2$$

and the **deviance statistic**

$$G = \sum_{i=1}^{n} d_i^2.$$

Both of these statistics are approximately $\chi_{n-p}^2$-distributed. When a test is rejected, there is a statistically significant lack of fit. Otherwise, there is no evidence of lack of fit.

These goodness-of-fit tests are analogous to the $F$-test in the analysis of variance table for ordinary regression. The null hypothesis is

$$H_0 : \beta_1 = \beta_2 = \ldots = \beta_{k-1} = 0.$$

A significant $p$-value means that at least one of the $X$ variables is a predictor of the probabilities of interest.

In general, one can also use the **likelihood ratio test** for testing the null hypothesis that any subset of the regression coefficients is equal to 0. Let us partition the vector of coefficients as $\beta^T = (\beta_1^T, \beta_2^T)$ with dimensions $p_1$ and $p_2$, respectively, such that $p = p_1 + p_2$. Suppose we wish to test

$$H_0 : \beta_2 = \mathbf{0}$$
$$H_A : \beta_2 \neq \mathbf{0}.$$

The likelihood ratio test statistic is given by

$$\Lambda^* = -2(\ell(\hat{\beta}_2) - \ell(\hat{\beta})),$$

where $\ell(\hat{\beta}_2)$ is the loglikelihood of the model specified by the null hypothesis evaluated at the maximum likelihood estimate of that reduced model. This test statistic follows a $\chi_{p_2}^2$ distribution.

The Wald test can also be extended for this purpose. The test statistic is

$$W - \hat{\beta}_2^T S_{\hat{\beta}_2}^{-1} \hat{\beta}_2,$$

where $S_{\hat{\beta}_2}$ is the estimated variance–covariance matrix of $\hat{\beta}_2$. The test statistic $W$ also follows a $\chi_{p_2}^2$ distribution.

Other goodness-of-fit tests include the **Osius and Rojek normal approximation test** (Osius and Rojek, 1992), **Stukel's score test** (Stukel, 1988), **Brown's test** (Brown, 1982), and the **Hosmer–Lemeshow test** (Hosmer et al., 2013). An extended study and discussion of goodness-of-fit tests in logistic regression models can be found in Hosmer et al. (1997).

**DFDEV and DFCHI**

**DFDEV** and **DFCHI** are statistics that measure the change in deviance and in Pearson's chi-square, respectively, that occurs when an observation is deleted from the dataset. Large values of either statistic indicate a poor fit for the respective observation. The formulas for these statistics are

$$\text{DFDEV}_i = d_i^2 + \bar{C}_i$$

and

$$\text{DFCHI}_i = \frac{\bar{C}_i}{h_{i,i}}.$$

**Pseudo-$R^2$**

The calculation of $R^2$ used in linear regression does not extend directly to logistic regression. Many notions of a **pseudo-$R^2$** measure exist in the literature; see Hu et al. (2006) for a broad discussion of this topic. The version we present is due to Maddala (1983), yet it is often referred to as the Cox–Snell pseudo-$R^2$. This pseudo-$R^2$ is defined as

$$R_{pseudo}^2 = 1 - \left( \frac{\mathcal{L}(\hat{\beta}_0)}{\mathcal{L}(\hat{\beta})} \right)^{2/n},$$

where $\mathcal{L}(\hat{\beta}_0)$ is the likelihood of the model when only the intercept is included and $\mathcal{L}(\beta)$ is the likelihood of the full model. This $R^2$ takes on values from 0 to 1 with 1 being a perfect fit.

**Probit Regression**

In Equation (20.3), we noted that the functional relationship between $\pi(\underline{X})$ and $\underline{X}^{\mathsf{T}}\beta$ is characterized through the logit function. However, one can also use

$$\text{probit}(\pi(\underline{X})) \equiv \Phi^{-1}(\pi(\underline{X})) = \underline{X}^{\mathsf{T}}\beta, \qquad (20.4)$$

where $\Phi^{-1}(\cdot)$ is the inverse of the standard normal cumulative distribution function. The function above is called the **probit function** and the corresponding regression is called **probit regression**. For all intents and purposes, the logistic regression with the logit function and probit regression are virtually the same, with the former being the most commonly used. One can assess the empirical fits of the two and choose the better, but again, the fits will usually be nearly identical. Traditional logistic regression with the logit function is often preferred because it has a natural interpretation with the odds ratio, which cannot be done with probit regression. These notions of the logit and probit functions fall under a bigger class of what we call **link functions**, which are taken up in greater detail in Chapter 21.

### 20.1.2 Nominal Logistic Regression

In binary logistic regression, we only have two possible outcomes. For nominal logistic regression, we allow for the possibility of having $k$ possible outcomes. When $k > 2$, such responses are known as **polytomous**.[3] The multiple nominal logistic regression model or **multinomial logistic regression model**[4] is given by the following:

$$
\pi_j(\underline{X}) = \begin{cases} \dfrac{e^{\underline{X}^T \beta_j}}{1 + \sum_{j=2}^{k} e^{\underline{X}^T \beta_j}}, & j = 2, \ldots, k; \\[3mm] \dfrac{1}{1 + \sum_{j=2}^{k} e^{\underline{X}^T \beta_j}}, & j = 1. \end{cases}
\tag{20.5}
$$

Notice that $k - 1$ of the groups have their own set of $\beta$ values. Furthermore, since $\sum_{j=1}^{k} \pi_j(\underline{X}) = 1$, we set the $\beta$ values for group 1 to be 0, which is called the **reference group**. Notice that when $k = 2$, we are back to binary logistic regression.

Suppose now that we have a sample of size $n$. $\pi_{i,j} \equiv \pi_j(\mathbf{x}_i)$ is the probability that an observation is in one of $k$ categories. The likelihood for the nominal logistic regression model is given by:

$$
\mathcal{L}(\beta; \mathbf{y}, \mathbf{X}) = \prod_{i=1}^{n} \prod_{j=1}^{k} \pi_{i,j}^{y_{i,j}} (1 - \pi_{i,j})^{1 - y_{i,j}},
$$

where the subscript $(i, j)$ means the $i^{\text{th}}$ observation belongs to the $j^{\text{th}}$ group. This yields the loglikelihood:

$$
\ell(\beta) = \sum_{i=1}^{n} \sum_{j=1}^{k} y_{i,j} \pi_{i,j}.
$$

Maximizing the likelihood (or loglikelihood) has no closed-form solution, so again iteratively reweighted least squares can be used to find an estimate of the regression coefficients, $\hat{\beta}$.

An odds ratio $(\theta)$ of 1 serves as the baseline for comparison. If $\theta = 1$, then there is no association between the response and predictors. If $\theta > 1$, then the odds of success are higher for the reference level of the factors (or for higher levels of continuous predictors). If $\theta < 1$, then the odds of success are less for

---

[3] The word **polychotomous** is sometimes used as well.
[4] This model is known by a variety of names, including **multiclass logistic regression** and **softmax regression**.

the reference level of the factor (or for higher levels of a continuous predictor). Values farther from 1 represent stronger degrees of association. For nominal logistic regression, the odds of success (at two different levels of the predictors, say $\underline{X}_{(1)}$ and $\underline{X}_{(2)}$) are:

$$\theta = \frac{\pi_j(\underline{X}_{(1)})/\pi_1(\underline{X}_{(1)})}{\pi_j(\underline{X}_{(2)})/\pi_1(\underline{X}_{(2)})}.$$

Many of the procedures discussed in binary logistic regression can be extended to nominal logistic regression with the appropriate modifications.

### 20.1.3   Ordinal Logistic Regression

For ordinal logistic regression, we again consider $k$ possible outcomes as in nominal logistic regression, except that the order matters. The multiple ordinal logistic regression model is the following:

$$\sum_{j=1}^{k^*} \pi_j(\underline{X}) = \frac{e^{\beta_{0,k^*} + \underline{X}^{\mathrm{T}}\boldsymbol{\beta}}}{1 + e^{\beta_{0,k^*} + \underline{X}^{\mathrm{T}}\boldsymbol{\beta}}}, \qquad (20.6)$$

such that $k^* \leq k$ and $\pi_1(\underline{X}) \leq \pi_2(\underline{X}) \leq \ldots \leq \pi_k(\underline{X})$. Notice that this model is a cumulative sum of probabilities, which involves just changing the intercept of the linear regression portion. So now $\boldsymbol{\beta}$ is $(p-1)$-dimensional and $\underline{X}$ is also $(p-1)$-dimensional such that the first entry of this vector is not a 1. Also, it still holds that $\sum_{j=1}^{k} \pi_j(\underline{X}) = 1$.

Suppose now that we have a sample of size $n$. $\pi_{i,j} \equiv \pi_j(\mathbf{x}_i)$ is still the probability that observation $i$ is in one of $k$ categories, but we are constrained by the model written in equation (20.6). The likelihood for the ordinal logistic regression model is given by:

$$\mathcal{L}(\boldsymbol{\beta}; \mathbf{y}, \mathbf{X}) = \prod_{i=1}^{n} \prod_{j=1}^{k} \pi_{i,j}^{y_{i,j}} (1 - \pi_{i,j})^{1 - y_{i,j}},$$

where the subscript $(i,j)$ means, again, that the $i^{\mathrm{th}}$ observation belongs to the $j^{\mathrm{th}}$ group. This yields the loglikelihood:

$$\ell(\boldsymbol{\beta}) = \sum_{i=1}^{n} \sum_{j=1}^{k} y_{i,j} \pi_{i,j}.$$

Notice that this is identical to the nominal logistic regression likelihood. Thus, maximization again has no closed-form solution, so we can use iteratively reweighted least squares.

For ordinal logistic regression, a proportional odds model is used to determine the odds ratio. Again, an odds ratio ($\theta$) of 1 serves as the baseline for comparison between the two predictor levels, say $\underline{X}_{(1)}$ and $\underline{X}_{(2)}$. Only one parameter and one odds ratio is calculated for each predictor. Suppose we are interested in calculating the odds of $\underline{X}_{(1)}$ to $\underline{X}_{(2)}$. If $\theta = 1$, then there is no association between the response and these two predictors. If $\theta > 1$, then the odds of success are higher for the predictor $\underline{X}_{(1)}$. If $\theta < 1$, then the odds of success are less for the predictor $\underline{X}_{(1)}$. Values farther from 1 represent stronger degrees of association. For ordinal logistic regression, the odds ratio utilizes cumulative probabilities and their complements and is given by:

$$\theta = \frac{\sum_{j=1}^{k^*} \pi_j(\underline{X}_{(1)}) / \left(1 - \sum_{j=1}^{k^*} \pi_j(\underline{X}_{(1)})\right)}{\sum_{j=1}^{k^*} \pi_j(\underline{X}_{(2)}) / \left(1 - \sum_{j=1}^{k^*} \pi_j(\underline{X}_{(2)})\right)}.$$

## 20.2   Poisson Regression

The **Poisson distribution** for a random variable $X$ has the following probability mass function for a given value $X = x$:

$$P(X = x|\lambda) = \frac{e^{-\lambda}\lambda^x}{x!},$$

for $x = 0, 1, 2, \ldots$. The Poisson distribution is characterized by the single parameter $\lambda$, which is the mean rate of occurrence for the event being measured. For the Poisson distribution, it is assumed that large counts with respect to the value of $\lambda$ are rare.

In **Poisson regression**, the dependent variable $Y$ is a count response variable that is assumed to follow a Poisson distribution, but the rate $\lambda$ is now determined by a set of predictors $\underline{X}$. The expression relating these quantities is

$$\lambda(\underline{X}) \equiv E(Y|\underline{X}) = \exp\{\underline{X}^T\beta\}.$$

The above is the most commonly used response function for Poisson regression.[5] Thus, the fundamental Poisson regression model is given by

$$P(Y|\underline{X}) = \frac{e^{-\exp\{\underline{X}^T\beta\}} \exp\{\underline{X}^T\beta\}^Y}{Y!}.$$

That is, for a given set of predictors, the categorical outcome follows a Poisson distribution with rate $\exp\{\underline{X}^T\beta\}$.

The development of goodness-of-fit measures and residual diagnostics for Poisson regression follows similarly to that for logistic regression. First, it is

---

[5]Other functions that ensure $\lambda$ is nonnegative can be employed, which falls under the discussion of link functions in Chapter 21.

necessary to at least define the likelihood. For a sample of size $n$, the likelihood for a Poisson regression is given by:

$$\mathcal{L}(\boldsymbol{\beta}; \mathbf{y}, \mathbf{X}) = \prod_{i=1}^{n} \frac{e^{-\exp\{\mathbf{x}_i^{\mathrm{T}}\boldsymbol{\beta}\}} \exp\{\mathbf{x}_i^{\mathrm{T}}\boldsymbol{\beta}\}^{y_i}}{y_i!}.$$

This yields the loglikelihood:

$$\ell(\boldsymbol{\beta}) = \sum_{i=1}^{n} y_i \mathbf{x}_i^{\mathrm{T}}\boldsymbol{\beta} - \sum_{i=1}^{n} \exp\{\mathbf{x}_i^{\mathrm{T}}\boldsymbol{\beta}\} - \sum_{i=1}^{n} \log(y_i!).$$

Maximizing the likelihood (or loglikelihood) has no closed-form solution, so like in logistic regression, we can again use iteratively reweighted least squares to find an estimate of the regression coefficients, $\hat{\boldsymbol{\beta}}$. Once this value of $\hat{\boldsymbol{\beta}}$ has been obtained, we can calculate fitted values from the estimated Poisson regression model:

$$\hat{y}_i = \hat{\lambda}(\mathbf{x}_i) = \exp\{\mathbf{x}_i^{\mathrm{T}}\hat{\boldsymbol{\beta}}\},$$

which are also used in calculating various goodness-of-fit measures and residuals. For the residuals we present, they serve the same purpose as in linear and logistic regression. When plotted versus the response, they will help identify suspect data points.

More details on Poisson regression, and more generally count regression models, can be found in the texts by Cameron and Trivedi (2013) and Hilbe (2014).

**Offset**
Often the count of interest is measured in terms of that unit's **exposure**, say, $t_i$. Examples of exposure are length of time (e.g., hours, days, or years), dimensional measurements (e.g., inches or square feet), and geographical levels (e.g., county-level, state-level, or country-level). The natural logarithm of the exposure is then incorporated into the response function relationship as an **offset term**, but with a regression parameter estimate that is fixed at 1. Namely, the equation is written as

$$\log(\lambda(\mathbf{x}_i)) = \exp\{\mathbf{x}_i^{\mathrm{T}}\boldsymbol{\beta}\} + t_i.$$

Note that we cannot explicitly have an exposure of $t_i = 0$; however, this usually is not an issue in practical applications.

**Overdispersion**
**Overdispersion** is when an observed dataset exhibits greater variability than what is expected based on the given model. For a regression setting, this means that the actual covariance matrix for the observed data exceeds that for the specified model for $Y|\underline{X}$. For a Poisson distribution, the mean and the variance are equal, which is infrequently observed in practice. This results in overdispersion in the Poisson regression model since the variance is oftentimes greater

than the mean. Various strategies can be employed to correct for or account for overdispersion. For example, one can remove potential outliers and check for appropriate transformations on the data to "correct" for overdispersion. If overdispersion is detected when fitting a Poisson regression model, one can incorporate an overdispersion parameter in the model, which is discussed in the next section on negative binomial regression. Note that overdispersion can also be measured in the logistic regression models that were discussed earlier, but can sometimes be tricky to detect.

**Goodness-of-Fit**
Goodness-of-fit for the fitted Poisson regression model can be measured by the Pearson chi-square statistic

$$P = \sum_{i=1}^{n} \frac{(y_i - \exp\{\mathbf{x}_i^T \hat{\boldsymbol{\beta}}\})^2}{\exp\{\mathbf{x}_i^T \hat{\boldsymbol{\beta}}\}}$$

and the deviance statistic

$$G = \sum_{i=1}^{n} \left[ y_i \log\left( \frac{y_i}{\exp\{\mathbf{x}_i^T \hat{\boldsymbol{\beta}}\}} \right) - (y_i \exp\{\mathbf{x}_i^T \hat{\boldsymbol{\beta}}\}) \right],$$

both of which are approximately $\chi^2_{n-p}$-distributed. As in the logistic regression setting, rejecting this test indicates there is a statistically significant lack-of-fit. The Pearson statistic can also be used as a test for overdispersion.

**Deviance**
Recall the measure of deviance introduced in the study of survival regressions (Chapter 18) and logistic regression. The deviance measure for Poisson regression is

$$D(\mathbf{y}, \hat{\boldsymbol{\beta}}) = 2\ell_S(\boldsymbol{\beta}) - \ell(\hat{\boldsymbol{\beta}}),$$

where $\ell_S(\boldsymbol{\beta})$ is the loglikelihood of the saturated model; i.e., where a model is fit perfectly to the data. This measure of deviance, which differs from the deviance statistic defined earlier, is a generalization of the sum of squares from linear regression. The deviance has an approximate $\chi^2_{n-p}$ distribution.

**Pseudo-$R^2$**
Just like with logistic regression, we define a pseudo-$R^2$ for the Poisson regression setting,

$$R^2_{pseudo} = \frac{\ell(\hat{\boldsymbol{\beta}}) - \ell(\hat{\beta}_0)}{\ell_S(\boldsymbol{\beta}) - \ell(\hat{\beta}_0)},$$

where $\ell(\hat{\beta}_0)$ is the loglikelihood of the model when only the intercept is included. The pseudo-$R^2$ goes from 0 to 1 with 1 being a perfect fit.

**Raw Residuals**

The raw residuals are calculated as

$$r_i = y_i - \exp\{\mathbf{x}_i^{\mathrm{T}}\hat{\boldsymbol{\beta}}\}.$$

Remember that the variance is equal to the mean for a Poisson random variable. Therefore, we expect that the variances of the residuals are unequal, which can lead to difficulties in the interpretation of the raw residuals. Regardless, they are still used, especially in the definition of other diagnostic measures.

**Pearson Residuals**

We can correct for the unequal variance in the raw residuals by dividing by the standard deviation, which leads to the Pearson residuals, given by

$$p_i = \frac{r_i}{\sqrt{\hat{\phi}\exp\{\mathbf{x}_i^{\mathrm{T}}\hat{\boldsymbol{\beta}}\}}},$$

where

$$\hat{\phi} = \frac{1}{n-p}\sum_{i=1}^{n}\frac{(y_i - \exp\{\mathbf{x}_i^{\mathrm{T}}\hat{\boldsymbol{\beta}}\})^2}{\exp\{\mathbf{x}_i^{\mathrm{T}}\hat{\boldsymbol{\beta}}\}}$$

is a dispersion parameter to help control overdispersion.

**Deviance Residuals**

The deviance residuals are given by

$$d_i = \mathbf{sgn}(r_i)\sqrt{2\left\{y_i\log\left(\frac{y_i}{\exp\{\mathbf{x}_i^{\mathrm{T}}\hat{\boldsymbol{\beta}}\}}\right) - (y_i - \exp\{\mathbf{x}_i^{\mathrm{T}}\hat{\boldsymbol{\beta}}\})\right\}},$$

which are analogous to the deviance residuals for logistic regression, but where we have incorporated the estimate of $\hat{\lambda}(\mathbf{x}_i) = \exp\{\mathbf{x}_i^{\mathrm{T}}\hat{\boldsymbol{\beta}}\}$. Note that the sum of squares of these residuals is the deviance statistic.

**Anscombe Residuals**

The **Anscombe residuals** (Anscombe, 1953) are given by

$$a_i = \frac{3(y_i^{2/3} - (\exp\{\mathbf{x}_i^{\mathrm{T}}\hat{\boldsymbol{\beta}}\})^{2/3})}{2(\exp\{\mathbf{x}_i^{\mathrm{T}}\hat{\boldsymbol{\beta}}\})^{1/6}}.$$

These residuals normalize the raw residuals and aid in the identification of heterogeneity and outliers.

**Hat Values**

The hat matrix serves the same purpose as in linear regression and logistic regression — to measure the influence of each observation on the overall fit of the model. The hat values, $h_{i,i}$, are the diagonal entries of the hat matrix

$$H = \mathbf{W}^{1/2}\mathbf{X}(\mathbf{X}^{\mathrm{T}}\mathbf{W}\mathbf{X})^{-1}\mathbf{X}^{\mathrm{T}}\mathbf{W}^{1/2},$$

where the weighting matrix is now $\mathbf{W} = \mathrm{diag}(\exp\{\mathbf{x}_i^{\mathrm{T}}\hat{\boldsymbol{\beta}}\})$. As before, a hat value is large if $h_{i,i} > 2p/n$.

## Studentized Residuals
Finally, we can again calculate Studentized versions of some of the earlier residuals. The Studentized Pearson residuals are given by

$$sp_i = \frac{p_i}{\sqrt{1 - h_{i,i}}}$$

and the Studentized deviance residuals are given by

$$sd_i = \frac{d_i}{\sqrt{1 - h_{i,i}}}.$$

## Rate Ratios
The interpretation of Poisson regression coefficients is not straightforward, so we can use **rate ratios**, which are analogous to odds ratios in logistic regression. Suppose we have two vectors of predictor variables $\underline{X}_1$ and $\underline{X}_2$, such that the vectors are identical except that one of the entries (say, $X_j$) is one unit higher in observation 2 than observation 1. The rate ratio is given by

$$\frac{\lambda_2}{\lambda_1} = \frac{\exp\{\underline{X}_2^{\mathrm{T}}\boldsymbol{\beta}\}}{\exp\{\underline{X}_1^{\mathrm{T}}\boldsymbol{\beta}\}} = \exp\{(\underline{X}_2 - \underline{X}_1)^{\mathrm{T}}\boldsymbol{\beta}\}.$$

Thus, the rate ratio for one unit increase in $X_j$ is $\exp\{\beta_j\}$. Of course, with observed data, we would replace the $\boldsymbol{\beta}$ and $\beta_j$ by their respective estimates, $\hat{\boldsymbol{\beta}}$ and $\hat{\beta}_j$.

## Inference
Inference in Poisson regression is nearly analogous to logistic regression. The estimated approximate variance–covariance matrix of the estimated regression coefficients for Poisson regression are calculated by

$$\mathrm{Var}(\hat{\boldsymbol{\beta}}) = [-H(\hat{\boldsymbol{\beta}})]^{-1},$$

where $H(\hat{\boldsymbol{\beta}})$ is the Hessian matrix. The estimated standard errors for each estimated regression coefficient (s.e.$(\hat{\beta}_k)$) is found by taking the square root of the diagonal of the above matrix. Then, the test statistic for

$$H_0 : \beta_k = 0$$
$$H_A : \beta_k \neq 0$$

follows a standard normal distribution and is calculated using a Wald statistic:

$$z^* = \frac{\hat{\beta}_k}{\text{s.e.}(\hat{\beta}_k)}.$$

We can then calculate $(1 - \alpha)$ confidence intervals for $\beta_k$ and the rate ratio $\exp\{\beta_k\}$ by

$$\hat{\beta}_k \pm z_{1-\alpha/2}\text{s.e.}(\hat{\beta}_k)$$

and

$$\exp\{\hat{\beta}_k \pm z_{1-\alpha/2}\text{s.e.}(\hat{\beta}_k)\},$$

respectively. If $g$ Poisson regression parameters are to be estimated with joint confidence of approximately $(1 - \alpha)$, then the joint Bonferroni confidence intervals can be constructed by

$$\hat{\beta}_k \pm z_{1-\alpha/2g}\text{s.e.}(\hat{\beta}_k).$$

Just as in logistic regression, we can develop tests of whether several $\beta_k = 0$. Let us partition the vector of coefficients as $\boldsymbol{\beta}^{\text{T}} = (\boldsymbol{\beta}_1^{\text{T}}, \boldsymbol{\beta}_2^{\text{T}})$ with dimensions $p_1$ and $p_2$, respectively, such that $p = p_1 + p_2$. Suppose we wish to test

$$H_0 : \boldsymbol{\beta}_2 = \mathbf{0}$$
$$H_A : \boldsymbol{\beta}_2 \neq \mathbf{0}.$$

The likelihood ratio test statistic is again given by

$$\Lambda^* = -2(\ell(\hat{\boldsymbol{\beta}}_2) - \ell(\hat{\boldsymbol{\beta}})),$$

which follows a $\chi^2_{p_2}$ distribution.

## 20.3   Negative Binomial Regression

The **negative binomial distribution** for a random variable $X$ has the following probability mass function for a given value $X = x$:

$$P(X = x | \mu, \theta) = \frac{\Gamma(x + \theta)}{x!\Gamma(\theta)} \frac{\mu^x \theta^\theta}{(\mu + \theta)^{x+\theta}},$$

for $x = 0, 1, 2, \ldots$, where $\Gamma(\cdot)$ is the gamma function defined earlier in (18.3); see Chapter 18. The negative binomial distribution is characterized by two parameters: the **mean parameter** $\mu > 0$ and the **dispersion parameter** $\theta > 0$. Unlike the Poisson distribution, the negative binomial distribution accommodates outcomes that are substantially larger than the mean.[6] A negative binomial random variable $X$ has mean $\mu$ and variance $\mu + (\mu^2/\theta)$ and is often presented as the distribution of the number of failures before the $r^{\text{th}}$ success occurs in a series of independent Bernoulli trials with probability of success $\pi$.

**Negative binomial regression** is similar to Poisson regression, but is more flexible and able to better characterize heavily right-skewed data. Let the dependent variable $Y$ be a count response that is assumed to follow a negative binomial distribution, but the mean $\mu$ is now determined by a set of predictors $\underline{X}$. The expression relating this quantities is

$$\mu(\underline{X}) \equiv \mathrm{E}(Y|\underline{X}) = \exp\{\underline{X}^{\mathrm{T}}\boldsymbol{\beta}\},$$

yielding the negative binomial regression model

$$P(Y|\underline{X}) = \frac{\Gamma(Y+\theta)}{Y!\,\Gamma(\theta)}\,\frac{\exp\{\underline{X}^{\mathrm{T}}\boldsymbol{\beta}\}^Y\,\theta^\theta}{(\exp\{\underline{X}^{\mathrm{T}}\boldsymbol{\beta}\}+\theta)^{Y+\theta}}.$$

The development of goodness-of-fit measures and residual diagnostics for negative binomial regression follows similarly to that for logistic and Poisson regression. First, we define the likelihood. For a sample of size $n$, the likelihood for a negative binomial regression is given by:

$$\mathcal{L}(\boldsymbol{\beta},\theta;\mathbf{y},\mathbf{X}) = \prod_{i=1}^n \frac{\Gamma(y_i+\theta)}{y_i!\,\Gamma(\theta)}\,\frac{\exp\{\mathbf{x}_i^{\mathrm{T}}\boldsymbol{\beta}\}^{y_i}\,\theta^\theta}{(\exp\{\mathbf{x}_i^{\mathrm{T}}\boldsymbol{\beta}\}+\theta)^{y_i+\theta}}.$$

This yields the loglikelihood:

$$\ell(\boldsymbol{\beta},\theta) = \sum_{i=1}^n \Bigg[ \log(\Gamma(y_i+\theta)) - \log(y_i!) + \log(\Gamma(\theta)) + y_i\mathbf{x}_i^{\mathrm{T}}\boldsymbol{\beta}$$
$$\theta\log(\theta) - (y_i+\theta)\log(\exp\{\mathbf{x}_i^{\mathrm{T}}\boldsymbol{\beta}\}+\theta) \Bigg].$$

Notice that the likelihood and loglikelihood are functions of *both* $\boldsymbol{\beta}$ and $\theta$, since these are the parameters of interest and $\mu$ is parameterized through $\boldsymbol{\beta}$. Maximizing the likelihood (or loglikelihood) has no closed-form solution, so we can again use iteratively reweighted least squares to find an estimate of the regression coefficients, $\hat{\boldsymbol{\beta}}$. Once this value of $\hat{\boldsymbol{\beta}}$ has been obtained, we can calculate fitted values from the estimated negative binomial regression model:

$$\hat{y}_i = \hat{\lambda}(\mathbf{x}_i) = \exp\{\mathbf{x}_i^{\mathrm{T}}\hat{\boldsymbol{\beta}}\},$$

---

[6]In fact, there are a few ways the negative binomial distribution can be parameterized. These distinct parameterizations are detailed in the text by Hilbe (2011). The way we have defined the negative binomial distribution actually arises from using a compound Poisson distribution where we place a gamma distribution on the rate parameter. This is called a **Poisson-gamma mixture**.

which are also used in calculating various goodness-of-fit measures and residuals. For the residuals we present, they again serve the purpose of helping us identify suspect data points.

Note that just as in Poisson regression, the count of interest may be in terms of that unit's exposure, $t_i$. We can then incorporate the natural logarithm of the exposure into the response function relationship as an **offset term**, yielding

$$\log(\lambda(\mathbf{x}_i)) = \exp\{\mathbf{x}_i^{\mathrm{T}}\boldsymbol{\beta}\} + t_i.$$

We next discuss a lot of the same goodness-of-fit measures and residuals as were presented for the logistic and Poisson regression models. We end this section with a discussion on underdispersion and another model related to the negative binomial regression model. More details on negative binomial regression can be found in the text by Hilbe (2011), which is devoted to the topic of negative binomial regression.

**Goodness-of-Fit**

Goodness-of-fit for the fitted negative binomial regression model can be measured by the Pearson chi-square statistic

$$P = \sum_{i=1}^{n} \frac{(y_i - \exp\{\mathbf{x}_i^{\mathrm{T}}\hat{\boldsymbol{\beta}}\})^2}{\exp\{\mathbf{x}_i^{\mathrm{T}}\hat{\boldsymbol{\beta}}\} + (\exp\{\mathbf{x}_i^{\mathrm{T}}\hat{\boldsymbol{\beta}}\})^2/\hat{\theta}}$$

and the deviance statistic

$$G = \sum_{i=1}^{n} \left[ y_i \log\left(\frac{y_i}{\exp\{\mathbf{x}_i^{\mathrm{T}}\hat{\boldsymbol{\beta}}\}}\right) - (y_i + \hat{\theta}) \log\left(\frac{y_i + \hat{\theta}}{\exp\{\mathbf{x}_i^{\mathrm{T}}\hat{\boldsymbol{\beta}}\} + \hat{\theta}}\right) \right],$$

both of which are approximately $\chi^2_{n-p-1}$-distributed. Notice that the degrees of freedom is $n - (p + 1)$ because we have $p$ parameters in the $\boldsymbol{\beta}$ vector and 1 parameter for $\theta$. As usual, when one of these tests is rejected, there is a statistically significant lack of fit; otherwise, there is no evidence of lack of fit.

**Deviance**

The deviance measure for negative binomial regression is

$$D(\mathbf{y}, \hat{\boldsymbol{\beta}}) = 2\ell_S(\boldsymbol{\beta}) - \ell(\hat{\boldsymbol{\beta}}),$$

where $\ell_S(\boldsymbol{\beta})$ is the loglikelihood of the saturated model; i.e., where a model is fit perfectly to the data. This measure of deviance, which differs from the deviance statistic defined earlier, is a generalization of the sum of squares from linear regression. The deviance has an approximate $\chi^2_{n-p}$ distribution.

**Pseudo-$R^2$**

According to Hilbe (2011), the most commonly used pseudo-$R^2$ for the negative binomial regression setting is given by

$$R^2_{pseudo} = 1 - \frac{\mathcal{L}(\hat{\boldsymbol{\beta}}_0)}{\mathcal{L}(\hat{\boldsymbol{\beta}})},$$

where $\mathcal{L}(\hat{\beta}_0)$ is the likelihood of the model when only the intercept is included and $\mathcal{L}(\beta)$ is the likelihood of the full model. This $R^2$ takes on values from 0 to 1 with 1 being a perfect fit. Note that this is similar to the pseudo-$R^2$ presented for the logistic regression model, but it is not raised to the power $2/n$. While we have presented different pseudo-$R^2$ formulas for logistic, Poisson, and negative binomial regression fits, one could use any of these. We refer to Heinzl and Mittlböck (2003) and Hu et al. (2006) for an overview of various pseudo-$R^2$.

### Raw Residuals

The raw residuals are calculated as

$$r_i = y_i - \exp\{\mathbf{x}_i^{\mathrm{T}}\hat{\beta}\}.$$

Just as in the Poisson regression setting, we expect that the variances of the residuals are unequal, due to the assumption of a negative binomial distribution.

### Pearson Residuals

We can correct for the unequal variance in the raw residuals by dividing by the standard deviation, which leads to the Pearson residuals, given by

$$p_i = \frac{r_i}{\sqrt{\exp\{\mathbf{x}_i^{\mathrm{T}}\hat{\beta}\} + (\exp\{\mathbf{x}_i^{\mathrm{T}}\hat{\beta}\})^2/\hat{\theta}}}.$$

Note the inclusion of the estimated dispersion parameter in the denominator.

### Deviance Residuals

The deviance residuals are given by

$$d_i = \mathbf{sgn}(r_i)\sqrt{2\left\{ y_i\log\left(\frac{y_i}{\exp\{\mathbf{x}_i^{\mathrm{T}}\hat{\beta}\}}\right) - (y_i + \hat{\theta})\log\left(\frac{y_i + \hat{\theta}}{\exp\{\mathbf{x}_i^{\mathrm{T}}\hat{\beta}\} + \hat{\theta}}\right)\right\}}.$$

Note that the sum of squares of these residuals is the deviance statistic.

### Anscombe Residuals

Following the definition in Hilbe (2011), the **Anscombe residuals** for negative binomial regression are given by

$$a_i = \frac{3\hat{\theta}\left\{(1 + y_i\hat{\theta}^{-1})^{2/3} - (1 + \hat{\theta}^{-1}\exp\{\mathbf{x}_i^{\mathrm{T}}\hat{\beta}\} + 3(y_i^{2/3} - (\exp\{\mathbf{x}_i^{\mathrm{T}}\hat{\beta}\})^{2/3}))\right\}}{2(\hat{\theta}^{-1}(\exp\{\mathbf{x}_i^{\mathrm{T}}\hat{\beta}\})^2 + \exp\{\mathbf{x}_i^{\mathrm{T}}\hat{\beta}\})^{1/6}}.$$

These residuals normalize the raw residuals and aid in the identification of heterogeneity and outliers.

**Hat Values**

Just like in Poisson regression, the hat values, $h_{i,i}$, are the diagonal entries of the hat matrix

$$H = \mathbf{W}^{1/2}\mathbf{X}(\mathbf{X}^\mathrm{T}\mathbf{W}\mathbf{X})^{-1}\mathbf{X}^\mathrm{T}\mathbf{W}^{1/2},$$

where $\mathbf{W} = \mathrm{diag}\left(\frac{\hat{\theta}\exp\{\mathbf{x}_i^\mathrm{T}\hat{\beta}\}}{\hat{\theta}+\exp\{\mathbf{x}_i^\mathrm{T}\hat{\beta}\}}\right)$. As before, a hat value is large if $h_{i,i} > 2p/n$.

**Studentized Residuals**

Finally, for the negative binomial regression model, we can also calculate the Studentized Pearson residuals

$$sp_i = \frac{p_i}{\sqrt{1 - h_{i,i}}}$$

and the Studentized deviance residuals

$$sd_i = \frac{d_i}{\sqrt{1 - h_{i,i}}}.$$

**Rate Ratios**

As with Poisson regression, we can interpret negative binomial regression coefficients using (incident) rate ratios. Suppose we have two vectors of predictor variables $\underline{X}_1$ and $\underline{X}_2$, such that the vectors are identical except that one of the entries (say, $X_j$) is one unit higher in observation 2 than observation 1. The rate ratio is given by

$$\frac{\lambda_2}{\lambda_1} = \frac{\exp\{\underline{X}_2^\mathrm{T}\beta\}}{\exp\{\underline{X}_1^\mathrm{T}\beta\}} = \exp\{(\underline{X}_2 - \underline{X}_1)^\mathrm{T}\beta\}.$$

Thus, the rate ratio for one unit increase in $X_j$ is $\exp\{\beta_j\}$. Again, with observed data, we replace $\beta$ and $\beta_j$ by their respective estimates, $\hat{\beta}$ and $\hat{\beta}_j$.

**Inference**

Inference in negative binomial regression is analogous to Poisson regression, but where we augment the regression coefficient vector to include the dispersion parameter — specifically, we write $\gamma = (\beta^\mathrm{T}, \theta)^\mathrm{T}$. The estimated approximate variance–covariance matrix of the estimated regression coefficients for negative binomial regression are calculated by

$$\mathrm{Var}(\hat{\gamma}) = [-H(\hat{\gamma})]^{-1}.$$

The estimated standard errors for each estimated regression coefficient (s.e.$(\hat{\beta}_k)$) and $\theta$ is found by taking the square root of the diagonal of the above matrix. Then, the test statistic for

$$H_0 : \beta_k = 0$$
$$H_A : \beta_k \neq 0$$

follows a standard normal distribution and is calculated using a Wald statistic:

$$z^* = \frac{\hat{\beta}_k}{\text{s.e.}(\hat{\beta}_k)}.$$

We are usually only interested in tests about the regression coefficients and not for $\theta$, but inference procedures on $\theta$ can be carried out. We can calculate $(1 - \alpha)$ confidence intervals for $\beta_k$ and the rate ratio $\exp\{\beta_k\}$ by

$$\hat{\beta}_k \pm z_{1-\alpha/2}\text{s.e.}(\hat{\beta}_k)$$

and

$$\exp\{\hat{\beta}_k \pm z_{1-\alpha/2}\text{s.e.}(\hat{\beta}_k)\},$$

respectively. If $g$ Poisson regression parameters are to be estimated with joint confidence of approximately $(1 - \alpha)$, then the joint Bonferroni confidence intervals can be constructed by

$$\hat{\beta}_k \pm z_{1-\alpha/2g}\text{s.e.}(\hat{\beta}_k).$$

For tests of whether several $\beta_k = 0$, use the partition $\boldsymbol{\beta}^{\mathrm{T}} = (\boldsymbol{\beta}_1^{\mathrm{T}}, \boldsymbol{\beta}_2^{\mathrm{T}})$ with dimensions $p_1$ and $p_2$, respectively, such that $p = p_1 + p_2$. For the test

$$H_0 : \boldsymbol{\beta}_2 = \mathbf{0}$$
$$H_A : \boldsymbol{\beta}_2 \neq \mathbf{0},$$

the likelihood ratio test statistic is again given by

$$\Lambda^* = -2(\ell(\hat{\boldsymbol{\beta}}_2) - \ell(\hat{\boldsymbol{\beta}})),$$

which follows a $\chi^2_{p_2}$ distribution.

## Underdispersion

As we noted earlier, overdispersion is often present when assuming a Poisson distribution for the count response. The negative binomial distribution is, perhaps, the most common alternative count distribution used when overdispersion is present in a Poisson model. One may encounter **underdispersion**, which means that the observed variability in the observed data is considerably less than that predicted under the assumed model. McCullagh and Nelder (1989) note that underdispersion is rare in practice, but Ridout and Besbeas (2004) present several real data examples demonstrating where underdispersion occurs, including data on the number of strikes that occurred in the United Kingdom's coal mining industry, clutch size for many species of birds, and counts of eggs fertilized by more than one sperm. More flexible count regressions have been developed to handle underdispersion and/or overdispersion. These models tend to be Poisson-based count regression models. Two such models are the **generalized Poisson regression model** (Famoye,

1993) and the **Conway–Maxwell–Poisson regression model** (Sellers and Shmueli, 2010).

## Quasi-Poisson Regression

Wedderburn (1974) introduced the **quasi-likelihood function**, which is a function with similar properties as the loglikelihood function, except that it does not correspond to an actual probability distribution. This framework allows one to develop and work with various quasi-distributions that help capture various features of the data that might not be accurately reflected by a particular parametric distribution. For example, one alternative to handle overdispersion of a Poisson regression model (besides the negative binomial regression model) is to use a **quasi-Poisson regression** model, where it is still assumed that $E(Y|\underline{X}) = \exp\{\underline{X}^{\mathrm{T}}\boldsymbol{\beta}\}$, but now $\mathrm{Var}(Y|\underline{X}) = \theta \exp\{\underline{X}^{\mathrm{T}}\boldsymbol{\beta}\}$, where $\theta$ is a dispersion parameter. In the classic Poisson regression model, we simply assume $\theta = 1$. One can compare the relative fits between a Poisson regression model and a negative binomial regression model. However, the fact that the latter is based on a parametric distribution and, thus, has a true loglikelihood function, typically makes it more appealing from an inference perspective. See Ver Hoef and Boveng (2007) for more on comparing quasi-Poisson regression and negative binomial regression model fits.

## Geometric Regression

The geometric distribution is a special case of the negative binomial distribution where the dispersion parameter is 1; i.e., $\theta = 1$. Thus, all of the discussion provided above for negative binomial regression can be analogously defined for the case where we set $\theta = 1$, which would pertain to a **geometric regression model**. Hilbe (2011) highlights some examples and data problems where using a geometric regression model appears appropriate since estimates of $\theta$ in the negative binomial regression model tend to be near 1.

## 20.4    Specialized Models Involving Zero Counts

With count data, there are certain issues that can arise with zero counts. For example, it might not be possible to observe zero counts, there may be more zero counts than expected with respect to underlying assumed count distribution, or the zero counts could come from their own generating process while all other positive counts come from their own generating process. The models for each of these settings are discussed below. These models can be developed using practically any discrete distribution, but most common are the Poisson and negative binomial distributions. Thus, we will use $f_{\gamma}(Y|\underline{X})$ to generically represent the probability mass function of the count distribution of interest, where $\gamma$ is the corresponding parameter vector.[7] We do not discuss how to numerically estimate their respective parameters, but typically Newton-Raphson or EM algorithms are used. More on estimation of these models is found in Zeileis et al. (2008) and Chapter 11 of Hilbe (2011).

### Zero-Truncated Count Regression Models

**Zero-truncated count regression models** are used when the count response structurally exclude zero counts. Hilbe (2011) gave an example where the response is the number of days a patient is in the hospital. The count begins at registration when a patient first enters a hospital. Thus, there can be no zero days. The zero-truncated count regression model is given by

$$P_{ZTR}(Y|\underline{X};\gamma) = \frac{f_\gamma(Y;\underline{X})}{1 - f_\gamma(Y;\underline{X})}, \quad Y = 1, 2, \ldots. \tag{20.7}$$

Note that the above distribution simply takes the original mass function of the count distribution and rescales the probabilities by excluding the possibility of zero counts.

### Hurdle Count Regression Models

**Hurdle count regression models** (Mullahy, 1986) are two-component models where a zero-truncated regression model is used for the positive count responses and a separate model is used for the zero counts versus larger counts. Hu et al. (2011) gave an example about studying cocaine users, where a secondary outcome could be the number of tobacco cigarettes smoked during last month. It is reasonable to assume that only non-smokers will have smoked zero cigarettes and smokers will report some positive (non-zero) number of cigarettes smoked. Hence the non-smokers would be the structural source of the zero counts. The hurdle count regression model is given by

$$P_{HUR}(Y|\underline{X},\gamma,\alpha) = \begin{cases} f_\alpha(0;\underline{Z}), & \text{if } Y = 0; \\ (1 - f_\alpha(0;\underline{Z}))\frac{f_\gamma(Y;\underline{X})}{f_\gamma(0;\underline{X})}, & \text{if } Y = 1, 2, \ldots. \end{cases} \tag{20.8}$$

In the above, $f_\alpha(\cdot;\underline{Z})$ is a separate model (e.g., a binomial regression model) used for the zero counts such that $\underline{Z}$ is a vector of predictor variables that can be comprised of different variables than those in $\underline{X}$. $\alpha$ is the corresponding parameter vector of the assumed zero count model.

### Zero-Inflated Count Regression Models

**Zero-inflated count regression models** (Lambert, 1992) are two-component mixture models where the zero count responses are assumed to come from one of two generating processes: either one that generates structural zeros or one where the zeros are generated according to some count

---

[7]Based on our earlier notation, $\gamma = \beta$ in the Poisson regression setting and $\gamma = (\beta^T, \theta)^T$ in the negative binomial regression setting.

regression model. For example, Lambert (1992) suggested the setting where one measures defects in a manufacturing process that can move between two states: (1) a *perfect* state where defects are extremely rare and (2) an *imperfect* state in which defects are possible. The zero-inflated count regression model is given by

$$P_{ZI}(Y|\underline{X}, \gamma, \delta) = \begin{cases} \omega_\delta(\underline{Z}) + (1 - \omega_\delta(\underline{Z}))f_\gamma(0; \underline{X}), & \text{if } Y = 0; \\ (1 - \omega_\delta(\underline{Z}))f_\gamma(Y; \underline{X}), & \text{if } Y = 1, 2, \ldots. \end{cases} \quad (20.9)$$

In the above, $0 \le \omega_\delta(\underline{Z}) \le 1$ are the mixing proportions and can be modeled (e.g., using a logit function) as a function of the predictors $\underline{Z}$ and some parameter vector $\alpha$. Again, the variables represented in $\underline{Z}$ can be uncoupled from those in the predictor vector $\underline{X}$.

## 20.5   Examples

**Example 20.5.1.** *(Subarachnoid Hemorrhage Data)*
Turck et al. (2010) analyzed data to facilitate early outcome prediction for $n = 113$ individuals suffering from irreversible brain damage from aneurysmal subarachnoid hemorrhage (aSAH). Each patient is given either a "good" or "poor" projected outcome. The measured variables considered for our analysis are:

- $Y$: outcome measure, where 1 means "good" and 0 means "poor"

- $X_1$: age of subject

- $X_2$: S100$\beta$ measurement from the drawn blood panel

- $X_3$: nucleoside diphosphate kinase A (NDKA) measurement from the drawn blood panel

- $X_4$: gender of the patient, where 1 is a female and 0 is a male

We proceed to fit a binary logistic regression model with the four predictor variables noted above. The output for this fit is as follows:

```
#############################
Coefficients:
              Estimate Std. Error z value Pr(>|z|)
(Intercept)    3.33222    1.01104   3.296 0.000981 ***
age           -0.03350    0.01826  -1.834 0.066631 .
s100b         -4.99552    1.27576  -3.916 9.01e-05 ***
ndka          -0.02904    0.01679  -1.729 0.083739 .
```

```
genderFemale  1.10249     0.50492    2.183 0.029001 *
---
Signif. codes:  0 *** 0.001 ** 0.01 * 0.05 . 0.1   1

(Dispersion parameter for binomial family taken to be 1)

    Null deviance: 148.04  on 112  degrees of freedom
Residual deviance: 113.08  on 108  degrees of freedom
AIC: 123.08
#############################
```

Based on the Wald test for each predictor, the $S100\beta$ measurement and gender of the subject are significant predictor. However, age and the NDKA measurement are only marginally significant. For the present time, we will retain all four of these predictors in the model.

We next look at the odds ratios of the different predictors:

```
#############################
                OR           2.5 %           97.5 %
(Intercept) 28.00038163 4.2420716318 231.45835787
age          0.96705650 0.9316391520   1.00147996
s100b        0.00676823 0.0004445675   0.06928629
ndka         0.97138001 0.9372892580   1.00306839
genderFemale 3.01164757 1.1394230699   8.39500298
#############################
```

These odds ratios are interpreted as follows:

- For a one-year increase in age, the odds of a good prognosis is multiplied by 0.967; i.e., an overall decrease of about 4% in the odds of receiving a good prognosis.

- For a one-unit increase in $S100\beta$, the odds of a good prognosis is multiplied by 0.007; i.e., an overall decrease of about 99.3% in the odds of receiving a good prognosis.

- For a one-unit increase in NDKA, the odds of a good prognosis is multiplied by 0.971; i.e., an overall decrease of about 2.9% in the odds of receiving a good prognosis.

- If the patient is a female, the odds of a good prognosis is multiplied by 3.012; i.e., an overall increase of over 300% in the odds of receiving a good prognosis when the patient is female instead of male.

Figure 20.1 gives scatterplots of various residuals versus the fitted values from the estimated logistic regression model. Typically, such residual plots are not very helpful because you will get two line patterns that emerge. This is a result of the observed response (which the residuals are based on) being binary. Thus, you don't get the random scatter when assessing, say, residual plots from a linear regression fit. Regardless, each scatterplot in Figure 20.1 has a LOESS curve overlaid. The LOESS curve tends to be relatively flat and about a value of 0 for the respective residual calculation, thus there do not appear to be any additional trends detected in these plots.

Below are the overall goodness-of-fit tests for this model:

```
#############################
         Statistic   p.value
Pearson   108.7579 0.4614418
Deviance  113.0783 0.3499458
#############################
```

Both of them have $p$-values, thus indicating that the model is a good fit.

Suppose we wish to test the joint significance of age and gender in this model; i.e.,

$$H_0 : \beta_1 = \beta_4 = 0$$

$$H_A : \text{at least one of } \{\beta_1, \beta_4\} \text{ is not 0.}$$

Below are the likelihood ratio test and Wald test for the above hypothesis:

```
#############################
Likelihood ratio test

Model 1: outcome == "Good" ~ s100b + ndka
Model 2: outcome == "Good" ~ age + s100b + ndka + gender
  #Df  LogLik Df Chisq Pr(>Chisq)
1   3 -59.879
2   5 -56.539  2  6.68    0.03544 *
---
Signif. codes:  0 *** 0.001 ** 0.01 * 0.05 . 0.1   1


Wald test

Model 1: outcome == "Good" ~ s100b + ndka
Model 2: outcome == "Good" ~ age + s100b + ndka + gender
  Res.Df Df      F  Pr(>F)
1    110
2    108  2 3.0747 0.05027 .
---
Signif. codes:  0 *** 0.001 ** 0.01 * 0.05 . 0.1   1
#############################
```

Based on the $\alpha = 0.05$ significance level, we would reject the null hypothesis with the likelihood ratio test and fail to reject the null hypothesis with the Wald test. Regardless, these results combined with the previous analysis would suggest that keeping both age and gender as predictors is appropriate.

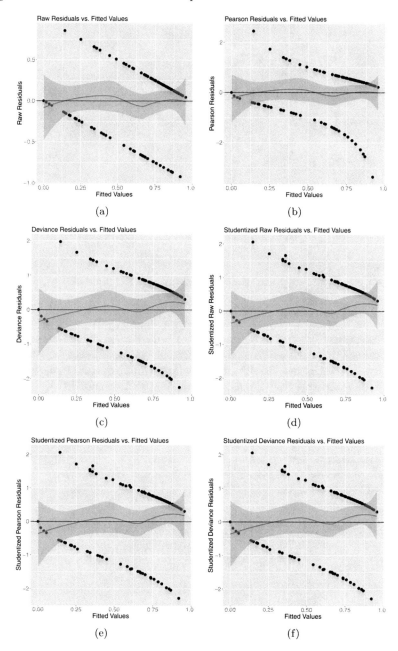

**FIGURE 20.1**

Plots of the following residuals versus the logistic regression fitted values for the subarachnoid hemorrhage data: (a) raw residuals, (b) Pearson residuals, (c) deviance residuals, (d) Studentized raw residuals, (e) Studentized Pearson residuals, and (f) Studentized deviance residuals.

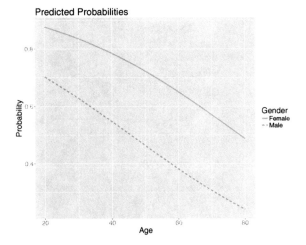

**FIGURE 20.2**
Profile plots of the predicted probabilities for each gender versus age.

The pseudo-$R^2$ value is about 0.266. While low with respect to the $R^2$ values we saw in linear regression settings, this value indicates the model fit is decent. When the pseudo-$R^2$ value gets very low, say, less than 0.10, then it might be indicative that the assumed model yields a poor fit.

Finally, Figure 20.2 provides a profile plot of the predicted probabilities versus age for each gender. Clearly, the profile for females indicates a better chance of having a good prognosis compared to the profile for males.

**Example 20.5.2.** *(Sportsfishing Survey Data)*
Herriges and Kling (1999) analyzed data from the 1989 Southern California Sportsfishing Survey. Various utility models were explored pertaining to the following variables:

- $Y$: preferred recreation mode for fishing: beach, pier, boat, or charter

- $X_1$: price of beach mode

- $X_2$: price of pier mode

- $X_3$: price of boat mode

- $X_4$: price of charter mode

- $X_5$: monthly income of fisherman

There is no baseline level for the categories, so we will choose the mode of beach as the baseline in building our nominal logistic regression model.

Below are the coefficient estimates, standard errors, and $p$-values for the Wald test for each of the coefficients:

```
#############################
$Coef
        price.beach price.pier price.boat price.charter income
pier        -0.0003     -0.0003    -0.0019        0.0013 -1e-04
boat         0.0204      0.0204    -0.0751        0.0460  0e+00
charter      0.0192      0.0192    -0.1501        0.1259 -1e-04

$SE
        price.beach price.pier price.boat price.charter income
pier         0.0018      0.0018     0.0083        0.0076  1e-04
boat         0.0018      0.0018     0.0089        0.0082  1e-04
charter      0.0018      0.0018     0.0087        0.0079  1e-04

$p.vals
        price.beach price.pier price.boat price.charter income
pier         0.8626      0.8626     0.8163        0.8646 0.0372
boat         0.0000      0.0000     0.0000        0.0000 0.5266
charter      0.0000      0.0000     0.0000        0.0000 0.1678
#############################
```

For example, a one dollar increase in the price of any mode has a non-significant effect on the log odds of choosing the pier mode versus the beach mode. The other $p$-values above have similar interpretations.

Below are the odds ratios for this analysis:

```
#############################
        price.beach price.pier price.boat price.charter income
pier         0.9997      0.9997     0.9981        1.0013 0.9999
boat         1.0206      1.0206     0.9277        1.0471 1.0000
charter      1.0194      1.0194     0.8606        1.1342 0.9999
#############################
```

For example, the relative risk ratio for a one dollar increase in the price to fish using a boat would be 0.8606 if someone chooses to charter versus using the beach to fish. Note that the relative risk for income is nearly 1 in all cases, which coincides with the mostly non-significant $p$-values obtained using the Wald test.

Finally, Figure 20.3 is a plot of the predicted probabilities for each chosen fishing mode versus income. The profiles show that there is little effect of income on the probability that one chooses the beach or pier as modes for their fishing. However, there are clearly different profiles for the choice of boat or chartering as a mode as the income level changes.

**Example 20.5.3.** *(Cheese-Tasting Experiment Data)*
McCullagh and Nelder (1989) presented data from an experiment concerning the effect on taste of various cheese additives. $n = 208$ subjects responded based on the 9-point hedonic scale, where a value of "1" corresponds to "strong

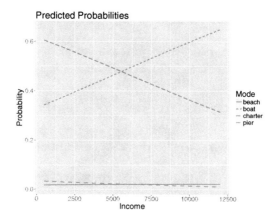

**FIGURE 20.3**
Profile plots of the predicted probabilities for each fishing mode versus the income.

dislike" and "9" corresponds to "excellent taste." Thus, the variables of interest are:

- $Y$: subject's response based on the 9-point hedonic scale

- $X$: cheese additive (four levels labeled A, B, C, and D)

We will analyze an ordinal logistic regression model.
    The estimated coefficients and intercepts for this model are:

```
#############################
Coefficients:
          Value Std. Error t value
CheeseB -3.352      0.4287   -7.819
CheeseC -1.710      0.3715   -4.603
CheeseD  1.613      0.3805    4.238

Intercepts:
      Value   Std. Error t value
1|2  -5.4674    0.5236  -10.4413
2|3  -4.4122    0.4278  -10.3148
3|4  -3.3126    0.3700   -8.9522
4|5  -2.2440    0.3267   -6.8680
5|6  -0.9078    0.2833   -3.2037
6|7   0.0443    0.2646    0.1673
7|8   1.5459    0.3017    5.1244
8|9   3.1058    0.4057    7.6547
```

```
Residual Deviance: 711.3479
AIC: 733.3479
##############################
```

The signs of the coefficients indicate that cheeses B and C (both negative coefficients) do not taste as good as cheese A, but cheese D (positive coefficient) tastes better than cheese A. In particular, there is an implied ordering that D>A>C>B. This can also be seen from the odds ratios and their 95% confidence intervals:

```
##############################
                OR      2.5 %      97.5 %
CheeseB 0.03501953 0.01479494  0.07962504
CheeseC 0.18088334 0.08621686  0.37084182
CheeseD 5.01673910 2.40947358 10.74767430
##############################
```

The other part of the output above gives various intercept quantities. These are the estimated log odds of falling into the categories on the left-hand side of the vertical bar versus all other categories when all other predictors are set to 0. So 1|2 is the estimated log odds of falling into category 1 versus all other categories when all predictors are set to 0, 2|3 is the estimated log odds of falling into categories 1 or 2 versus all other categories when all predictors are set to 0, etc.

Finally, the likelihood ratio test indicates that we have at least one significant predictor in this ordinal logistic regression model:

```
##############################
Likelihood ratio test

Model 1: Response ~ 1
Model 2: Response ~ Cheese
  #Df  LogLik Df  Chisq Pr(>Chisq)
1   8 -429.90
2  11 -355.67  3 148.45  < 2.2e-16 ***
---
Signif. codes:  0 *** 0.001 ** 0.01 * 0.05 . 0.1   1
##############################
```

**Example 20.5.4.** *(Biochemistry Publications Data)*
Long (1990) discussed the presence of gender differences in scientific productivity and position. This discussion included data on a sample of $n = 915$ biochemistry graduate students and focused on the count of articles produced in the last three years of their doctoral programs. The measured variables included the following:

- $Y$: count of articles produced in the last three years of their doctoral programs

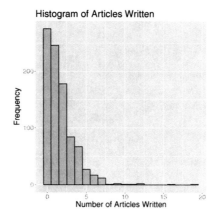

**FIGURE 20.4**
Histogram of the number of articles produced by the biochemistry graduate
students in the last three years of their doctoral program.

- $X_1$: gender of the student, where 1 is a female and 0 is a male

- $X_2$: marital status of the student, where 1 is married and 0 is single

- $X_3$: number of children the student has that are aged 5 or younger

- $X_4$: a measure of "prestige" of the doctoral program

- $X_5$: count of articles produced in the last three years of the student's doctoral
  advisor

We will start by fitting a Poisson regression and a negative binomial regression
model. Figure 20.4 is a histogram of the counts of articles produced by the
students. While not necessarily an extremely large proportion of the counts
are 0 (just over 30%), we will also investigate both zero-inflated and hurdle
versions of the Poisson regression and negative binomial regression models. For
brevity, we will omit the significance testing of each term in each model. Our
focus will be on a relative comparison of goodness-of-fit between the models.

Below are the coefficients for the Poisson regression and negative binomial
regression models. Clearly all of the coefficients have similar magnitude and
signs in both models. The final row gives the dispersion parameter estimated
for the negative binomial regression model.

```
###############################
            Poi.Reg  NB.Reg
(Intercept)  0.3046  0.2561
femWomen    -0.2246 -0.2164
marMarried   0.1552  0.1505
```

```
kid5          -0.1849 -0.1764
phd            0.0128  0.0153
ment           0.0255  0.0291
theta              NA  2.2644
##############################
```

These coefficients indicate that being a woman and having a larger number of children younger than 5 years of age result in, on average, the student having lesser number of publications, while being married, being in a more prestigious doctoral program, and having an advisor who publishes more, results in, on average, the student having a larger number of publications.

Below are the coefficients for the hurdle regression and zero-inflated regression models. Notice that the top-half of the output gives the estimated coefficients for the count portion of the regression models, while the bottom-half of the output gives the estimated coefficients for the zero-component portion of the models. All of the coefficients have similar magnitude and signs in the models for the count regression component. However, the zero-component portions do have some differences. Again, the final row gives the dispersion parameter estimated for the negative binomial regression models.

```
##############################
                    Poi.Hurdle NB.Hurdle ZIP.Reg ZINB.Reg
count_(Intercept)       0.6711    0.3551  0.6408   0.4167
count_femWomen         -0.2286   -0.2447 -0.2091  -0.1955
count_marMarried        0.0965    0.1034  0.1038   0.0976
count_kid5             -0.1422   -0.1533 -0.1433  -0.1517
count_phd              -0.0127   -0.0029 -0.0062  -0.0007
count_ment              0.0187    0.0237  0.0181   0.0248
zero_(Intercept)        0.2368    0.2368 -0.5771  -0.1917
zero_femWomen          -0.2512   -0.2512  0.1097   0.6359
zero_marMarried         0.3262    0.3262 -0.3540  -1.4995
zero_kid5              -0.2852   -0.2852  0.2171   0.6284
zero_phd                0.0222    0.0222  0.0013  -0.0377
zero_ment               0.0801    0.0801 -0.1341  -0.8823
theta                       NA    1.8285      NA   2.6548
##############################
```

Figure 20.5 provides scatterplots of the Pearson residuals versus the fitted values for each of the six models that were fit. For the most part, these plots look similar regardless of the fit. However, there are a few features worth noting. First is that the three negative binomial regression models that were fit all have smaller ranges of Pearson residuals compared to the corresponding Poisson regression models that were fit. This will likely yield better goodness-of-fit measures for the negative binomial regression models. Second, is that the largest fitted value appears to have less of an effect on the negative binomial regression fits compared to the Poisson regression fits. Finally, there are two

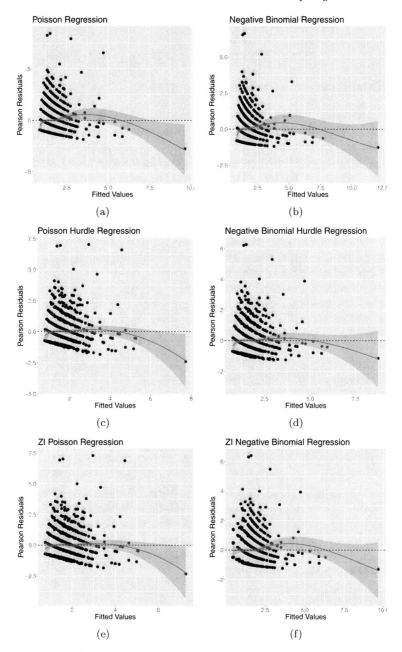

**FIGURE 20.5**
Plots of the Pearson residuals versus the fitted values for the following models fit to the biochemistry publications data: (a) Poisson regression, (b) negative binomial regression, (c) Poisson hurdle regression, (d) negative binomial hurdle regression, (e) zero-inflated Poisson regression, and (f) zero-inflated negative binomial regression.

values with fitted values near 1 with large Pearson residual from the negative binomial regression fits. These are not as prominent in the Poisson regression fits. This could be an indication that these points are outliers, which is a feature that is not as clear in the corresponding Poisson regression fits.

Finally, we calculated the BIC values for each of the six models under consideration, which are given below:

```
##############################
   Poi.Reg Poi.Hurdle    ZIP.Reg
   3343.026   3292.450   3291.373

    NB.Reg  NB.Hurdle   ZINB.Reg
   3169.649   3193.839   3188.628
##############################
```

Based on the above, the negative binomial regression model has the smallest BIC value. Thus, we would proceed to use that as our final model to perform any additional inference.

**Example 20.5.5.** *(Hospital Stays Data)*
Hilbe (2011) analyzed data from the U.S. national Medicare inpatient hospital database for patients in Arizona. These data are prepared yearly from hospital filing records. The data analyzed by Hilbe (2011) are limited to only one diagnostic group and the patients were randomly selected from the original database. We will investigate a model using the following variables:

- $Y$: length of the patient's hospital stay

- $X_1$: patient identifies primarily as Caucasian (1) or non-white (0)

- $X_2$: patient's status, where 1 is died and 0 is survived

- $X_3$: type of hospital admission: elective, urgent, or emergency

In these data, when a patient first enters the hospital, the count begins upon registration. Thus, the minimum length of stay is a value of 1. Due to this feature, the data are zero-truncated. So we will explore fitting zero-truncated Poisson regression and negative binomial regression models.

Figure 20.6 are **violin plots** (Hintze and Nelson, 1998) of the response variable versus each of the three categorical predictors. Violin plots are, essentially, boxplots that have been altered to reflect kernel densities. They help to visualize how the distribution of the response variable changes depending on the level of the categorical variable on the $x$-axis. Clearly each violin plot shows different shapes for the different levels of the respective predictor.

Below are the coefficients for the zero-truncated Poisson regression model:

```
##############################
Coefficients:
```

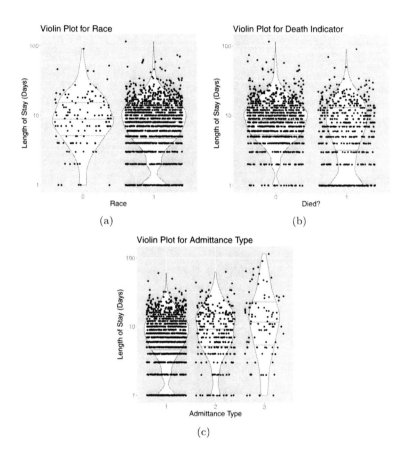

**FIGURE 20.6**
Violin plots of (a) patient's race, (b) whether or not the patient died, and (c) the admittance status of the patient to the hospital.

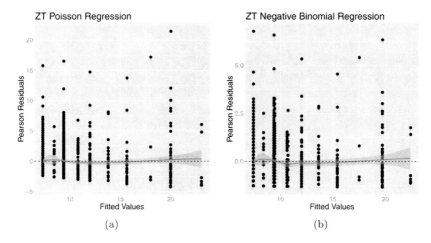

**FIGURE 20.7**
Partial residuals versus fitted values from the estimated (a) zero-truncated Poisson regression model and (b) zero-truncated negative binomial regression model.

```
               Estimate Std. Error  z value Pr(>|z|)
(Intercept)     2.38026    0.02731   87.161  < 2e-16 ***
white1         -0.14075    0.02741   -5.135 2.82e-07 ***
died1          -0.24517    0.01813  -13.520  < 2e-16 ***
type2           0.24427    0.02105   11.604  < 2e-16 ***
type3           0.75570    0.02605   29.006  < 2e-16 ***
---
Signif. codes:  0 *** 0.001 ** 0.01 * 0.05 . 0.1   1
##############################
```

Clearly, all predictors are found to be significant when modeling the length of stay of the patients. We next fit the zero-truncated negative binomial regression model:

```
##############################
Coefficients:
               Estimate Std. Error  z value Pr(>|z|)
(Intercept):1   2.33341    0.07402   31.524  < 2e-16 ***
(Intercept):2   0.63203    0.05468   11.559  < 2e-16 ***
white1         -0.13183    0.07450   -1.770   0.0768 .
died1          -0.25119    0.04475   -5.613 1.99e-08 ***
type2           0.26011    0.05505    4.725 2.30e-06 ***
type3           0.76917    0.08270    9.301  < 2e-16 ***
---
Signif. codes:  0 *** 0.001 ** 0.01 * 0.05 . 0.1   1
```

```
###############################
```

For the above output, the second (`Intercept`) line is the estimate of the dispersion parameter. For this model, we have mostly the same results, but the variable pertaining to the race of the patient has now become marginally significant.

We next look at scatterplots of the Pearson residuals versus the fitted values for the two zero-truncated regression fits. These are given in Figure 20.7. Generally, the residuals from the zero-truncated negative binomial regression fit are smaller than those from the zero-truncated Poisson regression fit. Moreover, the spread of the residuals from the zero-truncated negative binomial regression fit appear more constant compared to those from the zero-truncated Poisson regression fit. This indicates that the negative binomial distribution is likely a better distributional assumption for the zero-truncated regression model.

In addition to looking at the residual plots, we can compare the BIC values for both of the zero-truncated regression models, which are given below:

```
###############################
ZT.Poi.Reg   ZT.NB.Reg
 13714.556    9517.404
###############################
```

Based on the above, the negative binomial regression model has the smallest BIC value. Thus, we would proceed to use that as our final model to perform any additional inference.

# 21

## Generalized Linear Models

In this chapter, we consider **generalized linear models**, which are a generalization to ordinary linear regression that allow for different types of dependent variables to be modeled. In ordinary linear regression, the (conditional) distribution for the dependent variable is assumed to be normal. Here, we consider a broader class of distributions known as the exponential family.

## 21.1 The Generalized Linear Model and Link Functions

We have already seen special cases of generalized linear models, including multiple linear regression, logistic regression, Poisson regression, and negative binomial regression.[1] Generalized linear models provide a generalization of ordinary least squares regression that relates the **random term** (the $n$-dimensional vector of responses $\mathbf{Y}$) to the **systematic term** (the linear predictor $\mathbf{X}\boldsymbol{\beta}$) via a link function (denoted by $g(\cdot)$). Specifically, we have the relation

$$\mathrm{E}(\mathbf{Y}|\mathbf{X}) = \boldsymbol{\mu} = g^{-1}(\mathbf{X}\boldsymbol{\beta}),$$

so $g(\boldsymbol{\mu}) = \mathbf{X}\boldsymbol{\beta}$. The underlying idea is that the function $g(\cdot)$ "links" the mean vector $\boldsymbol{\mu}$ to the linear predictor $\mathbf{X}\boldsymbol{\beta}$. Some common link functions are:

- The **identity link**:

$$g(\boldsymbol{\mu}) = \boldsymbol{\mu} = \mathbf{X}\boldsymbol{\beta},$$

  which is used in traditional linear regression.

- The **logit link**:

$$g(\boldsymbol{\mu}) = \log\left(\frac{\boldsymbol{\mu}}{1 - \boldsymbol{\mu}}\right) = \mathbf{X}\boldsymbol{\beta}$$

$$\Rightarrow \boldsymbol{\mu} = \frac{e^{\mathbf{X}\boldsymbol{\beta}}}{1 + e^{\mathbf{X}\boldsymbol{\beta}}},$$

  which is used in logistic regression.

---

[1]In fact, a more "generalized" framework for regression models is called **general regression models**, which includes any parametric regression model.

- The **log link**:

$$g(\mu) = \log(\mu) = \mathbf{X}\boldsymbol{\beta}$$
$$\Rightarrow \mu = e^{\mathbf{X}\boldsymbol{\beta}},$$

which is used in Poisson regression. Closely related to this link function is the **log-log link**:

$$g(\mu) = -\log(-\log(\mu)) = \mathbf{X}\boldsymbol{\beta}$$
$$\Rightarrow \mu = \exp\{-e^{-\mathbf{X}\boldsymbol{\beta}}\}.$$

- The **probit link**:

$$g(\mu) = \Phi^{-1}(\mu) = \mathbf{X}\boldsymbol{\beta}$$
$$\Rightarrow \mu = \Phi(\mathbf{X}\boldsymbol{\beta}),$$

where $\Phi(\cdot)$ is the cumulative distribution function of the standard normal distribution. This link function is also sometimes called the **normit link**. This also can be used in logistic regression.

- The **complementary log-log link**:

$$g(\mu) = \log(-\log(1-\mu)) = \mathbf{X}\boldsymbol{\beta}$$
$$\Rightarrow \mu = 1 - \exp\{-e^{\mathbf{X}\boldsymbol{\beta}}\},$$

which can also be used in logistic regression. This link function is also sometimes called the **gompit link**.

- The **cauchit link**:

$$g(\mu) = \frac{1}{\pi}\arctan(\mu) + \frac{1}{2} = \mathbf{X}\boldsymbol{\beta}$$
$$\Rightarrow \mu = \tan\left[\pi\left(\mathbf{X}\boldsymbol{\beta} - \frac{1}{2}\right)\right],$$

which can be used for modeling binary data.

- The **power link**:

$$g(\mu) = \mu^{\lambda} = \mathbf{X}\boldsymbol{\beta}$$
$$\Rightarrow \mu = (\mathbf{X}\boldsymbol{\beta})^{1/\lambda},$$

where $\lambda \neq 0$. This is used in other regressions which we will shortly discuss, such as gamma regression and inverse Gaussian regression. When $\lambda = -1$, this yields the **inverse link**, which is used in exponential regression.

Also, the variance is typically a function of the mean and is often written as

$$\text{Var}(\mathbf{Y}|\mathbf{X}) = V(\boldsymbol{\mu}) = V(g^{-1}(\mathbf{X}\boldsymbol{\beta})).$$

A random variable $Y$ is assumed to belong to an exponential family distribution where the density can be expressed in the form

$$q(y; \theta, \phi) = \exp\left\{\frac{y\theta - b(\theta)}{a(\phi)} + c(y, \phi)\right\},$$

where $a(\cdot)$, $b(\cdot)$, and $c(\cdot)$ are specified functions, $\theta$ is called the **canonical parameter** of the distribution, and $\phi$ is the dispersion parameter, just like we introduced in Chapter 20. It can be shown that

$$\text{E}(Y) = b'(\theta) = \mu \tag{21.1}$$

and

$$\text{Var}(Y) = \phi b''(\theta) = \phi V(\mu). \tag{21.2}$$

Many probability distributions belong to the exponential family. For example, the normal distribution, the binomial distribution, and the Poisson distribution. When we also have a vector of predictors for which we incorporate a linear predictor $\underline{X}^{\mathrm{T}}\boldsymbol{\beta}$, we have, respectively, ordinary linear regression, logistic regression, and Poisson regression, respectively. The choice of link function is important to the particular generalized linear regression model. Previous research can suggest the appropriate link function to use. If not, we can use the **canonical link function**, where

$$g(\mu) = \theta = \underline{X}^{\mathrm{T}}\boldsymbol{\beta}.$$

For linear, logistic, and Poisson regression, the canonical link functions are, respectively, the identity link, the logit link, and the log link.

The unknown regression parameters, $\boldsymbol{\beta}$, are typically estimated by maximum likelihood using iteratively reweighted least squares. Suppose we have a sample of size $n$. First, we assume that

$$a_i(\phi) = \frac{\phi}{v_i},$$

where $v_i$ are known prior weights. Let $\eta_i = g(\mu_i)$.

---

**Iteratively Reweighted Least Squares**

1. Start with initial estimates $\mu_i^{(0)} = g^{-1}(\mathbf{x}_i^{\mathrm{T}}\boldsymbol{\beta}^{(0)})$.
2. For $t = 0, 1, \ldots$, calculate the **working responses**

$$z_i^{(t)} = \eta_i^{(t)} + (y_i - \mu_i^{(t)})g'\left(\mu_i^{(t)}\right)$$

and working weights

$$w_i^{(t)} = \frac{v_i}{V\left(\mu_i^{(0)}\right)\left(g'\left(\mu_i^{(t)}\right)\right)^2}.$$

Let $\mathbf{z}^{(t)} = (z_1^{(t)}, \ldots, z_n^{(t)})^{\mathrm{T}}$.

3. Using weighted least squares, calculate

$$\boldsymbol{\beta}^{(t+1)} = \left(\mathbf{X}^{\mathrm{T}}\mathbf{W}^{(t)}\mathbf{X}\right)^{-1}\mathbf{X}^{\mathrm{T}}\mathbf{W}^{(t)}\mathbf{z}^{(t)},$$

where $\mathbf{W}^{(t)} = \mathrm{diag}(w_1^{(t)}, \ldots, w_n^{(t)})$.

4. Repeat steps 2 and 3 until convergence. Denote the final estimate of $\boldsymbol{\beta}$ by $\hat{\boldsymbol{\beta}}$.

---

If using a canonical link, then the above is simply a Newton–Raphson algorithm.

The estimated variance–covariance matrix of $\boldsymbol{\beta}$ is given by

$$\hat{V}(\hat{\boldsymbol{\beta}}) = \phi(\mathbf{X}^{\mathrm{T}}\hat{\mathbf{W}}\mathbf{X})^{-1}, \tag{21.3}$$

where $\hat{\mathbf{W}}$ is the diagonal matrix with the final set of weights obtained from the iteratively reweighted least squares algorithm. Then, the square root of the diagonal entries of (21.3) gives the estimated standard errors for $\beta_j$, $j = 1, \ldots, p$. However, if $\phi$ is unknown for calculating (21.3), we can obtain an estimate using the following method of moments estimator:

$$\hat{\phi} = \frac{1}{n-p}\sum_{i=1}^{p}\frac{v_i(y_i - \hat{\mu}_i)^2}{V(\hat{\mu}_i)}.$$

$\boldsymbol{\beta}$ can also be estimated using Bayesian methods (which we briefly discuss in Chapter 25) or quasi-likelihood methods (which we discussed in the context of quasi-Poisson regression). The quasi-likelihood is a function which possesses similar properties to the loglikelihood function and is most often used with count or binary data. Specifically, for a realization $y$ of the random variable $Y$, it is defined as

$$Q(\mu; y) = \int_y^{\mu} \frac{y-t}{\sigma^2 V(t)} dt,$$

where $\sigma^2$ is a scale parameter. There are also tests using likelihood ratio statistics for model development to determine if any predictors may be dropped from the model.

Most of the diagnostic quantities discussed in Chapter 20 are diagnostic measures for those particular generalized linear models. Thus, for example, fitted values, raw residuals, Pearson residuals, deviance residuals, and hat values can all be calculated for the generalized linear model being fit. Moreover, inference can be conducted using Wald type tests for a single regression parameter or a likelihood ratio test can be used for several regression parameters. These quantities are easily constructed for any generalized linear model, but what changes is the assumed parametric distribution for the response variables. We briefly outline the form for these generalized residuals. More details on such measures, including a good introduction to generalized linear models, are found in the text by McCullagh and Nelder (1989).

**Raw Residuals**
The raw residuals for generalized linear models are calculated as

$$r_i = y_i - \hat{\mu}_i.$$

However, these residuals usually have unequal variance depending on the assumed model. These residuals are most informative for data where the response is normal (i.e., multiple linear regression models), but they are included in the definition of the other residuals we present.

**Pearson Residuals**
We can correct for the unequal variance in the raw residuals by dividing by the standard deviation. This leads to the Pearson residuals, given by

$$p_i = \frac{r_i}{\sqrt{V(\hat{\mu}_i)}}.$$

Note that

$$X^2 = \sum_{i=1}^{n} p_i^2,$$

which is the Pearson $X^2$ goodness-of-fit statistic. This is often used as a measure of residual variation in generalized linear model fits.

**Deviance Residuals**
The deviance residuals are given by

$$d_i = \mathbf{sgn}(y_i - \hat{\mu}_i)\sqrt{\mathrm{dev}_i},$$

where $\mathrm{dev}_i$ is the $i^{\mathrm{th}}$ observations contribution to the overall deviance measure

$$D = \sum_{i=1}^{n} \mathrm{dev}_i = \sum_{i=1}^{n} d_i^2.$$

**Anscombe Residuals**

The **Anscombe residuals** (Anscombe, 1953) tend to give values close to those of the standardized deviance. They are given by

$$a_i = \frac{A(y_i) - A(\hat{\mu}_i)}{A'(\hat{\mu}_i)\sqrt{V(\hat{\mu}_i)}},$$

where

$$A(\cdot) = \int_{y_i}^{\mu_i} \frac{d\mu_i}{V(\mu_i)^{1/3}}.$$

**Randomized Quantile Residuals**

While many regression models we have presented are specific cases of generalized linear models, there is one additional set of residuals that we define here. The **randomized quantile residuals** of Dunn and Smyth (1996) were developed as an alternative to the Pearson and deviance residuals, which are not always guaranteed to be close to normally distributed. Consider a generalized linear model set-up where $Y_1, \ldots, Y_n$ are responses each measured with a vector of predictors, $\underline{X}_1, \ldots, \underline{X}_n$. Let the conditional distribution of $Y_i | \underline{X}_i$ be characterized by the cumulative distribution function $F(\cdot; \boldsymbol{\mu}_i)$, where $\mu_i$ is related to the linear predictor $\underline{X}_i^T \boldsymbol{\beta}$ through a link function. Let $y_1, \ldots, y_n$ and $\mathbf{x}_1, \ldots, \mathbf{x}_n$ be the realizations of our quantities of interest, $a_i = \lim_{y \uparrow y_i} F(y; \hat{\boldsymbol{\mu}}_i)$, $b_i = F(y_i; \hat{\boldsymbol{\mu}}_i)$, and $u_i$ be a uniform random variable on $(a_i, b_i]$. Then, the randomized quantile residuals are defined by

$$rq_i = \begin{cases} \Phi^{-1}(b_i), & \text{if } F \text{ is continuous;} \\ \Phi^{-1}(u_i), & \text{if } F \text{ is discrete,} \end{cases} \tag{21.4}$$

where $\Phi(\cdot)$ is the cumulative distribution function of the standard normal distribution. The $rq_i$ are thus standard normal, apart from sampling variability in $\hat{\vartheta}_i$; see Dunn and Smyth (1996) for a more detailed discussion.

## 21.2 Gamma Regression

Suppose that the random variable $Y$ follows a gamma distribution. The gamma distribution can be parameterized various ways. Following McCullagh and Nelder (1989), the most convenient way to write the gamma density for our purposes is

$$f_Y(y) = \frac{y^{\nu-1}}{\Gamma(\nu)} \left(\frac{\nu}{\mu}\right)^{\nu} e^{-\frac{\nu y}{\mu}} \tag{21.5}$$

for $y > 0$ such that $\nu > 0$ and $\mu > 0$. Therefore, $\mathrm{E}(Y) = \mu$ and $\mathrm{Var}(Y) = \mu^2/\nu$.

Suppose we have the $n$ response values $Y_1, \ldots, Y_n$ and the corresponding $p$-dimensional vector of predictors $\underline{X}_1, \ldots, \underline{X}_n$. The gamma distribution can be used when the response values are positive and we require a more flexible distributional assumption than the normal. Since we also have a vector of predictors, we can construct a **gamma regression model** in a fairly straightforward manner. Assuming the gamma distribution as parameterized in (21.5) for the response, we can use the inverse link function, which is the canonical link for gamma regression, to give us

$$E(Y_i|\underline{X}_i) = \mu_i = (\underline{X}_i^T \beta)^{-1}$$

for the mean function and

$$\text{Var}(Y_i|\underline{X}_i) = \phi V(\mu_i) = \nu^{-1}(\underline{X}_i^T \beta)^{-2}$$

for the variance function. Note that these are obtained using the formulas given in (21.1) and (21.2), respectively.

As noted in the previous section, estimates for $\beta$ and the dispersion parameter $\phi$ can be obtained using the iteratively reweighted least squares routine. Residuals and goodness-of-fit measures are also easily obtained as before.

## 21.3 Inverse Gaussian (Normal) Regression

Suppose that the random variable $Y$ follows an inverse Gaussian distribution, which has the density

$$f_Y(y) = \sqrt{\frac{1}{2\pi\sigma^2 y^3}} \exp\left\{\frac{(y-\mu)^2}{2\sigma^2\mu^2 y}\right\} \tag{21.6}$$

for $y > 0$ such that $\mu > 0$ and $\sigma^2 > 0$. Therefore, $E(Y) = \mu$ and $\text{Var}(Y) = \mu^3\sigma^2$.

Suppose we have the $n$ response values $Y_1, \ldots, Y_n$ and the corresponding $p$-dimensional vector of predictors $\underline{X}_1, \ldots, \underline{X}_n$. Like the gamma distribution, the inverse Gaussian distribution can be used when the response values are positive and we require a more flexible distributional assumption than the normal. Since we also have a vector of predictors, we can construct an **inverse Gaussian regression model** in a fairly straightforward manner. Assuming the inverse Gaussian distribution as parameterized in (21.6) for the response, we can use the power link function with $\lambda = -2$, which is the canonical link for inverse Gaussian regression, to give us

$$E(Y_i|\underline{X}_i) = \mu_i = (\underline{X}_i^T \beta)^{-1/2}$$

for the mean function and

$$\text{Var}(Y_i|\underline{X}_i) = \phi V(\mu_i) = \sigma^2(\underline{X}_i^T \beta)^{-3/2}$$

for the variance function. These are, again, obtained using the formulas given in (21.1) and (21.2), respectively.

Again, estimates for $\boldsymbol{\beta}$ and the dispersion parameter $\phi$ can be obtained using the iteratively reweighted least squares routine. Residuals and goodness-of-fit measures are also easily obtained as before.

## 21.4    Beta Regression

Suppose now that our response variable $Y$ is constrained to be in the (open) unit interval (0,1), which includes rates and proportions. For example, $Y$ could be the unemployment rates of major European cities, the Gini index for each country, graduation rates of major universities, or percentage of body fat of medical subjects. Typically, we use a beta distribution for such random variables that have support on the (open) unit interval. A random variable $Y$ follows a beta distribution if it has the density

$$f_Y(y) = \frac{\Gamma(\phi)}{\Gamma(\mu\phi)\Gamma((1-\mu)\phi)} y^{\mu\phi-1}(1-y)^{(1-\mu)\phi-1}, \qquad (21.7)$$

for $0 < y < 1$ such that $0 < \mu < 1$ and $\phi > 0$ are mean and precision parameters (and $\phi^{-1}$ is called the dispersion parameter). Therefore, $\mathrm{E}(Y) = \mu$ and $\mathrm{Var}(Y) = \mu(1-\mu)/(1+\phi)$. We note that (21.7) is not the traditional formulation of the beta distribution, but is convenient for our following discussion.

Suppose we have the $n$ response values $Y_1, \ldots, Y_n$ and the corresponding $p$-dimensional vector of predictors $\underline{X}_1, \ldots, \underline{X}_n$. Suppose further that the response values take on values between 0 and 1. While logistic regression models can model binary responses that take on the values of 0 or 1, we have not presented a regression model that adequately handles continuous responses that fall between 0 and 1. Assuming the beta distribution for the response variable allows us to construct a **beta regression model**.

Following the setup in Ferrari and Cribari-Neto (2004), the beta regression model is defined as

$$g(\mu_i) = \underline{X}_i^{\mathrm{T}}\boldsymbol{\beta},$$

where $g(\cdot)$ is some link function. Thus far, the model has similar features as a generalized linear model. However, this is not the case. As noted in Cribari-Neto and Zeileis (2010), while the beta regression model shares some properties with generalized linear models, it is actually not a special case of a generalized linear model, even in the fixed dispersion setting. Moreover, there is no canonical link function for beta regression, but Ferrari and Cribari-Neto (2004) suggests using one of the logit, probit, complementary log-log, log-log, or cauchit links.

The conditional expectation of $Y_i$ is

$$\mathrm{E}(Y_i|\underline{X}_i) = \mu_i = g^{-1}(\underline{X}_i^{\mathrm{T}}\boldsymbol{\beta}),$$

while the conditional variance of $Y_i$ is

$$\text{Var}(Y_i|\underline{X}_i) = \frac{\mu_i(1 - \mu_i)}{1 + \phi} = \frac{g^{-1}(X_i^T\boldsymbol{\beta})(1 - g^{-1}(X_i^T\boldsymbol{\beta}))}{1 + \phi}.$$

The above variance clearly depends on $\mu_i$, which results in the beta regression model being naturally heteroskedastic.

Let $y_1, \ldots, y_n$ and $\mathbf{x}_1, \ldots, \mathbf{x}_n$ be our realizations of the response variables and predictor vectors, respectively. The loglikelihood function is

$$\ell(\boldsymbol{\beta}, \phi) = \sum_{i=1}^{n} \ell_i(\mu_i, \phi)$$

$$= \sum_{i=1}^{n} \{\log(\Gamma(\phi)) - \log(\Gamma(\mu_i\phi)) - \log(\Gamma((1 - \mu_i)\phi)) + (\mu_i\phi - 1)\log(y_i)$$

$$+ ((1 - \mu_i)\phi - 1)\log(1 - y_i)\},$$

where we emphasize that $\mu_i$ is a function of $\boldsymbol{\beta}$. Estimation of the parameters can then be performed using maximum likelihood.

Let $\hat{\mu}_i = g^{-1}(\mathbf{x}_i^T\hat{\boldsymbol{\beta}})$, which is based on the maximum likelihood estimate of $\boldsymbol{\beta}$. Standard residuals as calculated for generalized linear models can also be calculated for beta regression models. In particular, we can calculate the raw residuals

$$r_i = y_i - \hat{\mu}_i,$$

the Pearson residuals

$$p_i = \frac{r_i}{\sqrt{\frac{\hat{\mu}_i(1 - \hat{\mu}_i)}{1 + \hat{\phi}}}},$$

and the deviance residuals

$$d_i = \text{sgn}(r_i)\sqrt{-2(\ell_i(\hat{\mu}_i, \hat{\phi}) - \ell_i(y_i, \hat{\phi}))}.$$

Note that for the deviance residuals, $\ell_i(y_i, \hat{\phi})$ is the quantity that appears in the formulation of the loglikelihood function, but is analyzed at $\mu_i = y_i$. In other words, this quantifies the loglikelihood of the saturated model. While we do not discuss them here, there are numerous alternatives to the above residuals that were developed in Espinheira et al. (2008).

An extension of the beta regression model is to further apply a link function to the dispersion parameter, which relates it to a linear combination of another set of predictors. This model is called the **varying dispersion beta regression model** and was introduced in Simas et al. (2010) along with other extensions to the beta regression model. This model can provide greater flexibility over the traditional beta regression model that we presented in this section.

## 21.5    Generalized Estimating Equations

**Generalized estimating equations (GEE) models** were developed by Liang and Zeger (1986) as an extension to generalized linear models with correlated data. Our set-up of GEE models follows closely to that presented in Chapter 8 of Hedeker and Gibbons (2006). Suppose that we have a sample of size $n$ where for each observation $i$, there are measurements at different time points, $j = 1, \ldots, n_i$. Thus, $\mathbf{y}_i = (y_{i,1}, y_{i,2}, \ldots, y_{i,n_i})^{\mathrm{T}}$, which is called **longitudinal data** or **trajectory data**. Moreover, suppose that at each time $j$, the subject also has a $p$-dimensional vector of predictors measured, which in matrix form is an $n_i \times k$ matrix $\mathbf{X}_i = (\mathbf{x}_{i,1}, \mathbf{x}_{i,2}, \ldots, \mathbf{x}_{i,n_i})^{\mathrm{T}}$. One feature of GEE models is that only the marginal distribution of $y_{i,j}$ at each time point needs to be specified and not the joint distribution of a response vector $\mathbf{y}_i$. Keeping this in mind, we specify the GEE models as an extension of generalized linear models, but with a slightly different set-up.

First, specify the linear predictor

$$\eta_{i,j} = \mathbf{x}_{i,j}^{\mathrm{T}} \boldsymbol{\beta}$$

and a link function

$$g(\mu_{i,j}) = \eta_{i,j},$$

which is chosen with consideration to the assumed distribution of the response. The variance is then

$$\mathrm{Var}(y_{i,j}) = \phi V(\mu_{i,j}),$$

which is characterized as a function of the mean. As before, $V(\mu_{i,j})$ is a known variance function and $\phi$ is a dispersion parameter. However, we need to reflect the dependency structure of the longitudinal responses, which we denote by $\mathbf{R}$ and is called the **working correlation structure**. Each observation has its own $n_i \times n_i$ correlation matrix $\mathbf{R}_i$. The matrix depends on a vector of association parameters, $\mathbf{a}$, which is assumed the same for all subjects.

Define $\mathbf{D}_i = \mathrm{diag}(V(\mu_{i,1}), \ldots, V(\mu_{i,n_i}))$, which is an $n_i \times n_i$ matrix. The working variance–covariance matrix of $\mathbf{y}_i$, which reflects the within-subject dependencies, is written as

$$V_i(\mathbf{a}) = \phi \mathbf{D}_i^{1/2} \mathbf{R}_i(\mathbf{a}) \mathbf{D}_i^{1/2}. \tag{21.8}$$

The working correlation matrix in the above can be arbitrarily parameterized, but there are some standard structures that are commonly considered. In order to easily define some of these structures, let us write the generic working correlation matrix as $\mathbf{R}(\mathbf{a}) = [a_{j,j'}]_{n \times n}$, where $j, j' = 1, \ldots, n$ and $a_{j,j'}$ represents the $(j, j')$ entry of the correlation matrix. Note that we are constructing correlation matrices, so $a_{j,j'} = 1$ for $j = j'$. We now briefly outline six common correlation structures for GEE models:

- The **independence structure** is given by

$$a_{j,j'} = 0, \quad \text{for all } j \neq j'.$$

This structure assumes that the longitudinal data are not correlated, which is usually not a reasonable assumption.

- The **exchangeable structure** is given by

$$a_{j,j'} = \rho, \quad \text{for all } j \neq j'.$$

This is one of the most common structures and it assumes that all correlations are the same, regardless of the time interval.

- The **AR(1) structure** is given by

$$a_{j,j'} = \rho^{|j-j'|}.$$

This structure assumes that there is a marked decrease in the within-subject correlation with respect to the order of the lag. Autoregressive structures with higher orders can also be constructed.

- For $m < n$, the $m$-**dependent non-stationary structure** is given by

$$a_{j,j'} = \begin{cases} \rho_{j,j'}, & \text{if } 0 < |j - j'| \leq m; \\ 0, & \text{otherwise.} \end{cases}$$

This structure is a banded correlation matrix and assumes there is some natural order to the data.

- For $m < n$, the $m$-**dependent stationary structure** is given by

$$a_{j,j'} = \begin{cases} \rho_{|j-j'|}, & \text{if } 0 < |j - j'| \leq m; \\ 0, & \text{otherwise.} \end{cases}$$

This structure is also a banded correlation matrix and, mathematically, is what is called a **Toeplitz matrix**. Within a time lag, all of the correlations are assumed to be the same and do not have a functional relationship between the lags of different orders as in the AR(1) structure.

- The **unstructured** correlation matrix is given by

$$a_{j,j'} = \rho_{j,j'}.$$

This means that all correlation coefficients are assumed to be different and are freely estimated from the data. Computationally, this can become difficult for large matrices.

After choosing a working correlation structure, the GEE estimator of $\beta$ is then found as a solution to

$$\sum_{i=1}^{n} \mathbf{S}_i^{\mathrm{T}}[V(\hat{\mathbf{a}})]^{-1}(\mathbf{y}_i - \boldsymbol{\mu}_i) = \mathbf{0},$$

where $\mathbf{S}_i = \frac{\partial \boldsymbol{\mu}_i}{\partial \boldsymbol{\beta}}$. The above is a set of score equations. Since the GEE model is not based on a probability distribution, it is considered a quasi-likelihood model. Thus, solving for this GEE iterates between the quasi-likelihood solution for $\beta$ and a robust method for estimating $\mathbf{a}$ as a function $\beta$. We iterate the following two steps for $t = 0, 1, \ldots$, where $t = 0$ indicates initial estimates, until convergence:

1. Conditioned on the estimates of $\mathbf{R}_i(\mathbf{a}^{(t)})$ and $\phi^{(t)}$, calculate estimates $\beta^{(t+1)}$ using iteratively reweighted least squares as outlined at the beginning of this chapter.

2. Conditioned on the estimates of $\beta^{(t+1)}$, obtain estimates of $\mathbf{a}^{(t+1)}$ and $\phi^{(t)}$ using the Pearson residuals

$$p_{i,j} = \frac{y_{i,j} - \mu_{i,j}^{(t+1)}}{\sqrt{V(\mathbf{a}^{(t)})_{j,j}}} \tag{21.9}$$

to consistently estimate these quantities. Again, note that $\mu_{i,j}^{(t+1)}$ is a function of $\beta^{(t+1)}$.

Upon convergence of the above algorithm, let $\hat{\beta}$, $\hat{\phi}$, and $\hat{\mathbf{a}}$ denote our final estimates. We can obtain standard errors of the estimated regression coefficients to conduct inference. The standard errors are the square root of the diagonal entries of the matrix $V(\hat{\beta})$. Below are the two variance estimators used for GEE models:

1. The **empirical estimator** (also called the **robust estimator** or the **sandwich estimator**) is given by

$$V(\hat{\beta}) = \mathbf{M}_0^{-1} \mathbf{M}_1 \mathbf{M}_0^{-1},$$

where

$$\mathbf{M}_0 = \sum_{i=1}^{n} \mathbf{S}_i^{\mathrm{T}} V_i(\hat{\mathbf{a}}) \mathbf{S}$$

and

$$\mathbf{M}_1 = \sum_{i=1}^{n} \|(\mathbf{y}_i - \hat{\boldsymbol{\mu}}_i)^{\mathrm{T}} V_i(\hat{\mathbf{a}})^{-1} \mathbf{S}_i\|^2.$$

This estimator is consistent when the mean model is correctly specified. This type of "sandwich" estimation is usually attributed to Huber (1967).

2. The **model-based estimator** (also called the **naïve estimator**) is given by

$$V(\hat{\boldsymbol{\beta}}) = \left[\sum_{i=1}^{n} \mathbf{S}_i^{\mathrm{T}} V_i(\hat{\mathbf{a}})^{-1} \mathbf{S}_i\right]^{-1},$$

which is consistent when both the mean model and the covariance model are correctly specified.

Finally, model selection can also be performed using a variation of AIC using the quasi-likelihood. The **quasi-likelihood under the independence model criterion** (or **QIC**) was developed by Pan (2001), which makes and adjustment to the penalty term in AIC. Specifically, it is defined as

$$\text{QIC} = -2Q(\hat{\boldsymbol{\mu}}; I) + 2\text{tr}(\hat{\Omega}_I \hat{V}_{\mathbf{R}}(\hat{\boldsymbol{\beta}})),$$

where $\hat{\boldsymbol{\mu}}$, again, depends on $\hat{\boldsymbol{\beta}}$, $I$ represents the independent covariance structure used to calculate the quasi-likelihood, $\hat{\Omega}_I$ is a variance estimator obtained under the assumption of an independence correlation structure, and $\hat{V}_{\mathbf{R}}(\hat{\boldsymbol{\beta}})$ is the empirical estimator obtained using a general working correlation structure **R**.

The QIC can be used to select the best correlation structure and the best fitting GEE model. A smaller QIC indicates that the corresponding correlation matrix is the preferred correlation matrix. When $\text{tr}(\hat{\Omega}_I \hat{V}_{\mathbf{R}}(\hat{\boldsymbol{\beta}})) \approx \text{tr}(I) = p$, then Pan (2001) presented a simplified version of QIC:

$$\text{QIC}_u = -2Q(\hat{\boldsymbol{\mu}}; I) + 2p.$$

$\text{QIC}_u$ can be potentially useful in variable selection; however, it is easy to see that it cannot be applied to select the working correlation matrix **R**.

Crowder (1995) noted how some of the parameters in the working correlation matrix are "subject to an uncertainty of definition," which has implications on the asymptotic properties of the estimators. An alternative to GEE is to use another two-stage approach called **quasi-least squares regression** (Chaganty, 1997; Shults and Chaganty, 1998; Shults and Hilbe, 2014). Like GEE, quasi-least squares iterates between updating $\boldsymbol{\beta}$ and $\rho$. However, GEE uses the Pearson residuals in (21.9), whereas quasi-least squares solves estimating equations for $\rho$ that are orthogonal to $\boldsymbol{\beta}$.

More on GEE models in the context of longitudinal data analysis can be found in the texts by Hedeker and Gibbons (2006) and Fitzmaurice et al. (2011). We also refer the reader to the text by Hardin and Hilbe (2013), which focuses solely on GEE models. Quasi-likelihood regression is discussed in detail in the text by Shults and Hilbe (2014).

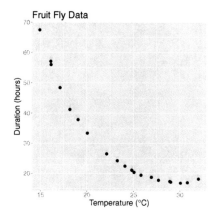

**FIGURE 21.1**
A scatterplot of duration of embryonic period versus the temperature of the
incubation for the fruit fly data.

## 21.6    Examples

**Example 21.6.1.** *(Fruit Fly Data)*
Powsner (1935) presented a study on the effects of temperature on development of the common fruit fly. While Powsner (1935) considered the four stages
in development — embryonic, egg-larval, larval, and pupal — only the embryonic stage is considered for our analysis. A total of $n = 23$ batches were
studied. While numerous quantities were measured, the variables we use in
our analysis are:

- $Y$: the average duration of the embryonic period (in hours) starting at the
  time the eggs were laid

- $X$: the experimental temperature (in degrees Celsius) where the eggs were
  incubated after remaining for about 30 minutes at a controlled temperature

- $W$: the number of eggs in each batch; i.e., batch size

A plot of the average duration versus temperature is given in Figure 21.1.
   These data were also analyzed in Chapter 8 of McCullagh and Nelder
(1989), who motivated their chosen model based on an empirical version of
the Arrhenius law for the rate of simple chemical reactions. In their analysis,
they developed an empirical model that was fit using a gamma regression
with a log link and weighted by the sample batch size. Here, we also analyze a
weighted gamma regression model using the sample batch size $(W)$, but with

the canonical link:

$$\mu^{-1} = \beta_0 + \beta_1 X + \beta_2 (X - \delta)^{-1}. \tag{21.10}$$

As noted in McCullagh and Nelder (1989), the above functional form is used because no embryonic development occurs below a certain critical temperature. Thus, the function should reflect temperatures that have asymptotes. Moreover, note that while (21.10) is nonlinear in temperature, it is linear in $\beta$ for a fixed $\delta$.

Figure 21.2(a) is, again, a scatterplot of the average duration versus temperature, but with the fitted gamma regression curve overlaid. Clearly this appears to provide a good fit. Thus, we proceed to look at some diagnostic plots. First we look at plots of the deviance and Pearson residuals versus the fitted values, given in Figures 21.2(b) and 21.2(c), respectively. Both of these residual plots look very similar. No noticeable trends are present in either plot, so this does not indicate any issues with the model fit that should be investigated. Figure 21.2(d) is a plot of the Cook's distance values. The first observation appears to have a large Cook's $D_i$ value relative to the others. However, the value is only 0.81 which is less than the typical rule of thumb threshold value of 1. Thus, we would not claim that any of the observations appear to be overly influential for this model fit. Overall, this indicates that the mean function in (21.10) is a justifiable model for these data.

Below is the output for the fitted model:

```
#############################
Coefficients:
                          Estimate Std. Error t value Pr(>|t|)
(Intercept)              -3.844e-02  8.819e-04  -43.59  < 2e-16 ***
temp                      3.595e-03  5.721e-05   62.85  < 2e-16 ***
I(1/(temp - delta.hat))   3.421e-02  3.534e-03    9.68 5.45e-09 ***
---
Signif. codes:  0 *** 0.001 ** 0.01 * 0.05 . 0.1   1

(Dispersion parameter for Gamma family taken to be 0.06914058)

    Null deviance: 617.8643  on 22  degrees of freedom
Residual deviance:   1.4044  on 20  degrees of freedom
AIC: 4688.8

Number of Fisher Scoring iterations: 3
#############################
```

Thus, the parameter estimates for (21.10) are $\hat{\beta}_0 = -0.0384$, $\hat{\beta}_1 = 0.0036$, and $\hat{\beta}_2 = 0.0034$. All of these are significant according to the output above. The estimated dispersion parameter is 0.0691. Moreover, the estimate of $\delta$ in (21.10) is 33.56 with an estimated standard error of 1.78, which was estimated using a separate nonlinear least squares routine.

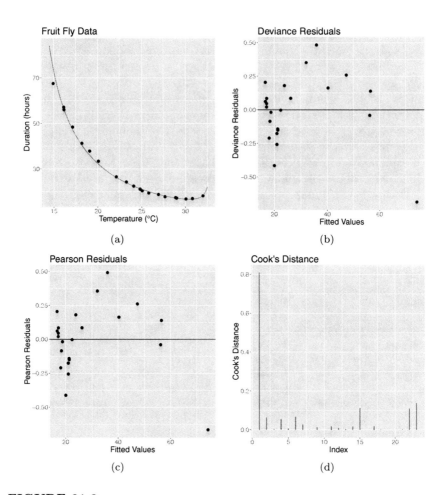

**FIGURE 21.2**
(a) A scatterplot of the fruit fly data with the estimated mean gamma regression curve. (b) Deviance residuals versus the fitted values from the estimated fruit fly model. (c) Pearson residuals versus the fitted values from the estimaed fruit fly model. (d) Cook's $D_i$ values versus observation number for the fruit fly data.

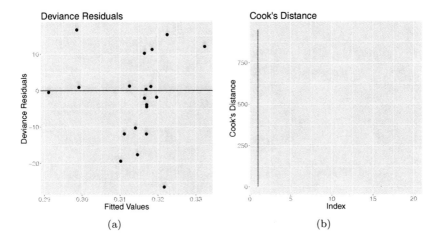

**FIGURE 21.3**
(a) Deviance residuals versus the fitted values from the estimated Canadian auto insurance model. (b) Cook's $D_i$ values versus observation number for the Canadian auto insurance data.

**Example 21.6.2.** *(Canadian Auto Insurance Data)*
The Statistical Unit of the Canadian Underwriters' Association collated automobile insurance policies for private passenger automobile liability for nonfarmers in Canada excluding those in the province of Saskatchewan. The data for policy years 1956 and 1957 were reported and analyzed in Bailey and Simon (1960). The policy statistics were summarized according to various policyholder classifications and a total of $n = 20$ observations are included in this dataset. We will focus on the following variables regarding the policyholder's risk exposure:

- $Y$: the average cost of claims for policy years 1956 and 1957 (in 1000's of Canadian dollars)

- $X_1$: earned adjusted premium (in 1000's of Canadian dollars)

- $X_2$: earned car years under the policy

- $W$: total cost of the claims (in 1000's of Canadian dollars)

These data are also available from and analyzed by Smyth (2011), who used a gamma regression model for the analysis. We proceed to analyze these data using a weighted inverse Gaussian regression model with the canonical link. The weighting variable is the total cost of the claims ($W$).

Figure 21.3(a) is a plot of the deviance residuals versus fitted values from the estimated inverse Gaussian model. No noticeable trends are present in this plot, so this does not indicate any issues with the model fit. However, the plot

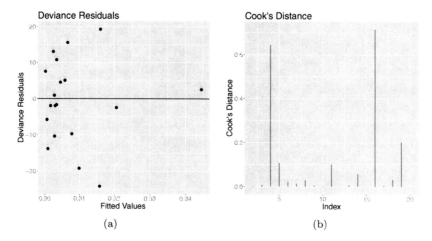

**FIGURE 21.4**
(a) Deviance residuals versus the fitted values from the estimated Canadian
auto insurance model with the largest influential value removed. (b) Cook's
$D_i$ values versus observation number for the Canadian auto insurance data
with the largest influential value removed.

of Cook's distance values due to this fit (Figure 21.3(b)) indicates that the
first observation is highly influential with a value of 950. We will drop this
value from the dataset and rerun the analysis.

Figure 21.4 gives plots of the deviance residuals and the Cook's distance
values for the new fitted model with the first observation omitted. Again, the
plot of the deviance residuals appears good (Figure 21.4(a)), but now the plot
of the Cook's distance values is much improved (Figure 21.4(b)) as all of the
Cook's $D_i$ values are less than 1.

Below is the output for the fitted model with the highly-influential value
omitted:

```
##############################
Coefficients:
              Estimate Std. Error t value Pr(>|t|)
(Intercept)  1.095e+01  7.788e-01  14.062    2e-10 ***
Premium      1.139e-03  4.175e-04   2.727   0.0149 *
Insured     -7.195e-05  2.734e-05  -2.632   0.0181 *
---
Signif. codes:  0 *** 0.001 ** 0.01 * 0.05 . 0.1   1

(Dispersion parameter for inverse.gaussian family
 taken to be 148.0018)
```

```
    Null deviance: 3483.0   on 18   degrees of freedom
Residual deviance: 2412.4   on 16   degrees of freedom
AIC: -937631

Number of Fisher Scoring iterations: 4
###############################
```

Thus, the parameter estimates for the mean function are $\hat{\beta}_0 = 10.95$, $\hat{\beta}_1 = 1.14 \times 10^{-3}$, and $\hat{\beta}_2 = -7.20 \times 10^{-5}$. All of these are significant regression coefficients according to the output above. The estimated dispersion parameter is 148.00.

**Example 21.6.3.** *(Credit Loss Data)*
In the analysis of credit portfolio losses, *loss given default* (LGD) is the proportion of exposure that will be lost if a default occurs. Various credit models have been built and studied that relate the probability of default to LGD and other credit-related variables. Bruche and González-Aguado (2010) provided credit loss data that were extracted from the Altman-NYU Salomon Center Corporate Bond Default Database. The years studied were from 1982 to 2005, giving us $n = 24$ observations. The variables are:

- $Y$: probability of default

- $X_1$: year the statistics were collected (1982–2005)

- $X_2$: number of defaults

- $X_3$: mean LGD

- $X_4$: LGD volatility

These data were subsequently analyzed by Huang and Oosterlee (2011) using various beta regression models. In that paper, they developed more complex credit loss models. Here, we will proceed with development of a simpler beta regression model using the variables outlined above.

We first fit a beta regression model with a canonical link using the four predictors stated above. The output for the fitted beta regression model is as follows:

```
###############################
Coefficients (mean model with logit link):
            Estimate Std. Error z value Pr(>|z|)
(Intercept) 61.656314  21.245642   2.902  0.00371 **
year        -0.033760   0.010628  -3.176  0.00149 **
defs         0.012500   0.002254   5.546 2.93e-08 ***
LGD.mean     0.009160   0.010229   0.895  0.37053
LGD.vol      0.010037   0.017184   0.584  0.55914
```

```
Phi coefficients (precision model with identity link):
     Estimate Std. Error z value Pr(>|z|)
(phi)      977.4         284.0    3.442 0.000578 ***
---
Signif. codes:  0 '***' 0.001 '**' 0.01 '*' 0.05 '.' 0.1 ' ' 1

Type of estimator: ML (maximum likelihood)
Log-likelihood: 101.5 on 6 Df
Pseudo R-squared: 0.8079
Number of iterations: 15 (BFGS) + 8 (Fisher scoring)
#############################
```

The variables for mean LGD and LGD volatility are not statistically significant predictors for the mean equation, with $p$-values of 0.371 and 0.559, respectively. While we could drop LGD volatility and then rerun the analysis, we will proceed to drop both LGD variables and refit the model. The fitted beta regression model after dropping these two LGD variables is as follows:

```
#############################
Coefficients (mean model with logit link):
            Estimate Std. Error z value Pr(>|z|)
(Intercept) 65.61240   20.67647   3.173 0.001507 **
year        -0.03540    0.01040  -3.404 0.000665 ***
defs         0.01396    0.00141   9.901  < 2e-16 ***

Phi coefficients (precision model with identity link):
     Estimate Std. Error z value Pr(>|z|)
(phi)      943.2         274.1    3.441  0.00058 ***
---
Signif. codes:  0 '***' 0.001 '**' 0.01 '*' 0.05 '.' 0.1 ' ' 1

Type of estimator: ML (maximum likelihood)
Log-likelihood: 101.1 on 4 Df
Pseudo R-squared: 0.7999
Number of iterations: 199 (BFGS) + 10 (Fisher scoring)
#############################
```

Clearly, the predictors of year and number of defaults are both statistically significant. Moreover, we can compare the two models by their BIC values. The model fit with the LGD variables has a BIC value of $-183.97$ and the model fit without these variables has a BIC value of $-189.51$. Since the smaller the BIC value the "better" the fit, this indicates that the model without the LGD variables is better.

Figure 21.5(a) is a scatterplot of the probability of default versus the number of defaults. The estimated beta regression curves are overlaid for the years 1985, 1995, and 2005. These fitted mean curves appear appropriate for the

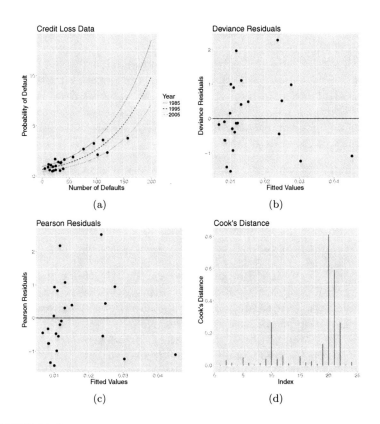

**FIGURE 21.5**

(a) A scatterplot of the credit loss data with the estimated mean beta regression curves for years 1985, 1995, and 2005. (b) Deviance residuals versus the fitted values from the estimated beta regression model. (c) Pearson residuals versus the fitted values from the estimaed beta regression model. (d) Cook's $D_i$ values versus observation number from the estimated beta regression model.

given data. We next look at some diagnostic plots. First we look at plots of the deviance and Pearson residuals versus the fitted values, given in Figures 21.5(b) and 21.5(c), respectively. Both of these residual plots look similar and no noticeable trends are present in either plot. Figure 21.5(d) is a plot of the Cook's distance values. Based on this plot, there do not appear to be any influential observations for this model fit. Overall, this indicates that the beta regression model fit appears appropriate for these data.

**Example 21.6.4.** *(Epilepsy Data)*
Leppik et al. (1985) presented data from a longitudinal study on $n = 59$ epileptic patients. The patients suffered from simple or complex partial seizures. They were randomized to receive either the antiepileptic drug progabide or a placebo. Four successive postrandomization visits were scheduled at 2-week increments and the number of seizures during that timeframe were recorded. The variables in this dataset are:

- $Y$: the number of seizures during each 2-week treatment period

- $X_1$: the number of seizures before the experiment began

- $X_2$: the age of the patient at the beginning of the study

- $X_3$: the treatment group during the respective treatment period

Thall and Vail (1990) and Hothorn and Everitt (2014) both analyzed these data using GEE models for count data with overdispersion present. We'll proceed with a similar strategy by fitting quasi-Poisson regression models with various working correlation structures.

Figures 21.6(a) and 21.6(b) show plots of the seizure rates per base rate of seizures versus the time period for the placebo and progabide groups, respectively. Clearly there are some general trends over the four time periods for most of the subjects. Moreover, Hothorn and Everitt (2014) noted that the variances differ across each treatment by time combination. A plot of the variances versus means for these different levels is given in Figure 21.7. Clearly there are some differences regarding the variance trends, especially with the first two time periods generally having larger variances in the number of seizures relative to those in the second two time periods, which tend to have lower variance in the number of seizures. Moreover, the variances are substantially larger than their respective means. All of these features suggest that overdispersion is present and, thus, using a quasi-Poisson model for our GEE is appropriate.

Without any direct guidance on what the working correlation structure should be, a reasonable assumption is that one other than an independence structure should be used. This is based on the general trends noted in Figures 21.6(a) and 21.6(b). For comparison, we do fit the general quasi-Poisson regression model with an independence structure for the correlation matrix:

##############################

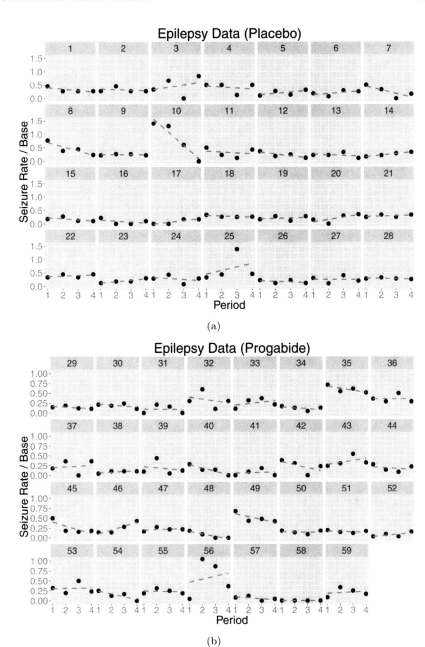

**FIGURE 21.6**
Plots of seizure rate per base rate of seizures versus the time period for the (a) placebo group and (b) progabide group.

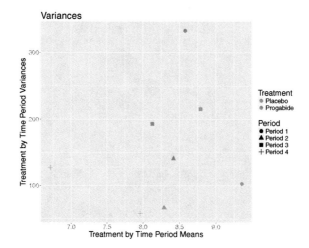

**FIGURE 21.7**
Plot of the variances versus the means for each treatment by time combination.

```
Coefficients:
                    Estimate Naive S.E. Naive z Robust S.E. Robust z
(Intercept)         -0.1306    0.30596  -0.427     0.36515   -0.358
base                 0.0227    0.00115  19.714     0.00124   18.332
age                  0.0227    0.00908   2.505     0.01158    1.964
treatmentProgabide  -0.1527    0.10785  -1.416     0.17111   -0.892

Estimated Scale Parameter:  5.09
Number of Iterations:  1

Working Correlation
      [,1] [,2] [,3] [,4]
[1,]    1    0    0    0
[2,]    0    1    0    0
[3,]    0    0    1    0
[4,]    0    0    0    1
#############################
```

An exchangeable working correlation structure could be more appropriate, albeit the exchangeability condition is a fairly strong assumption. The output using this working correlation matrix is as follows:

```
#############################
Coefficients:
               Estimate Naive S.E. Naive z Robust S.E. Robust z
(Intercept)    -0.1306     0.4522  -0.289     0.36515   -0.358
base            0.0227     0.0017  13.339     0.00124   18.332
age             0.0227     0.0134   1.695     0.01158    1.964
```

```
treatmentProgabide  -0.1527     0.1594  -0.958     0.17111   -0.892

Estimated Scale Parameter:  5.09
Number of Iterations:  1

Working Correlation
       [,1]   [,2]   [,3]   [,4]
[1,]  1.000  0.395  0.395  0.395
[2,]  0.395  1.000  0.395  0.395
[3,]  0.395  0.395  1.000  0.395
[4,]  0.395  0.395  0.395  1.000
#############################
```

Note in the above that the näive standard errors are slightly larger than those under the independence correlation structure, with the exception of the estimate for the base number of seizures before the experiment began. Another alternative could be to use an AR(1) correlation structure, which gives the following results:

```
#############################
Coefficients:
                     Estimate Naive S.E. Naive z Robust S.E. Robust z
(Intercept)          -0.2388    0.43788  -0.545     0.36452   -0.655
base                  0.0231    0.00163  14.149     0.00124   18.703
age                   0.0256    0.01294   1.981     0.01176    2.179
treatmentProgabide   -0.1648    0.15376  -1.072     0.16042   -1.027

Estimated Scale Parameter:  5.15
Number of Iterations:  3

Working Correlation
       [,1]   [,2]   [,3]   [,4]
[1,]  1.000  0.497  0.247  0.123
[2,]  0.497  1.000  0.497  0.247
[3,]  0.247  0.497  1.000  0.497
[4,]  0.123  0.247  0.497  1.000
#############################
```

In the above, we get slightly different results overall. But, the standard error estimates are comparable to those in the previous GEE fits. Moreover, the estimated scale parameter (5.15) is very similar to the previous two fits (5.09). We also calculate the QIC values for the three models above:

```
#############################
          QIC
gee_fit1 -5131
gee_fit2 -5131
gee_fit3 -5132
#############################
```

Clearly, these three QIC values are all close, with a slightly smaller value favoring the third model with the AR(1) structure. Overall, based on our initial assessment of the general trends over time period for each subject, a working correlation like the exchangeable or AR(1) structure is likely most appropriate.

# 22

## Multivariate Multiple Regression

Up until now, we have only concentrated on univariate responses; i.e., the case where the response $Y$ is simply a single value for each observation. Sometimes, however, you may have multiple responses measured for each observation, whether it be different characteristics or perhaps measurements taken over time. When our regression setting must accommodate multiple responses for a single observation, the technique is called multivariate regression.

## 22.1 The Model

A **multivariate multiple regression model** is a multivariate linear model that describes how a *vector* of responses relates to a set of predictors.[1] For example, you may have a newly machined component which is divided into four sections (or sites). Various experimental predictors may be the temperature and amount of stress induced on the component. The responses may be the average length of the cracks that develop at each of the four sites.

The general structure of a multivariate multiple regression model is as follows:

- A set of $p-1$ predictors, or independent variables, are measured for each of the $i = 1, \ldots, n$ observations:

$$\mathbf{X}_i = \begin{pmatrix} X_{i,1} \\ \vdots \\ X_{i,p-1} \end{pmatrix}.$$

- A set of $m$ responses, or dependent variables, are measured for each of the $i = 1, \ldots, n$ observations:

$$\mathbf{Y}_i = \begin{pmatrix} Y_{i,1} \\ \vdots \\ Y_{i,m} \end{pmatrix}.$$

---

[1] In fact, the extension to a matrix response is fairly straightforward and leads to **matrix-variate regression** (Viroli, 2012).

- Each of the $j = 1, \ldots, m$ responses has its own regression model:

$$Y_{i,j} = \beta_{0,j} + \beta_{1,j} X_{i,1} + \beta_{2,j} X_{i,2} + \ldots + \beta_{p-1,j} X_{i,p-1} + \epsilon_{i,j}.$$

Vectorizing the above model for a single observation yields:

$$\mathbf{Y}_i = \mathbf{B}^{\mathrm{T}} (1 \ \mathbf{X}_i^{\mathrm{T}})^{\mathrm{T}} + \boldsymbol{\epsilon}_i, \tag{22.1}$$

where

$$\mathbf{B} = \begin{pmatrix} \boldsymbol{\beta}_1^{\mathrm{T}} \\ \boldsymbol{\beta}_2^{\mathrm{T}} \\ \vdots \\ \boldsymbol{\beta}_m^{\mathrm{T}} \end{pmatrix}^{\mathrm{T}} = \begin{pmatrix} \beta_{0,1} & \beta_{0,2} & \cdots & \beta_{0,m} \\ \beta_{1,1} & \beta_{1,2} & \cdots & \beta_{1,m} \\ \vdots & \vdots & \ddots & \vdots \\ \beta_{p-1,1} & \beta_{p-1,2} & \cdots & \beta_{p-1,m} \end{pmatrix}$$

and

$$\boldsymbol{\epsilon}_i = \begin{pmatrix} \epsilon_{i,1} \\ \vdots \\ \epsilon_{i,m} \end{pmatrix}.$$

Notice that $\boldsymbol{\epsilon}_i$ is the vector of errors for the $i^{\mathrm{th}}$ observation and is assumed to be normally distributed with mean vector $\mathbf{0}$ and variance–covariance matrix $\Sigma \geq 0$.

- Finally, we may explicitly write down the multivariate multiple regression model:

$$\underline{\mathbf{Y}}_{n \times m} = \begin{pmatrix} \mathbf{Y}_1^{\mathrm{T}} \\ \vdots \\ \mathbf{Y}_n^{\mathrm{T}} \end{pmatrix}$$

$$= \begin{pmatrix} 1 & \mathbf{X}_1^{\mathrm{T}} \\ \vdots & \vdots \\ 1 & \mathbf{X}_n^{\mathrm{T}} \end{pmatrix} \begin{pmatrix} \beta_{0,1} & \beta_{0,2} & \cdots & \beta_{0,m} \\ \beta_{1,1} & \beta_{1,2} & \cdots & \beta_{1,m} \\ \vdots & \vdots & \ddots & \vdots \\ \beta_{p-1,1} & \beta_{p-1,2} & \cdots & \beta_{p-1,m} \end{pmatrix} + \begin{pmatrix} \boldsymbol{\epsilon}_1^{\mathrm{T}} \\ \vdots \\ \boldsymbol{\epsilon}_n^{\mathrm{T}} \end{pmatrix}$$

$$= \underline{\mathbf{X}}_{n \times p} \mathbf{B}_{p \times m} + \boldsymbol{\varepsilon}_{n \times m}.$$

Or more compactly, without the dimensional subscripts, we will write:

$$\underline{\mathbf{Y}} = \underline{\mathbf{X}} \mathbf{B} + \boldsymbol{\varepsilon}. \tag{22.2}$$

Of course, we can extend or modify the multivariate linear regression model in (22.2) to many similar settings that we discussed in this book. For example, we could have censored responses, temporal dependencies, or random effects. One specific example is where we also include latent variables as discussed with SEMs in Chapter 11. Letting these latent variables be denoted by the

$q$-dimensional random vectors $\mathbf{Z}_1, \ldots, \mathbf{Z}_n$, we are then interested in a model of the form

$$\underline{\mathbf{Y}} = \underline{\mathbf{Z}}\mathbf{A} + \underline{\mathbf{X}}\mathbf{B} + \varepsilon,$$

where $\underline{\mathbf{Z}}$ is an $n \times q$ matrix of the latent variables, $\mathbf{A}$ is a $q \times m$ factor loading matrix, and all other quantities are as defined in (22.2). This particular model is called the **factor regression model** and has been heavily used in the analysis of gene expression data (Carvalho, 2008).

## 22.2   Estimation and Statistical Regions

In this section, we present a selection of estimation topics for the multivariate multiple linear regression model. Generally speaking, most routines for the univariate multiple linear regression model extend to the multivariate setting without much difficulty. We refer to the text by Johnson and Wichern (2007) for additional details on the multivariate multiple linear regression model and multivariate analysis in general.

**Least Squares**
Extending the least squares theory from the multiple regression setting to the multivariate multiple regression setting is fairly intuitive. The biggest hurdle is dealing with the matrix calculations, which statistical packages perform for you anyhow. We can also formulate similar assumptions for the multivariate model.
    Let

$$\epsilon_{(j)} = \begin{pmatrix} \epsilon_{1,j} \\ \vdots \\ \epsilon_{n,j} \end{pmatrix},$$

which is the vector of errors for the $j^{\text{th}}$ trial of all $n$ observations.[2] We assume that $\mathrm{E}(\epsilon_{(j)}) = \mathbf{0}$ and $\mathrm{Cov}(\epsilon_{(i)}, \epsilon_{(k)}) = \sigma_{i,k}\mathbf{I}_m$ for each $i, k = 1, \ldots, n$. Notice that the $j^{\text{th}}$ trial of the $n$ observations have variance–covariance matrix $\Sigma = \{\sigma_{i,k}\}$, but observations from different entries of the vector are uncorrelated.
    The least squares estimate for $\mathbf{B}$ is simply given by:

$$\hat{\mathbf{B}} = (\underline{\mathbf{X}}^{\mathrm{T}}\underline{\mathbf{X}})^{-1}\underline{\mathbf{X}}^{\mathrm{T}}\underline{\mathbf{Y}}.$$

Using $\hat{\mathbf{B}}$, we can calculate the predicted values as:

$$\hat{\underline{\mathbf{Y}}} = \underline{\mathbf{X}}\hat{\mathbf{B}}$$

---

[2]Note: The parentheses used in the subscript does not imply some sort of ordering as when we define order statistics.

and the residuals as:

$$\hat{\varepsilon} = \mathbf{Y} - \hat{\mathbf{Y}}.$$

Furthermore, an estimate of $\Sigma$ (which is the maximum likelihood estimate of $\Sigma$) is given by:

$$\hat{\Sigma} = \frac{1}{n}\hat{\varepsilon}^{\mathrm{T}}\hat{\varepsilon}.$$

### Hypothesis Testing

Suppose we are interested in testing the hypothesis that our multivariate responses do not depend on the predictors $X_{i,q+1}, \ldots, X_{i,p-1}$. We can partition $\mathbf{B}$ to consist of two matrices: one with the regression coefficients of the predictors we assume will remain in the model and one with the regression coefficients we wish to test. Similarly, we can partition $\mathbf{X}$ in a similar manner. Formally, the test is

$$H_0 : \boldsymbol{\beta}_{(2)} = \mathbf{0},$$

where

$$\mathbf{B} = \left( \frac{\boldsymbol{\beta}_{(1)}}{\boldsymbol{\beta}_{(2)}} \right)$$

and

$$\underline{\mathbf{X}} = \left( \ \underline{\mathbf{X}}_1 \ | \ \underline{\mathbf{X}}_2 \ \right).$$

Here $\underline{\mathbf{X}}_2$ is an $n \times (p - q - 1)$ matrix of predictors corresponding to the null hypothesis and $\underline{\mathbf{X}}_1$ is an $n \times (q)$ matrix of predictors we assume will remain in the model. Furthermore, $\boldsymbol{\beta}_{(2)}$ and $\boldsymbol{\beta}_{(1)}$ are $(p-q-1) \times m$ and $q \times m$ matrices, respectively, for these predictor matrices.

Under the null hypothesis, we can calculate

$$\hat{\boldsymbol{\beta}}_{(1)} = (\underline{\mathbf{X}}_1^{\mathrm{T}}\underline{\mathbf{X}}_1)^{-1}\underline{\mathbf{X}}_1^{\mathrm{T}}\underline{\mathbf{Y}}$$

and

$$\hat{\Sigma}_1 = (\underline{\mathbf{Y}} - \underline{\mathbf{X}}_1\hat{\boldsymbol{\beta}}_{(1)})^{\mathrm{T}}(\underline{\mathbf{Y}} - \underline{\mathbf{X}}_1\hat{\boldsymbol{\beta}}_{(1)})/n.$$

These values (which are maximum likelihood estimates under the null hypothesis) can be used to calculate one of four commonly used multivariate test statistics:

$$\textbf{Wilks' Lambda} = \frac{|n\hat{\Sigma}|}{|n(\hat{\Sigma}_1 - \hat{\Sigma})|}$$

$$\textbf{Pillai's Trace} = \mathrm{tr}[(\hat{\Sigma}_1 - \hat{\Sigma})\hat{\Sigma}_1^{-1}]$$

$$\textbf{Hotelling--Lawley Trace} = \mathrm{tr}[(\hat{\Sigma}_1 - \hat{\Sigma})\hat{\Sigma}^{-1}]$$

$$\textbf{Roy's Greatest Root} = \frac{\lambda_1}{1 + \lambda_1}.$$

In the above, $\lambda_1$ is the largest nonzero eigenvalue of $(\hat{\Sigma}_1 - \hat{\Sigma})\hat{\Sigma}^{-1}$. Also, the value $|\Sigma|$ is the determinant of the variance–covariance matrix $\Sigma$ and is called the **generalized variance**, which assigns a single numerical value to express the overall variation of this multivariate problem. All of the above test statistics have approximate $F$ distributions with degrees of freedom that are more complicated to calculate than what we have seen. Most statistical packages will report at least one of the above if not all four. For large sample sizes, the associated $p$-values will likely be similar, but various situations (such as many large eigenvalues of $(\hat{\Sigma}_1 \hat{\Sigma})\hat{\Sigma}^{-1}$ or a relatively small sample size) will lead to a discrepancy between the results. In this case, it is usually accepted to report the Wilks' lambda value as this is the likelihood ratio test.

**Confidence Regions**

One problem is to predict the mean responses corresponding to fixed values $\mathbf{x}_h$ of the predictors. Using various distributional results concerning $\hat{\mathbf{B}}^T \mathbf{x}_h$ and $\hat{\Sigma}$, it can be shown that the $100 \times (1 - \alpha)\%$ **simultaneous confidence intervals** for $E(Y_i | \mathbf{X} = \mathbf{x}_h) = \mathbf{x}_h^T \hat{\beta}_i$ are

$$\mathbf{x}_h^T \hat{\beta}_i \pm \sqrt{\left(\frac{m(n-p-2)}{n-p-1-m}\right) F_{m,n-p-1-m;1-\alpha}} \\ \times \sqrt{\mathbf{x}_h^T (\underline{\mathbf{X}}^T \underline{\mathbf{X}})^{-1} \mathbf{x}_h \left(\frac{n}{n-p-2} \hat{\sigma}_{i,i}\right)}, \tag{22.3}$$

for $i = 1, \ldots, m$. Here, $\hat{\beta}_i$ is the $i^{\text{th}}$ column of $\hat{\mathbf{B}}$ and $\hat{\sigma}_{i,i}$ is the $i^{\text{th}}$ diagonal element of $\hat{\Sigma}$. Also, notice that the simultaneous confidence intervals are constructed for each of the $m$ entries of the response vector, thus why they are considered *simultaneous*. Furthermore, the collection of these simultaneous intervals yields what we call a $100 \times (1 - \alpha)\%$ confidence region for $\hat{\mathbf{B}}^T \mathbf{x}_h$.

**Prediction Regions**

Another problem is to predict new responses $\mathbf{Y}_h = \mathbf{B}^T \mathbf{x}_h + \varepsilon_h$. Again, skipping over a discussion on various distributional assumptions, it can be shown that the $100 \times (1 - \alpha)\%$ **simultaneous prediction intervals** for the individual responses $Y_{h,i}$ are

$$\mathbf{x}_h^T \hat{\beta}_i \pm \sqrt{\left(\frac{m(n-p-2)}{n-p-1-m}\right) F_{m,n-p-1-m;1-\alpha}} \\ \times \sqrt{(1 + \mathbf{x}_h^T (\underline{\mathbf{X}}^T \underline{\mathbf{X}})^{-1} \mathbf{x}_h) \left(\frac{n}{n-p-2} \hat{\sigma}_{i,i}\right)}, \tag{22.4}$$

for $i = 1, \ldots, m$. The quantities here are the same as those in the simultaneous confidence intervals. Furthermore, the collection of these simultaneous prediction intervals are called a $100 \times (1 - \alpha)\%$ prediction region for $\hat{\mathbf{y}}_h$.

## MANOVA

The **multivariate analysis of variance (MANOVA)** table is similar to its univariate counterpart. The sum of squares values in a MANOVA are no longer scalar quantities, but rather matrices. Hence, the entries in the MANOVA table are called **sum of squares and cross-products (SSCPs)**. These quantities are described in a little more detail below:

- The **sum of squares and cross-products for total** is SSCPTO $= \sum_{i=1}^{n}(\mathbf{Y}_i - \bar{\mathbf{Y}})(\mathbf{Y}_i - \bar{\mathbf{Y}})^{\mathrm{T}}$, which is the sum of squared deviations from the overall mean vector of the $\mathbf{Y}_i$s. SSCPTO is a measure of the overall variation in the $\mathbf{Y}$ vectors. The corresponding total degrees of freedom are $n - 1$.

- The **sum of squares and cross-products for the errors** is SSCPE $= \sum_{i=1}^{n}(\mathbf{Y}_i - \hat{\mathbf{Y}}_i)(\mathbf{Y}_i - \hat{\mathbf{Y}}_i)^{\mathrm{T}}$, which is the sum of squared observed errors (residuals) for the observed data vectors. SSE is a measure of the variation in $\mathbf{Y}$ that is not explained by the multivariate regression. The corresponding error degrees of freedom are $n - p$.

- The **sum of squares and cross-products due to the regression** is SSCPR = SSCPTO − SSCPE, and it is a measure of the total variation in $Y$ that can be explained by the regression with the predictors. The corresponding model degrees of freedom are $p - 1$.

A MANOVA table is given in Table 22.1.

**TABLE 22.1**

MANOVA table for the multivariate multiple linear regression model

| Source | df | SSCP |
|---|---|---|
| **Regression** | $p - 1$ | $\sum_{i=1}^{n}(\hat{\mathbf{Y}}_i - \bar{\mathbf{Y}})(\hat{\mathbf{Y}}_i - \bar{\mathbf{Y}})^{\mathrm{T}}$ |
| **Error** | $n - p$ | $\sum_{i=1}^{n}(\mathbf{Y}_i - \hat{\mathbf{Y}}_i)(\mathbf{Y}_i - \hat{\mathbf{Y}}_i)^{\mathrm{T}}$ |
| **Total** | $n - 1$ | $\sum_{i=1}^{n}(\mathbf{Y}_i - \bar{\mathbf{Y}})(\mathbf{Y}_i - \bar{\mathbf{Y}})^{\mathrm{T}}$ |

Notice in the MANOVA table that we do not define any mean square values or an $F$-statistic. Rather, a test of the significance of the multivariate multiple regression model is carried out using a Wilks' lambda quantity similar to

$$\Lambda^* = \frac{\left| \sum_{i=1}^{n}(\mathbf{Y}_i - \hat{\mathbf{Y}}_i)(\mathbf{Y}_i - \hat{\mathbf{Y}}_i)^{\mathrm{T}} \right|}{\left| \sum_{i=1}^{n}(\mathbf{Y}_i - \bar{\mathbf{Y}})(\mathbf{Y}_i - \bar{\mathbf{Y}})^{\mathrm{T}} \right|},$$

which follows a $\chi^2$ distribution. However, depending on the number of variables and the number of trials, modified versions of this test statistic must be used, which will affect the degrees of freedom for the corresponding $\chi^2$ distribution.

## Dimension Reduction and Envelopes

We can also be faced with the issue of fitting a multivariate linear regression model when the number of predictors $p$ is large. Fortunately, most of the dimension reduction techniques discussed in Chapter 16 can be applied in the multivariate regression setting. Another dimension reduction method that is used in the multivariate regression setting involves the notion of **envelopes** as established in Cook et al. (2007, 2010). To briefly define envelopes, we use the subspace construction as discussed in Chapter 16 and Appendix B.

The $m$-dimensional subspace $\mathcal{Q}$ is an **invariant subspace** of the $m \times m$ matrix $\mathbf{M}$ if $\mathbf{M}\mathcal{Q} \subseteq \mathcal{Q}$. $\mathcal{Q}$ is a **reducing subspace** of $\mathbf{M}$ if $\mathbf{M}\mathcal{Q}^{\perp} \subseteq \mathcal{Q}^{\perp}$, where $\mathcal{Q}^{\perp}$ is the orthogonal complement of $\mathcal{Q}$. Now, suppose that the subspace $\mathcal{S}$ is contained within the span of $\mathbf{M}$. We then define the **M-envelope** of $\mathcal{S}$, written as $\mathcal{E}_{\mathbf{M}}(\mathcal{S})$, as the intersection of all reducing subspaces of $\mathbf{M}$ that contain $\mathcal{S}$. This way of constructing an envelope is analogous to how the intersection of all DRSs and the intersection of all mean DRSs define the CS and CRS, respectively, in Chapter 16.

In order to assist in the rest of our discussion and to closely follow the notation in Cook et al. (2010), let us rewrite the multivariate linear regression model in (22.1) as

$$\mathbf{Y} = \boldsymbol{\alpha} + \mathbf{BX} + \boldsymbol{\epsilon},$$

where $\mathbf{X}$ is $p$-dimensional but without the first entry set to equal to 1 and $\boldsymbol{\alpha}$ is an $m$-dimensional parameter vector for the intercept term. This model has a total of $m + pm + m(m+1)/2$ unknown real parameters.

Let $\mathcal{B}$ be the subspace spanned by $\mathbf{B}$ and let $d = \dim(\mathcal{B})$. Consider the $\boldsymbol{\Sigma}$-envelope of $\mathcal{B}$, $\mathcal{E}_{\boldsymbol{\Sigma}}(\mathcal{B})$, which has dimension $u$, $0 < d \leq u \leq m$. We next let the $m \times u$ matrix $\boldsymbol{\Gamma}$ be a semi-orthogonal basis matrix for $\mathcal{E}_{\boldsymbol{\Sigma}}(\mathcal{B})$ and let $(\boldsymbol{\Gamma}, \boldsymbol{\Gamma}_0)$ be an $r \times r$ orthogonal matrix. Then, there is a $u \times p$ matrix $\mathbf{N}$ such that $\mathbf{B} = \boldsymbol{\Gamma}\mathbf{N}$. Additionally, let $\boldsymbol{\Omega} = \boldsymbol{\Gamma}^{\mathrm{T}}\boldsymbol{\Sigma}\boldsymbol{\Gamma}$ and $\boldsymbol{\Omega}_0 = \boldsymbol{\Gamma}_0^{\mathrm{T}}\boldsymbol{\Sigma}\boldsymbol{\Gamma}_0$ be $u \times u$ and $(m-u) \times (m-u)$ symmetric matrices, respectively. Then, Cook et al. (2010) give the following **envelope regression model**:

$$\mathbf{Y} = \boldsymbol{\alpha} + \boldsymbol{\Gamma}\mathbf{NX} + \boldsymbol{\epsilon}$$
$$\boldsymbol{\Sigma} = \boldsymbol{\Gamma}^{\mathrm{T}}\boldsymbol{\Omega}\boldsymbol{\Gamma} + \boldsymbol{\Gamma}_0^{\mathrm{T}}\boldsymbol{\Omega}_0\boldsymbol{\Gamma}_0,$$

where $\boldsymbol{\epsilon}$ is normally distributed with mean vector $\mathbf{0}$ and variance–covariance matrix $\boldsymbol{\Sigma}$. It can then be shown that the number of parameters needed to estimate the above model is $m + pu + m(m+1)/2$. In particular, the number of parameters in an envelope regression model is maximally reduced with respect to the original multivariate regression model. Maximum likelihood estimation can then be carried out following the details in Cook et al. (2010).

## 22.3 Reduced Rank Regression

**Reduced rank regression** (Izenman, 1975) is a way of constraining the multivariate linear regression model so that the rank of the regression coefficient matrix has less than full rank. The objective in reduced rank regression is to minimize the sum of squared residual subject to a reduced rank condition. Without the rank condition, the estimation problem is an ordinary least squares problem. Reduced rank regression is important in that it contains as special cases the classical statistical techniques of principal component analysis, canonical variate and correlation analysis, linear discriminant analysis, exploratory factor analysis, multiple correspondence analysis, and other linear methods of analyzing multivariate data. It is also heavily utilized in neural network modeling (Aoyagi and Watanabe, 2005) and econometrics (Johansen, 1995), just to name a couple of applied areas.

Recall that the multivariate regression model is

$$\underline{\mathbf{Y}} = \underline{\mathbf{X}}\mathbf{B} + \varepsilon,$$

where $\underline{\mathbf{Y}}$ is an $n \times m$ matrix, $\underline{\mathbf{X}}$ is an $n \times p$ matrix, and $\mathbf{B}$ is a $p \times m$ matrix of regression parameters. A reduced rank regression is when we have the rank constraint

$$\text{rank}(\mathbf{B}) = t, \quad 0 \leq t \leq \min\{p, m\},$$

where $t = 0$ implies $\underline{\mathbf{X}}$ and $\underline{\mathbf{Y}}$ are independent and $t = \min\{p, m\}$ implies the full rank model setting where we implement traditional ordinary least squares. When the rank condition above holds, then there exist two non-unique full rank matrices $\mathbf{A}_{p \times t}$ and $\mathbf{C}_{t \times m}$, such that

$$\mathbf{B} = \mathbf{AC}.$$

Moreover, there may be an additional set of predictors, say $\underline{\mathbf{W}}$, such that $\underline{\mathbf{W}}$ is an $n \times q$ matrix. Letting $\mathbf{D}$ denote a $q \times m$ matrix of regression parameters, we can then write the reduced rank regression model as follows:

$$\underline{\mathbf{Y}} = \underline{\mathbf{X}}\mathbf{AC} + \underline{\mathbf{W}}\mathbf{D} + \varepsilon.$$

In order to get estimates for the reduced rank regression model, first note that $\mathrm{E}(\epsilon_{(j)}) = \mathbf{0}$ and $\mathrm{Var}(\epsilon_{(j)}^{\mathrm{T}}) = \mathbf{I}_m \otimes \Sigma$, where $\otimes$ represents the Kronecker product of the two matrices (see Appendix B). For simplicity in the following, let $\underline{\mathbf{Z}}_0 = \underline{\mathbf{Y}}$, $\underline{\mathbf{Z}}_1 = \underline{\mathbf{X}}$, and $\underline{\mathbf{Z}}_2 = \underline{\mathbf{W}}$. Next, we define the moment matrices $M_{i,j} = \mathbf{Z}_i^{\mathrm{T}}\mathbf{Z}_j/m$ for $i, j = 0, 1, 2$ and $S_{i,j} = M_{i,j} - M_{i,2}M_{2,2}^{-1}M_{2,i}$, $i, j = 0, 1$. Then, the parameter estimates for the reduced rank regression model are as follows:

$$\hat{\underline{\mathbf{A}}} = (\hat{\nu}_1, \ldots, \hat{\nu}_t)\Phi$$

$$\hat{\underline{\mathbf{C}}}^{\mathrm{T}} = S_{0,1}\hat{\mathbf{A}}(\hat{\mathbf{A}}^{\mathrm{T}}S_{1,1}\hat{\mathbf{A}})^{-1}$$

$$\hat{\underline{\mathbf{D}}} = M_{0,2}M_{2,2}^{-1} - \hat{\underline{\mathbf{C}}}^{\mathrm{T}}\hat{\mathbf{A}}^{\mathrm{T}}M_{1,2}M_{2,2}^{-1},$$

where $(\hat{\nu}_1, \ldots, \hat{\nu}_t)$ are the eigenvectors corresponding to the $t$ largest eigenvalues $\hat{\lambda}_1, \ldots \hat{\lambda}_t$ of $|\lambda S_{1,1} - S_{1,0} S_{0,0}^{-1} S_{0,1}| = 0$ and where $\Phi$ is an arbitrary $t \times t$ matrix with full rank.

Typically, the appropriate rank is chosen by fitting the reduced rank regression models, $\mathcal{M}_1, \mathcal{M}_2, \ldots, \mathcal{M}_{t+1}$, for ranks $t = 0, 1, \ldots, t$, respectively. Then, the results are compared using a model selection criterion like BIC. However, another criterion applied to this specific problem is the **singular BIC (sBIC)** of Drton and Plummer (2017). Using their notation, let $\mathcal{I} = \{1, \ldots, t+1\}$ be the index set of our models such that $i \preceq j$ is used to represent the ordering of the indices when $\mathcal{M}_i \subseteq \mathcal{M}_j$. Let $\mathcal{L}(\mathcal{M}_i)$ be the marginal likelihood of model $\mathcal{M}_i$ and $\hat{\mathcal{L}}(\mathcal{M}_i)$ an appropriately computed approximation. Then,

$$\mathrm{sBIC}(\mathcal{M}_i) = \log\{\hat{\mathcal{L}}(\mathcal{M}_i)\},$$

where $\hat{\mathcal{L}}(\mathcal{M}_i)$ is the unique solution to the system of equations

$$\sum_{j \preceq i} \{\hat{\mathcal{L}}(\mathcal{M}_i) - \hat{L}_{ij}\} \hat{\mathcal{L}}(\mathcal{M}_i) = 0,$$

for $i \in \mathcal{I}$ and known constants $\hat{L}_{ij}$. More on the computational details can be found in Drton and Plummer (2017).

## 22.4 Seemingly Unrelated Regressions

Suppose we have $M$ regression equations

$$\mathbf{Y}_m = \mathbf{X}_m \boldsymbol{\beta}_m + \boldsymbol{\epsilon}_m, \quad m = 1, \ldots, M,$$

where for the $m^{\mathrm{th}}$ equation, $\mathbf{Y}_m$ is an $n \times 1$ vector of responses, $\mathbf{X}_m$ is a (fixed) $n \times p_m$ predictor matrix of rank $p_m$, $\boldsymbol{\beta}_m$ is a $p_m$-dimensional vector of unknown regression coefficients, and $\boldsymbol{\epsilon}_m$ is an $n$-dimensional vector of errors. We can rewrite the above system of regression equations as

$$
\begin{aligned}
\mathbf{Y}_S &= \begin{pmatrix} \mathbf{Y}_1 \\ \mathbf{Y}_2 \\ \vdots \\ \mathbf{Y}_M \end{pmatrix} \\
&= \begin{pmatrix} \mathbf{X}_1 & 0 & \cdots & 0 \\ 0 & \mathbf{X}_2 & \cdot & 0 \\ \vdots & \vdots & \ddots & \vdots \\ 0 & 0 & \cdots & \mathbf{X}_M \end{pmatrix} \begin{pmatrix} \boldsymbol{\beta}_1 \\ \boldsymbol{\beta}_2 \\ \vdots \\ \boldsymbol{\beta}_M \end{pmatrix} + \begin{pmatrix} \boldsymbol{\epsilon}_1 \\ \boldsymbol{\epsilon}_2 \\ \vdots \\ \boldsymbol{\epsilon}_M \end{pmatrix} \\
&= \underline{\mathbf{X}}_S \boldsymbol{\beta}_S + \boldsymbol{\epsilon}_S.
\end{aligned}
$$

Letting $q = \sum_{m=1}^{M} p_m$, we see that $\mathbf{Y}_S$ and $\epsilon_S$ are of dimension $nM \times 1$, $\mathbf{X}_S$ is $nM \times q$, and $\boldsymbol{\beta}_S$ is $q \times 1$. The manner of combining such a system of regression equations as above is called **seemingly unrelated regressions** or (**SUR**) and was first proposed by Zellner (1961). For estimating the above model, note that $E[\epsilon_S \epsilon_S^T] = \Sigma \otimes \mathbf{I}_n$, where $\Sigma$ is an $M \times M$ variance–covariance matrix. Defining $\hat{\Sigma}_S^{-1} = \hat{\Sigma}^{-1} \otimes \mathbf{I}_n$, where $\hat{\Sigma}$ is a consistent estimator for $\Sigma$, the least squares estimator for SUR due to Zellner (1961) is

$$\hat{\boldsymbol{\beta}}_S = (\mathbf{X}_S^T \hat{\Sigma}_S^{-1} \mathbf{X}_S)^{-1} \mathbf{X}_S^T \hat{\Sigma}_S^{-1} \mathbf{Y}_S.$$

Recall that we first mentioned SUR in Chapter 11 as part of the 3SLS algorithm for estimation in instrumental variables regression.

Inference and goodness-of-fit measures for SUR are fairly straightforward to develop. For example, letting $\mathbf{C}$ be a $r \times q$ contrast matrix of rank $r$ and $\mathbf{c}$ an $r$-dimensional vector, the general linear hypothesis

$$H_0 : \mathbf{C}\boldsymbol{\beta}_S = \mathbf{c}$$
$$H_A : \mathbf{C}\boldsymbol{\beta}_S \neq \mathbf{c}$$

can be tested using the test statistic

$$U = \left( \frac{(\mathbf{C}\hat{\boldsymbol{\beta}}_S - \mathbf{c})^T [\mathbf{C}(\mathbf{X}_S^T \hat{\Sigma}_S^{-1} \mathbf{X}_S)^{-1} \mathbf{C}^T]^{-1} (\mathbf{C}\hat{\boldsymbol{\beta}}_S - \mathbf{c})}{\mathbf{e}_S^T \hat{\Sigma}_S^{-1} \mathbf{e}_S} \right) \left( \frac{nM - q}{r} \right), \quad (22.5)$$

which follows an $F_{r, nM-q}$ distribution. In the above, $\mathbf{e}_S$ is the $nM \times 1$ vector of residuals from the fitted SUR model.

A common goodness-of-fit measure that is often calculated is **McElroy's $R^2$** (McElroy, 1977), which is similar to the traditional $R^2$, but is calculated based on the (vectorized) residuals of the SUR fit and the variance–covariance matrix of the residuals. Consider the special case of testing the general linear hypothesis where all regression coefficients (except the intercept terms) are 0. We can rewrite the SUR model to separate the slope coefficients from the intercept terms as follows:

$$\mathbf{Y}_S = \mathbf{W}_S \boldsymbol{\beta}_{S,0} + \mathbf{Z}\boldsymbol{\beta}_{S,1} + \epsilon_S,$$

where $\mathbf{W}_S$ and $\mathbf{Z}_S$ are matrices with certain structures to allow the above separation; see McElroy (1977) for details. Then, McElroy's $R^2$ is defined as

$$\text{McElroy's } R^2 = 1 - \frac{\mathbf{e}_S^T \hat{\Sigma}_S^{-1} \mathbf{e}_S}{\mathbf{Y}_S^T ([\mathbf{I}_{nM} - n^{-1} \mathbf{W}_S \mathbf{W}_S^T] \hat{\Sigma}_S^{-1}) \mathbf{Y}_S}.$$

This measure retains a similar interpretation for SUR as $R^2$ does when interpreting the traditional multiple linear regression fit.

## 22.5 Examples

**Example 22.5.1.** *(Amitriptyline Data)*
Rudorfer (1982) presented a study on the side effects of amitriptyline - a drug some physicians prescribe as an antidepressant. Some of the various physiological side effects that can occur when being administered this drug are irregular heartbeat and abnormal blood pressure readings. Data were gathered on $n = 17$ patients admitted to a hospital with an overdose of amitriptyline. The variables in this dataset are:

- $Y_1$: total TCAD plasma level

- $Y_2$: amount of amitriptyline present in TCAD plasma level

- $X_1$: gender, where 0 is for a male subject and 1 is for a female subject

- $X_2$: amount of the drug taken at the time of overdose

- $X_3$: PR wave measurement

- $X_4$: diastolic blood pressure

- $X_5$: QRS wave measurement

Johnson and Wichern (2007) proposed fitting a multivariate multiple linear regression model to these data. We present an analysis using such a multivariate regression model.

First we obtain the regression estimates for each response:

```
##############################
              TOT        AMI
(Intercept) -2879.4782 -2728.7085
GEN1          675.6508   763.0298
AMT             0.2849     0.3064
PR             10.2721     8.8962
DIAP            7.2512     7.2056
QRS             7.5982     4.9871
##############################
```

Then we obtain individual ANOVA tables for each response and see that the multiple regression model for each response is statistically significant:

```
##############################
 Response TOT :
           Df  Sum Sq Mean Sq F value    Pr(>F)
Regression  5 6835932 1367186  17.286 6.983e-05 ***
Residuals  11  870008   79092
---
```

Signif. codes:  0 '***' 0.001 '**' 0.01 '*' 0.05 '.' 0.1 ' ' 1


 Response AMI :
             Df  Sum Sq Mean Sq F value    Pr(>F)
Regression    5 6669669 1333934  15.598 0.0001132 ***
Residuals    11  940709   85519
---
Signif. codes:  0 '***' 0.001 '**' 0.01 '*' 0.05 '.' 0.1 ' ' 1
##############################

The following also gives the SSCP matrices for this fit:

##############################
$SSCPR
          TOT     AMI
TOT 6835932 6709091
AMI 6709091 6669669


$SSCPE
        TOT     AMI
TOT 870008 765676
AMI 765676 940709


$SSCPTO
          TOT     AMI
TOT 7705940 7474767
AMI 7474767 7610378
##############################

We can also see which predictors are statistically significant for each response:

##############################
Response TOT :

Coefficients:
              Estimate Std. Error t value Pr(>|t|)
(Intercept) -2.88e+03   8.93e+02   -3.22  0.00811 **
GEN1         6.76e+02   1.62e+02    4.17  0.00156 **
AMT          2.85e-01   6.09e-02    4.68  0.00068 ***
PR           1.03e+01   4.25e+00    2.41  0.03436 *
DIAP         7.25e+00   3.23e+00    2.25  0.04603 *
QRS          7.60e+00   3.85e+00    1.97  0.07401 .
---
Signif. codes:  0 *** 0.001 ** 0.01 * 0.05 . 0.1   1

Residual standard error: 281 on 11 degrees of freedom
Multiple R-squared:  0.887,Adjusted R-squared:  0.836

**TABLE 22.2**

95% confidence regions and 95% prediction regions for the new observation in the amitriptyline study

| Response | 95% Confidence Region | 95% Prediction Region |
|:---:|:---:|:---:|
| $Y_1$ | $(1402, 1683)$ | $(1256, 1829)$ |
| $Y_2$ | $(1336, 1750)$ | $(1121, 1964)$ |

```
F-statistic: 17.3 on 5 and 11 DF,  p-value: 6.98e-05


Response AMI :

Coefficients:
              Estimate Std. Error t value Pr(>|t|)
(Intercept) -2.73e+03   9.29e+02   -2.94  0.01350 *
GEN1         7.63e+02   1.69e+02    4.53  0.00086 ***
AMT          3.06e-01   6.33e-02    4.84  0.00052 ***
PR           8.90e+00   4.42e+00    2.01  0.06952 .
DIAP         7.21e+00   3.35e+00    2.15  0.05478 .
QRS          4.99e+00   4.00e+00    1.25  0.23862
---
Signif. codes:  0 *** 0.001 ** 0.01 * 0.05 . 0.1   1

Residual standard error: 292 on 11 degrees of freedom
Multiple R-squared:  0.876,Adjusted R-squared:  0.82
F-statistic: 15.6 on 5 and 11 DF,  p-value: 0.000113
#############################
```

We can proceed to drop certain predictors from the model in an attempt to improve the fit as well as view residual plots to assess the regression assumptions. Figure 22.1 gives the Studentized residual plots for each of the responses. Notice that the plots have a fairly random pattern, but there is one value high with respect to the fitted values. We could also formally test (e.g., with a Levene's test) to see if this affects the constant variance assumption and also to study pairwise scatterplots for any potential multicollinearity in this model. However, we do not perform such analyses here.

Next, we want to construct 95% confidence and prediction regions for a new male observation with predictor values $x_h = (1, 2000, 200, 50, 100)^T$. These regions are reported in Table 22.2. Thus, we are 95% confident that a male patient with the values given in $x_h$ will have mean total TCAD plasma levels between 1402 and 1683 and mean amount of amitriptyline present between 1336 and 1750. Moreover, we are 95% confident that a future male observation with these values given in $x_h$ will have a total TCAD plasma levels between 1256 and 1829 and an amount of amitriptyline present between 1121 and 1964.

Finally, we explore the appropriateness of using a reduced rank regression

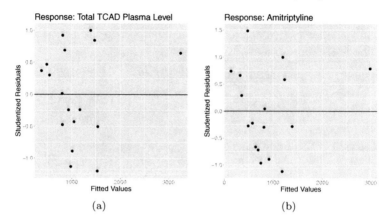

**FIGURE 22.1**
Plots of the Studentized residuals versus fitted values for the response (a) total
TCAD plasma level and the response (b) amount of amitriptyline present in
TCAD plasma level.

model. Below are the model selection results for fitting multivariate regression
models of rank 0, 1, and 2:

```
##############################
$logLike
[1] -24947168  -1764686  -1685320

$sBIC
[1]      -Inf      -Inf -1685337

$BIC
[1] -24947176  -1764700  -1685337
##############################
```

The above output shows that the largest BIC and sBIC values are for a rank
of $t = 2$. Therefore, using the ordinary least squares estimates are the best to
use. But for comparison, we did fit a reduced rank regression model for $t = 1$.
The coefficient estimates for this model are given below:

```
##############################
A matrix:
        latvar
TOT 1.0000000
AMI 0.9876859


C matrix:
```

```
              latvar
GEN1  723.4966160
AMT     0.2973651
PR      9.6474593
DIAP    7.2730097
QRS     6.3395246
```

```
B1 matrix:
                    TOT       AMI
(Intercept) -2709.676 -2900.628
#############################
```

**Example 22.5.2.** *(Investment Data)*
Grunfeld (1958) reported data on 11 large U.S. manufacturing firms over 20 years from 1935–1954. The variables in this dataset are:

- $Y$: gross investment to the firm

- $X_1$: market value of the firm

- $X_2$: capital stock of plant and equipment

These data were used by Zellner (1961) to illustrate SUR, who analyzed the data for two of the firms: General Electric and Westinghouse. We take this same strategy due to the large number of quantities that would be estimated for an analysis on all 11 firms. For our analysis, we look at the firms of General Motors and Goodyear.

First we obtain goodness-of-fit statistics about the SUR and the individual regression fits obtained by ordinary least squares:

```
#############################
        N DF    SSR detRCov OLS-R2 McElroy-R2
system 40 34 145050  664650  0.921      0.862

                N DF    SSR    MSE RMSE   R2 Adj R2
General.Motors 20 17 143641 8449.5 91.9 0.921  0.912
Goodyear       20 17   1409   82.9  9.1 0.665  0.626
#############################
```

The top portion of the output gives some goodness-of-fit quantities for the SUR fit. In particular, there is the traditional $R^2$ obtained using ordinary least squares, which is 0.921. This is a large value and indicative that there is a high proportion of variation explained in the gross investment of the firms based on the two predictors. McElroy's $R^2$ is also high, indicating that the SUR fit is good. The bottom portion of the output gives the goodness-of-fit measures for the ordinary least squares fit to each firm. Each of these have fairly large $R^2$ values for the corresponding regression fit.

Next we obtain the SUR fits for the two firms:

```
###############################
SUR estimates for 'General.Motors' (equation 1)
Model Formula: General.Motors_invest ~ General.Motors_value
   + General.Motors_capital
<environment: 0x111223180>

              Estimate Std. Error t value Pr(>|t|)
(Intercept) -142.6947    104.8361   -1.36  0.19125
value          0.1164      0.0256    4.55  0.00028 ***
capital        0.3799      0.0369   10.30    1e-08 ***
---
Signif. codes:  0 *** 0.001 ** 0.01 * 0.05 . 0.1  1

Residual standard error: 91.921 on 17 degrees of freedom
Number of observations: 20 Degrees of Freedom: 17
SSR: 143641.153 MSE: 8449.48 Root MSE: 91.921
Multiple R-Squared: 0.921 Adjusted R-Squared: 0.912


SUR estimates for 'Goodyear' (equation 2)
Model Formula: Goodyear_invest ~ Goodyear_value + Goodyear_capital
<environment: 0x111223180>

              Estimate Std. Error t value Pr(>|t|)
(Intercept)  -6.8618      9.3125   -0.74   0.4713
value         0.0704      0.0336    2.10   0.0514 .
capital       0.0848      0.0278    3.05   0.0072 **
---
Signif. codes:  0 *** 0.001 ** 0.01 * 0.05 . 0.1  1

Residual standard error: 9.104 on 17 degrees of freedom
Number of observations: 20 Degrees of Freedom: 17
SSR: 1409.155 MSE: 82.891 Root MSE: 9.104
Multiple R-Squared: 0.665 Adjusted R-Squared: 0.626
###############################
```

The above output shows that the predictors are significant in each of the estimated SUR models. These results, combined with the results from the goodness-of-fit measures, indicate that there is substantial variation among the firms. Thus, a single-equation model that pools these two firms is inadequate for characterizing this variability.

# 23

## Semiparametric Regression

In Chapter 17, we discussed nonparametric regression models as a flexible non-linear modeling paradigm in settings where identifying a (reasonable) parametric form would be too difficult. Sometimes, however, we may want to include more of a structure in what we are modeling if the nonparametric regression fit is not very good. Fortunately, there is a middle-ground of regression methodology that allows us to include some structure, but also have some of the flexibility of nonparametric regression. These are called **semiparametric regression models** and they contain both parametric and nonparametric components. In fact, we have already discussed some semiparametric regression models, such as the Cox proportional hazards model in Chapter 18. In this model, there is the baseline hazard, which is nonparametric, and the hazards ratio, which is parametric.

We can outline the general context of three types of regression models:

**Parametric Models:** These models are fully determined up to a parameter vector. If the underlying assumptions are correct, then the estimated model is valid for inference, the fitted model generally has a meaningful interpretation, and the estimated model will tend to have good predictive ability on independent datasets. If the assumptions are violated, then fitted parametric estimates may provide inconsistencies and misleading interpretations.

**Nonparametric Models:** These models provide flexibility over (possibly) restrictive parametric forms. However, they may be difficult to interpret and the curse of dimensionality can be very problematic when faced with a large number of predictors.

**Semiparametric Models:** These models combine components of parametric and nonparametric regressions. They allow for easy interpretation of the parametric component while providing the flexibility of the nonparametric component.

In this chapter, we present some of the more common semiparametric regression models. Two texts that provide somewhat different treatments of these models are Ruppert et al. (2003) and Härdle et al. (2004b), both of which were used as references for the material in this chapter.

## 23.1　Single-Index Models

Suppose we have the $p$-dimensional vector of variables $\underline{X}$. An **index** is a summary of those $p$ variables into a single number; e.g., the Consumer Price Index, the House Price Index, or the Urban Health Index. Such a single-index term can be used for statistical modeling with the added benefit that it has reduced the dimensionality of the problem considerably. The **single-index model**, or **SIM** (Ichimura, 1993), is of the form

$$Y = m(\underline{X}^{\mathrm{T}}\boldsymbol{\beta}) + \epsilon$$
$$\Rightarrow \mathrm{E}(Y|\underline{X}) = m(\underline{X}^{\mathrm{T}}\boldsymbol{\beta}), \tag{23.1}$$

where $m$ is a smooth, but unspecified function, $\boldsymbol{\beta}$ is the vector of coefficients in the single-index term $\underline{X}^{\mathrm{T}}\boldsymbol{\beta}$, and $\epsilon$ is an error term with mean 0.

Suppose we have a sample of size $n$ and that we measured a response variable and $p$ predictor variables. The general steps for estimating the SIM in (23.1) are as follows:

1.　Estimate $\boldsymbol{\beta}$ by $\hat{\boldsymbol{\beta}}$.

2.　Compute the index values $\hat{\eta}_i = \mathbf{x}_i^{\mathrm{T}}\hat{\boldsymbol{\beta}}$.

3.　Estimate the function $m$ nonparametrically for the regression of $y_i$ on $\hat{\eta}_i$.

In Step 1, an estimator for $\boldsymbol{\beta}$ can be obtained either via an iterative approximation procedure or direct estimation. Iterative procedures available include **semiparametric least squares** (Ichimura, 1993) or **pseudo-maximum likelihood estimation** (see Härdle et al., 2004b, Chapter 6). Direct estimation, which requires that all predictor variables be continuous, can be accomplished using a **(weighted) average derivatives estimator** (Härdle and Stoker, 1989). In Step 3, Ichimura (1993) proposed to nonparametrically estimate the function $m$ using kernel regression methods and the Nadaraya-Watson estimator (17.6):

$$\hat{m}_h(\eta_0) = \frac{\sum_{i=1}^{n} K\left(\frac{\eta_0 - \hat{\eta}_i}{h}\right) y_i}{\sum_{i=1}^{n} K\left(\frac{\eta_0 - \hat{\eta}_i}{h}\right)}.$$

Testing is also possible for comparing the SIM to a fully parametric competitor. Horowitz and Härdle (1994) developed a test regarding the following hypotheses:

$$H_0 : \mathrm{E}(Y|\underline{X} = \mathbf{x}_0) = f(\mathbf{x}_0^{\mathrm{T}}\boldsymbol{\beta})$$
$$H_A : \mathrm{E}(Y|\underline{X} = \mathbf{x}_0) = m(\mathbf{x}_0^{\mathrm{T}}\boldsymbol{\beta}),$$

where $f$ is a known function and $m$ is, again, the unspecified function for the SIM in (23.1). The authors derived a test statistic for the above, which they show follows a normal distribution.

## 23.2 (Generalized) Additive Models

The structure assumed in a fully parametric regression model will certainly introduce bias if the structure does not accurately describe reality; however, it can result in a dramatic reduction in variance. The nonparametric regression methods of Chapter 17 were aimed at balancing enough structure to make the model fit stable, but not so much structure as to bias the fit. This, of course, becomes challenging when the number of predictors $p$ is large and there are clear nonlinear effects. One popular class of flexible semiparametric regression models available for such settings uses an additive structure. An **additive model**, or **AM** (Hastie and Tibshirani, 1990), is of the form

$$Y = \alpha + \sum_{j=1}^{p} m_j(X_j) + \epsilon$$

$$\Rightarrow E(Y|\underline{X}) = \alpha + \sum_{j=1}^{p} m_j(X_j), \tag{23.2}$$

where the $m_1, \ldots, m_p$ are unspecified smooth functions, $\alpha$ is an intercept term, and $\epsilon$ is an error term with mean 0. Notice that instead of estimating a single function of several variables, the AM in (23.2) requires estimation of $p$ functions of one-dimensional variables $X_j$.

Suppose we have a sample of size $n$ and that we measured a response variable and $p$ predictor variables. The additive components in an AM are usually approximated using a **backfitting algorithm** (Breiman and Friedman, 1985), the general steps for which are as follows:

---

Backfitting Algorithm

1. Initialize the algorithm by setting $\hat{\alpha} = \bar{y}$ and $\hat{m}_j^{(0)} \equiv 0$, for $j = 1, \ldots, p$.

2. For $t = 1, 2, \ldots$, repeat for $j = 1, \ldots, p$ the cycles

$$\mathbf{r}_{i,j} = y_i - \hat{\alpha} - \sum_{k<j} \hat{m}_j^{(t+1)}(x_{i,j}) - \sum_{k>j} \hat{m}_j^{(t)}(x_{i,j})$$

$$\hat{m}_j^{(t+1)} = \mathcal{S}_j(\mathbf{r}_j) - \frac{1}{n}\sum_{i=1}^{n} \hat{m}_j(x_{i,j}),$$

until $|\hat{m}_j^{(t+1)} - \hat{m}_j^{(t)}|$ is less than some prespecified threshold for each $j = 1, \ldots, p$. $\mathbf{r}_j = (r_{1,j}, \ldots, r_{n,j})^{\mathrm{T}}$ is a vector of the current residuals for component $j$ and $\mathcal{S}_j$ is a smoother that targets the residuals $\mathbf{r}_j$ as a function of the $x_{i,j}$.

---

The AM in (23.2) can also be generalized to allow for inclusion of a known parametric link function, $g$. Such a model is called a **generalized additive model**, or **GAM** (Hastie and Tibshirani, 1990), and is of the form

$$g(\mathrm{E}(Y|\underline{X})) = \alpha + \sum_{j=1}^{p} m_j(X_j). \tag{23.3}$$

For example, if $g$ is the logit link and the responses are binary, then we can use (23.3) to construct an additive logistic regression model. If $g$ is the log link and the responses are counts, then we can use (23.3) to construct an additive Poisson regression model.

Estimation of GAMs combines the backfitting procedure with a likelihood maximizer. The approach recasts a Newton–Raphson routine for maximum likelihood estimation in generalized linear models as an IRLS algorithm (see Chapter 12). It involves replacing the weighted linear regression in the IRLS algorithm by a weighted backfitting algorithm. This algorithm is called a **local scoring algorithm** and is discussed in Chapter 6 of Hastie and Tibshirani (1990).

Backfitting algorithms are, perhaps, the most common approaches for estimating AMs and GAMs. One other method is to use **marginal integration estimators**, which estimate the marginal influence of a particular predictor variable on a multiple regression fit. When the real, underlying model to the data is additive, then backfitting and marginal integration estimators perform similarly. Otherwise, there are some differences. More discussion on marginal integration estimators can be found in Chapter 8 of Härdle et al. (2004b) and the references therein.

---

## 23.3 (Generalized) Partial Linear Models

Suppose we want to model the response as a linear function of some of the predictors, while the relationship to the other predictors is characterized non-parametrically. Specifically, suppose that the $p$-dimensional predictor vector $\underline{X} = (\underline{U}^{\mathrm{T}}, \underline{V}^{\mathrm{T}})^{\mathrm{T}}$, where $\underline{U} = (U_1, \ldots, U_r)^{\mathrm{T}}$ and $\underline{V} = (V_1, \ldots, V_s)^{\mathrm{T}}$ such that $p = r + s$. Let $\underline{U}$ be the predictors for the linear relationship with $Y$ and $\underline{V}$ be the predictors handled nonparametrically when being related to $Y$. Regressing

$Y$ on $\underline{X}$ leads to a **partial linear model**, or **PLM** (Green and Yandell, 1985; Engle et al., 1986; Härdle et al., 2000), which is of the form

$$Y = \underline{U}^{\mathrm{T}}\boldsymbol{\beta} + m(\underline{V}) + \epsilon$$
$$\Rightarrow \mathrm{E}(Y|\underline{U}, \underline{V}) = \underline{U}^{\mathrm{T}}\boldsymbol{\beta} + m(\underline{V}), \tag{23.4}$$

where the $m$ is an unspecified smooth function of $\underline{V}$, $\boldsymbol{\beta}$ is an $r$-dimensional vector of regression coefficients, and $\epsilon$ is an error term with mean 0.

Just like AMs and GAMs, we can generalize PLMs to allow for inclusion of a known parametric link function, $g$. Such a model is called a **generalized partial linear model**, or **GPLM** (Severini and Staniswalis, 1994), and is of the form

$$g(\mathrm{E}(Y|\underline{U}, \underline{V})) = \underline{U}^{\mathrm{T}}\boldsymbol{\beta} + m(\underline{V}). \tag{23.5}$$

The same type of link functions and distributional assumptions used in generalized linear models and GAMs can be used here. Note that one could also assume that the component $m(\underline{V})$ in (23.5) is actually a SIM, say, $m(\underline{V}^{\mathrm{T}}\boldsymbol{\gamma})$. This model is called a **generalized partial linear single-index model** (Carroll et al., 1997).

Estimation for PLMs and GPLMs is a bit more complicated than for AMs and GAMs, but two primary approaches are used in the literature. First, we can use a backfitting algorithm with the appropriate modifications for either PLMs or GPLMs. The backfitting algorithm alternates between parametric estimation for $\boldsymbol{\beta}$ and nonparametric estimation for $m$. Second, we can use profile likelihood methods.[1] For PLMs, the profile likelihood approach leverages the fact that the conditional distribution of $Y$ given $U$ and $V$ is parametric. For a fixed value of $\boldsymbol{\beta}$, it is then used to estimate the least favorable nonparametric function $m_{\boldsymbol{\beta}}$ independent of $\boldsymbol{\beta}$. This least favorable function is then used to construct the profile likelihood for $\boldsymbol{\beta}$, which we maximize to find $\hat{\boldsymbol{\theta}}$. Another approach based on the profile likelihood involves the **Speckman estimator** (Speckman, 1988). This estimator is constructed by first determining the parametric component by applying an ordinary least squares estimator to a nonparametrically modified design matrix and response vector. Then, the nonparametric component is estimated by smoothing the residuals with respect to the parametric part. For GPLMs, a generalized Speckman estimator is necessary, which is based on the results from a modified IRLS algorithm. Details on all of these estimation methods are given in Chapter 7 of Härdle et al. (2004b).

---

[1] The profile likelihood was briefly discussed in the context of Box-Cox transformations in Example 4.7.3.

## 23.4    (Generalized) Partial Linear Partial Additive Models

Assume the same partitioning of the predictor vector $\underline{X} = (\underline{U}^{\mathrm{T}}, \underline{V}^{\mathrm{T}})^{\mathrm{T}}$ as in the PLM and GPLM settings. In both of these models, if $\underline{V}$ is high-dimensional, then nonparametrically estimating the smooth function $m$ suffers from the curse of dimensionality as well as challenges with its interpretation. Thus, we can impose an additive structure in the nonparametric component to get a **partial linear partial additive model**, or **PLPAM**, which has the form

$$
\begin{aligned}
Y &= \alpha + \underline{U}^{\mathrm{T}}\beta + \sum_{j=1}^{s} m_j(V_j) + \epsilon \\
&\Rightarrow \mathrm{E}(Y|\underline{U}, \underline{V}) = \alpha + \underline{U}^{\mathrm{T}}\beta + \sum_{j=1}^{s} m_j(V_j),
\end{aligned}
\tag{23.6}
$$

where the $m_1, \ldots, m_s$ are unspecified smooth functions, $\beta$ is an $r$-dimensional vector of regression coefficients, $\alpha$ is an intercept term, and $\epsilon$ is an error term with mean 0. The PLPAM can also be generalized to allow for inclusion of a known parametric link function, $g$. Such a model is called a **generalized partial linear partial additive model**, or **GPLPAM**, and is of the form

$$
g(\mathrm{E}(Y|\underline{X})) = \alpha + \underline{U}^{\mathrm{T}}\beta + \sum_{j=1}^{s} m_j(V_j).
\tag{23.7}
$$

Estimation of PLPAMs and GPLPAMs is more difficult, especially since the theory is underdeveloped compared to the other semiparametric regression models already discussed. Estimation can be performed using a backfitting algorithm (Hastie and Tibshirani, 1990) or a marginal integration approach (Härdle et al., 2004a).

## 23.5    Varying-Coefficient Models

In the previous sections, the semiparametric regression models were introduced as a way to increase the flexibility of linear regression models. Basically, we were replacing subsets of the predictors with smooth nonparametric functions of the predictors or of a single-index term. Suppose we again partition the predictor vector as $\underline{X} = (\underline{U}^{\mathrm{T}}, \underline{V}^{\mathrm{T}})^{\mathrm{T}}$. A different generalization compared to the previous semiparametric regression models is a class of models that are linear in the predictors $\underline{U}$, but the regression coefficients are allowed to change smoothly with the variables $\underline{V}$, which are called **effect modifiers**. Assuming this structure for the regression leads to a **varying-coefficient model** (Hastie and Tibshirani, 1993), which is a conditionally linear model of the form

$$Y = \alpha(\underline{V}) + \underline{U}^{\mathrm{T}}\boldsymbol{\beta}(\underline{V}) + \epsilon$$
$$\Rightarrow \mathrm{E}(Y|\underline{U},\underline{V}) = \alpha(\underline{V}) + \underline{U}^{\mathrm{T}}\boldsymbol{\beta}(\underline{V}). \tag{23.8}$$

In the above model, the intercept and slope coefficients are, respectively, $\alpha(\underline{V})$ and $\boldsymbol{\beta}(\underline{V}) = (\beta_1(\underline{V}), \ldots, \beta_r(\underline{V}))^{\mathrm{T}}$, which are unknown functions of $\underline{V}$, and $\epsilon$ is an error term with mean 0. Thus, for a given $\underline{V}$, (23.8) is a linear model, but the coefficients can vary with $\underline{V}$. The varying-coefficient model can also be generalized to allow for inclusion of a known parametric link function, $g$. Such a model is called a **generalized varying-coefficient model** (Fan and Zhang, 2008) and is of the form

$$g(\mathrm{E}(Y|\underline{X})) = \alpha(\underline{V}) + \underline{U}^{\mathrm{T}}\boldsymbol{\beta}(\underline{V}). \tag{23.9}$$

The varying-coefficient model in (23.8) is a local linear model. Thus, approaches for estimating $\alpha(\underline{V})$ and $\boldsymbol{\beta}(\underline{V})$ include some of those discussed in Chapter 17, like local least squares and spline-based methods. The generalized varying-coefficient model in (23.9) can be estimated using a local likelihood approach. Details for most of these estimation procedures as well as development of relevant hypothesis tests, are discussed in Fan and Zhang (2008).

## 23.6  Projection Pursuit Regression

An exploratory method introduced by Friedman and Stuetzle (1981) called **projection pursuit regression** (or **PPR**) attempts to reveal possible nonlinear and interesting structures in

$$Y = m(\underline{X}) + \epsilon$$

by looking at an expansion of $m(\underline{X})$:

$$m(\underline{X}) = \sum_{j=1}^{M} \phi_j(\underline{X}^{\mathrm{T}}\boldsymbol{\beta}_j), \tag{23.10}$$

where $\phi_j$ and $\boldsymbol{\beta}_j$ are unknowns. The above is more general than stepwise regression and we can easily obtain graphical summaries of such models. When the number of predictors $p$ is large, PPR can be used instead to mitigate the effects of the curse of dimensionality that is typically an issue when using nonparametric smoothers. Thus, PPR can be characterized at the intersection of the dimension reduction techniques discussed in Chapter 16 and the smoothing methods discussed in Chapter 17.

Suppose we have a sample of size $n$ and $p$ predictor variables. To estimate the response function from the data, Friedman and Stuetzle (1981) proposed the following algorithm for PPR:

---

PPR Algorithm

1. Initialize the algorithm by specifying the current residuals: $r_i^{(0)} = y_i$, $i = 1, \ldots, n$. It is assumed that the response is centered; i.e., $\bar{y} = 0$.

2. For $j = 1, \ldots,$ define

$$R_{(j)}^2 = 1 - \frac{\sum_{i=1}^n \left( r_i^{(j-1)} - \phi_j(\mathbf{x}_i^{\mathsf{T}} \boldsymbol{\beta}_j) \right)^2}{\sum_{i=1}^n \left( r_i^{(j-1)} \right)^2}.$$

Find the parameter vector $\hat{\boldsymbol{\beta}}_j$ that maximizes $R_{(j)}^2$ (this is called **projection pursuit**) and the corresponding smoother $\hat{\phi}_j$.

3. If $R_{(j)}^2$ is smaller than a user-specified threshold, then do not include the $j^{\text{th}}$ term in the estimate of (23.10) and set $M = j-1$. Otherwise, compute new residuals

$$r_i^{(j)} = r_i^{(j-1)} - \hat{\phi}_j(\mathbf{x}_i \hat{\boldsymbol{\beta}}_j)$$

and repeat Step 2 until $R_{(j)}^2$ becomes small. A small $R_{(j)}^2$ implies that $\hat{\phi}_j(\mathbf{x}_i^{\mathsf{T}} \hat{\boldsymbol{\beta}}_j)$ is approximately the zero function and we will not find any other useful direction.

---

The advantages of using PPR for estimation is that we are using univariate regressions which are quick and easy to estimate. Also, PPR is able to approximate a fairly rich class of functions as well as ignore variables providing little to no information about $m(\cdot)$. Some disadvantages of using PPR include having to examine a $p$-dimensional parameter space to estimate $\hat{\boldsymbol{\beta}}_j$ and interpretation of a single term may be difficult.

## 23.7 Examples

**Example 23.7.1.** *(The 1907 Romanian Peasant Revolt Data)*
The 1907 Romanian peasants' revolt was sparked by the discontent of peasants regarding the inequity of land ownership, which was primarily controlled by

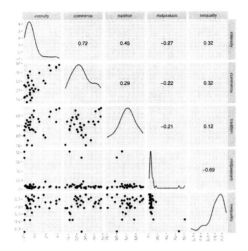

**FIGURE 23.1**
Matrix of pairwise scatterplots of the 1907 Romanian peasant revolt data. The diagonal of this matrix gives a density estimate of the respective variable, while the strictly upper triangular portion of this matrix gives the correlations between the column and row variables.

just a few large landowners. The revolt ended violently with thousands of peasants killed. Chirot and Ragin (1975) developed two models of peasant rebellion, which were examined using historical census data relevant to the event. The data is at the county-level of Romania at the time, so $n = 32$. The variables analyzed in this study are:

- $Y$: a measure of the intensity of the rebellion in the county

- $X_1$: the percent commercialization of agriculture

- $X_2$: a measure of traditionalism within the county

- $X_3$: strength of the county's middle peasantry

- $X_4$: rate of inequality of land tenure

Figure 23.1 is a scatterplot matrix of all the pairwise scatterplots for the four predictors and the response. Clearly there are nonlinearities that we could characterize when building a regression model between the intensity of the revolt as a function of these four other predictors. We will explore various semiparametric regression fits to these data.

First we fit a linear regression model to these data for comparison:

```
###############################
Coefficients:
```

```
              Estimate Std. Error t value Pr(>|t|)
(Intercept) -12.919018    5.507499  -2.346 0.026587 *
commerce      0.091140    0.020268   4.497 0.000118 ***
tradition     0.116787    0.060688   1.924 0.064906 .
midpeasant   -0.003342    0.017695  -0.189 0.851625
inequality    1.137970    2.850304   0.399 0.692853
---
Signif. codes:  0 *** 0.001 ** 0.01 * 0.05 . 0.1   1

Residual standard error: 1.227 on 27 degrees of freedom
Multiple R-squared:  0.5836,Adjusted R-squared:  0.5219
F-statistic: 9.462 on 4 and 27 DF,  p-value: 6.476e-05
##############################
```

This model shows that the strength of the county's middle peasantry and the rate of inequality of land tenure are highly non-significant predictors of the intensity of the revolt. We investigate greater flexibility by fitting an additive model where the basis for a thin-plate smoothing spline is applied to each predictor:

```
##############################
Parametric coefficients:
             Estimate Std. Error t value Pr(>|t|)
(Intercept) 0.0003125  0.1830571   0.002    0.999

Approximate significance of smooth terms:
                edf Ref.df      F  p-value
s(tradition)  1.336  1.555  1.817    0.244
s(midpeasant) 1.775  1.948  1.605    0.262
s(inequality) 1.000  1.000  2.552    0.123
s(commerce)   1.888  1.982 16.171 1.33e-05 ***
---
Signif. codes:  0 *** 0.001 ** 0.01 * 0.05 . 0.1   1

R-sq.(adj) =  0.659   Deviance explained = 72.5%
GCV = 1.3725  Scale est. = 1.0723    n = 32
##############################
```

The smoothing term for commerce is statistically significant, but the smoothing terms for each of the other three predictors are not statistically significant. However, there is a noticeable improvement in the $R^2_{adj}$, which increased from 58.4% to 65.9%. The components of the additive fits with partial residuals overlaid are given in Figure 23.2. This gives a sense of where the curves appear to provide a good fit. For example, curvature in the inequality predictor appears to be reasonably captured, whereas the smoothness of the strength of the middle peasantry predictor could be improved upon. However, there is also a sparse region in the observed values of this latter predictor.

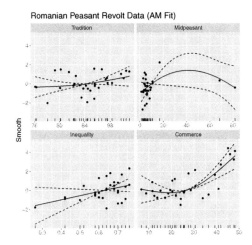

**FIGURE 23.2**
Additive model fits of each predictor for the 1907 Romanian peasant revolt data with partial residuals overlaid.

We next fit a partial linear partial additive model where the basis for a thin-plate smoothing spline is applied to inequality and strength of the middle peasantry predictors, while the measures of tradition and commerce enter as linear predictors:

```
#############################
Parametric coefficients:
            Estimate Std. Error t value Pr(>|t|)
(Intercept) -16.37869    5.36535  -3.053  0.00628 **
tradition     0.17160    0.06645   2.582  0.01779 *
commerce      0.07288    0.02200   3.313  0.00347 **
---
Signif. codes:  0 *** 0.001 ** 0.01 * 0.05 . 0.1   1

Approximate significance of smooth terms:
                edf Ref.df     F p-value
s(midpeasant) 7.053  7.853 1.463   0.233
s(inequality) 1.946  2.433 1.281   0.317

R-sq.(adj) =  0.639   Deviance explained = 76.7%
GCV = 1.8178  Scale est. = 1.1362    n = 32
#############################
```

Clearly the linear terms are both statistically significant, whereas those modeled using a smoother are not statistically significant. The components of the

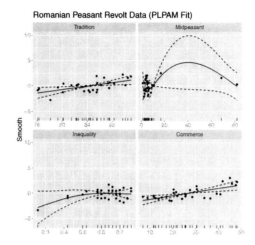

**FIGURE 23.3**
Partial linear partial additive model fits of each predictor for the 1907 Romanian peasant revolt data with partial residuals overlaid.

additive fits with partial residuals overlaid are given in Figure 23.3. There is a decrease in the $R^2_{adj}$ from the additive model fit. So this provides evidence that, perhaps, it is unnecessary to model smooth terms of the inequality and strength of the middle peasantry predictors. We might consider removing the predictors altogether from the model. However, there does appear to be some improvement by modeling the smooth terms of the remaining two predictors based on the results from the previous additive model fit.

We also fit a partial linear model, where the commerce variable is modeled as a linear predictor and the other three variables comprise the nonparametric part. The estimate for the linear coefficient (i.e., the coefficient for the commerce variable) is 0.0988 with a standard error of 0.0158. The components of the fits for both the parametric and nonparametric components are given in Figure 23.4 with partial residuals overlaid. The biggest contributor to the nonparametric component's fit appears to be the midpeasant predictor, but again, the regions where the data is sparse could be affecting this fit. While the linear effect of commerce on the intensity of the rebellion is adequately captured, it does not appear that the effects of the tradition and inequality measures are contributing much to the nonparametric component of the fit.

We next fit a single index model to these data with only the traditionalism measure and the commercialization of agriculture measure as predictors:

```
##############################
Single Index Model
Regression Data: 32 training points, in 3 variable(s)
```

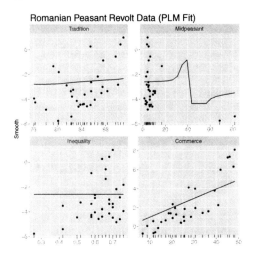

Romanian Peasant Revolt Data (PLM Fit)

**FIGURE 23.4**
Partial linear model fits of each predictor for the 1907 Romanian peasant revolt data with partial residuals overlaid.

```
        Int tradition   commerce
Beta:     1 0.1580466 0.1426025
Bandwidth: 0.505307
Kernel Regression Estimator: Local-Constant

Residual standard error: 0.8536655
R-squared: 0.7641328

Continuous Kernel Type: Second-Order Gaussian
No. Continuous Explanatory Vars.: 1
#############################
```

The above shows the estimated regression coefficients for the single index model as well as the regular $R^2$. A plot of the single index fit is also provided in Figure 23.5. In general, this appears to be a good fit since noticeable features of the average relationship between the intensity of the revolt versus the index appear to be adequately captured.

Finally, we also fit a varying-coefficient regression model using all of the predictors:

```
#############################
Smooth Coefficient Model
Regression data: 32 training points, in 4 variable(s)
```

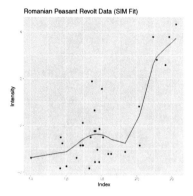

**FIGURE 23.5**
Single index model fit of the 1907 Romanian peasant revolt data.

```
                 commerce tradition midpeasant inequality
Bandwidth(s): 13.61683    40.9569    2.366861  0.1359696

Bandwidth Type: Fixed

Residual standard error: 0.7395195
R-squared: 0.8218895

Continuous Kernel Type: Second-Order Gaussian
No. Continuous Explanatory Vars.: 4
##############################
```

The above gives the bandwidth estimates as well as the $R^2$, which is fairly large. This could be a reasonable model to use given the larger value of $R^2$. For comparison, we can calculate the average values of the varying-coefficient regression estimates with the estimated regression coefficients obtained from the multiple linear regression fit:

```
##############################
    Intercept commerce tradition midpeasant inequality
lm  -13.9645   0.0771   0.1181    0.0315    2.0259
vc  -12.9190   0.0911   0.1168   -0.0033    1.1380
##############################
```

Clearly they are both similar estimates, although the sign for the strength of the middle peasantry predictor differs between the two methods.

**Example 23.7.2.** *(Ragweed Data)*
Stark et al. (1997) used meteorologic data to develop models for predicting daily ragweed pollen levels. The data are for $n = 355$ days from the city of Kalamazoo, Michigan. The variables analyzed in this study are:

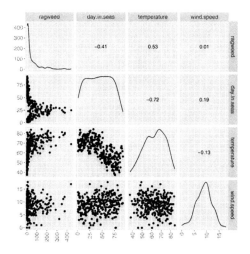

**FIGURE 23.6**
Matrix of pairwise scatterplots of the ragweed data. The diagonal of this matrix gives a density estimate of the respective variable, while the strictly upper triangular portion of this matrix gives the correlations between the column and row variables.

- $Y$: ragweed level (grains per cubic meter)
- $X_1$: day number in the current ragweed pollen season
- $X_2$: temperature of the following day, in degrees Fahrenheit
- $X_3$: an indicator where 1 indicates at least 3 hours of intense rain the following day and 0 otherwise
- $X_4$: wind speed forecast (in knots) for the following day

Figure 23.6 provides a scatterplot matrix of all the pairwise scatterplots for the three continuous predictors and the response. Clearly there are nonlinearities that we could characterize when building a regression model between the intensity of the revolt as a function of these four other predictors. These data were also thoroughly analyzed in Ruppert et al. (2003) using various semiparametric regression models. We also proceed to analyze various semiparametric regression models using various combinations of the predictors.

First, note that the ragweed level is a count variable. We begin by fitting a Poisson regression model to these data for comparison:

```
#############################
Coefficients:
                Estimate Std. Error z value Pr(>|z|)
```

```
(Intercept)     -3.7509832   0.1101586   -34.05    <2e-16 ***
day.in.seas     -0.0112381   0.0005049   -22.26    <2e-16 ***
temperature      0.0928500   0.0012571    73.86    <2e-16 ***
rain0therwise    1.1227606   0.0515481    21.78    <2e-16 ***
wind.speed       0.0572064   0.0027120    21.09    <2e-16 ***
---
Signif. codes:  0 *** 0.001 ** 0.01 * 0.05 . 0.1   1


(Dispersion parameter for poisson family taken to be 1)

    Null deviance: 29360  on 334  degrees of freedom
Residual deviance: 14846  on 330  degrees of freedom
AIC: 16133

Number of Fisher Scoring iterations: 6
#############################
```

This model shows that all predictors are significant in this model. However, we noted the nonlinearities in the scatterplots of Figure 23.6. Thus, we turn to fitting a generalized additive model where we assume a Poisson distribution for the response and relate the sum of functions of independent variables to the response via the canonical log link. We build a model with temperature as a predictor, but also the day number in the ragweed pollen season as a predictor. A thin-plate smoothing spline is applied to each of these predictors:

```
#############################
Parametric coefficients:
            Estimate Std. Error z value Pr(>|z|)
(Intercept)  2.54957    0.02984   85.44   <2e-16 ***
---
Signif. codes:  0 *** 0.001 ** 0.01 * 0.05 . 0.1   1

Approximate significance of smooth terms:
                 edf Ref.df Chi.sq p-value
s(day.in.seas) 8.876  8.991   6588  <2e-16 ***
s(temperature) 8.682  8.942   1276  <2e-16 ***
---
Signif. codes:  0 *** 0.001 ** 0.01 * 0.05 . 0.1   1

R-sq.(adj) =  0.677   Deviance explained = 79.6%
UBRE = 16.946  Scale est. = 1          n = 335
#############################
```

Smooth terms for the day number in the season and the temperature are both significant for this generalized additive model fit. The $R^2_{adj}$ and percentage deviance explained are both relatively high, with values of 67.7% and 79.6%,

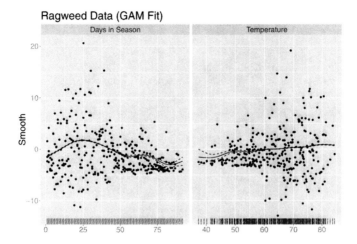

**FIGURE 23.7**
Generalized additive model fits of each predictor for the ragweed data with
partial residuals overlaid. Notice the curvature identified in each plot.

respectively. The components of the additive fits with partial residuals overlaid
are given in Figure 23.7. These curves provide reasonable fits given the general
trend of the partial residuals with respect to each predictor.

We next fit a generalized partial linear partial additive model, where we
use the previous generalized additive model and include the rain indicator and
wind speed as linear predictors:

```
##############################
Parametric coefficients:
             Estimate Std. Error z value Pr(>|z|)
(Intercept)  0.885402   0.063516   13.94   <2e-16 ***
rainOtherwise 0.929259  0.052481   17.71   <2e-16 ***
wind.speed   0.086296   0.002875   30.01   <2e-16 ***
---
Signif. codes:  0 *** 0.001 ** 0.01 * 0.05 . 0.1   1

Approximate significance of smooth terms:
                edf Ref.df Chi.sq p-value
s(day.in.seas) 8.937  8.998   6572  <2e-16 ***
s(temperature) 8.596  8.908   1168  <2e-16 ***
---
Signif. codes:  0 *** 0.001 ** 0.01 * 0.05 . 0.1   1

R-sq.(adj) =  0.749   Deviance explained = 84.4%
```

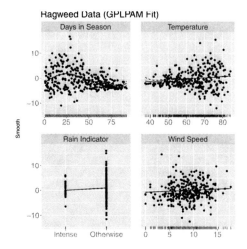

**FIGURE 23.8**
Generalized partial linear partial additive model fits of each predictor for the ragweed data with partial residuals overlaid.

```
UBRE = 12.772  Scale est. = 1          n = 335
##############################
```

Clearly the linear terms and smooth terms are all statistically significant. The $R^2_{adj}$ and percentage deviance explained are 74.9% and 84.4%, respectively. These are both larger than what was obtained from the generalized additive model fit. The components of the generalized partial linear partial additive fits with partial residuals overlaid are given in Figure 23.8. The fits for the smooth terms in this figure are similar to what were observed for the generalized additive model fits in Figure 23.7.

Finally, it may also be informative to predict the probability of having an intense day of rain, which would require us to treat the $X_3$ variable as the response. We fit a generalized partial linear model where we assume a Bernoulli distribution for the response and relate the sum of functions of independent variables to the response via the probit link. The estimates for the linear coefficients — i.e., the coefficient for the ragweed count and the wind speed — are given below:

```
##############################
            Coef     SE
ragweed    -0.0241 0.0084
wind.speed  0.0219 0.0358
##############################
```

The coefficient for wind speed in this model is within one standard error of

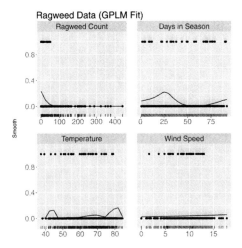

**FIGURE 23.9**
Generalized partial linear model fits of each predictor for the ragweed data
with partial residuals overlaid.

0 and the coefficient for ragweed count is about three standard errors from
0. Thus, neither of these might be considered particular informative linear
predictors for predicting an intense day of rainfall. Regardless, the fits for
both the parametric and nonparametric components are given in Figure 23.9
with partial residuals overlaid. Some smooth nonlinear trends appear to be
indicated by the fits for the nonparametric components, but overall there are
no clear and definitive statements that we might be willing to make about
these relationships.

**Example 23.7.3.** *(Boston Housing Data)*
Harrison and Rubinfeld (1978) analyzed data on the median value of homes
in Boston, Massachusetts as a function of various demographic and socioeco-
nomic predictors. The sample size of this dataset is $n = 506$. The variables
analyzed in this study are:

- $Y$: median home price (in $1000 U.S. dollars)

- $X_1$: per capita crime rate by town

- $X_2$: proportion of residential land zoned for lots

- $X_3$: proportion of non-retail business acres per town

- $X_4$: a river boundary indicator where 1 indicates the tract bounds the river

- $X_5$: nitrogen oxides concentration (parts per 10 million)

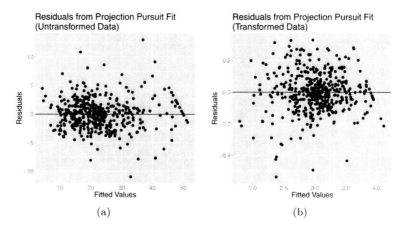

**FIGURE 23.10**
(a) Raw residuals versus fitted values from the projection pursuit regression fit for the model with the untransformed predictors. (b) Raw residuals versus fitted values from the projection pursuit regression fit for the model studied by Harrison and Rubinfeld (1978) and Aldrin et al. (1993).

- $X_6$: average number of rooms per dwelling

- $X_7$: proportion of owner-occupied units built prior to 1940

- $X_8$: weighted mean of distances to five Boston employment centers

- $X_9$: index of accessibility to radial highways

- $X_{10}$: full-value property-tax rate per \$10,000 U.S. dollars

- $X_{11}$: pupil-teacher ratio by town

- $X_{12}$: the proportion of black individuals by town

- $X_{13}$: the percentage of the population classified as lower status

Harrison and Rubinfeld (1978) did a thorough study of an appropriate model to use for relating $Y$ to the 13 predictors. After studying various transformations, the model they identified is as follows:

$$\log(Y) = \beta_0 + \beta_1 X_1 + \beta_2 X_2 + \beta_3 X_3 + \beta_4 X_4 + \beta_5 X_5^2 + \beta_6 X_6^2$$
$$+ \beta_7 X_7 + \beta_8 \log(X_8) + \beta_9 \log(X_9) + \beta_{10} X_{10}$$
$$+ \beta_{11} X_{11} + \beta_{12}(X_{12} - 0.63)^2 + \beta_{13} \log(X_{13}) + \epsilon.$$

Aldrin et al. (1993) also analyzed these data using projection pursuit regression as an effective way to handle moderate nonlinearities in the relationship. They analyzed both the model identified in Harrison and Rubinfeld

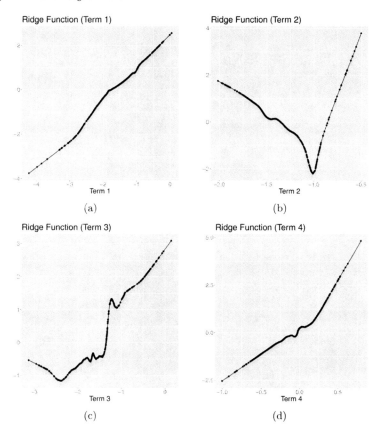

**FIGURE 23.11**
Ridge functions for (a) term 1, (b) term 2, (c) term 3, and (d) term 4 in the projection pursuit regression fit of the Boston housing data.

(1978) as well as the same model, but where all of the variables are left untransformed. We analyze both the untransformed variables and transformed variables using projection pursuit with four terms retained for the final model. Figure 23.10 show plots of the residuals versus fitted values for both of these fits. While the fit using the untransformed variables produces a reasonable looking residual plot (Figure 23.10(a)), there is a clear improvement in the residuals based on the fit using the transformed variables (Figure 23.10(b)). These fits have, respectively, MSE values of 6.519 and 0.014, which further illustrates the better fit achieved using the transformed variables.

Using the projection pursuit regression fit for the transformed variables, we can also construct plots of the ridge functions for each of the four terms in the final model. These are given in Figure 23.11 to help visualize the general nonlinear behavior characterized by each of the terms in this fit.

# 24

## Data Mining

The field of statistics is constantly being presented with larger and more complex datasets. This notion of **big data** presents a formidable challenge for statisticians and practitioners to collect, store, and curate the data, produce informative visualizations, and find meaningful trends. The general tools and the approaches for dealing with these challenges in massive datasets is referred to as **data mining**, **statistical learning**, or **machine learning**.

Data mining problems typically involve an outcome measurement which we wish to predict based on a set of **feature measurements**, which is another name for predictors. The set of these observed measurements is called the **training data**, from which we attempt to build a **learner**, which is a model used to predict the outcome for new subjects. These learning problems are (roughly) classified as either supervised or unsupervised.[1] A **supervised learning** problem is one where the goal is to predict the value of an outcome measure based on a number of input measures, such as classification with labeled samples from the training data. An **unsupervised learning** problem is one where there is no outcome measure and the goal is to describe the associations and patterns among a set of input measures, which involves clustering unlabeled training data by partitioning a set of features into a number of statistical classes. The regression problems that are the focus of this text are (generally) supervised learning problems.

Data mining is an extensive field in and of itself. In fact, many of the methods utilized in this field are regression-based. For example, smoothing splines, shrinkage methods, and multivariate regression methods are all often found in data mining. The purpose of this chapter will not be to revisit these methods, but rather to add to our toolbox additional regression methods, which are methods that happen to be utilized more in data mining problems. Hastie et al. (2009) is one of the most comprehensive texts on the topic, which we used to determine the topics discussed in this chapter.

---

[1] A third class of problems is called **semi-supervised learning**, which falls between unsupervised learning and supervised learning. Semi-supervised learning techniques make use of supervised learning techniques where unlabeled data is also used for training purposes. We do not formally discuss such techniques here, but the interested reader can refer to the edited text by Chapelle et al. (2006).

## 24.1   Classification and Support Vector Regression

**Classification** uses a training set of data containing observations whose subpopulation is known to identify the subpopulation to which new observations belong such that the labels of the subpopulation are unknown. An algorithm that performs classification is called a **classifier**. The classification problem is often contrasted with **clustering**, where the problem is to analyze a dataset and determine how (or if) the dataset can be divided into subgroups. In data mining, classification is a supervised learning problem whereas clustering is an unsupervised learning problem.

In this chapter, we focus on a special classification technique which has many regression applications in data mining. **Support Vector Machines** (or **SVMs**) perform classification by constructing an $N$-dimensional hyperplane that optimally separates the data into two categories. The algorithm is a supervised learning algorithm. SVM models are also closely related to neural networks, which we discuss later in this chapter. The predictor variables are called **attributes** and a transformed attribute that is used to define the hyperplane is the feature. The task of choosing the most suitable representation is known as **feature selection**, which is analogous to variable selection that we have discussed throughout the text. A set of features that describes one case (i.e., a row of predictor values) is called a **vector**. So the goal of SVM modeling is to find the optimal hyperplane that separates clusters of vectors in such a way that cases with one category of the target variable are on one side of the plane and cases with the other category are on the other size of the plane. The vectors near the hyperplane are the support vectors.

Suppose we wish to perform classification with the data shown in Figure 24.1(a) and our data has a categorical target variable with two categories. Some examples could be whether or not a person has coronary heart disease, if an e-mail is "spam" or "not spam," and if a manufactured unit "passes" or "fails" quality control specifications. Also assume that the attributes have continuous values. Figure 24.1(b) provides a snapshot of how we perform SVM modeling. In this example, SVMs attempt to find a 1-dimensional hyperplane (i.e., a line) that separates the cases based on their target categories. There are an infinite number of possible lines and we show only one in Figure 24.1(b). The question is which line is optimal and how do we define that line?

The dashed lines drawn parallel to the separating line mark the distance between the dividing line and the closest vectors to the line. The distance between the dashed lines is called the **margin**. The vectors (i.e., points) that constrain the width of the margin are the **support vectors**. An SVM analysis finds the line (or, more generally, the hyperplane) that is oriented so that the margin between the support vectors is maximized. Unfortunately, real data problems will typically not be as simple as the setting in Figure 24.1. The

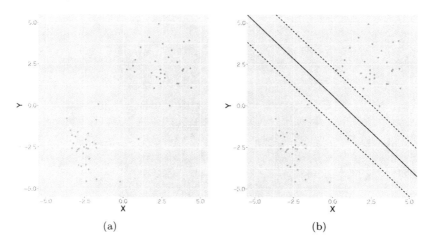

**FIGURE 24.1**

(a) A plot of the data where classification is to be performed. (b) The data where a support vector machine has been used. The points near the parallel dashed lines are the support vectors. The region between the parallel dashed lines is called the margin, which is the region we want to optimize.

challenge is to develop an SVM model that accommodates such characteristics as:

1. handling more than two attributes;

2. being able to separate points with nonlinear curves;

3. having minimum classification error on new points;

4. handling of cases where the clusters cannot be completely separated; and

5. handling classification with more than two categories.

The setting with nonlinear curves and where clusters cannot be completely separated is illustrated in Figure 24.2. Without loss of generality, our discussion will mainly be focused on the one-attribute and one-feature setting. Moreover, we will be utilizing support vectors in order to build a regression relationship that fits our data adequately.

We next seek some mechanism for minimizing any penalties (or costs) associated with incorrect classification. This can be accomplished through **loss functions**, which represent the cost associated with an inaccurate prediction; i.e., a prediction different from either a desired or a true value. When discussing SVM modeling in the regression setting, the loss function must incorporate a distance measure as well.

We provide a quick illustration of some common loss functions in Figure

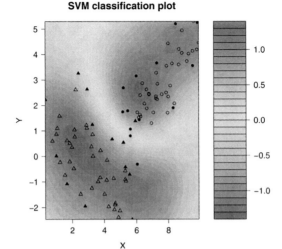

**SVM classification plot**

**FIGURE 24.2**
A plot of data where a support vector machine has been used for classification.
The data were generated where we know that the circles belong to group 1 and
the triangles belong to group 2. The white contours show where the margin
is; however, there are clearly some values that have been misclassified since
the two clusters are not well-separated. The points that are solid were used
as the training data.

24.3. Figure 24.3(a) is a quadratic loss function, which is what we use in
ordinary least squares. Figure 24.3(b) is a Laplacian loss function, which is less
sensitive to outliers than the quadratic loss function. Figure 24.3(c) is Huber's
loss function, which is a robust loss function that has optimal properties when
the underlying distribution of the data is unknown. Finally, Figure 24.3(d) is
called the $\epsilon$-insensitive loss function, which enables a sparse set of support
vectors to be obtained.

**Support Vector Regressions** (or **SVRs**), as discussed in Vapnik (2000),
first maps the input onto an $N$-dimensional feature space using some fixed
(nonlinear) mapping, and then a linear model is constructed in this feature
space. Using mathematical notation, the linear model in the feature space is
given by

$$f(\mathbf{x}, \boldsymbol{\omega}) = \sum_{j=1}^{N} \omega_j g_j(\mathbf{x}) + b,$$

where $g_j(\cdot)$, $j = 1, \ldots, N$, denotes a set of nonlinear transformations, $\boldsymbol{\omega}$ is
a vector of weights, and $b$ is a bias term. If the data is assumed to be of
zero mean, as it usually is, then the bias term is dropped. Note that $b$ is

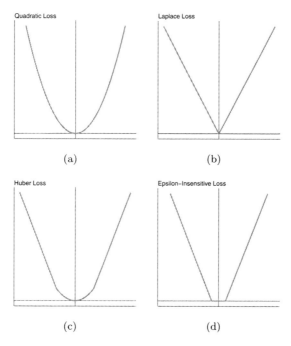

**FIGURE 24.3**
Plots of the (a) quadratic loss, (b) Laplace loss, (c) Huber's loss, and (d) $\epsilon$-insensitive loss functions.

not considered stochastic in this model and is not akin to the error terms in previous regression models.

The goal is to find a function $f(\mathbf{x})$ that has at most a deviation of $\epsilon$ from the actual obtained targets for the training data and simultaneously be as flat as possible. The optimal regression function is given by the minimum of the functional

$$\Phi(\boldsymbol{\omega}, \xi) = \frac{1}{2}\|\boldsymbol{\omega}\|^2 + C\sum_{i=1}^{n}(\xi_i^- + \xi_i^+),$$

where $C > 0$ is a pre-specified constant that determines the tradeoff between the flatness of $f(\cdot)$ and the amount up to which deviations larger than $\epsilon$ are tolerated and $\xi_i^-$ and $\xi_i^+$ are **slack variables** (i.e., variables added to an inequality constraint to transform it to an equality) representing upper and lower constraints, respectively, on the output of the system. In other words, we have the following constraints:

$$y_i - f(\mathbf{x}_i, \boldsymbol{\omega}) - b \leq \epsilon + \xi_i^+$$

$$f(\mathbf{x}_i, \boldsymbol{\omega}) + b - y_i \leq \epsilon + \xi_i^-$$

$$\xi_i^-, \xi_i^+ \geq 0, \quad i = 1, \ldots, n,$$

where $y_i$ is defined through the loss function we are using. The four loss functions we show in Figure 24.3 are as follows:

**Quadratic Loss:**

$$\mathcal{L}_2 = (f(\mathbf{x}) - y)^2$$

**Laplace Loss:**

$$\mathcal{L}_1 = |f(\mathbf{x}) - y|$$

**Huber's Loss:**

$$\mathcal{L}_H = \begin{cases} \frac{1}{2}(f(\mathbf{x}) - y)^2, & \text{for } |f(\mathbf{x}) - y| < \delta; \\ \delta|f(\mathbf{x}) - y| - \frac{\delta^2}{2}, & \text{otherwise} \end{cases}$$

**$\epsilon$-Insensitive Loss:**

$$\mathcal{L}_\epsilon = \begin{cases} 0, & \text{for } |f(\mathbf{x}) - y| < \epsilon; \\ |f(\mathbf{x}) - y| - \epsilon, & \text{otherwise} \end{cases}$$

Depending on which loss function is chosen, then an appropriate optimization problem can be specified. This can involve kernel methods as discussed in Chapter 17. Moreover, specification of the kernel type as well as values like $C$, $\epsilon$, and $\delta$ all control the complexity of the model in different ways.[2] There are many subtleties depending on which loss function is used and the investigator should become familiar with the loss function being employed. Regardless, this a convex optimization approach that requires the use of numerical methods. For a more detailed tutorial on support vector regression, see Smola and Schölkopf (2004).

It is also desirable to strike a balance between complexity and the error that is present with the fitted model. **Test error**, or **generalization error**, is the expected prediction error over an independent test sample $\mathcal{T}$ and is given by

$$Err_\mathcal{T} = \mathrm{E}[\mathcal{L}(Y, \hat{f}(\mathbf{X}))|\mathcal{T}],$$

where $\mathbf{X}$ and $Y$ are drawn randomly from their joint distribution. **Training error** is the average loss over the training sample and is given by

$$\overline{err} = \frac{1}{n} \sum_{i=1}^{n} \mathcal{L}(y_i, \hat{f}(\mathbf{x}_i)).$$

We would like to know the test error of our estimated model $\hat{f}(\cdot)$. As the model increases in complexity, it is able to capture more complicated underlying structures in the data, which thus decreases bias. But then the estimation error increases, which thus increases variance. This is known as the **bias-variance tradeoff**. In between there is an optimal model complexity that gives minimum test error.

---

[2]The quantity $\delta$ is a specified threshold constant.

## 24.2 Prediction Trees and Related Methods

Suppose we have a complicated relationship between a continuous response $Y$ and $p$ predictors $X_1, \ldots, X_p$, where it is difficult to identify a particular functional model that adequately characterizes the true relationship. We can move away from a parametric model and use a strategy that closely follows the human decision-making process. **Tree-based methods** involve partitioning the predictor space into a number of simple regions where we take a summary measure (e.g., mean or median) of a response variable to make predictions. The initial portion of this section focuses solely on tree-based algorithms for regression and classification problems. We then discuss other methods that are loosely related to tree-based approaches.

### Prediction Trees

When there are lots of complicated interactions and nonlinear relationships present, one strategy is to use **regression trees**, which are nonparametric methods that partition the predictor space into a set of rectangles and then a simple model is fit, like a constant, in each one. If fitting the model is still complicated, we then further partition these rectangles into smaller regions, which allow interactions to be handled in an easier manner. This strategy of identifying conceptually simpler models through partitioning the predictor space is called **recursive partitioning**.

More generally, regression trees can be thought of as one of two different tree-based methods belonging to a class of predictive modeling strategies called **decision trees** or **decision tree learning**. The other tree-based method is **classification trees**, which differs from regression trees in that the response $Y$ is now a class variable. Thus, classification trees are used to build a model where the response (or target) variable is a finite set of values representing a finite number of classes while regression trees are used to build a model where the response (or target) variable has continuous values. Conceptually these two types of decision trees are similar, so we give the general steps for most tree-based algorithms:

---

Sketch of Tree-Based Algorithms

1. Start at the **root node** of the tree, which is simply the entire dataset.

2. Let $S_j$ represent a binary split of the predictor variable $X_j$. If $X_j$ is a continuous variable (or an ordered discrete variable), then $S_j$ is an interval of the form $(-\infty, s]$. If $X_j$ is a categorical variable, then $S_j$ is a subset of the possible values taken by $X_j$. For each $X_j$, find the set $S_j$ that minimizes the sum of a specified measure (called the **node impurity measure**) of the observed $Y$ values in

those two nodes, which are also called **child nodes**. Then, choose the split and predictor variable that minimizes the sum of this node impurity measure over all predictor $X_j$ and splits $S_j$.

3. Repeat the previous step until some stopping criterion is employed, such as when the subset at a node has all of the same values as the response variable, a specified number of regions (or **terminal nodes**) is met, or splitting no longer adds values to the predicted values. All nodes between the root node and the terminal nodes are called **internal nodes**.

---

An example of the above is given in Figure 24.4. We consider two predictors, $X_1$ and $X_2$, each taking on values between 0 and 10. Four binary splits are illustrated, resulting in five regions labeled as $R_1, \ldots, R_5$. Figure 24.4(a) shows what the partitioning looks like in the rectangular predictor space and Figure 24.4(b) shows the corresponding decision tree.

Notice that the "growing" of trees is really more algorithmic than it is statistical in the sense that we are not making any sort of distributional assumptions nor are we working with some sort of likelihood function. But, how much should we grow the trees? If we have a small tree, it might be too simple. If we have a large tree, we risk overfitting the data. We can take one of two primary strategies. First, we can decide whether or not the split should be based on a hypothesis test of whether or not it significantly improves the fit. Second, we can view the tree size as a tuning parameter, and, thus, choose the size using methods such as crossvalidation.

For the first strategy, we simply carry out an appropriate hypothesis test for each variable. If the lowest $p$-value is significant after adjusting for the number of comparisons, partition the data using the optimal split for the variable with the lowest $p$-value. We then stop growing the tree when no significant variables are found. One appealing feature of this approach is that you get $p$-values for each split and they are guaranteed to be significant. However, you may determine that a split is not important, but later on it actually is very important. To mitigate this potential problem, we can grow our tree to be very large (i.e., a large number of terminal nodes) and then employ the second strategy of viewing the tree size as a tuning parameter.

We can use some common rules to stop growing our tree, such as those in Step 3 of the algorithm presented earlier. We denote this largest tree under consideration as $T_0$. Furthermore, consider a subtree $T$ that can be obtained by collapsing any number of the internal nodes of $T_0$, which is called **pruning**. Let $|T|$ denote the number of terminal nodes in subtree $T$ and index those nodes with $m$, with node $m$ corresponding to region $R_m$. We then use a node impurity measure, $Q_m(T)$, which will differ depending on the type of decision

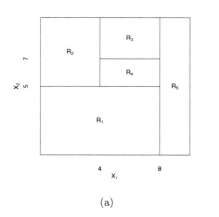
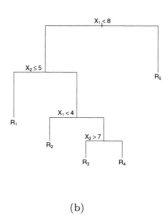

(a)                                        (b)

**FIGURE 24.4**
(a) A partition of a two-dimensional predictor space based on some hypothetical binary splits. (b) The decision tree corresponding to the partition.

tree you are growing. Finally, we define the *cost-complexity criterion* as

$$C_\alpha(T) = \sum_{m=1}^{|T|} N_m Q_m(T) + \alpha|T|, \qquad (24.1)$$

where $N_m$ is the number of observations in node $m$. The tuning parameter $\alpha$ behaves like other regularization parameters by balancing stability (tree size) with goodness-of-fit. For any given $\alpha$, there is a tree $T_\alpha$ which minimizes the above criterion, for example, $\alpha = 0$ returns the full tree $T_0$. As we alluded to earlier, $\alpha$ is usually chosen via cross-validation.

For some of the mathematical details of decision trees, we follow the general set-up in Chapter 9 of Hastie et al. (2009). Suppose we have a sample of size $n$ with one continuous response variable, $y_i$, and $p$ predictors, $\mathbf{x}_i = (x_{i,1}, \ldots, x_{i,p})^\mathrm{T}$. First, suppose we have the partition of the predictor space into $M$ regions, say, $R_1, \ldots, R_M$. For regression trees (i.e., where the response is a continuous variable), we model the response as a constant $c_m$ in each region:

$$f(\mathbf{X}) = \sum_{m=1}^{M} c_m \mathrm{I}\{\mathbf{X} \in R_m\}.$$

Minimizing the sum of squares for the above model with our observed data yields

$$\hat{c}_m = \sum_{i=1}^{n} (y_i - f(\mathbf{x}_i))^2$$
$$= \frac{\sum_{i=1}^{n} y_i \mathrm{I}\{\mathbf{x}_i \in R_m\}}{\sum_{i=1}^{n} \mathrm{I}\{\mathbf{x}_i \in R_m\}}.$$

We proceed to grow the tree by finding the best binary partition in terms of the $\hat{c}_m$ values. Pruning can be accomplished by using the cost-complexity criterion in (24.1) such that

$$Q_m(T) = \frac{1}{N_m} \sum_{i:x_i \in R_m} (y_i - \hat{c}_m)^2.$$

For classification trees (where the response is a classification variable), we model the classification variable as a constant $\pi_{m,k}$, for $k = 1, \ldots, K$ classes, in each region:

$$\mathrm{P}(Y = k|\mathbf{X}) = \sum_{m=1}^{M} \pi_{m,k} \mathrm{I}\{\mathbf{X} \in R_m\}.$$

While regression trees use a squared-error loss approach with the cost-complexity criterion $Q_m(T)$, different criteria need to be employed for classification trees. Some common measures according to Hastie et al. (2009) are

- Misclassification error: $\frac{1}{N_m} \sum_{i \in R_m} \mathrm{I}\{y_i \neq \underset{k}{\mathrm{argmax}}\ \hat{\pi}_{m,k}\}$

- Gini index: $\sum_{k=1}^{K} \hat{\pi}_{m,k}(1 - \hat{\pi}_{m,k})$

- Deviance: $-\sum_{k=1}^{K} \hat{\pi}_{m,k} \log(\hat{\pi}_{m,k})$

All three criteria are similar, but the Gini index and deviance are differentiable which makes them easier to optimize numerically.

The majority of what we have discussed is the foundation for implementing tree-based algorithms and is often generically referred to as **classification and regression trees** or **CART** (Breiman et al., 1984). There are numerous related tree-based algorithms, some of which are direct modifications of CART. These algorithms include **CHAID** (Kass, 1980), **C4.5** (Quinlan, 1993a), and **M5'** (Witten et al., 2011). An accessible overview and comparison of these and many other tree-based algorithms is given in Loh (2011).

### PRIM

We can also move away from the binary splits performed in regression trees to one where we attempt to find boxes, cubes, or hypercubes in which the response average is high. This approach to finding such regions is called **bump hunting**. One of the major algorithms to accomplish this is the **patient rule induction method** or **PRIM** due to Friedman and Fisher (1999).

Unlike tree-based methods, PRIM constructs boxes that are not the result of binary splits. PRIM starts with a main box comprised of all of the data, compresses one face of the box, and sets aside observations falling outside the box in a process called **peeling**. The face for compression is determined by the largest box mean post-compression. This process continues until the current box contains some minimum number of data points. After this top-down sequence is computed, PRIM reverses the process and expands along any edge where such an expansion increases the box mean. This part of the process is called **pasting**. As noted by Hastie et al. (2009), CART and similar algorithms partition the data quickly, whereas PRIM is more "patient" in how it obtains a solution.

## MARS

**Multivariate adaptive regression splines (MARS)** due to Friedman (1991) is another nonparametric regression method that can be viewed as a modification of CART and is well-suited for high-dimensional (large number of predictors) problems.[3] MARS uses expansions in piecewise linear basis functions of the form $(X - t)_+$ and $(t - X)_+$ such that the "+" again means the positive part. These two functions together are called a **reflected pair** and, individually, are just the truncated power function from Chapter 17.

In MARS, each function is piecewise linear with a knot at $t$. The idea is to form a reflected pair for each predictor $X_j$ with knots at each observed value $x_{i,j}$ of that predictor. These form a set of basis functions that we denote by $\mathcal{C}$.

The objective with building a model via MARS is to end up with

$$f(\mathbf{X}) = \beta_0 + \sum_{m=1}^{M} \beta_m h_m(\mathbf{X}),$$

where each $h_m(\mathbf{X})$ is a function in $\mathcal{C}$ or a product of two or more such functions in $\mathcal{C}$. MARS can build such a model in two primary ways. The **forward pass approach** – which is similar to forward selection in stepwise regression – starts with an intercept term and then systematically adds functions from the set $\mathcal{C}$ and their products. The **backward pass approach** – which is similar to backward elimination in stepwise regression – starts with an **overfit model**, which produces a good fit to the data, but likely does not generalize well to new data. Then, at each step of the backward pass, the model is pruned such that the model subsets are often compared using a generalized crossvalidation criterion. Regardless of the model-building strategy employed, the result is that MARS automatically determines the most important independent variables as well as the most significant interactions among them.

---

[3]MARS is actually a licensed and trademarked term, so many open source implementations of MARS are called **Earth**.

**HME Models**

The **hierarchical mixture-of-experts** (**HME**) procedure (Jacobs et al., 1991) can be viewed as a parametric tree-based method that recursively splits the function of interest at each node. However, the splits are done probabilistically and the probabilities are functions of the predictors.

Figure 24.5 depicts the architecture of the HME model with two levels as presented in Young and Hunter (2010). The HME model involves a set of functions (the leaves in the tree structure) called **expert networks** that are combined together by a classifier (the non-leaf nodes) called **gating networks**. The gating networks split the feature space into regions where one expert network seems more appropriate than the other. The general HME model is then written as

$$f(Y) = \sum_{j_1=1}^{k_1} \lambda_{j_1}(\mathbf{X}, \boldsymbol{\tau}) \sum_{j_2=1}^{k_2} \lambda_{j_2}(\mathbf{X}, \boldsymbol{\tau}_{j_1})$$

$$\cdots \sum_{j_r=1}^{k_r} \lambda_{j_r}(\mathbf{X}, \boldsymbol{\tau}_{j_1,j_2,\dots,j_{r-1}}) g(Y; \mathbf{X}, \boldsymbol{\theta}_{j_1,j_2,\dots,j_r}),$$

which has a tree structure with $r$ levels; i.e., $r$ levels where probabilistic splits occur. The $\lambda(\cdot)$ functions provide the probabilities for the splitting and, in addition to depending on the predictors, they also have their own set of parameters. These parameters are the different $\boldsymbol{\tau}$ values, which need to be estimated. These mixing proportions are typically modeled using logistic regressions, but other parametric and nonparametric approaches have been proposed; cf. Yau et al. (2003) and Young and Hunter (2010). Finally, $\boldsymbol{\theta}$ is simply the parameter vectors for the regression modeled at each terminal node of the tree constructed using the HME structure.

Jordan and Xu (1995) point out that the mixture-of-linear-regressions model,[4] which assigns probabilities of class membership based only on residuals, can be viewed as one end of a continuum. At the other end are tree-based methods like CART and MARS, which rigidly partition the predictor space into categories by using a "hard" split at each node. Viewed in this way, HME models lie between these extremes because it incorporates gating networks that allow the probability of subgroup membership to depend on the predictors and, essentially, use "soft" probabilistic splits at each node.

## 24.3 Some Ensemble Learning Methods for Regression

**Ensemble methods** combine predictions from multiple learning algorithms in order to improve predictions from a single learning algorithm. While quite

---

[4]We discuss mixtures of regressions in Chapter 25.

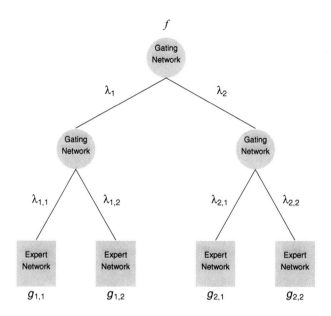

**FIGURE 24.5**
A diagram of an HME model with two levels as presented in Young and Hunter (2010). The density $f$ is approximated by the weighted sum of lower gating networks, which are the weighted sum of expert networks: $f = \sum_{j=1}^{2} \lambda_j \sum_{l=1}^{2} \lambda_{j,l} g_{j,l}$. We simplified notation by writing $g(Y; \mathbf{X}, \boldsymbol{\theta}_{j,l})$ as $g_{j,l}$, $\lambda_j(\mathbf{X}, \boldsymbol{\tau})$ as $\lambda_j$, and so on.

powerful, ensemble methods do require more computation when evaluating the predictions compared to evaluating the predictions from a single model. Most ensemble methods are available for both classification and regression.

**Boosted Regression**
The first approach we discuss is called **boosting** or **boosted regression** (Freund and Schapire, 1997). Boosted regression is highly flexible in that it allows the researcher to specify the feature measurements without specifying their functional relationship to the outcome measurement. Because of this flexibility, a boosted model will tend to fit better than a linear model and therefore inferences made based on the boosted model may have more credibility.

Boosting can be done for both classification trees and regression trees. Below we present the boosting algorithm for regression trees, which closely follows Algorithm 8.2 of James et al. (2013):

---

Boosting for Regression Trees

For a training dataset of size $n$:

1. Set $\hat{f}^{(0)} = 0$ and $r_i^{(0)} = y_i$ for all $i = 1, \ldots, n$. Let $\mathbf{r}^{(0)} = (r_1^{(0)}, \ldots, r_n^{(0)})^{\mathrm{T}}$.

2. For $m = 1, \ldots, M$, repeat the following:

   (a) Fit a tree $\hat{g}^{(m)}$ with $s$ splits ($s+1$ terminal nodes) to the training data consisting of $\mathbf{X}$ (the matrix of observed predictors) and the current residual vector $\mathbf{r}^{(m-1)}$.

   (b) Obtain the updated tree

   $$\hat{f}^{(m)}(\mathbf{X}) = \hat{f}^{(m-1)}(\mathbf{X}) + \lambda \hat{g}^{(m)}(\mathbf{X}),$$

   which updates the current tree by adding a shrunken version of the new tree (controlled through the shrinkage parameter $\lambda$).

   (c) Obtain the updated residuals:

   $$r_i^{(m)} = r_i^{(m-1)} - \lambda \hat{g}^{(m)}(\mathbf{x}_i).$$

3. The boosted model is then calculated as

$$\hat{f}(\mathbf{x}) = \sum_{m=1}^{M} \hat{f}^{(m)}(\mathbf{X}).$$

---

The above is a general algorithm for boosting regression trees. There are multiple variants on how to perform boosting for decision trees, just like with the algorithms for constructing decision trees. Some of the more common algorithms for boosting classifiers include **AdaBoost** (Freund and Schapire, 1997), **ExpBoost** (Rettinger et al., 2006), and **TrAdaBoost** (Dai et al., 2007). Pardoe and Stone (2010) discuss the extensions of each of these algorithms to the regression setting.

**Bagging**
In Chapter 17, we introduced the bootstrap as a way to get confidence intervals and estimates for the regression parameters. However, the notion of the bootstrap can also be extended to fitting a regression model. **Bootstrap aggregating** or **bagging** (Breiman, 1996) is used to improve the stability and accuracy of algorithms used in classification and regression. Bagging averages predictions over a collection of bootstrap samples, thus reducing variance. Although bagging is usually used with decision trees, it can be used with many other modeling approaches.

Bagging for Regression Trees

For a training dataset of size $n$:

1. Generate $B$ bootstrap samples of the training data, which we write as $\{\mathbf{Y}_1^*, \mathbf{X}_1^*\}, \ldots, \{\mathbf{Y}_B^*, \mathbf{X}_B^*\}$.
2. Calculate the trees $\hat{f}^{*1}(\mathbf{X}_1^*), \ldots, \hat{f}^{*B}(\mathbf{X}_B^*)$.
3. Average all of the predictions to get the estimate:

$$\hat{f}_{\text{bag}}(\mathbf{X}) = \sum_{b=1}^{B} \hat{f}^{*b}(\mathbf{X}_b^*). \tag{24.2}$$

In fact, as $B \to \infty$, (24.2) is a Monte Carlo estimate of the true bagging estimate. The true bagging estimate would be the quantity calculated in (24.2), but based on $B$ different training datasets if such training data were available.

### Random Forests

Suppose we again have a training dataset of size $n$ with one response and $p$ predictors. Suppose further that a small number of the predictors are actually very strong predictors for the response variable. When bagging, these predictors will be selected in many of the $B$ trees constructed from the bootstrap samples, thus causing the trees to become correlated. We can slightly modify bagging to decorrelate the trees, which is a procedure known as **random forests** (Ho, 1995). The difference is that at each candidate split in the decision tree, a random subset of the predictors is used. Below is the algorithm:

Random Forests for Regression Trees

For a training dataset of size $n$:

1. Generate $B$ bootstrap samples of the training data, which we write as $\{\mathbf{Y}_1^*, \mathbf{X}_1^*\}, \ldots, \{\mathbf{Y}_B^*, \mathbf{X}_B^*\}$.
2. Calculate the trees $\hat{f}^{\dagger 1}(\mathbf{X}_1^*), \ldots, \hat{f}^{\dagger B}(\mathbf{X}_B^*)$, such that each candidate split in the tree learning algorithm, a random subset of the predictors is chosen.
3. Average all of the predictions to get the estimate:

$$\hat{f}_{\text{rforest}}(\mathbf{X}) = \sum_{b=1}^{B} \hat{f}^{\dagger b}(\mathbf{X}_b^*). \tag{24.3}$$

A typical rule of thumb for the number of predictors to select in each split of the regression tree is $\lfloor p/3 \rfloor$. If using random forests for classification trees, then a typical rule of thumb is $\lfloor \sqrt{p} \rfloor$.

## 24.4   Neural Networks

In the era of big data, advanced computing power has become essential to handle more complex data problems. Researchers in areas like statistics, artificial intelligence, and data mining have all been faced with the challenge to develop simple, flexible, powerful procedures for modeling these large datasets. One such model is the **neural network** approach, which attempts to model the response as a nonlinear function of various linear combinations of the predictors, much like the objective with some of the semiparametric regression models in Chapter 23. As one might expect, neural networks were first used as models for the human brain as well as for applications in artificial intelligence.

There are *many* different types of neural networks in the literature; see the text by Bishop (1995) for a thorough treatment of some of these neural networks. We focus on one of the more commonly used neural networks called the **single-hidden-layer feedforward neural network** or the **single-layer perceptron**. Assume we have a response $Y$ and a $p$-dimensional vector of predictors $\underline{X}$, where the first entry is set to 1 for an intercept term. We define the $m$-dimensional vector of **derived predictor values** $\underline{S} = (S_1, \ldots, S_m)^{\mathrm{T}}$ such that

$$S_h = \begin{cases} g_h(\underline{X}^{\mathrm{T}}\beta_h), & h = 2, \ldots, m; \\ 1, & h = 1. \end{cases}$$

Then, the single-hidden-layer feedforward neural network model relates the response $Y$ to the vector of derived predictor variables $\underline{S}$ by

$$Y = f_Y(\underline{S}^{\mathrm{T}}\boldsymbol{\alpha}) + \epsilon,$$

where the nonlinear function $f_Y$ is called an **activation function**. Combining the above, we can form the neural network model as:

$$Y = f_Y(\underline{S}^{\mathrm{T}}\boldsymbol{\alpha}) + \epsilon$$

$$= f_Y \left( \alpha_0 + \sum_{h=2}^{m} \alpha_{h-1} g_h(\underline{X}^{\mathrm{T}}\beta_h) \right) + \epsilon.$$

Note that if we are doing regression (i.e., where $Y$ is a continuous variable), then the $g_h$ functions are usually taken as the identity function. If we are doing classification (i.e., where $Y$ is a discrete class variable), then the $g_h$ functions are usually taken as the **softmax function**

$$g_h(\underline{X}^{\mathrm{T}}\beta_h) = \frac{\exp\{\underline{X}^{\mathrm{T}}\beta_h\}}{\sum_{k=1}^{m} \exp\{\underline{X}^{\mathrm{T}}\beta_k\}}.$$

There are various numerical optimization algorithms for fitting neural networks, such as quasi-Newton methods and conjugate-gradient algorithms. One important thing to note is that parameter estimation in neural networks often utilizes penalized least squares to control the level of overfitting. Suppose we have a sample of size $n$. The penalized least squares criterion is given by:

$$Q = \sum_{i=1}^{n} \left[ y_i - f_Y \left( \alpha_0 + \sum_{h=2}^{m} \alpha_{h-1} g_h(\mathbf{x}_i^{\mathrm{T}} \boldsymbol{\beta}_h) \right) \right]^2 + p_\lambda(\boldsymbol{\alpha}, \boldsymbol{\beta}_2, \ldots, \boldsymbol{\beta}_m),$$

where the penalty term is given by:

$$p_\lambda(\boldsymbol{\alpha}, \boldsymbol{\beta}_2, \ldots, \boldsymbol{\beta}_m) = \lambda \left[ \sum_{k=0}^{m-1} \alpha_k^2 + \sum_{h=2}^{m} \sum_{j=0}^{p-1} \beta_{h,j}^2 \right].$$

## 24.5 Examples

**Example 24.5.1.** *(Simulated Motorcycle Accident Data)*
Schmidt et al. (1981) conducted an experiment from a simulated collision involving different motorcycles for the purpose of testing crash helmets. Data from $n = 133$ tests are available. The variables analyzed in this study are:

- $Y$: g-force measurement of acceleration

- $X$: time (in milliseconds) after impact

These data were also analyzed by Silverman (1985) for illustrating important advances in taking a spline smoothing approach to nonparametric regression estimation. These data have become a classic dataset for researchers to demonstrate new approaches to nonparametric and semiparametric regression curve fitting. The data are plotted in Figure 24.6(a). Given the complex nonlinear trend present that is clearly present, we will fit SVR models to these data.

An SVR using an $\epsilon$-insensitive loss function is fit to this data. The fits based on $\epsilon \in \{0.01, 0.10, 0.70\}$ are shown in Figure 24.6(b). As $\epsilon$ decreases, different characteristics of the data are emphasized, but the level of complexity of the model is increased. As noted earlier, we want to try and strike a good balance regarding the model complexity. For the training error, we get values of 0.209, 0.202, and 0.273 for the three levels of $\epsilon$. Since our objective is to minimize the training error, the value of $\epsilon = 0.10$, which has a training error of 0.202, is chosen. This corresponds to the green line in Figure 24.6(b).

**Example 24.5.2.** *(Pima Indian Diabetes Data)*
The Pima Indian population near Phoenix, Arizona has a high incidence rate of diabetes. Smith et al. (1988) analyzed data on women who were at least 21

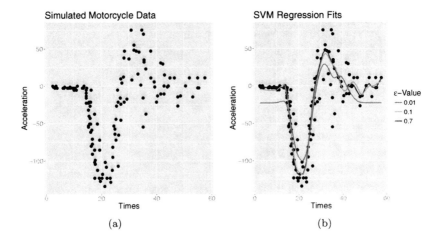

**FIGURE 24.6**
(a) The simulated motorcycle accident data where the time until impact (in milliseconds) is plotted versus the recorded head acceleration (in g). (b) The data with different values of $\epsilon$ used for the estimated SVR with an $\epsilon$−insensitive loss function. Note how the smaller the $\epsilon$, the more features that are picked up by the fit, however, the complexity of the model also increases.

years old in this population. The data were collected by the National Institute of Diabetes and Digestive and Kidney Diseases. Diabetes in these women was diagnosed by the World Health Organization (WHO). The variables analyzed in this study are:

- $Y$: an indicator where 1 indicates the subject is diabetic per the WHO criteria and 0 otherwise

- $X_1$: number of times the subject was pregnant

- $X_2$: plasma glucose concentration at two hours in an oral glucose tolerance test

- $X_3$: diastolic blood pressure (mm Hg)

- $X_4$: triceps skin fold thickness (mm)

- $X_5$: body mass index

- $X_6$: diabetes pedigree function

- $X_7$: age (in years)

The dataset we will analyze consists of $n = 532$ complete records, which was obtained after dropping records with missing values. The dataset is partitioned into a training set containing a randomly selected set of 200 patients and a test set containing the remaining 332 patients.

First we build a regression tree using the Pima Indian diabetes training data. The resulting regression tree is given in Figure 24.7(a). Using the test data, the classification table based on this regression tree fit is as follows:

```
#############################
        truth
predicted    No    Yes
     No   0.5482 0.1446
     Yes  0.1235 0.1837
#############################
```

Therefore, the misclassification rate is 0.2681. In an attempt to improve upon this rate, we prune the above tree using an estimated complexity parameter of (approximately) 0.015. The pruned regression tree is given in Figure 24.7(b). Using the test data, the classification table based on this pruned regression tree fit is as follows:

```
#############################
        truth
predicted    No    Yes
     No   0.5813 0.1536
     Yes  0.0904 0.1747
#############################
```

Therefore, the misclassification rate is 0.2440, a slight improvement over the original tree.

We next use MARS to build a regression model:

```
#############################
GLM Yes =
  8.5
+  0.04 * pmax(0,  glu -  109)
-  0.12 * pmax(0,   42 -  bmi)
-  0.53 * pmax(0,  bmi -   42)
-    70 * pmax(0,  ped -  0.2)
+    71 * pmax(0,  ped - 0.29)
-    34 * pmax(0, 0.38 -  ped)
-   0.1 * pmax(0,   51 -  age)
-  0.29 * pmax(0,  age -   51)

Earth selected 9 of 16 terms, and 4 of 7 predictors
Termination condition: Reached nk 21
Importance: glu, age, ped, bmi, npreg-unused, bp-unused, skin-unused
Number of terms at each degree of interaction: 1 8 (additive model)
Earth GCV 0.16    RSS 27    GRSq 0.3    RSq 0.4

GLM null.deviance 256 (199 dof)   deviance 159 (191 dof)   iters 5
#############################
```

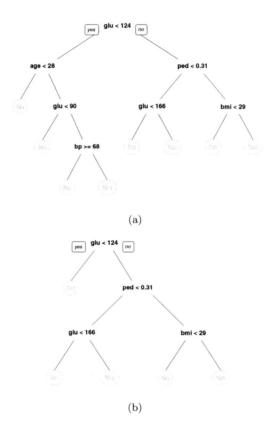

(a)

(b)

**FIGURE 24.7**
(a) Regression tree built using the Pima Indian diabetes training data. (b) The pruned regression tree based on the complexity parameter of 0.015.

Based on the above output, the predictors for number of times the subject was pregnant $(X_1)$, diastolic blood pressure $(X_3)$, and triceps skin fold thickness $(X_4)$ are not included. The $R^2$ is also only 0.4046, which is not very large. Using the test data, the classification table based on this MARS model is as follows:

```
############################
        truth
predicted    No    Yes
     No   0.5783 0.1596
     Yes  0.0934 0.1687
############################
```

Therefore, the misclassification rate is 0.2530, which is slightly worse than the pruned regression tree fit from above. Curves for the partial components of the fitted MARS model are given in Figure 24.8. For each figure, the curve is estimated while holding the other three predictors at their (approximate) mean values. For example, we see a general increase in the probability of being diagnosed with diabetes as a patient's glucose and pedigree increase, whereas the probability, surprisingly, decreases past a BMI of about 42 and an age of about 52.

Finally, we fit a single hidden-layer and double hidden-layer neural network model to these data. Dependency graphs of these neural networks with the estimated weights are given in Figure 24.9. Clearly greater complexity is being modeled with the double hidden-layer neural network in Figure 24.9(b). Using the test data, the classification table based on the single hidden-layer neural network fit is as follows:

```
############################
        truth
predicted    No    Yes
     No   0.5964 0.1325
     Yes  0.0753 0.1958
############################
```

Therefore, the misclassification rate is 0.2078. This is noticeably lower than the misclassification rates from the other methods discussed. The classification table based on double hidden-layer neural network fit is as follows:

```
############################
        truth
predicted    No    Yes
     No   0.5934 0.1265
     Yes  0.0783 0.2018
############################
```

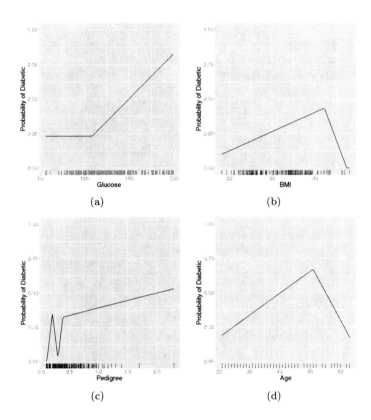

**FIGURE 24.8**
Curves for the partial components of the fitted MARS model for the predictors of (a) glucose, (b) BMI, (c) pedigree, and (d) age, where the other three predictors are held at (approximately) their mean values.

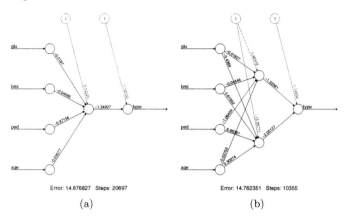

Error: 14.876827  Steps: 20697                Error: 14.762351  Steps: 10355

(a)                                                        (b)

**FIGURE 24.9**
(a) The fitted single hidden-layer neural network model and (b) the fitted double hidden-layer neural network model to the Pima Indian diabetes training data.

Therefore, the misclassification rate is 0.2048, which is comparable to what was obtained with the single hidden-layer neural network. Overall, the neural network approach provided the lowest misclassification rates out of all of the approaches we considered.

**Example 24.5.3.** *(Boston Housing Data, cont'd)*
We return to the Boston housing data analyzed in Example 23.7.3. As noted earlier, Harrison and Rubinfeld (1978) performed a thorough study to develop a meaningful theoretical model and Aldrin et al. (1993) used projection pursuit regression to achieve good predictions for these data. We first apply various ensemble learning methods for the purpose of regression and compare the predictions between these methods.

Without applying any transformations on the original data, we consider all of the original 13 predictors for our model. We fit the regular multiple linear regression model to these data, as well as grow a random forest, fit a boosted regression model, and perform bagging for regression trees. Figure 24.10 is a matrix of scatterplots comparing the predictions from these methods to the actual response of median home value from the Boston housing data. Based on these results, the random forest approach appears to provide the best predictions. But overall, the ensemble methods demonstrate fairly good agreement and especially provide better predictions than those produced by the multiple linear regression fit.

Next, we only consider the predictors of per capita crime rate by town ($X_1$), the average number of rooms per dwelling ($X_6$), and the full-value property-tax rate per \$10,000 ($X_{10}$). We proceed to implement PRIM, where we set

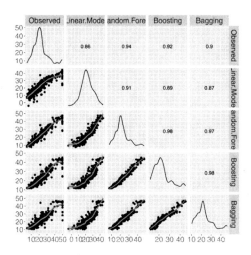

**FIGURE 24.10**
Matrix of scatterplots of the Boston housing data fit using a linear model and
various ensemble learning methods for regression.

each of the quantiles to determine the amount of peeling and pasting at 0.01.
The results are given below:

```
###############################
        box-fun box-mass threshold.type
box1       27.9    0.484              1
box2*      17.4    0.516             NA
overall    22.5    1.000             NA

Box limits for box1
     crim   rm tax
min -8.89 6.13 135
max  6.54 9.30 700

Box limits for box2
     crim   rm tax
min -8.89 3.04 135
max 97.87 9.30 763
###############################
```

The above shows that two boxes result after implementing PRIM. The first
part of the output gives a summary of the mean in each box and the mass
of observations in each box. The second part of the output gives the range of
each predictor. The first box, for example, contains 48.4% of the towns where
the average median home price is $27,900. This region has the bulk of the

homes with high median prices and is described in terms of per capita crime rates between 0 and 6.54 (where we have set the lower value to 0 since it is a rate), the average number of rooms per dwelling is between 6 and 9, and the full-value property-tax rate is between $135 and $700 per $10,000 of the property value. The second box, which has an asterisk, is comprised of all the additional observations not processed by PRIM.

# 25

## Miscellaneous Topics

This chapter presents a collection of many additional topics in regression. We present a much higher-level overview of these topics compared to the previous topics in this handbook. By no means are we minimizing the utility of these topics. In fact, most of the topics have books or monographs devoted specifically to them. We have supplemented each discussion with at least one major reference to the respective topic.

## 25.1 Multilevel Regression Models

Assume that we have data from $J$ classes with a different number of subjects, $n_j$, from each of the $j = 1, \ldots, J$ classes. Thus, the units of analysis are the subjects, which are nested within a higher-level of aggregated units. We can model such hierarchical data using a **multilevel regression models**,[1] which enable us to model a single response variable (measured at the lowest level) to predictor variables that occur at all levels in the hierarchy. While technically there can be many levels in the hierarchical structure of the data, we simply present two levels — a lower level (which we call *Level 1*) and a higher level (which we call *Level 2*) — to illustrate how to construct multilevel regression models. There are a number of very comprehensive and lucid texts on multilevel regression models, including those by Raudenbush and Bryk (2002), Gelman and Hill (2006), and Goldstein (2011).

Before formulating some common multilevel regression models, let us first present some notation corresponding to the data in this setup. Let

$$
\mathbf{Y}_j = \begin{pmatrix} Y_1 \\ \vdots \\ Y_{n_j} \end{pmatrix} \quad \text{and} \quad \mathbf{X}_j = \begin{pmatrix} X_{1,1} & X_{1,2} & \cdots & X_{1,p} \\ \vdots & \vdots & \ddots & \vdots \\ X_{n_j,1} & X_{n_j,2} & \cdots & X_{n_j,p} \end{pmatrix}
$$

---

[1] In the literature, multilevel regression models are referred to by a variety of different names. Some of these names include **hierarchical regression models, nested effects regression models, random coefficients regression models**, and **variance components regression models**. With the exception of the usage of hierarchical regression, there are some slight differences with how the other models are sometimes presented. We do not discuss such differences here, but simply note that all of these models are basically characterizing similar hierarchical data structures as we present in our discussion.

represent, respectively, the response vector and the predictor matrix of $p$ predictors from class $j$, for $j = 1, \ldots, J$. These are the measurements taken at Level 1 of the hierarchy. Next, let

$$
\mathbf{Z} = \begin{pmatrix} Z_{1,1} & Z_{1,2} & \cdots & Z_{1,q} \\ \vdots & \vdots & \ddots & \vdots \\ Z_{J,1} & Z_{J,2} & \cdots & Z_{J,q} \end{pmatrix}
$$

be a $J \times q$ matrix of predictors measured at Level 2 of the hierarchy. In other words, this is a set of $q$ predictors that are measured for each class at the higher level of the hierarchy. In order to simplify notation in the models that we present, the first column of $\mathbf{Z}$ could be all 1s to accommodate an intercept term in the Level 2 regression models.

With the notation for the data established, we now present the four basic multilevel regression models that are discussed in Gelman and Hill (2006):

- The **varying-intercept model with no predictors** is given by

$$
\mathbf{Y}_j = \alpha_j \mathbf{1}_{n_j} + \boldsymbol{\epsilon}_j,
$$

where $\alpha_j$ is the intercept for class $j$ and $\boldsymbol{\epsilon}_j$ is an $n_j$-dimensional vector of error terms for those measurements in class $j$. This Level 1 regression simply indicates that we have a constant term that is allowed to vary for each class. The error terms are assumed to be normally distributed with mean 0 and variance $\sigma_\epsilon^2$. One could also model the error terms such that each class has its own variance, but we forego adding that additional complexity in our setup. Letting $\boldsymbol{\alpha} = (\alpha_1, \ldots, \alpha_J)^{\mathrm{T}}$, the hierarchical structure appears through the following Level 2 regression:

$$
\boldsymbol{\alpha} = \gamma \mathbf{1}_J + \mathbf{u},
$$

where $\gamma$ is the overall mean of the different varying-intercepts and $\mathbf{u}$ is a $J$-dimensional vector of error terms that quantify the deviation of each intercept from their overall mean. Again, the entries in the $\mathbf{u}$ vector are assumed to be normally distributed with mean 0 and variance $\sigma_u^2$. Moreover, the errors at the two levels of the multilevel regression model, $\mathbf{u}$ and $\boldsymbol{\epsilon}_j$, are assumed to be independent. As noted in Gelman and Hill (2006), this is not a particularly interesting model; however, it serves as a good foundation for presenting the more intricate multilevel regression models.

- The **varying-intercept model** takes the previous model and incorporates the predictors $\mathbf{X}_j$. The Level 1 regression model is given by

$$
\mathbf{Y}_j = \alpha_j \mathbf{1}_{n_j} + \mathbf{X}_j \boldsymbol{\beta} + \boldsymbol{\epsilon}_j,
$$

where now $\boldsymbol{\beta}$ is the vector of slopes for the $p$ predictor variables, which will be the same vector for all of the $J$ groups. We further assume that a set of higher-level predictors are available for each $j$; i.e., those defined in the predictor matrix $\mathbf{Z}$. Thus, the Level 2 regression model on the intercepts is:

$$\boldsymbol{\alpha} = \mathbf{Z}\boldsymbol{\gamma} + \mathbf{u},$$

where now $\boldsymbol{\gamma}$ is a $q$-dimensional vector of Level 2 regression coefficients. Thus, this multilevel regression model says we have random intercepts which are allowed to vary across the groups, but those intercepts are further relating the response variable at Level 1 to an aggregated Level 2 predictor.

- The **varying-slope model** has the Level 1 regression model

$$\mathbf{Y}_j = \alpha\mathbf{1}_{n_j} + \mathbf{X}_j\boldsymbol{\beta}_j + \boldsymbol{\epsilon}_j,$$

where now $\boldsymbol{\beta}_j$ is the vector of slopes for the $p$ predictor variables in group $j$, but a single intercept $\alpha$ is the same for all of the $J$ groups. We again assume that $\mathbf{Z}$ is a set of higher-level predictors that are available. Let $\boldsymbol{\beta}_k^* = (\beta_{k,1}, \ldots, \beta_{k,J})^\mathrm{T}$ be the slopes for each of the $J$ classes for predictor $X_k$, $k = 1, \ldots, p$. The Level 2 regression model can be characterized on these slope vectors as follows:

$$\boldsymbol{\beta}_k^* = \mathbf{Z}\boldsymbol{\eta}_k + \mathbf{v}_k,$$

where now $\boldsymbol{\eta}_k$ is a $q$-dimensional vector of Level 2 regression coefficients characterizing the relationship between the slopes and the Level 2 predictors in $\mathbf{Z}$ and $\mathbf{v}_k$ is a $J$-dimensional vector of error terms that quantify the deviation of each slope from their overall mean. The entries in the $\mathbf{v}_k$ vectors are assumed to be normally distributed with mean 0 and variance $\sigma_v^2$. This multilevel regression model says we have random slopes which are allowed to vary across the groups.

- The **varying-intercept, varying-slope model** combines the features of the two previous multilevel regression models. It has the Level 1 regression model

$$\mathbf{Y}_j = \alpha_j\mathbf{1}_{n_j} + \mathbf{X}_j\boldsymbol{\beta}_j + \boldsymbol{\epsilon}_j,$$

where all of the components of this model are as defined in the previous two models. Then, the Level 2 regression model is characterized as follows:

$$\boldsymbol{\alpha} = \mathbf{Z}\boldsymbol{\gamma} + \mathbf{u},$$
$$\boldsymbol{\beta}_k^* = \mathbf{Z}\boldsymbol{\eta}_k + \mathbf{v}_k.$$

All of the error terms in the Level 1 and Level 2 regression models are assumed to be independent. This multilevel regression model says we have random intercepts and random slopes which are allowed to vary across the groups.

The four models presented above are essentially the basic multilevel regression models with standard assumptions. One could, of course, consider more complex variants on these models. For example, one might need to consider a dependency structure between error terms or perhaps consider nonlinear functional forms for either the Level 1 regression, the Level 2 regression, or both.

Estimation of multilevel regression models can be accomplished using least squares or maximum likelihood estimation. In the case of the errors being independent and with equal variance — as we assumed for our discussion — then these estimates will be the same. Depending on the complexity of the multilevel regression model's structure, other estimation approaches include iteratively reweighted maximum marginal likelihood methods, the use of EM algorithms, and empirical or hierarchical Bayes estimation methods. The general structure of multilevel regression models allows for estimation to be easily performed in a Bayesian framework, which is an approach emphasized in Gelman and Hill (2006). We do not discuss this approach here, but we give a brief introduction to Bayesian regression later in this chapter.

**Example 25.1.1.** *(Radon Data)*
Price et al. (1996) conducted a statistical analysis to identify areas with higher-than-average indoor radon concentrations based on $n = 919$ owner-occupied homes in 85 Minnesota counties. The subset of variables we use for our analysis are:

- $Y$: log radon measurements (in log picocurie per liter)

- $X$: an indicator for the floor level of the home where the measurement was taken, where 1 indicates the first floor and 0 indicates the basement (level 1 predictor)

- $W$: the county name (level 2 predictor)

By treating the indicator variable of floor level $(X)$ as the level 1 predictor and the county name factor variable $(Y)$ as the level 2 predictor, Gelman and Hill (2006) analyzed these data using various multilevel regression models. We follow some of their initial analysis of these data, but focus on fitting the basic multilevel regression models discussed in this section.

First we fit the varying-intercepts model where no predictor is present. In other words, we have one intercept for each of the $n$ counties, but no slope since no predictor is included. The output for this estimated model is as follows:

```
#############################
Random effects:
 Groups    Name          Variance Std.Dev.
 county    (Intercept)   0.09581  0.3095
 Residual                0.63662  0.7979
Number of obs: 919, groups:  county, 85

Fixed effects:
```

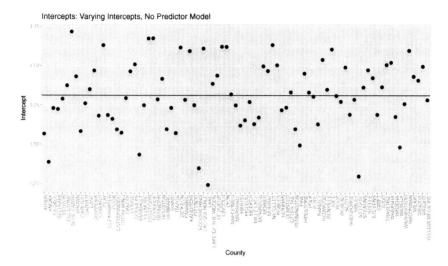

Intercepts: Varying Intercepts, No Predictor Model

**FIGURE 25.1**
Intercepts for each county from the varying intercepts, no predictor model for
the radon data.

```
              Estimate Std. Error t value
(Intercept)   1.31258     0.04891   26.84
#############################
```

While not very interesting, these results do allow us to see how the intercepts
vary by county. A scatterplot of these random intercepts is given in Figure
25.1.

Next we fit the varying-intercepts model with the level 2 predictor present.
In other words, we have one intercept for each of the $n$ counties, a single slope
due to the predictor of the floor indicator. The output for this estimated model
is as follows:

```
#############################
Random effects:
 Groups    Name         Variance Std.Dev.
 county    (Intercept)  0.1077   0.3282
 Residual               0.5709   0.7556
Number of obs: 919, groups:  county, 85

Fixed effects:
             Estimate Std. Error t value
(Intercept)  1.46160     0.05158  28.339
floor       -0.69299     0.07043  -9.839
```

```
Correlation of Fixed Effects:
     (Intr)
floor -0.288
##############################
```

The top part of the output gives the variance estimates due to the random effects, which in this case is the random intercepts. The middle part of the output gives the inference about the intercept and slope terms for the models, averaged over the counties. Both of these are highly significant at reasonable significance levels given the large $t$-values. The bottom part of the output gives the expected correlation of the estimated slope and intercept term. It is roughly a measure of collinearity. In this case, it would mean that in running this experiment again, we would expect, on average, the slope to decrease as the intercept increases. A plot of the radon data from 9 of the 85 Minnesota counties is given in Figure 25.2. The estimated multilevel regression lines with random intercepts are given by the solid black lines.

Next we fit the varying-slopes model with the level 2 predictor present. In other words, we have one intercept for all of the $n$ counties, and a slope due to the predictor of the floor indicator for each of the counties. The output for this estimated model is as follows:

```
##############################
Random effects:
 Groups     Name   Variance Std.Dev.
 county    floor 0.1154    0.3396
 Residual         0.6586    0.8115
Number of obs: 919, groups:  county, 85

Fixed effects:
            Estimate Std. Error t value
(Intercept)  1.32674    0.02932   45.25
floor       -0.55462    0.08920   -6.22

Correlation of Fixed Effects:
     (Intr)
floor -0.329
##############################
```

The above output is structured like the previous output. However, the top part for the variance estimates due to the random effects is now in terms of the random slopes. Interpretations for the rest of these results are analogous to the previous output. The estimated multilevel regression lines with random slopes are given by the dashed red lines in Figure 25.2.

Finally we fit the varying-intercepts, varying-slopes model with the level 2 predictor present. In other words, we have an intercept and a slope for each of the $n$ counties. The output for this estimated model is as follows:

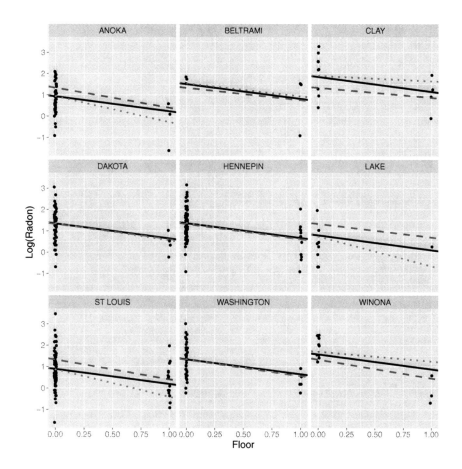

**FIGURE 25.2**
Scatterplots of the log radon measurements versus floor level of the measurements from 9 of the 85 Minnesota counties with multilevel regression lines for the varying-intercepts model (solid black lines), varying-slopes model (dashed red lines) and varying-intercepts, varying-slopes model (dotted green lines). Colors are in e-book version.

```
#############################
Random effects:
  Groups    Name           Variance Std.Dev. Corr
  county    (Intercept) 0.1216    0.3487
            floor          0.1181    0.3436  -0.34
  Residual                 0.5567    0.7462
Number of obs: 919, groups:  county, 85

Fixed effects:
             Estimate Std. Error t value
(Intercept)  1.46277    0.05387   27.155
floor       -0.68110    0.08758   -7.777

Correlation of Fixed Effects:
      (Intr)
floor -0.381
#############################
```

The above output is structured like the previous output. However, the top part for the variance estimates due to the random effects is now in terms of both the random intercepts and the random slopes. Interpretations for these results are analogous to the previous output. The estimated multilevel regression lines with random intercepts and random slopes are given by the dotted green lines in Figure 25.2.

## 25.2 Functional Linear Regression Analysis

**Functional data** are observations that can be treated as functions rather than univariate or multivariate quantities. Such data provide us with information about shapes like curves, surfaces, or any other type of measurement varying over a continuum. Longitudinal data can be considered a type of functional data, where repeated measurements are taken over time on the same sample; e.g., blood pressure readings, wind speed readings, and psychological testing results.

Regression models where we have functional responses and functional predictors can also be developed. The mathematical details, however, are more advanced than what we are able to present here; see Yao et al. (2005) for the details. **Functional linear regression models** are of the form

$$Y(t) = \alpha(t) + \int_{\mathcal{S}} \beta(s,t)\underline{X}(s)ds + \epsilon(t), \qquad (25.1)$$

where $Y(t)$, $\underline{X}(t)$, and $\epsilon(t)$ represent the functional response, vector of functional predictors, and the error process, respectively. $\beta(s,t)$ is the bivariate regression function representing the functional regression coefficients. It is smooth and square integrable, which means $\int_{\mathcal{T}} \int_{\mathcal{S}} \beta^2(s,t) ds dt < \infty$ for appropriately defined spaces $\mathcal{S}$ and $\mathcal{T}$.

An alternative to the functional linear regression model in (25.1) is the **pointwise regression model**

$$Y(t) = \alpha(t) + \beta(t)\underline{X}(t) + \epsilon(t), \qquad (25.2)$$

where the regression function only depends on $t$. Observations used to estimate models like (25.1) and (25.2) are trajectories or functions, and are observed at points $t_1, \ldots, t_k$ in time, where $k$ is large. In other words, we are trying to fit a regression surface to a collection of functions.

Estimation of $\beta(t)$ is beyond the scope of this discussion as it requires knowledge of Fourier transformations, functional principal components, and more advance multivariate techniques. Furthermore, estimation of $\beta(t)$ is intrinsically an infinite-dimensional problem. However, estimates found in the literature have been shown to possess desirable properties of an estimator. There are also inference procedures available for such models as presented in Yao et al. (2005).

We can also consider models with measurement errors in a functional linear regression framework using what is called **covariate-adjusted regression** (Sentürk and Müller, 2005). Covariate-adjusted regression models assume that any predictors observed are actually contaminated versions of their true value. When estimating in the presence of measurement errors, we traditionally assume that there is an additive error to each observed predictor value as with the models presented in Chapter 11. However, covariate-adjusted regression can include, and is often discussed from the perspective of, multiplicative errors on both the predictors *and* the response. An example where a covariate-adjusted regression model has been applied was given in Sentürk and Müller (2006) and involves serum creatinine levels that were regressed on cholesterol levels and serum albumin. The observed response and the two predictors are known to depend on body mass index (BMI), which has a confounding effect on the regression relationship. Normalization by weight or BMI is common in medical data, so this implicitly implies that the confounding is of a multiplicative nature.

Suppose that the measured values of the response $Y$ and the $p$ predictors $X_1, \ldots, X_p$ are contaminated with a multiplicative error. Suppose that this multiplicative error is characterized as an unknown function of $U$. Thus, we would actually observe the distorted versions

$$\tilde{X}_j = \phi_j(U)X_j$$

and

$$\tilde{Y} = \psi(U)Y,$$

where $\phi_j(\cdot)$ and $\psi(\cdot)$ are unknown, smooth functions — called **distorting**

**functions** — of the contaminating covariate $U$. Then, the covariate-adjusted regression model can be written as

$$\tilde{\mathbf{Y}} = \underline{\tilde{X}}^{\mathrm{T}} \boldsymbol{\beta} + \boldsymbol{\epsilon},$$

where the $(X_j, U, \epsilon)$ are mutually independent. The class of covariate-adjusted regression models is quite general and flexible depending on how the distorting functions are defined. Estimation of covariate-adjusted regression models falls within a similar framework as estimation of functional linear regression models and, thus, go beyond the level of mathematical rigor presented in this book.

More on functional regression analysis — and, more generally, functional data analysis — can be found in the text by Ramsay and Silverman (2005). The text by Ramsay and Silverman (2002) is also devoted to functional data analysis, but where all concepts are presented through case studies and only high-level mathematical details are included.

**Example 25.2.1.** *(Gait Data)*
Olshen et al. (1989) presented and analyzed data collected at the Motion Analysis Laboratory at Children's Hospital in San Diego. The data consists of certain angular measurements based on the gait of $n = 39$ children. The angles are measured over 20 time points across one gait cycle. The measurements in this experiment are:

- $Y$: angle made by the knee of the child during the gait cycle

- $X$: angle made by the hip of the child during the gait cycle

These data were thoroughly analyzed using various functional data models in Ramsay and Silverman (2005). We will proceed to analyze these data using a concurrent functional regression model.

Figure 25.3(a) gives the observed predictor function of the hip angle. Treating the functional relationship between hip and gait time as a predictor in our model is clearly appropriate. Similarly, Figure 25.3(b) gives the observed response function of the knee angle. Again, it is clearly appropriate that the response is a function. Figure 25.4 is a 3D plot of the knee angle versus hip angle and time in the gait cycle. From this figure, we see the general path of the gait of the child in terms of the angular measurements on their knee and hip.

We create a Fourier basis for the functional data with 21 basis functions and smooth the data. These smoothed values are then used to fit the concurrent functional regression model. The smoothed predicted functional responses are plotted in Figure 25.5. The observed data (of which there are 20 points per child) is overlaid on this plot as well. Overall, these fitted functional regression curves look appropriate for the data.

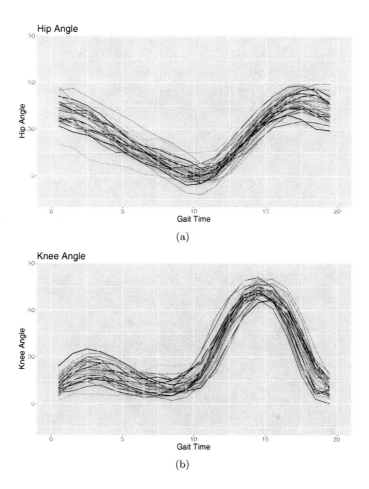

**FIGURE 25.3**
(a) Observed predictor function of hip angle for the gait data. (b) Observed response function of knee angle for the gait data.

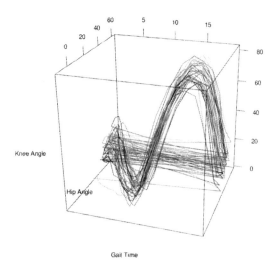

**FIGURE 25.4**
3D plot of the functional relationship between knee angle and the predictor of hip angle at the normalized gait times.

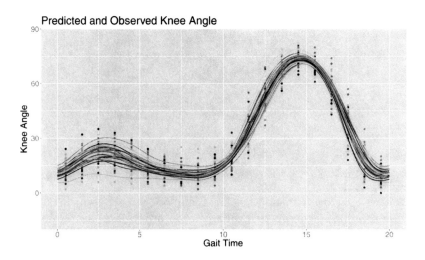

**FIGURE 25.5**
Predicted functions of knee angles for the gait data. Observed values are overlaid as solid points.

## 25.3 Regression Depth

We have discussed the notion of order statistics, such as in the context of resistant regression methods of Chapter 12. The applications we have presented with ordered data have all inherently been univariate data problems. Ordering of multivariate data is a more complex problem, however, one technique uses **statistical depth functions**, which provide a center-outward ordering of multivariate observations. Statistical depth functions allow one to define reasonable analogues of univariate order statistics by utilizing various geometrical notions of how *deep* a data point is relative to the rest of the multivariate data. There are numerous depth functions available in the literature. We do not discuss these depth functions here, but a good general overview of some common depth functions is found in Serfling (2006). The notion of statistical depth is also available in the regression setting. Specifically, **regression depth**, due to Rousseeuw and Hubert (1999), is a quality measure for robust linear regression.

Statistically speaking, the regression depth of a hyperplane $\mathcal{H}$ is the smallest number of residuals that need to change sign to make $\mathcal{H}$ a **nonfit**.[2] A regression hyperplane is called a nonfit if it can be rotated to vertical (i.e., parallel to the axis of any of the dependent variables) without passing through any data points. Included in this count are any points that lie exactly on the hyperplane. A nonfit is a very poor regression hyperplane, because it is combinatorially equivalent to a vertical hyperplane, which posits no relationship between independent and dependent variables. The regression depth of a hyperplane (say, $\mathcal{L}$) is the minimum number of points whose removal makes $\mathcal{H}$ into a nonfit. For example, consider the simulated data of size $n = 10$ in Figure 25.6. Removing the blue triangles and rotating the regression line until it is horizontal (i.e., the dashed blue line) demonstrates that the black line has regression depth of $\mathcal{L} = 3$.[3] Hyperplanes with high regression depth behave well in general error models, including skewed errors or distributions with heteroskedastic errors.

The regression depth of $n$ points in $p$ dimensions (i.e., number of regression coefficients) has an upper bound of $\lceil n/(p+1) \rceil$. In other words, there exist point sets for which no hyperplane has regression depth larger than this bound. For simple linear regression, like the example in Figure 25.6, this means there is always a line with regression depth of at least $\lceil n/3 \rceil$.

When confronted with outliers, then you may be confronted with the choice of other regression lines or hyperplanes to consider for your data. Some of these regressions may be biased or altered from the traditional ordinary least

---

[2]This definition also has convenient statistical properties, such as invariance under affine transformations, which we do not discuss in greater detail.

[3]Colors are in e-book version.

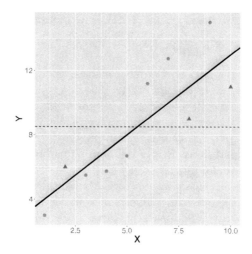

**FIGURE 25.6**
A dataset showing how the regression depth is determined.

squares line. In such cases, regression depth can help provide a measure of a fitted line that best captures the effects due to outliers.

**Example 25.3.1.** *(Animal Weight Data)*
Rousseeuw and Leroy (2003) presented and analyzed data on $n = 28$ species of animals. For the purpose of illustrating robust regression, they analyzed the relationship between the following variables:

- $Y$: brain weight (kg)

- $X$: body weight (kg)

The data are transformed on the $\log_{10}$ scale before being analyzed.

There are three observations that appear to be possible outliers and/or influential. A scatterplot of the data along with the simple linear regression fit and the deepest regression fit are presented in Figure 25.7. Clearly there is some influence of those three observations on the ordinary least squares fit, which do not influence the deepest regression fit. The intercept and slope estimates from the ordinary least squares fit are $(1.110, 0.496)$, while the corresponding estimates from the deepest regression fit are $(0.981, 0.703)$. This fit has a regression depth of $\mathcal{L} = 11$.

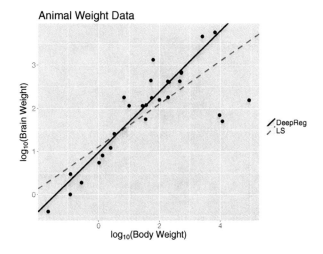

**FIGURE 25.7**
A scatterplot of the animal weight data with the ordinary least squares fit (solid black line) and deepest regression line (dashed red line) overlaid. Colors are in e-book version.

## 25.4 Mediation and Moderation Regression

Consider the following research questions:

- Will the increase in technology-related jobs cause demographic shifts in major metropolitan cities?

- Do changes in dietary guidelines affect obesity rates in children?

- Does trauma affect brain stem activation in a way that inhibits memory?

Such questions suggest a chain of relations where a predictor variable affects another variable, which then affects the response variable. A **mediation regression model** attempts to identify and characterize the mechanism that underlies an observed relationship between an independent variable and a dependent variable via the inclusion of a third explanatory variable called a **mediator variable**.

Instead of modeling a direct, causal relationship between the independent and dependent variables, a mediation model hypothesizes that the independent variable causes the mediator variable which, in turn, causes the dependent variable. Mediation models are heavily utilized in the area of psychometrics (MacKinnon et al., 2007), while other scientific disciplines (including Statistics) have criticized the methodology. One such criticism is that sometimes

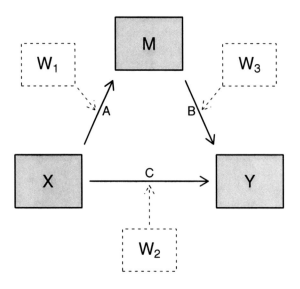

**FIGURE 25.8**
Diagram showing the basic flow of a mediation regression model and possible
moderator variables, $W_1$, $W_2$, and $W_3$. $X$, $Y$, and $M$ are the independent,
dependent, and mediator variables, respectively.

the roles of the mediator variable and the dependent variable can be switched
and yield a model which explains the data equally well, thus causing identifiability issues (Robins and Greenland, 1992). We present the classical model
for mediation regression due to Baron and Kenny (1986), which simply has
one independent variable, one dependent variable, and one mediator variable
in the model. Extensions with more than one of any of these variables can
also be constructed.

The following three regression models are used in our discussion:

1. $Y = \beta_0 + \beta_1 X + \epsilon$

2. $M = \theta_0 + \theta_1 X + \gamma$

3. $Y = \alpha_0 + \alpha_1 X + \alpha_2 M + \delta.$

Figure 25.8 gives a diagram showing these general relationships, which we
reference in the discussion below. The first model is the classic simple linear
regression model, which is the relationship between $X$ and $Y$ that we typically
wish to study. In causal analysis, this is written as $X \rightarrow Y$ and follows path

$C$ in the diagram. The second and third models show how we incorporate the mediator variable into this framework so that $X$ causes the mediator $M$ and $M$ causes $Y$; i.e., $X \rightarrow M \rightarrow Y$. The first segment of this relationship follows path $A$ in the diagram and then path $B$. So $\alpha_1$ is the coefficient relating $X$ to $Y$ adjusted for $M$, $\alpha_2$ is the coefficient relating $M$ to $Y$ adjusted for $X$, $\theta_1$ is the coefficient relating $X$ to $M$, and $\epsilon$, $\gamma$, and $\delta$ are error terms for the three relationships.

By rearranging the three models above, we see that the mediated effect can be estimated by either $\hat{\theta}_1 \hat{\alpha}_2$ or $\hat{\beta}_1 - \hat{\alpha}_1$. These estimates can be obtained via traditional estimation procedures, like ordinary least squares and maximum likelihood. The test statistic for each quantity is

$$T^* = \frac{\hat{\theta}_1 \hat{\alpha}_2}{\hat{S}} \quad \text{and} \quad T^* = \frac{\hat{\beta}_1 - \hat{\alpha}_1}{\hat{S}},$$

where

$$\hat{S} = \sqrt{\hat{\theta}_1^2 \widehat{\text{Var}}(\hat{\alpha}_2) + \hat{\alpha}_2^2 \widehat{\text{Var}}(\hat{\theta}_1)}$$

is the pooled standard error term. These test statistics are then compared to a normal distribution and, thus, $p$-values and confidence intervals are easily obtained for the mediation effect. This testing is formally known as **Sobel's test** (Sobel, 1982).

The strength and form of mediated effects may depend on, yet, another variable. These variables, which affect the hypothesized relationship amongst the variables already in our model, are called **moderator variables** and are often characterized as an interaction effect. Now consider the following regression equations as formulated in Muller et al. (2005):

4.  $Y = \xi_0 + \xi_1 X + \xi_2 W + \xi_3 XW + \upsilon$
5.  $M = \vartheta_0 + \vartheta_1 X + \vartheta_2 W + \vartheta_3 XW + \eta$
6.  $Y = \phi_0 + \phi_1 X + \phi_2 W + \phi_3 XW + \phi_4 M + \phi_5 MW + \varphi.$

These three models above are for **moderated mediation**, which means there is the effect of a mediator variable $M$ on the relationship between $X$ and $Y$, but $M$ depends on the level of the moderator variable $W$. Thus, the mediational mechanism differs for subgroups (or subregions) of the study. More specifically, Model 4 assesses moderation by $W$ of the effect of $X$ on $Y$, Model 5 assesses moderation by $W$ of the effect of $X$ on $M$, and Model 6 assesses moderation by $W$ of both the effects of $X$ on $Y$ and $M$ on $Y$. Moderated mediation occurs when statistical testing provides evidence of $\xi_3 = 0$, which indicates that there is an overall effect of $X$ on $Y$ that does not depend on $W$. If statistical testing provides evidence of $\xi_3 \neq 0$, then the process is called **mediated moderation**. While we used $W$ to denote the moderator variable in the models above, we use $W_1$, $W_2$, and $W_3$ in Figure 25.8 to illustrate the different locations where a moderator variable (or potentially multiple moderator variables) could be located.

Finally, there are two other types of variables worth defining. Consider the two linear regression models:

7.  $Y = \beta_1 + \beta_1 X + \epsilon$

8.  $Y = \psi_1 + \psi_1 X + \psi_2 Z + \omega.$

In these models, $X$ and $Z$ are both dependent variables. If the inclusion of $Z$ in the model strengthens the effect of $X$ on $Y$ (i.e., the magnitude of the effect measured by $\psi_1$ is much greater than that measured by $\beta_1$), then $Z$ is called a **suppressor variable**. In terms of extra sums of squares, this implies $\text{SSR}(X|Z) >> \text{SSR}(Z)$. However, if the relationship between $X$ and $Y$ is found to be highly significant, but inclusion of another variable $Z$ (which may not have been part of the research study and, hence, may not have been measured) indicates that conclusions made about the relationship between $X$ and $Y$ are erroneous, then $Z$ is called a **confounding variable**. In terms of extra sums of squares, this implies $\text{SSR}(Z) >> \text{SSR}(X|Z)$. Confounding variables can usually be controlled through properly-designed experiments, like those underlying the models discussed in Chapter 15. We also briefly alluded to the notion of confounding in Chapter 7.

MacKinnon (2008) is a very accessible introductory text on mediation analysis. Additional discussion distinguishing confounding and suppressor variables from mediation effects is given in MacKinnon et al. (2000). Several examples of real studies involving suppressor variables are given in Thompson and Levine (1997).

**Example 25.4.1.** *(Job Stress Data)*
Bliese and Halverson (1996) presented an analysis of job stress based on how individuals perceive and respond to their work environment. The variables from their study that we will investigate are as follows:

- $Y$: the respondent's numeric score about overall well-being at work

- $X_1$: the respondent's number of daily work hours

- $X_2$: a numeric score of the leadership climate at the respondent's place of employment

- $M$: a numeric score of the level of cohesion at the respondent's place of employment

The study consisted of $n = 7382$ workers from 99 groups. However, to illustrate moderated and mediated regression, we will not account for the grouping variable.

We first run a moderated regression analysis where we treat $X_1$ as the predictor that is mediated by the variable $M$. $X_2$ is treated as another predictor variable, one in which it is not mediated by $M$. This is identical to fitting a multiple linear regression model with an interaction term where the predictors have been centered. Below are the fitted results for these data:

```
#############################
Coefficients:
                Estimate Std. Error t value Pr(>|t|)
(Intercept)     2.777962   0.009522 291.731  < 2e-16 ***
HRS.c          -0.042700   0.004234 -10.086  < 2e-16 ***
COHES.c         0.072833   0.012107   6.016 1.88e-09 ***
LEAD.c          0.450032   0.013625  33.029  < 2e-16 ***
HRS.c:COHES.c  -0.009508   0.004572  -2.079   0.0376 *
---
Signif. codes:  0 *** 0.001 ** 0.01 * 0.05 . 0.1   1

Residual standard error: 0.8143 on 7377 degrees of freedom
Multiple R-squared:  0.1969,Adjusted R-squared:  0.1964
F-statistic:    452 on 4 and 7377 DF,  p-value: < 2.2e-16
#############################
```

In the above, the interaction is statistically significant. Thus, a worker's well-being tends to be adversely affected by a combined increase in hours worked and decrease in work environment cohesion.

We next fit a mediated regression model in the setup of Sobel's test for mediation. For this model, we do not include the $X_2$ score on leadership. Below are the estimates for the three models discussed earlier in this section:

```
#############################
$'Mod1: Y~X'
                Estimate    Std. Error    t value      Pr(>|t|)
(Intercept)   3.51693620   0.052902697   66.47934   0.000000e+00
pred         -0.06523285   0.004590274  -14.21110   3.078129e-45

$'Mod2: Y~X+M'
                Estimate    Std. Error    t value      Pr(>|t|)
(Intercept)   2.68753214   0.066005670   40.71669   0.000000e+00
pred         -0.05595744   0.004493628  -12.45262   3.053679e-35
med           0.23627275   0.011755705   20.09856   1.617929e-87

$'Mod3: M~X'
                Estimate    Std. Error    t value      Pr(>|t|)
(Intercept)   3.51036694   0.051010192  68.816971   0.000000e+00
pred         -0.03925723   0.004426065  -8.869556   9.091143e-19

$Indirect.Effect
[1] -0.009275413

$SE
[1] 0.001143062
```

```
$z.value
[1] -8.114535

$N
[1] 7382
##############################
```

In the above, the indirect mediating effect is -0.0093 with an estimated standard error of 0.0011. Therefore, a 95% confidence interval is, roughly $-0.0093 \pm (1.9600 \times 0.0011)$, or $(-0.0115, -0.0071)$, indicating that the cohesion level does have a significant mediating effect on the relationship between the respondent's overall well-being at work and their number of daily hours worked.

## 25.5   Meta-Regression Models

In Statistics, a **meta-analysis** combines the results of several scientific studies that address a set of related research hypotheses. This is normally accomplished by identification of a common measure of **effect size**, which is a descriptive statistic that quantifies the estimated magnitude of a relationship between variables without making any inherent assumption about if such a relationship in the sample reflects a true relationship for the population. In a meta-analysis, a weighted average from the results of the individual studies might be used as the output. While there are other differences between the studies that the researcher should try to characterize as best as possible, the general aim of a meta-analysis is to more powerfully estimate the true effect size as opposed to a smaller effect size derived in a single study under a given set of assumptions and conditions.

**Meta-regressions** are similar in essence to classic regressions, in which a response variable is predicted according to the values of one or more predictor variables. In meta-regression, the response variable is the effect estimate, for example, a mean difference, a risk difference, a log odds ratio, or a log risk ratio. The predictor variables are characteristics of studies that might influence the size of intervention effect, which are called **potential effect modifiers** or covariates. Meta-regressions differ from traditional linear regressions in two ways. First, meta-regressions incorporate weighting of the studies by the precision of their respective effect estimate since larger studies have more influence on the relationship than smaller studies. Second, meta-regressions typically allow for residual heterogeneity among intervention effects not modeled by the predictor variables. Meta-regressions have varied uses in areas such as clinical trials (Freemantle et al., 1999), econometrics (Stanley and Jarrell, 1989), and ecology (Gates, 2002).

The regression coefficient obtained from a meta-regression analysis will describe how the response variable — the intervention effect — changes with a unit increase in the predictor variable – the potential effect modifier or moderator variables. The statistical significance of the regression coefficient is a test of whether there is a linear relationship between intervention effect and the predictor variable. If the intervention effect is a ratio measure, the log-transformed value of the intervention effect should always be used in the regression model, and the exponential of the regression coefficient will give an estimate of the relative change in intervention effect with a unit increase in the predictor variable. More pragmatic details about meta-regression analyses are discussed in Thompson and Higgins (2002).

Let us first introduce some notation common to all of the meta-regression models that we will present. For $i = 1, \ldots, n$ studies, we let $Y_i$ denote the effect size in study $i$ and $\underline{X}_i$ be a $p$-dimensional vector for study $i$ composed of different characteristics measured across all $n$ studies. We also let $\boldsymbol{\beta}$ be a $p$-dimensional vector of regression coefficients as explained in the previous paragraph and $\alpha$ will be an intercept term that estimates the overall effect size. With this notation established, the three general types of meta-regression models found in the literature are as follows:

**Simple meta-regression:**

$$Y_i = \alpha + \underline{X}_i^{\mathrm{T}} \boldsymbol{\beta} + \epsilon,$$

$\epsilon$ specifies the between-study variation. Note that this model does not allow specification of within-study variation.

**Fixed-effect meta-regression:**

$$Y_i = \alpha + \underline{X}_i^{\mathrm{T}} \boldsymbol{\beta} + \omega_i,$$

where $\sigma_{\omega_i}^2$ is the variance of the effect size in study $i$.[4] In this model, we assume that the true effect size $\delta_\theta$ is distributed as $\mathcal{N}(\theta, \sigma_\theta^2)$, where $\sigma_\theta^2$ is the within-study variance of the effect size. Thus, this model allows for within-study variability, but not between-study variability because all studies have the identical expected fixed effect size $\delta_\theta$; i.e., $\epsilon = 0$. Moreover, if between-study variation cannot be ignored, then parameter estimates are biased.

**Random-effects meta-regression:**

$$Y_i = \alpha + \underline{X}_i^{\mathrm{T}} \boldsymbol{\beta} + \omega + \epsilon_i,$$

where $\sigma_{\epsilon_i}^2$ is the variance of the effect size in study $i$ and $\sigma_\omega^2$ is the between-study variance. In this model, we assume that $\theta$ in $\mathcal{N}(\theta, \sigma_\theta)$ is a random variable following a hyper-distribution $\mathcal{N}(\mu_\theta, \varsigma_\theta)$. Because of the random

---

[4]Note that for the "fixed-effect" model, no plural is used as in the experimental design models discussed in Chapter 15. This is because only *one* true effect across all studies is assumed.

component in this model, an estimate for $\sigma_\omega^2$ is typically found using estimation procedures for random-effects models, such as restricted maximum likelihood.

**Example 25.5.1.** *(Tuberculosis Vaccine Literature Data)*
Colditz et al. (1994) presented the data and meta-analysis from $n = 13$ studies on the effectiveness of the Bacillus Calmette-Guerin (BCG) vaccine against tuberculosis. The treatment is allocated in one of three ways: randomly, systematically assigned, or alternately. The variables that we will investigate are as follows:

- $Y$: the logarithm of the relative risk of developing tuberculosis in the vaccinated versus non-vaccinated groups

- $T_1$: number of tuberculosis positive cases in the treated (vaccinated) group

- $T_2$: number of tuberculosis negative cases in the treated (vaccinated) group

- $T_3$: number of tuberculosis positive cases in the control (non-vaccinated) group

- $T_4$: number of tuberculosis negative cases in the control (non-vaccinated) group

- $X_1$: a treatment allocation indicator where 1 means random allocation and 0 otherwise

- $X_2$: a treatment allocation indicator where 1 means systematically assigned and 0 otherwise

- $M_1$: publication year of the study

- $M_2$: absolute latitude of the study location (in degrees)

The variables $T_1$, $T_2$, $T_3$, and $T_4$ are not modeled directly, but rather used to calculate the relative risk $Y$. Various meta-regression models were also fit to these data by Viechtbauer (2010). We proceed to fit similar fixed-effect and random-effects meta-regression models here.

First we fit a fixed-effect meta-regression model where the relative risk is modeled as a function of the treatment allocation, which is characterized by the group indicator variables $X_1$ and $X_2$. A plot of the estimated relative risk versus the observed risk is given in Figure 25.9(a). The output from the fixed-effect meta-regression fit is below:

```
#############################
 Model Results:

               estimate    se      z    ci.l   ci.u     p
 intrcpt         -0.719  0.078 -9.211 -0.871 -0.566 0.000
```

```
allocrandom          0.424  0.095  4.453  0.237  0.611 0.000
allocsystematic      0.305  0.123  2.479  0.064  0.547 0.013
```

Heterogeneity & Fit:

```
           QE    QE.df      QEp       QM   QM.df QMp
[1,] 132.368   10.000    0.000   19.865   2.000   0
############################
```

The top-half of the output gives the summary of the estimated regression coefficients for this model. Clearly, based on the 95% confidence intervals and the $p$-values (which are the values given in the last three columns) all of these coefficients are significantly different from 0. Moreover, this indicates that the treatment allocation method is a significant predictor of the relative risk on the log scale.

The second-half of the output gives various measures about the measure of the error in the model and the model fit. The result of the test for residual heterogeneity is significant ($Q_E = 132.4$, df$_E = 10$, $p < 0.001$), suggesting that other moderator variables could be influencing the effectiveness of the vaccine. The result of the omnibus test for the regression coefficients is also significant ($Q_M = 19.9$, df$_E = 2$, $p < 0.001$), suggesting that at least one of the group indicator variables explains a significant amount of variability in the relative risk on the log scale.

Next we fit a random-effects meta-regression model where the relative risk is modeled as a function of the moderator variables of the study's publication year and the absolute latitude of the original study location, which are the variables $M_1$ and $M_2$, respectively. A plot of the estimated relative risk versus the observed risk is given in Figure 25.9(b). The output from the random-effects meta-regression fit is below:

```
############################
 Model Results:

         estimate      se       z     ci.l     ci.u      p
intrcpt    -3.546  29.096  -0.122  -60.572   53.481  0.903
year        0.002   0.015   0.130   -0.027    0.031  0.897
ablat      -0.028   0.010  -2.737   -0.048   -0.008  0.006
```

Heterogeneity & Fit:

```
          QE   QE.df      QEp      QM   QM.df    QMp
[1,]  28.325  10.000   0.002  12.204   2.000  0.002
############################
```

The output is summarized similar to the fixed-effect meta-regression fit. Based on the 95% confidence intervals and the $p$-values, the location of the study is

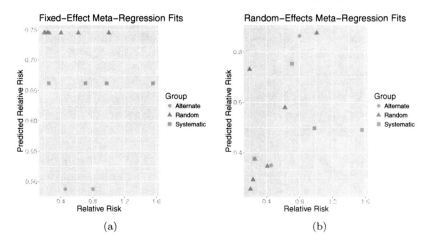

**FIGURE 25.9**
(a) Scatterplot of the predicted relative risk versus the observed relative risk
for the estimated fixed-effect meta-regression model. Different colors and plotting symbols are used for the different allocation methods. (b) Scatterplot of
the predicted relative risk versus the observed relative risk for the estimated
random-effects meta-regression model. Colors are in e-book version.

a significant predictor of the relative risk on the log scale, whereas the year
of the study is not a significant predictor. The result of the test for residual
heterogeneity is significant ($Q_E = 28.3$, $df_E = 10$, $p = 0.002$), suggesting that
other moderator variables could be influencing the effectiveness of the vaccine.
The result of the omnibus test for the regression coefficients is also significant
($Q_M = 12.2$, $df_E = 2$, $p < 0.002$), suggesting that at least one of the mediator
variables explains a significant amount of variability in the relative risk on the
log scale. Based on the individual tests, this is likely due to the location of
the study.

## 25.6  Regression Methods for Analyzing Survey Data

Statistical inference is often conducted using **model-based approaches**,
such as regression modeling, to test a particular hypothesis about the relationship between two variables. As we have seen, there are numerous assumptions
in order for the estimation of the particular regression model to be valid. Regression models (and statistical models in general) are probability models of
the random process used to generate the sample from the population of interest. In contrast, the analysis of data from complex surveys is treated using

**design-based approaches**, where we specify a population such that the data values are unknown, but treated as fixed, not random as in the model-based approach. The **sample design** is the random selection of individuals from the population of interest. Examples of traditional sample designs include **simple random sampling, stratified sampling, cluster sampling**, and **systematic sampling**. See the texts by Lohr (2010) and Thompson (2012) for a thorough overview of sampling designs and survey methods.

Common estimates obtained in survey analysis include population means and totals. It is usually necessary to also report standard errors for these estimates, especially in publications involving official statistics. In survey research, we are assuming to sample without replacement from a finite population of size $N$ according to some specified sample design. The standard error calculations need to reflect these assumptions, which usually involves multiplying the traditional standard error formula by a **finite population correction factor** or a **design effect** (Kish, 1965).

Even though complex surveys are design-based, we can still use regression modeling in the analysis stage of a survey. In survey applications, predictor variables are often called **auxiliary variables** and the response variable represents the quantity of interest, such as the number of people in a housing unit, the number of acres on a piece of farmland, or the number of robberies committed in a city. The auxiliary variables are correlated with the variable of interest and are used to improve the precision of estimators of the mean and total of a population. We briefly discuss some topics found at the intersection of regression methods and survey analysis.

**Survey Weighted Least Squares**

In the analysis of survey data, one can use a **working model** that is intended to reasonably describe the structure of the population of interest. For a sample of size $n$, one might consider the traditional linear regression model

$$\mathbf{Y} = \mathbf{X}\boldsymbol{\beta} + \boldsymbol{\epsilon}, \tag{25.3}$$

where $\mathrm{Var}(\boldsymbol{\epsilon}) = \sigma^2 \mathbf{I}_n$. Suppose, however, that our sample of size $n$ has a set of survey weights $w_1, \ldots, w_n$ and that the $\epsilon_i$ are independently normal with mean 0 and variance $\nu_i \sigma^2$, where the $\nu_i$ are known constants. Letting $\mathbf{V} = \mathrm{diag}(\nu_1, \ldots, \nu_n)$ and $\mathbf{W} = \mathrm{diag}(w_1, \ldots, w_n)$, the **survey weighted least squares estimator** is then

$$\hat{\boldsymbol{\beta}}_{\mathrm{SWLS}} = (\mathbf{X}^{\mathsf{T}}\mathbf{W}\mathbf{V}^{-1}\mathbf{X})^{-1}\mathbf{X}^{\mathsf{T}}\mathbf{W}\mathbf{V}^{-1}\mathbf{Y}, \tag{25.4}$$

where if we assume that $\mathbf{V} = \mathbf{I}_n$, then the above is the usual weighted least squares estimator. See Li and Valliant (2009) for more details on survey weighted least squares.

**Ratio Estimation and Linear Regression**

Following similar notation as that used in Chapter 4 of Lohr (2010), suppose that we have a population of size $N$ and that for each sample unit, we measure a quantity of interest $Y_i$ and some auxiliary variable $X_i$, for $i = 1, \ldots, N$. Then, their totals are

$$t_Y = \sum_{i=1}^{N} Y_i \quad \text{and} \quad t_X = \sum_{i=1}^{N} X_i$$

and their ratio is

$$B = \frac{t_Y}{t_X} = \frac{\bar{Y}_N}{\bar{X}_N}.$$

The simplest form of ratio estimation involves taking a simple random sample of size $n < N$ and computing the estimates for the above, respectively, as

$$\hat{t}_y = \sum_{i=1}^{n} y_i, \quad \hat{t}_x = \sum_{i=1}^{n} x_i, \quad \text{and} \quad \hat{B} = \frac{\hat{t}_y}{\hat{t}_x} = \frac{\bar{y}}{\bar{x}},$$

where $\hat{B}$ is the **ratio estimator**. For known levels of $t_X$ and $\bar{X}_N$, one can then obtain ratio-based estimators of the $t_Y$ and $\bar{Y}_N$:

$$\hat{t}_{y;\text{ratio}} = \hat{B} t_X \quad \text{and} \quad \hat{\bar{y}}_{\text{ratio}} = \hat{B} \bar{X}_N.$$

While at first one might question the availability of quantities like $t_X$ and $\bar{X}_N$, such quantities are usually assumed known from official statistics, like those produced by census agencies throughout the world.

Ratio estimation works best if the data are well fit by a regression through the origin model, but this is not always the case. Assuming a simple linear regression model relating the variable of interest to the auxiliary variable, the **regression estimators** of $t_Y$ and $\bar{Y}_N$ are given by

$$\hat{t}_{y;\text{reg}} = \hat{t}_y + \hat{\beta}_1(t_X - \hat{t}_x) \quad \text{and} \quad \hat{\bar{y}}_{\text{reg}} = \hat{\beta}_0 + \hat{\beta}_1 \bar{X} = \bar{y} + \hat{\beta}_1(\bar{X}_N - \bar{x}),$$

respectively, where $\hat{\beta}_0$ and $\hat{\beta}_1$ are the ordinary least squares estimates from regressing the $y$-values on the $x$-values. Note that the regression estimator above adjusts $\bar{y}$ by the quantity $\hat{\beta}_1(\bar{X}_N - \bar{x})$. Both the ratio estimator and the regression estimator of $\bar{Y}_N$ are biased. See Chapter 4 of Lohr (2010) for details on the bias of these estimators as well as how to obtain their respective standard errors.

### Generalized Regression Estimators

Suppose that we are interested in population totals for a variable, which is measured with $p$ auxiliary variables. Assume the multiple linear regression model in (25.3) as our working model, but where $\sigma^2$ is also known. Let $\mathbf{t_X}$ be the $p$-dimensional vector of true population totals for the $p$ auxiliary variables. $\mathbf{t_X}$ is assumed known so that it can be used to adjust $\hat{t}_y$. Letting $\Sigma_N = \text{diag}(\sigma_1^2, \ldots, \sigma_N^2) = \sigma^2 \text{diag}(\nu_1, \ldots, \nu_N)$, we can define the survey weighted least squares estimate of $\beta$ using (25.4) by:

$$\mathbf{B} = (\mathbf{X}_N^{\mathrm{T}} \Sigma_N^{-1} \mathbf{X}_N)^{-1} \mathbf{X}_N^{\mathrm{T}} \Sigma_N^{-1} \mathbf{Y}_N,$$

where $\mathbf{X}_N$ is an $N \times p$ matrix of the auxiliary variables and $\mathbf{Y}_N$ is an $N$-dimensional vector of the variable of interest. We can then estimate $\mathbf{B}$ by replacing $\mathbf{X}_N$, $\Sigma_N$, and $\mathbf{Y}_N$ with analogous quantities based on our sample of size $n < N$:

$$\hat{\mathbf{B}} = \left( \sum_{i=1}^{n} \frac{w_i}{\sigma_i^2} \mathbf{x}_i^{\mathrm{T}} \mathbf{x}_i \right)^{-1} \sum_{i=1}^{n} \frac{w_i}{\sigma_i^2} \mathbf{x}_i y_i.$$

Then, the **generalized regression estimator** of the population total is

$$\hat{t}_{y;\mathrm{GREG}} = \hat{t}_y + (\mathbf{t_X} - \hat{\mathbf{t}}_\mathbf{x})\hat{\mathbf{B}},$$

which is a regression adjustment of the famous **Horvitz–Thompson estimator** of the population total (Horvitz and Thompson, 1952). An important property of the generalized regression estimator is that for any choice of constants $\sigma_i^2$, it calibrates the sample to the population total for the $p$ auxiliary variables used in the regression.

## Small Area Models

In addition to quantities like population totals and population means, sample surveys can be used to provide estimates for a variety of subpopulations or domains. In the context of sample surveys, a domain estimator is said to be **direct** if it is based only on the domain-specific sample data, which may include a number of auxiliary variables. A domain (area) is regarded as *large* if the domain-specific sample is large enough to yield direct estimates of adequate precision. A domain is regarded as *small* if it is not large enough to yield direct estimates of adequate precision. Such domains are called **small areas**. A small area is usually a small geographic unit, such as a county or a city, but it can also refer to a particular demographic within an area, such as individuals living below the poverty line or children below the age of 10. Comprehensive treatment of both the theory and applications of small area estimation in sample surveys can be found in the texts by Mukhopadhyay (1998) and Rao (2003).

Following the notation and terminology in Chapter 5 of Rao (2003), we define two of the most common small area regression models used in practice. These models are essentially variants on mixed effects models, like we discussed in Chapter 15 and in the context of multilevel regression models earlier in this chapter. The first small area model is the **basic area level (type A) model**, which relates the small area means to area-specific auxiliary variables. Let $\bar{Y}_i$ be the mean of our variable of interest for small area $i$, $i = 1, \ldots, m$. Assume that $\theta_i = g(\bar{Y}_i)$ for some specified $g(\cdot)$, which is related to area-specific auxiliary data $\mathbf{z}_i = (z_{i,1}, \ldots, z_{i,p})^{\mathrm{T}}$ through the model

$$\theta_i = \mathbf{z}_i^{\mathrm{T}} \boldsymbol{\beta} + b_i \nu_i, \tag{25.5}$$

where the $b_i$s are known positive constants. Furthermore, the $\nu_i$s are area-specific random effects assumed to be independent and identically distributed with mean 0 and variance $\sigma_v^2$. For making inferences about the $\bar{Y}_i$s under (25.5), we assume that direct estimators $\hat{\bar{Y}}_i$ are available and that

$$\hat{\theta}_i = g(\hat{\bar{Y}}_i) = \theta_i + \epsilon_i, \qquad (25.6)$$

where the sampling errors $\epsilon_i$ are conditionally independent with mean 0 and known variances $\psi_i$. Finally, combining (25.5) with (25.6) then gives a special case of a linear mixed model known as the **Fay–Herriot model** (Fay and Herriot, 1979):

$$\hat{\theta}_i = \mathbf{z}_i^{\mathrm{T}} \boldsymbol{\beta} + b_i \nu_i + \epsilon_i,$$

where the $\nu_i$ and $\epsilon_i$ are independent. Fay and Herriot (1979) developed this model for estimating per capita income for small places in the United States with population less than 1000.

The second small area model is the **basic unit level (type B) model**, which relates the unit values of the variable of interest to unit-specific auxiliary variables. For this model, we assume that unit-specific auxiliary data — given by the $p$-dimensional vector $\mathbf{x}_{i,j}$ — are available for each population element $j$ in each small area $i$ and that only population means $\bar{X}_i$ are known. Then, the variable of interest $y_{i,j}$ is assumed to be related to the auxiliary variables through a nested area unit level regression model:

$$y_{i,j} = \mathbf{x}_{i,j}^{\mathrm{T}} \boldsymbol{\beta} + \nu_i + \epsilon_{i,j},$$

where the $\nu_i$s are again area-specific random effects assumed to be independent and identically distributed with mean 0 and variance $\sigma_v^2$, $\epsilon_{i,j} = c_{i,j} \tilde{\epsilon}_{i,j}$ with known constants $c_{i,j}$, and $\tilde{\epsilon}_{i,j}$ are independent and identically distributed random variables with mean 0 and variance $\sigma_e^2$. The $\tilde{\epsilon}_{i,j}$s are also independent of the $\nu_i$s. This model was first used by Battese et al. (1988) to model corn and soybean crop areas in Iowa.

**Example 25.6.1.** *(Census of Agriculture Data)*
Lohr (2010) analyzed data from the United States Census of Agriculture, which is conducted every five years. This census collects data on all farms in the United States from which $1000 or more of agricultural products were produced and sold. The variables that we will investigate, which are reported at the county level, are as follows:

- $Y$: the number of acres devoted to farms

- $X_1$: number of farms with 1000+ acres

- $X_2$: number of farms with $\leq 9$ acres

- $X_3$: total number of farms

As with most survey data, the variables of interest will often by highly correlated. In a regression model, this means multicollinearity will be issue. We

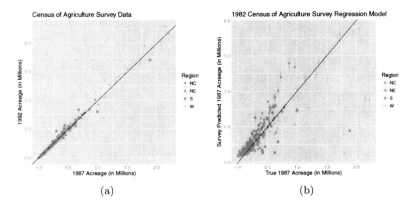

(a) (b)

**FIGURE 25.10**
(a) The plot of millions of acres from the Census of Agriculture data (1992 versus 1987) with a line going through the origin and slope $\hat{B}$. (b) Predicted values of farm acreage in 1987 using the survey-weighted regression model fit to the 1982 Census of Agriculture data plotted against the true values of farm acreage.

ignore such effects, but one can assess and handle multicollinearity in regression models for survey data using the strategies presented in, for example, Liao and Valliant (2012).

Lohr (2010) used data from a stratified sample of size $n = 300$ out of $N = 3078$ counties from the 1992 Census of Agriculture. Four strata are used based on the region of the United States where the farm is located: south, west, north central, and northeast. The variables discussed above are available for the censuses conducted in 1982, 1987, and 1992.

First we compute the ratio and regression estimators for the number of acres in 1992 (the variable of interest) based on the number of acres in 1987 (the auxiliary variable). According to Lohr (2010), a total of $t_X = 964,470,625$ acres were devoted to farms in the United States in 1987, resulting in $\bar{X}_N = t_X/N = 313,343$ acres of farm per county. Using the sample, we have $\hat{t}_x = 89,578,186$ and $\hat{t}_y = 88,683,801$, which gives $\hat{B} = 0.990$. The sample is plotted in Figure 25.10(a) with the the line going through the origin and having slope $\hat{B}$ overlaid. The ratio estimators are

$$\hat{t}_{y;\text{ratio}} = 954,840,958.466 \quad \text{and} \quad \hat{\bar{y}}_{\text{ratio}} = 310,214.736.$$

The respective regression estimators are

$$\hat{t}_{y;\text{ratio}} = 967,132,262.622 \quad \text{and} \quad \ddot{\bar{y}}_{\text{ratio}} = 310,421.949.$$

We next build a survey-weighted regression model that takes into consideration the complex design of this sample. We initially use the data from the

1982 Census to build this model and then specify new levels of the auxiliary variables from the 1987 Census to predict the acreage for that year.

Below are the results when fitting the model with all three auxiliary variables:

```
#############################
Coefficients:
            Estimate Std. Error t value Pr(>|t|)
(Intercept) 72446.21   32630.27   2.220   0.0272 *
largef82     3662.85     262.31  13.964   <2e-16 ***
smallf82      385.38     150.79   2.556   0.0111 *
farms82        24.31      35.21   0.690   0.4904
---
Signif. codes:  0 *** 0.001 ** 0.01 * 0.05 . 0.1   1
#############################
```

The auxiliary variable of total number of farms ($X_3$) is not a statistically significant predictor ($p = 0.4904$) of total acreage given that the other two auxiliary variables are in the model. Thus, we drop this auxiliary variable from the model and refit it using the two remaining auxiliary variables:

```
#############################
Coefficients:
            Estimate Std. Error t value Pr(>|t|)
(Intercept)  86287.4    16912.9   5.102 6.04e-07 ***
largef82      3666.5      261.6  14.018  < 2e-16 ***
smallf82       449.6      112.6   3.993 8.24e-05 ***
---
Signif. codes:  0 *** 0.001 ** 0.01 * 0.05 . 0.1   1
#############################
```

The remaining auxiliary variables are both statistically significant.

Using the model above, we next specify new levels of the auxiliary variables from the 1987 Census to predict the acreage for that year. These are plotted in Figure 25.10(b) against their true values as reported on the 1987 Census of Agriculture. The vertical lines extending from each observation is 1 standard error. Overall, there is good agreement between these predicted values and the true values given the simplicity of the model we developed. It also appears that the model tends to perform better in some of the strata (the west and northeast regions) compared to other strata (the north central and south regions). Additional variables from this census, as well as data from other national surveys, could be used to improve the model fits.

## 25.7 Regression with Missing Data and Regression Imputation

Missing data occur with many datasets analyzed in practice. Regardless if recording data from an observational study, a properly designed experiment, or collecting survey data, one is bound to eventually face the problem of missing data. There are two primary types of missing data that can occur:

- Data can be **completely missing**. If we are measuring a single variable (i.e., the univariate setting), then this means an observation is simply missing. In the regression setting, this means that the response and any covariates measured are missing for a particular observation. In survey analysis, this type of missingness is called **unit nonresponse**.

- Data can be **partially missing**. This occurs when each observation has many variables measured, but some of the observations do not have values recorded for some variables. This could be done by design (e.g., a condition might not be met for certain observations), a subject could terminate during a longitudinal study, or a respondent in a survey may (intentionally or unintentionally) omit answering certain questions. In survey analysis, this type of missingness is called **item nonresponse**.

These two types of missing data are more generally part of a class of missing-data patterns. The text by Little and Rubin (2002) provides an accessible and thorough treatment of missing data analysis.

Assume that all of the data in our study are arranged in a rectangular array; i.e., a matrix. If we observe $p$ variables on $n$ subjects, then we would have an $(n \times p)$ data matrix, which we will denote as $\mathbf{X}$, where $x_{i,j}$ is the $(i,j)^{\text{th}}$ entry of the matrix. We next define the $(n \times p)$ **missing-data indicator matrix M**, such that $m_{i,j}$ is the indicator of the missingness of $x_{i,j}$ (1 implies missing, 0 implies present). **M** then can be used to characterize the missing-data patterns as well as the missing-data mechanism, which concerns the approximate relationship between the missingness and the values in the data matrix. There are three basic missing-data mechanisms generally considered in the literature (Rubin, 1976), which we briefly define:

- **Missing completely at random** or **MCAR** occurs when the probability of missingness is the same for all units.

- **Missing at random** or **MAR** occurs when the probability of missingness depends on some available information.

- **Not missing at random** or **NMAR** occurs when the probability of missing depends on the value of a missing response variable and cannot be completely explained by values in the observed data.

One common strategy for handling missing data is to use **imputation**, which is the process of replacing missing data with reasonably-determined values. Following Chapter 4 of Little and Rubin (2002), they note that one can consider imputations as "means or draws from a predictive distribution of the missing values." This yields two general classes for generating a predictive distribution for the imputation given the observed data. First is the class of **explicit models**, where a predictive distribution is based on a formal statistical model, e.g., a regression model. Second is the class of **implicit models**, which takes a more algorithmic approach to imputation, albeit there is an implied underlying model. While numerous imputation methods fall within these two classes, we briefly present some of the imputation methods for regression settings. Again, we refer to Little and Rubin (2002) for a broader discussion of imputation methods.

In a regression analysis, partially missing data could occur with the response variable, some (or all) of the predictor variables, or with both the response and predictor variables. If only some response variables are missing, then we could simply fit a regression model to the complete data cases and predict the missing response values.

If missing data occur for a certain predictor variable, say, $X_j$, we could simply impute using the mean of all of the available values for that predictor. This is called **mean imputation**. A variant on this procedure is if we have some way to partition our data into classes or cells. For any observations having a missing predictor variable within a given cell, we could then impute the cell mean of all of the available values for that predictor.

In **regression imputation**, we estimate a regression model where the variable of interest is regressed on variables observed for all cases; i.e., complete and incomplete cases. We then impute the missing values for the variable of interest using the fitted values from this estimated regression model. In fact, mean imputation is a special case of regression imputation where indicator variables are constructed for the classes or cells within which the means are imputed. However, one problem with regression imputation is that the imputed data do not have an error term included in their estimation. Thus, no residual variance is reflected. We can modify this procedure by adding a residual drawn to reflect uncertainty in the fitted value, which is called **stochastic regression imputation**. Typically, the residual is drawn from a normal distribution with mean 0 and variance equal to the MSE from the estimated regression model. While we do not present the details here, another approach to regression imputation that handles general patterns of missingness can be found in Buck (1960).

One final imputation method we mention is **deductive imputation**, which often occurs in the data editing stage by employing logical relations among the variables. For example, suppose in a survey that an individual responded that they have been admitted to the emergency room at a hospital twice in the past year. However, they did not respond to the yes/no question asking if they have ever been admitted to an emergency room. Clearly, we could impute the answer of "yes" based on their response to their number of times admitted.

**Example 25.7.1.** *(Mammal Sleep Data)*
Allison and Cicchetti (1976) analyzed the interrelationships between sleep, ecological, and constitutional variables for different mammals. In total, their data consisted of $n = 62$ mammals, but the presence of missing items results in only $n_c = 42$ complete cases. The variables of interest are as follows:

- $Y$: the total number of hours of sleep at night

- $X_1$: the body weight of the mammal subject (in kilograms)

- $X_2$: the brain weight of the mammal subject (in grams)

- $X_3$: the amount of paradoxical sleep achieved (in hours)

- $X_4$: the average life span of the mammal (in years)

- $X_5$: the average gestation time of the mammal (in days)

- $X_6$: an index rating of the extent to which the mammal is preyed upon (scores of 1 to 5)

- $X_7$: an index rating of the extent to which the mammal is safe during sleep (scores of 1 to 5)

- $X_8$: an index rating of the general predatory danger of the mammal (scores of 1 to 5)

A total of 12 observations have a missing value for the paradoxical sleep variable $(X_3)$, while the total number of hours slept $(Y)$, life span $(X_4)$, and gestation time $(X_5)$ each have 4 observations with missing values. We will use different imputation strategies for these missing values. But first, for comparison, we fit a multiple linear regression model to the complete-case data:

```
#############################
Coefficients:
            Estimate Std. Error t value Pr(>|t|)
(Intercept) 12.928769   2.370019   5.455 4.81e-06 ***
BodyWgt      0.003186   0.005694   0.560   0.5796
BrainWgt    -0.001326   0.003393  -0.391   0.6986
Dream        1.106606   0.528930   2.092   0.0442 *
Span         0.000247   0.044702   0.006   0.9956
Gest        -0.013270   0.007165  -1.852   0.0730 .
Pred         1.318790   1.145813   1.151   0.2580
Exp          0.186037   0.680124   0.274   0.7861
Danger      -2.613785   1.587452  -1.647   0.1091
---
```

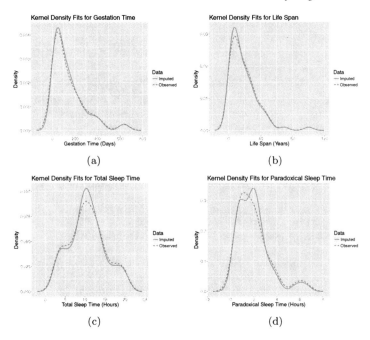

**FIGURE 25.11**
Kernel density fits for the observed data (dashed line) and imputed data
(solid line) for (a) gestation time, (b) life span, (c) total sleep time, and (d)
paradoxical sleep time.

```
Signif. codes:  0 *** 0.001 ** 0.01 * 0.05 . 0.1   1

Residual standard error: 2.898 on 33 degrees of freedom
  (20 observations deleted due to missingness)
Multiple R-squared:  0.6952,Adjusted R-squared:  0.6213
F-statistic: 9.409 on 8 and 33 DF,  p-value: 1.167e-06
#############################
```

The above shows that only paradoxical sleep $(X_3)$ is statistically significant
$(p = 0.0442)$ and gestation time $(X_5)$ is marginally significant. We will see how
much these results change when we employ different imputation strategies to
utilize all 62 mammals.

First we perform regression imputation for gestation time by regressing it
against body weight and brain weight. The estimated regression coefficients
are as follows:

```
#############################
Coefficients:
(Intercept)        BodyWgt        BrainWgt
```

```
   104.43948      -0.05812       0.16684
##############################
```

For the life span variable, we use deductive imputation. We sort the data on body weight and then brain weight. We then take the average of the life spans for the mammals whose records immediately precede and immediately succeed the mammal with a missing life span value.

For the total sleep variable, we simply use mean imputation. In other words, we replaced the 4 missing values with the overall mean of the observed sleep times.

For the paradoxical sleep variable, we again use a type of deductive imputation. Since the paradoxical sleep time is a proportion of the total sleep time, we simply computed the average proportion of time (based on the complete observations) that the mammals spend in paradoxical sleep. This is approximately 0.1815. Thus, for the mammals with missing paradoxical sleep values, we multiplied their total sleep time by 0.1815.

We can visualize the impact of the above assumptions by comparing the kernel density fits of the observed values with the imputed dataset for each of the four variables; see Figure 25.11. For the gestation time (Figure 25.11(a)) and life span (Figure 25.11(b)), the density fits for the observed values and for the imputed dataset look quite similar. For the total sleep time (Figure 25.11(c)) and paradoxical sleep time (Figure 25.11(d)), there are noticeable differences in the fits around the respective peaks. Different imputation strategies will affect such comparisons.

Below is the multiple linear regression fit to the imputed dataset using the imputed values we calculated:

```
##############################
Coefficients:
              Estimate Std. Error t value Pr(>|t|)
(Intercept)  9.4524861  1.7571735   5.379 1.72e-06 ***
BodyWgt      0.0002020  0.0015134   0.133    0.894
BrainWgt    -0.0006139  0.0016842  -0.365    0.717
Dream        1.7122923  0.3857686   4.439 4.62e-05 ***
Span        -0.0145852  0.0362912  -0.402    0.689
Gest        -0.0040640  0.0051721  -0.786    0.436
Pred         0.6638043  0.7216907   0.920    0.362
Exp         -0.0359587  0.5060170  -0.071    0.944
Danger      -1.1774339  0.9730461  -1.210    0.232
---
Signif. codes:  0 *** 0.001 ** 0.01 * 0.05 . 0.1   1

Residual standard error: 2.856 on 53 degrees of freedom
Multiple R-squared:  0.6426,Adjusted R-squared:  0.5886
F-statistic: 11.91 on 8 and 53 DF,  p-value: 1.567e-09
##############################
```

We still obtain similar results as we did when using the observed data; however, gestation time is no longer marginally significant. Again, the choice of imputation strategy will impact the overall estimates, but based on the kernel density fits in Figure 25.11, the imputation strategies we used appear reasonable.

## 25.8   Bayesian Regression

**Bayesian inference** is a type of statistical inference conducted after updating our model, which is based on previous beliefs, as a result of receiving more information or incoming data. Bayesian inference is based on **Bayes' Theorem**, which says for two events, $A$ and $B$,

$$P(A|B) = \frac{P(B|A)P(A)}{P(B)}, \tag{25.7}$$

where $P(B) \neq 0$. Even though the inference methods we present in this book are likelihood-based (also called the **frequentist's view**), Bayesian methods are widely used and accepted in practice. There is a copious amount of literature on Bayesian methods. We refer to the texts by Carlin and Louis (2008) and Gelman et al. (2013) and the references therein.

In a Bayesian analysis, we start with our model, which is the distribution of the observed data conditioned on its parameters. This is also termed the likelihood and, for most practical purposes, can simply be viewed as equivalent to the traditional likelihood function. In order to update our model, we define a distribution for the parameter(s) in this model, which is based on previous beliefs. This distribution is called a **prior distribution** and the parameters in this distribution are called **hyperparameters**. We multiply the likelihood function of our model by the prior distribution and then divide by the **marginal density function**, which is the joint density function with the parameter integrated out. The result is called the **posterior distribution**. Fortunately, the marginal density function is just a normalizing constant and does not usually have to be calculated in practice.

For multiple linear regression, the ordinary least squares estimator (and maximum likelihood estimator) of $\beta$,

$$\hat{\beta} = (\mathbf{X}^{\mathrm{T}}\mathbf{X})^{-1}\mathbf{X}^{\mathrm{T}}\mathbf{y},$$

and the maximum likelihood estimator of $\sigma^2$,

$$\hat{\sigma}^2 = \left(\frac{n-p}{p}\right) \mathrm{MSE},$$

are constructed in a frequentist setup. In a Bayesian setup, we only make assumptions about the observed data and attempt to describe the plausibility of possible outcomes by "borrowing" information from a larger set of similar observations.

In **Bayesian linear regression**, the conditional likelihood is given by:

$$\mathcal{L}(\mathbf{y}|\mathbf{X},\boldsymbol{\beta},\sigma^2) = (2\pi\sigma^2)^{-n/2} \exp\left(-\frac{1}{2\sigma^2}\|\mathbf{y} - \mathbf{X}\boldsymbol{\beta}\|^2\right).$$

We define a **conjugate prior**, which is a prior that yields a joint density that is of the same functional form as the likelihood. Since the likelihood is quadratic in $\boldsymbol{\beta}$, we can rewrite

$$\|\mathbf{y} - \mathbf{X}\boldsymbol{\beta}\|^2 = \|\mathbf{y} - \mathbf{X}\hat{\boldsymbol{\beta}}\|^2 + (\boldsymbol{\beta} - \hat{\boldsymbol{\beta}})^{\mathrm{T}}(\mathbf{X}^{\mathrm{T}}\mathbf{X})(\boldsymbol{\beta} - \hat{\boldsymbol{\beta}})$$

to give us the following form of the likelihood:

$$\mathcal{L}(\mathbf{y}|\mathbf{X},\boldsymbol{\beta},\sigma^2) \propto (\sigma^2)^{-v/2} \exp\left(-\frac{vs^2}{2\sigma^2}\right)$$
$$\times (\sigma^2)^{-(n-v)/2} \exp\left(-\frac{1}{2\sigma^2}\|\mathbf{X}(\boldsymbol{\beta} - \hat{\boldsymbol{\beta}})\|^2\right),$$

where $vs^2 = \|\mathbf{y} - \mathbf{X}\hat{\boldsymbol{\beta}}\|^2$ and $v = n - p$. While skipping over some additional details, the above form of the likelihood indicates that we can use some form for the prior distribution like

$$\pi(\boldsymbol{\beta},\sigma^2) = \pi(\boldsymbol{\beta}|\sigma^2)\pi(\sigma^2),$$

which breaks the prior distribution up into the product of two prior distributions as justified through an application of Bayes' theorem. $\pi(\sigma^2)$ is the prior for $\sigma^2$, which is an inverse gamma distribution with shape hyperparameter $\alpha$ and scale hyperparameter $\gamma$. $\pi(\boldsymbol{\beta}|\sigma^2)$ is the prior for $\boldsymbol{\beta}$, which is a multivariate normal distribution with location and dispersion hyperparameters $\bar{\boldsymbol{\beta}}$ and $\Sigma$, respectively. This yields the joint posterior distribution:

$$p(\boldsymbol{\beta},\sigma^2|\mathbf{y},\mathbf{X}) \propto \mathcal{L}(\mathbf{y}|\mathbf{X},\boldsymbol{\beta},\sigma^2)\pi(\boldsymbol{\beta}|\sigma^2)\pi(\sigma^2)$$
$$\propto \sigma^{-n-\alpha} \exp\left\{-\frac{1}{2\sigma^2}(\tilde{s} + (\boldsymbol{\beta} - \tilde{\boldsymbol{\beta}})^{\mathrm{T}}(\Sigma^{-1} + \mathbf{X}^{\mathrm{T}}\mathbf{X})(\boldsymbol{\beta} - \tilde{\boldsymbol{\beta}}))\right\},$$

where

$$\tilde{\boldsymbol{\beta}} = (\Sigma^{-1} + \mathbf{X}^{\mathrm{T}}\mathbf{X})^{-1}(\Sigma^{-1}\bar{\boldsymbol{\beta}} + \mathbf{X}^{\mathrm{T}}\mathbf{X}\hat{\boldsymbol{\beta}})$$
$$\tilde{s} = 2\gamma + \hat{\sigma}^2(n - p) + (\bar{\boldsymbol{\beta}} - \tilde{\boldsymbol{\beta}})^{\mathrm{T}}\Sigma^{-1}\bar{\boldsymbol{\beta}} + (\hat{\boldsymbol{\beta}} - \tilde{\boldsymbol{\beta}})^{\mathrm{T}}\mathbf{X}^{\mathrm{T}}\mathbf{X}\hat{\boldsymbol{\beta}}.$$

Finally, it can be shown that the distribution of $(\boldsymbol{\beta}|\mathbf{X},\mathbf{y})$ is a multivariate-$t$ distribution with $(n + \alpha - p - 1)$ degrees of freedom such that:

$$\mathrm{E}(\boldsymbol{\beta}|\mathbf{X},\mathbf{y}) = \tilde{\boldsymbol{\beta}}$$
$$\mathrm{Var}(\boldsymbol{\beta}|\mathbf{X},\mathbf{y}) = \frac{\tilde{s}(\Sigma^{-1} + \mathbf{X}^{\mathrm{T}}\mathbf{X})^{-1}}{n + \alpha - p - 3}.$$

Furthermore, the distribution of $(\sigma^2|\mathbf{X},\mathbf{y})$ is an inverse gamma distribution with shape parameter $(n + \alpha - p)$ and scale parameter $\hat{\sigma}^2(n + \alpha - p)/2$.

One can also construct statistical intervals based on draws simulated from a Bayesian posterior distribution. A $100 \times (1 - \alpha)\%$ **credible interval** is constructed by taking the middle $100 \times (1 - \alpha)\%$ of the values simulated from the posterior distribution. The interpretation of these intervals is that there is a $100 \times (1 - \alpha)\%$ probability that the true population parameter is in the calculated $100 \times (1 - \alpha)\%$ credible interval. This interpretation is often how many people initially try to interpret confidence intervals.

**Example 25.8.1.** *(Animal Weight Data, cont'd)*
We continue with the animal weight data of Example 25.3.1. We will perform a Bayesian analysis of these data, where we illustrate the impact of the prior specification on the resulting posterior sample. We use the conjugate prior, but specify two different sets of hyperparameters. For the prior on $\boldsymbol{\beta}$, we use a strong prior where the dispersion hyperparameter is $50\mathbf{I}_2$ and a diffuse prior where the dispersion hyperparameter is $0.01\mathbf{I}_2$. For both MCMC chains, the location hyperparameter used for the prior on $\boldsymbol{\beta}$ is $\bar{\boldsymbol{\beta}} = (0.981, 0.703)^{\mathrm{T}}$, which are the intercept and slope values obtained for the deepest regression fit in Example 25.3.1. For both MCMC chains, we also use the same hyperparameters for the prior on $\sigma^2$, which are 3 degrees of freedom and the sample variance of the $\log_{10}$-transformed values of the animal's brain weight, which is 1.086. Each MCMC chain consists of 10000 draws from the posterior.

Figure 25.12 gives various diagnostic plots for the two MCMC chains. Figure 25.12(a) gives the kernel density fits to each posterior distribution for each chain. Clearly, we see that the first chain, which has the stronger prior, yields posterior means for $\boldsymbol{\beta}$ that are closer to the location hyperparameter of $\bar{\boldsymbol{\beta}}$. These differences are not quite as visible in Figure 25.12(b) due to the scale of the posterior draws; however, the plot of the running means again helps us to visualize these different posterior means. The corresponding plots for the posterior sample for $\sigma^2$ are nearly identical for each MCMC chain since we used the same hyperparameters for its prior in each chain. Finally, the autocorrelation plots for each posterior sample in each chain (Figure 25.12(d)) do not indicate that the draws were correlated with each other, which would have required us to employ other strategies to ensure that our posterior draws were, indeed, independent.

Since the MCMC diagnostic plots do not indicate any problems, we plot the regression lines based on the posterior means from each chain over the data (Figure 25.13). The line resulting from the first chain with the stronger prior structure has an intercept of 0.997 and slope of 0.557, while the line results from the second chain with the diffuse prior structure has an intercept of 1.110 and slope of 0.496. Notice that these latter estimates are the same as the maximum likelihood estimates (i.e., the ordinary least squares estimates) for the simple linear regression model. Finally, we also report the posterior means and 95% credible intervals for each of the three parameters in Table 25.1. The results for both chains are reported.

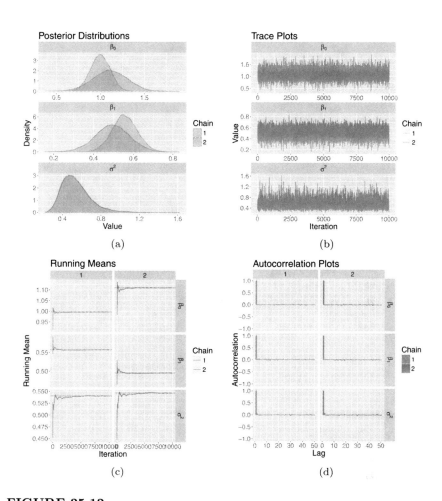

**FIGURE 25.12**
Plots from the posterior distributions for the two MCMC chains ran under
the different prior structures. (a) Density plots of the posterior distribution
for each parameter. (b) Trace plots where the chains are plotted as a time
series. (c) Running means plot where the chains are plotted as a time series.
(d) Autocorrelation plots of the posterior.

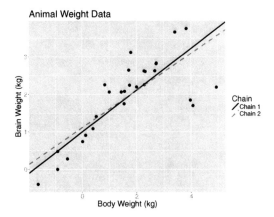

**FIGURE 25.13**
Scatterplot of the animal weight data with the posterior means of the regression lines from each chain.

**TABLE 25.1**
Posterior means (PMs) and 95% credible intervals (CIs) for the animal weight data

| Parameter | Chain 1 | | Chain 2 | |
|:---:|:---:|:---:|:---:|:---:|
| | PM | 95% CI | PM | 95% CI |
| $\beta_0$ | 0.997 | $(0.780, 1.213)$ | 1.110 | $(0.720, 1.497)$ |
| $\beta_1$ | 0.557 | $(0.435, 0.684)$ | 0.496 | $(0.323, 0.667)$ |
| $\sigma^2$ | 0.540 | $(0.319, 0.912)$ | 0.546 | $(0.320, 0.928)$ |

## 25.9   Quantile Regression

For a random variable $X$ with probability density function $F_X(x) = P(X \leq x)$, the $\tau^{\text{th}}$ **quantile**, $\tau \in [0, 1]$, is given by

$$Q_X(\tau) = F_X^{-1}(\tau) = \inf\{x : F_X(x) \geq \tau\}.$$

For example, if $\tau = \frac{1}{2}$, then the corresponding value of $x$ would be the median. For an observed random sample $x_1, \ldots, x_n$ from $F_X$, the $\tau^{\text{th}}$ sample quantile is the solution

$$\hat{q}_\tau = \operatorname*{argmin}_q \sum_{i=1}^n \rho_\tau(x_i - q),$$

where

$$\rho_\tau(z) = z(\tau - \mathrm{I}\{z < 0\})$$

is called the **linear check function**. This notion of (sample) quantiles also extends to the regression setting.

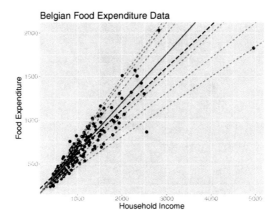

**FIGURE 25.14**
Various quantile regression fits for the Belgian food expenditure data.

In **quantile regression**, for a dataset of size $n$ with $p - 1$ predictors, we seek a solution to the criterion

$$\hat{\boldsymbol{\beta}}_\tau = \underset{\boldsymbol{\beta}}{\operatorname{argmin}} \sum_{i=1}^{n} \rho_\tau(y_i - \mathbf{x}_i^\mathrm{T}\boldsymbol{\beta}),$$

which estimates the $\tau^{\text{th}}$ linear conditional quantile function $Q_{Y|X}(\tau) = \mathbf{X}^\mathrm{T}\boldsymbol{\beta}_\tau$. The calculated quantity $\hat{\boldsymbol{\beta}}_\tau$ is called the $\tau^{\text{th}}$ **regression quantile**.

We actually encountered quantile regression in Chapter 12. Least absolute deviations regression is just the case of quantile regression where $\tau = 1/2$. Estimation for quantile regression can usually be accomplished through linear programming or other optimization procedures. Thus, statistical intervals are also easily computed. For a thorough discussion on quantile regression, we refer to the text by Koenker (2005).

**Example 25.9.1.** *(Belgian Food Expenditure Data)*
Koenker and Bassett, Jr. (1982) presented and analyzed data on $n = 235$ working class households in Belgian to study Engel's law, which is a theory studied by economists that the percentage of income for food purchases decreases as income rises. Thus, the variables in this study are:

- $Y$: annual household food expenditure (in Belgian francs)

- $X$: annual household income (in Belgian francs)

Figure 25.14 gives a plot of these data. Overlaid on the plot is a solid blue line, which is the least absolute deviation fit; i.e., $\tau = 0.50$. The dotted red lines (from bottom to top) are the quantile regression fits for

**TABLE 25.2**

Sample quantile regression coefficients

| $\tau$ | $\hat{\beta}_{\tau,0}$ | $\hat{\beta}_{\tau,1}$ |
|--------|-----------|-----------|
| 0.05 | 124.88 | 0.34 |
| 0.10 | 110.14 | 0.40 |
| 0.25 | 95.48 | 0.47 |
| 0.50 | 81.48 | 0.56 |
| 0.75 | 62.40 | 0.64 |
| 0.90 | 67.35 | 0.69 |
| 0.95 | 64.10 | 0.71 |
| OLS | 147.48 | 0.49 |

$\tau = 0.05, 0.10, 0.25, 0.75, 0.90$, and $0.95$, respectively. Essentially, this says is that if we looked at those households with the highest food expenditures, they will likely have larger regression coefficients, such as the $\tau = 0.95$ regression quantile, while those with the lowest food expenditures will likely have smaller regression coefficients, such as the $\tau = 0.05$ regression quantile. The estimated quantile regression coefficients for all the values of $\tau$ we considered are given in Table 25.2. For reference, the ordinary least squares estimates are provided at the bottom of this table, which is the dashed black line overlaid on Figure 25.14.

## 25.10 Monotone Regression

Suppose we have a set of data $(x_1, y_1), \ldots, (x_n, y_n)$. For ease of notation, let us assume there is already an ordering on the predictor variable. Specifically, we assume that $x_1 \leq \ldots \leq x_n$. **Monotonic regression** is a nonparametric technique where we attempt to find a weighted least squares fit of the responses $y_1, \ldots, y_n$ to a set of scalars $a_1, \ldots, a_n$ with corresponding weights $w_1, \ldots, w_n$, subject to monotonicity constraints giving a simple or partial ordering of the responses. In other words, the responses are suppose to strictly increase (or decrease) as the predictor increases and the regression line we fit is piecewise constant (which resembles a step function). The weighted least squares problem for monotonic regression is:

$$\underset{\mathbf{a}}{\mathrm{argmin}} \sum_{i=1}^{n} w_i(y_i - a_i)^2 \qquad (25.8)$$

and is subject to one of two possible constraints:

1. If the direction of the trend is to be monotonically increasing, then the process is called **isotonic regression** and the constraint is $y_i \geq y_j$ for all $i > j$ where this ordering is true.

2. If the direction of the trend is to be monotonically decreasing, then the process is called **antitonic regression** and the constraint is $y_i \leq y_j$ for all $i > j$ where this ordering is true.

Two other constraints that can be used when minimizing Equation (25.8) lead to the following types of regressions:

1. If minimizing (25.8) with respect to the constraint

$$(a_i - a_{i-1})(x_{i+1} - x_i) \geq (a_{i+1} - a_i)(x_i - x_{i-1})$$

for $i = 1, \ldots, n$ and $a_0 \equiv a_{n+1} = -\infty$, then we refer to the problem as **concave regression**.

2. If minimizing (25.8) with respect to the constraint

$$(a_i - a_{i-1})(x_{i+1} - x_i) \leq (a_{i+1} - a_i)(x_i - x_{i-1})$$

for $i = 1, \ldots, n$ and $a_0 \equiv a_{n+1} = \infty$, then we refer to the problem as **convex regression**.

One can also estimate the above monotonic, concave, and convex regression problems under $L_p$ for $p > 0$:

$$\underset{\mathbf{a}}{\text{argmin}} \sum_{i=1}^{n} w_i |y_i - a_i|^p,$$

where the appropriate constraints are imposed for the respective regression problem. All of these regression problems are sometimes referred to as **restricted least squares regression** (Dykstra, 1983).

Classic optimization for these restricted least squares regression problems where $w_i \equiv 1$ are due to an iterative algorithm called **Dykstra's algorithm** (Dykstra, 1983). The algorithm is designed to find the projection of a point onto a finite intersection of closed convex sets. While Dykstra's algorithm can be employed when the weights are different than 1, Lange (2013) notes that this setting is handled better by the **pool adjacent violators algorithm**. More information on the optimization details of monotone regression and the pool adjacent violators algorithm is found in de Leeuw et al. (2009).

**Example 25.10.1.** *(Animal Weight Data, cont'd)*
We continue with the animal weight data of Example 25.3.1. As we saw with these data, there is a clear increasing trend between the brain weight and body weight (both transformed to be on the $\log_{10}$ scale) of the $n = 28$ species of animals. Thus, we proceed to obtain an isotonic regression fit on these data. For simplicity, we take $w_i \equiv 1$ for all $i$.

The isotonic regression fit is given in Figure 25.15. Figure 25.15(a) is the actual isotonic regression fit. The horizontal lines represent the values of the scalars minimizing the weighted least squares problem given earlier. Figure

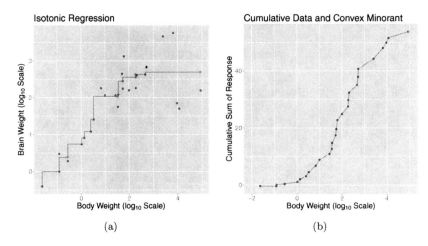

(a)             (b)

**FIGURE 25.15**
(a) An isotonic regression fit to the animal weight data. (b) The convex minorant from the isotonic regression overlaid on the cumulative sum of the brain
weight versus the ordered body weight, both on the $\log_{10}$ scale.

25.15(b) shows the cumulative sums of the brain weight plotted against the
ordered body weight. The piecewise regression line that is plotted is called the
**convex minorant**.[5] Each value of brain weight where this convex minorant
intersects at the value of the cumulative sum is the same value of the brain
weight where the slope changes in the isotonic regression plot.

## 25.11    Generalized Extreme Value Regression Models

**Extreme value analysis** (or **EVA**) is a branch of statistics that focuses on
quantifying the stochastic behavior of a process at unusually large or unusually small levels. In the observed data, these levels are characterized by the
observed maxima and minima, or more simply, the extrema. EVA is used in
a wide range of important applications, including the analysis of sea levels
(Tawn, 1992), financial applications (Gilli and Këllezi, 2006), and engineering
applications (Castillo et al., 2005).

    Perhaps the most important model in EVA is the generalized extreme value
distribution. A random variable $X$ follows a **generalized extreme value
distribution** (written $X \sim \text{GEV}(\mu, \sigma, \xi)$) if it has cumulative distribution

---

[5]When fitting an antitonic regression, this plot would be with respect to the **concave
majorant**.

function

$$G_X(x) = \exp\left\{-\left[1 + \xi\left(\frac{x-\mu}{\sigma}\right)\right]^{-1/\xi}\right\}, \qquad (25.9)$$

which is defined on the set $\{x : 1 + \xi(x - \mu)/\sigma > 0\}$ with parameter space $\{\mu, \sigma, \xi : -\infty < \mu, \xi < \infty, \sigma > 0\}$. The parameters $\mu$, $\sigma$, and $\xi$, are location, scale, and shape parameters, respectively. The family of generalized extreme value distributions is quite flexible in that it yields three important classes of distributions depending on the behavior of the shape parameter:

- Type I: Gumbel family ($\xi \to 0$)

- Type II: Fréchet family ($\xi > 0$)

- Type III: Weibull family ($\xi < 0$)

The above extreme value distribution relates to the maxima. An analogous distribution can be developed for the minima by substituting $(-x)$ for $x$ in the above distribution and then subtracting the distribution from 1. More on extreme value theory is found in Coles (2001).

Most applications in EVA involve time-ordered sequences of measurements. For the remainder of our discussion, let $t = 1, 2, \ldots$ denote a time-ordered index as we used in time series analysis in Chapter 13. Suppose that $\{Y_t : t \in T\}$ is a non-stationary sequence of generalized extreme value random variables, where $T$ is again an index set. Suppose further that we also measure three sets of predictor vectors at each $t$. Call these vectors $\mathbf{X}_t$, $\mathbf{Z}_t$, and $\mathbf{W}_t$, which are $p$-dimensional, $q$-dimensional, and $r$-dimensional, respectively.[6] Analogous to generalized linear models, the generalized extreme value model parameters $\{\mu, \sigma, \xi\}$ depend on a linear combination of unknown regression parameters $\{\boldsymbol{\gamma}, \boldsymbol{\beta}, \boldsymbol{\delta}\}$ through mean functions (i.e., inverse link functions) $\{g_1(\cdot), g_2(\cdot), g_3(\cdot)\}$:

$$\mu(\mathbf{X}_t) = g_1(\mathbf{X}_t^{\mathsf{T}}\boldsymbol{\gamma})$$
$$\sigma(\mathbf{Z}_t) = g_2(\mathbf{Z}_t^{\mathsf{T}}\boldsymbol{\beta})$$
$$\xi(\mathbf{W}_t) = g_3(\mathbf{W}_t^{\mathsf{T}}\boldsymbol{\delta}).$$

The above setup is formally a **generalized extreme value regression model**, which we write as

$$(Y_t|\mathbf{X}_t, \mathbf{Z}_t, \mathbf{W}_t) \sim \mathrm{GEV}(\mu(\mathbf{X}_t), \sigma(\mathbf{Z}_t), \xi(\mathbf{W}_t)).$$

Appropriate numerical techniques can be employed to find the maximum likelihood estimates of the regression coefficients, $\hat{\boldsymbol{\gamma}}$, $\hat{\boldsymbol{\beta}}$, and $\hat{\boldsymbol{\delta}}$. Model diagnostics are also available when fitting a generalized extreme value regression model. Refer to Chapter 6 of Coles (2001) for more details on estimation and inference about these models.

---

[6]Note that some or all variables may be shared across the three predictor vectors. In the latter case, we would simply have a predictor vector $\mathbf{X}_t$.

**Example 25.11.1.** *(Annual Maximum Sea Levels Data)*
Coles (2001) presented and analyzed data on $n = 83$ annual maximum sea
levels recorded at Fremantle, located near Perth in Western Australia. This
dataset consists of records from most of the years between 1897 and 1989. The
variables in this study are:

- $Y$: annual sea level maxima

- $X_1$: year

- $X_2$: annual mean values of the Southern Oscillation Index (SOI)

We proceed to fit various generalized extreme value regression models to
these data. In particular, we explore fitting three different models where we
(1) model $\mu$ as a linear function of both predictors, (2) model $\mu$ and $\sigma$ as linear
functions of both predictors, and (3) model $\mu$, $\sigma$, and $\xi$ as linear functions of
both predictors. The BIC values for these three models are $-85.5$, $-84.6$, and
$-76.3$, respectively. Since the first model has the lowest BIC value, we will
thus focus on the fitted model with $\mu$ modeled as a linear function of $X_1$ and
$X_2$. Below are the parameter estimates and their standard error estimates for
this fitted model:

```
############################
 Negative Log-Likelihood Value:   -53.9



 Estimated parameters:
     mu0      mu1       mu2     scale      shape
  1.38221  0.00211  0.05452  0.12073 -0.14999


 Standard Error Estimates:
     mu0      mu1       mu2     scale      shape
 0.029084 0.000496 0.019644 0.010027 0.064636
############################
```

Figure 25.16 is a plot of the annual maximum sea level versus year. Overlaid
is the fit for $\hat{\mu}(X_1, X_2)$ for fixed values of $X_2 \in \{-1, 0, 1, 2\}$. Clearly these
values of $\hat{\mu}(X_1, X_2)$ appear to capture the increasing trend in these data. The
quality of this fit is also supported by the Q-Q plot of the residuals (Figure
25.17). There are no noticeable patterns in the Q-Q plot, which suggests that
the proposed model is appropriate for these data.

---

## 25.12   Spatial Regression

Suppose a demographer is studying the growth (measured in terms of housing
units) of a neighborhood over the last year. In doing so, they will surely want

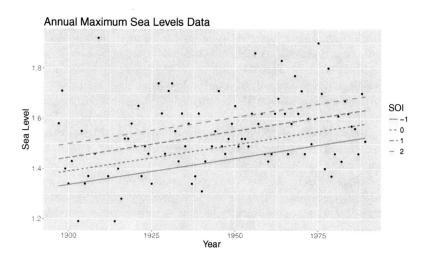

**FIGURE 25.16**
Estimate of $\mu(t)$ for the annual maximum sea levels data at different values of the SOI.

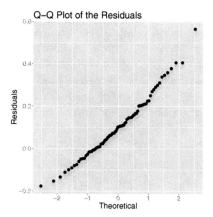

**FIGURE 25.17**
Q-Q plot of the residuals from the estimated generalized extreme value model.

to look at the growth of adjacent neighborhoods and other neighborhoods within the same county. However, the number of housing units measured in, say, a different state likely will not have any impact on the growth in the neighborhood of interest. The framework for such modeling is likely to incorporate some sort of spatial effect as neighborhoods nearest to the neighborhood of interest are likely to have a greater impact on the growth measurement while neighborhoods farther away will have a smaller or negligible impact.

**Spatial regression** deals with the specification, estimation, and diagnostic analysis of regression models which incorporate spatial effects. Two broad classes of spatial effects are often distinguished: **spatial heterogeneity** and **spatial dependency**. We will provide a brief overview of both types of effects, but it should be noted that we will only skim the surface of what is a very rich area. We refer to the texts by Cressie (1993), Cressie and Wikle (2011), and Banerjee et al. (2014) for a thorough treatment of spatial analysis.

A spatial (linear) regression model reflecting spatial heterogeneity is written locally as

$$\mathbf{Y} = \mathbf{X}\boldsymbol{\beta}(\mathbf{s}) + \boldsymbol{\epsilon}, \tag{25.10}$$

where $\mathbf{s}$ indicates that the regression coefficients are to be estimated locally at the coordinates specified by $\mathbf{s}$ and $\boldsymbol{\epsilon}$ is an error term distributed with mean 0 and variance $\sigma^2$. $\mathbf{s}$ could, for example, be the two dimensional vector of longitude and latitude for a geographic location. The response and predictor variables also depend on $\mathbf{s}$, but we have suppressed that dependency to simplify notation except when it helps clarify the discussion. The model in (25.10) is called **geographically weighted regression** or **GWR**. The estimation of $\boldsymbol{\beta}(\mathbf{s})$ is found using a weighting scheme such that

$$\hat{\boldsymbol{\beta}}(\mathbf{s}) = (\mathbf{X}^{\mathrm{T}}\mathbf{W}(\mathbf{s})\mathbf{X})^{-1}\mathbf{X}^{\mathrm{T}}\mathbf{W}(\mathbf{s})\mathbf{Y}.$$

The weights in the geographic weighting matrix $\mathbf{W}(\mathbf{s})$ are chosen such that those observations near the point in space where the parameter estimates are desired have more influence on the result than those observations further away.

The GWR model is similar to local regression models, like those discussed in Chapter 17. While the choice of a geographic (or spatially) weighted matrix is a blend of art and science, one commonly used weight is the Gaussian weight function:

$$w_i(\mathbf{s}) = \exp\{-d_i/h\},$$

where $d_i$ is the Euclidean distance between observation $i$ and location $\mathbf{s}$ and $h$ is the bandwidth. Thus, $\mathbf{W}(\mathbf{s}) = \mathrm{diag}(w_1(\mathbf{s}), \ldots, w_n(\mathbf{s}))$.

The resulting parameter estimates or standard errors for the spatial heterogeneity model may be mapped in order to examine local variations in the parameter estimates. Hypothesis tests are also possible regarding this model.

Spatial regression models accommodate spatial dependency in two primary ways:

1. A **spatial lag** dependency is where the spatial correlation manifests through the dependent variable.

2. A **spatial error** dependency is where the spatial correlation manifest through the error term.

A spatial lag model is a spatial regression model which models the response as a function of not only the predictors, but also values of the response observed at other (likely neighboring) locations $\mathbf{s}^* \neq \mathbf{s}$:

$$\mathbf{Y}(\mathbf{s}) = g(\mathbf{Y}(\mathbf{s}^*); \boldsymbol{\theta}) + \mathbf{X}(\mathbf{s})\boldsymbol{\beta}(\mathbf{s}) + \boldsymbol{\epsilon},$$

where the function $g$, which depends on a parameter vector $\boldsymbol{\theta}$, can be very general, but is often simplified by using a spatially weighted matrix.

Assuming a spatially weighted matrix $\mathbf{W}(\mathbf{s})$ that has row-standardized spatial weights, i.e., $\sum_{j=1}^{n} w_{i,j} = 1$, we obtain a **mixed regressive spatial autoregressive model** (Anselin, 2009):

$$\mathbf{Y} = \rho \mathbf{W}(\mathbf{s})\mathbf{Y} + \mathbf{X}\boldsymbol{\beta} + \boldsymbol{\epsilon}, \qquad (25.11)$$

where $\rho$ is the **spatial autoregressive coefficient**, $\mathbf{W}(\mathbf{s})\mathbf{Y}$ is the spatial lag, and the error terms in the vector $\boldsymbol{\epsilon}$ are *iid* $\mathcal{N}(0, \sigma^2)$. The proper solution to the equation for all observations requires (after some matrix algebra)

$$\mathbf{Y} = (\mathbf{I}_n - \rho \mathbf{W}(\mathbf{s}))^{-1}\mathbf{X}\boldsymbol{\beta} + (\mathbf{I}_n - \rho \mathbf{W}(\mathbf{s}))^{-1}\boldsymbol{\epsilon}$$

to be solved simultaneously for $\boldsymbol{\beta}$ and $\rho$. When no spatial predictors are present, the model in (25.11) is called a **(first-order) spatial autoregressive model**.

The inclusion of a spatial lag is similar to autoregressive terms in time series models, although with a fundamental difference. Unlike time dependency, a spatial dependency is multidirectional, implying feedback effects and simultaneity (Anselin, 2009). More precisely, if $\mathbf{s}$ and $\mathbf{s}^*$ are neighboring locations, then the response $Y(\mathbf{s}^*)$ enters on the right-hand side in the equation for $Y(\mathbf{s})$, but $Y(\mathbf{s})$ also enters on the right-hand side in the equation for $Y(\mathbf{s}^*)$.

In spatial error models, the spatial autocorrelation does not enter as an additional variable in the model, but through the covariance structure of the random error. In other words, $\text{Var}(\boldsymbol{\epsilon}) = \Sigma$ such that the off-diagonals of $\Sigma$ are not 0. One common way to model the error structure is through **direct representation**, which is similar to the weighting scheme used in GWR. In this setting, the off-diagonals of $\Sigma$ are given by $\sigma_{i,j} = \sigma^2 g(d_{i,j}, \boldsymbol{\phi})$, where again $d_{i,j}$ is the Euclidean distance between locations $i$ and $j$ and $\boldsymbol{\phi}$ is a vector of parameters which may include a bandwidth parameter.

Another way to model the error structure is through a **spatial process**, such as specifying the error terms to have a spatial autoregressive structure as in the spatial lag model from earlier:

$$\boldsymbol{\epsilon} = \lambda \mathbf{W}(\mathbf{s})\boldsymbol{\epsilon} + \mathbf{u},$$

where $\mathbf{u}$ is a vector of random error terms. Other spatial processes include

**conditional autoregressive processes** and **spatial moving average processes**, both which resemble similar time series processes.

Estimation of these spatial regression models can be accomplished through various techniques, but they differ depending on if you have a spatial lag dependency or a spatial error dependency. Such estimation methods include maximum likelihood estimation, the use of instrumental variables, and semiparametric methods. Moreover, prediction at a new location $\mathbf{s}^*$ using the estimated models can be performed, which is formally known as **kriging**.

There are also tests for the spatial autocorrelation coefficient, of which the most notable uses **Moran's $I$ statistic** (Moran, 1948, 1950). Moran's $I$ statistics is calculated as

$$I = \frac{\mathbf{e}^T\mathbf{W}(\mathbf{s})\mathbf{e}/S_0}{\mathbf{e}^T\mathbf{e}/n},$$

where $\mathbf{e}$ is a vector of ordinary least squares residuals, $\mathbf{W}(\mathbf{s})$ is a geographic weighting matrix, and $S_0 = \sum_{i=1}^n \sum_{j=1}^n w_{i,j}(\mathbf{s})$ is a normalizing factor. Then, Moran's $I$ test can be based on a normal approximation using a standardized value $I$ statistic such that

$$\mathrm{E}(I) = \mathrm{tr}(\mathbf{MW}(\mathbf{s})/(n-p))$$

and

$$\mathrm{Var}(I) = \frac{\mathrm{tr}(\mathbf{MW}(\mathbf{s})\mathbf{MW}(\mathbf{s})^T) + \mathrm{tr}(\mathbf{MW}(\mathbf{s})\mathbf{MW}(\mathbf{s})) + [\mathrm{tr}(\mathbf{MW}(\mathbf{s}))]^2}{(n-p)(n-p+2)},$$

where $\mathbf{M} = \mathbf{I}_n - \mathbf{X}(\mathbf{X}^T\mathbf{X})^{-1}\mathbf{X}^T$.

**Example 25.12.1.** *(Boston Housing Data, cont'd)*
Let us return to the Boston housing data of Example 23.7.3. Our original analysis did not include a spatial component, but we will now use spatial weights based on a *sphere-of-influence* spatial measure that was used for these data. These spatial weights are used when fitting a spatial regression model with spatial error dependency.

First we run a test for the spatial autocorrelation coefficient on the residuals from the ordinary least squares fit:

```
#############################
Global Moran I for regression residuals

data:
model: lm(formula = log(medv) ~ crim + zn + indus + chas +
I(nox^2) + I(rm^2) + age + log(dis) + log(rad) + tax +
ptratio + black + log(lstat), data = Boston)
weights: boston.w

Moran I statistic standard deviate = 10, p-value <2e-16
alternative hypothesis: two.sided
```

```
sample estimates:
Observed Moran I       Expectation         Variance
      0.436430           -0.016887         0.000976
#############################
```

As can be seen, the *p*-value is very small and so the spatial autocorrelation coefficient is significant. Thus, we fit a spatial regression model with spatial error dependency based. The functional form of the model is the one established by Harrison and Rubinfeld (1978), which was used in Example 23.7.3:

```
#############################
Type: error
Coefficients: (asymptotic standard errors)
                Estimate  Std. Error   z value  Pr(>|z|)
(Intercept)   3.85706007  0.16083868   23.9809 < 2.2e-16
crim         -0.00545832  0.00097262   -5.6120 2.000e-08
zn            0.00049195  0.00051835    0.9491 0.3425904
indus         0.00019244  0.00282240    0.0682 0.9456390
chas         -0.03303431  0.02836929   -1.1644 0.2442462
I(nox^2)     -0.23369324  0.16219196   -1.4408 0.1496289
I(rm^2)       0.00800078  0.00106472    7.5145 5.707e-14
age          -0.00090974  0.00050116   -1.8153 0.0694827
log(dis)     -0.10889418  0.04783715   -2.2764 0.0228250
log(rad)      0.07025729  0.02108181    3.3326 0.0008604
tax          -0.00049870  0.00012072   -4.1311 3.611e-05
ptratio      -0.01907769  0.00564160   -3.3816 0.0007206
black         0.00057442  0.00011101    5.1744 2.286e-07
log(lstat)   -0.27212778  0.02323159  -11.7137 < 2.2e-16

Lambda: 0.702, LR test value: 212, p-value: < 2.22e-16
Asymptotic standard error: 0.0327
    z-value: 21.5, p-value: < 2.22e-16
Wald statistic: 461, p-value: < 2.22e-16

Log likelihood: 256 for error model
ML residual variance (sigma squared): 0.0181, (sigma: 0.135)
Number of observations: 506
Number of parameters estimated: 16
AIC: -480, (AIC for lm: -270)
#############################
```

The interpretations of the coefficients are similar to those estimates from a fitted multiple linear regression model. For example, the interpretation of first slope coefficient in the above output is that the average change in the log median home price decreases by about 0.005 for each unit increase when holding the other variables constant and accounting for spatial variability. In the last line of the above output, the values for AIC are given for this model's fit and the corresponding model without spatial errors. Clearly the model with spatial errors produces a better fit according to AIC. However, note there are five predictors that are not significant at the 0.05 significance level. This suggests that further variable selection could be employed to potentially find a more parsimonious model.

Figure 25.18 provides **hex maps** — a map where hexagons represent aggregations of two-dimensional data on a third dimension — of the the observed median housing prices (Figure 25.18(a)) and predicted median housing prices (Figure 25.18(b)). Each hexagon represents the mean of the median housing prices that fall within that region. Even though we stated above possible improvements to the model we built, these figures show good agreement between the observed and predicted median housing prices.

## 25.13   Circular Regression

A **circular random variable** is one which takes values on the circumference of a circle; i.e., the angle is in the range of $(0, 2\pi)$ radians or $(0°, 360°)$. A **circular-circular regression** is used to determine the relationship between a circular predictor variable $X$ and a circular response variable $Y$. Circular data occurs when there is periodicity to the phenomena at hand or where there are naturally angular measurements. An example could be determining the relationship between wind direction measurements (the response) on an aircraft and wind direction measurements taken by radar (the predictor). If only the response is a circular variable, then the relationship is characterized using **circular-linear regression**. If only the predictor is a circular variable, then the relationship is characterized using **linear-circular regression**.

Both types of circular regression models can be given by

$$y_i = g(x_i, \boldsymbol{\beta}) + \epsilon_i (\text{mod } 2\pi),$$

where $g$ is some meaningful function (e.g. a trigonometric polynomial) and the $q$-dimensional parameter vector $\boldsymbol{\beta}$ could have some elements that are complex-valued. The expression $\epsilon_i(\text{mod } 2\pi)$ is read as $\epsilon_i$ modulus $2\pi$ and is a way to express the remainder of the quantity $\epsilon_i/(2\pi)$.[7] In this model, $\epsilon_i$ is a circular random error assumed to follow a von Mises distribution with circular mean 0 and concentration parameter $\kappa$. The von Mises distribution is the circular analog of the univariate normal distribution, but has a more complex form and its tails are truncated. A random variable $X$ follows the von Mises distribution with circular mean $\mu$ and concentration parameter $\kappa$ such that the probability

---

[7]For example, $11(\text{mod } 7) = 4$ because 11 divided by 7 leaves a remainder of 4.

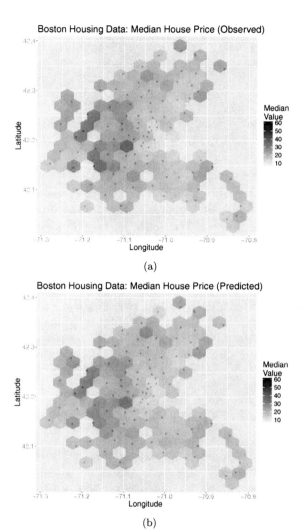

**FIGURE 25.18**

Hex map for (a) observed median housing prices and (b) predicted median housing prices. The approximate geographical centers of the 92 towns in Boston are overlaid as points.

density function is

$$f_X(x) = \frac{e^{\kappa \cos(x-\mu)}}{2\pi I_0(\kappa)}$$

and cumulative distribution function is

$$F_X(x) = \frac{1}{2\pi I_0(\kappa)} \left\{ x I_0(\kappa) + 2 \sum_{j=1}^{\infty} \frac{I_j(\kappa) \sin(j(x-\mu))}{j} \right\},$$

both of which are defined on the range $x \in [0, 2\pi)$. In the above, $I_p(\cdot)$ is called a **modified Bessel function of the first kind of order** $p$. The Bessel function is the contour integral

$$I_p(z) = \frac{1}{2\pi i} \oint e^{(z/2)(t-1/t)} t^{-(p+1)} dt,$$

where the contour encloses the origin and traverses in a counterclockwise direction in the complex plane such that $i = \sqrt{-1}$. Maximum likelihood estimates can be obtained for the circular regression models with minor differences in the details when dealing with a circular predictor or linear predictor variable. Clearly, such formulas do not lend themselves to closed-form solutions, requiring us to use numerical methods. For a thorough treatment on the topic, see the text by Chernov (2010) and the references therein.

Given the more geometric nature of circular regression models, there are a number of helpful data visualizations that can be produced for assessing the adequacy of the fitted models. While we do not provide an overview of all such visualizations, one important plot is a **rose diagram**, which is basically a circular version of a histogram that allows you to display circular data as well as the frequency of each bin. It can also be used to visualize the distribution of the residuals from the estimated circular regression model.

**Example 25.13.1.** *(Wind Direction Data)*
Oliveira et al. (2014) presented and analyzed data on the hourly observations of wind direction measured at a weather station in Texas from May 20, 2003 to July 31, 2003 inclusive. We are interested in the same subset of the measurements analyzed in Oliveira et al. (2014). The variables we are interested in analyzing are:

- $Y$: the wind direction measurements at 12:00 p.m. (in radians)

- $X$: the wind direction measurements at 6:00 a.m. (in radians)

Therefore, there are $n = 73$ pairs of direction measurements we will model.

We fit a circular-circular regression where the function $g$ represents a trigonometric polynomial of order 2 fitting the 12:00 p.m. wind direction measurement against the immediately preceding 6:00 a.m. wind direction measurement. The coefficients and relevant output from this fit are given below:

```
###############################
$coefficients
                  [,1]         [,2]
(Intercept)   0.3132750  -0.05772414
cos.x1        0.5759673   0.25468282
cos.x2       -0.0851367  -0.12376941
sin.x1       -0.0918941   0.47877083
sin.x2        0.1187634  -0.19018101


$p.values
            [,1]       [,2]
[1,] 0.8826034 0.9253103


$kappa
[1] 1.821278
###############################
```

The two columns of the estimated coefficients correspond to those coefficients of the model predicting the cosine and sine of the response, respectively. The rows correspond to the coefficients according to increasing trigonometric order. The two $p$-values are for testing whether or not a one-degree increase in the order is necessary. Since both of these $p$-values are large, we would claim that there is no statistically significant evidence of needing to fit a one-degree larger trigonometric polynomial. Finally, the error terms are assumed to follow a von Mises distribution with circular mean 0 and concentration parameter $\kappa$. The estimate of $\kappa$ is the third quantity above, which is 1.821.

The error terms from the circular-circular regression fit are plotted on a rose diagram in Figure 25.19(a). This clearly shows far fewer residuals falling within the $[\pi/2, 3\pi/2)$ range than those falling within the $[0, \pi/2)$ and $[3\pi/2, 2\pi)$ ranges. A scatterplot of the 12:00 p.m. wind direction versus the previous wind direction measurement at 6:00 a.m. along with the circular-circular regression fit is given in Figure 25.19(b). Overall, the use of a second-order trigonometric polynomial in the circular-circular regression model appears to be a good fit to these directional measurements.

## 25.14   Rank Regression

Chapter 17 discussed different approaches for nonparametric regression analysis. In the broader field of nonparametrics, classic approaches focus on rank-based methods as emphasized in the text by Lehmann (2006). Taking a rank-based approach to regression is a straightforward extension of these classic methods.

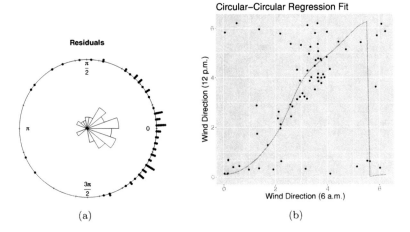

**FIGURE 25.19**
(a) Rose diagram of the circular residuals assuming a von Mises error distribution for the circular-circular regression fit of the wind direction data. (b) Scatterplot of the 12:00 p.m. wind direction versus the previous wind direction measurement at 6:00 a.m. along with the circular-circular regression fit overlaid.

Suppose we have the multiple linear regression model

$$\mathbf{Y} = \mathbf{X}\boldsymbol{\beta} + \boldsymbol{\epsilon},$$

but where the only assumption made on the error terms in the vector $\boldsymbol{\epsilon}$ is that the distribution is continuous. **Rank regression** (Jurečková, 1971; Jaeckel, 1972) entails finding a rank-based estimator to the regression coefficient $\boldsymbol{\beta}$. The rank-based estimator of $\boldsymbol{\beta}$ is

$$\hat{\boldsymbol{\beta}}_{\psi} = \underset{\boldsymbol{\beta}}{\operatorname{argmin}} \|\mathbf{y} - \mathbf{X}\boldsymbol{\beta}\|_{\psi},$$

where $\| \cdot \|_{\psi}$ is a pseudo-norm defined as

$$\|\mathbf{u}\|_{\psi} = \sum_{i=1}^{n} a(\mathcal{R}(u_i))u_i$$

such that $\mathcal{R}$ is the rank function, $a(t) = \psi\left(\frac{t}{n+1}\right)$, and $\psi(\cdot)$ (called a **score function**) is a non-decreasing, square integrable function defined on the open unit interval $(0, 1)$.

Based on asymptotic properties of $\hat{\boldsymbol{\beta}}_{\psi}$ (Hettmansperger and McKean, 2011), an approximate $100 \times (1 - \alpha)\%$ confidence interval for $\beta_j$ is

$$\hat{\beta}_{\psi;j} \pm t_{n-p;1-\alpha/2}\text{s.e.}(\hat{\beta}_{\psi;j}).$$

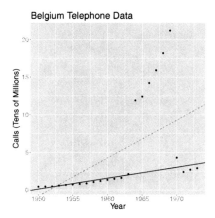

**FIGURE 25.20**
Scatterplot of the Belgium telephone data with the rank regression fit (solid black line) and the ordinary least squares fit (dashed red line) overlaid. Colors are in e-book version.

Kloke and McKean (2012) provide details on how to calculate s.e.$(\hat{\beta}_{\psi;j})$ as well as how to test the general linear hypothesis

$$H_0 : \mathbf{M}\beta = \mathbf{0}$$
$$H_A : \mathbf{M}\beta \neq \mathbf{0}$$

for some contrast matrix $\mathbf{M}$ in the rank regression setting.

**Example 25.14.1.** *(Belgium Telephone Data)*
Rousseeuw and Leroy (2003) presented data on telephone calls placed in Belgium spanning the years 1950 to 1973. Thus, the dataset consists of $n = 24$ observations. The variables in this study are:

- $Y$: the number of telephone calls placed (in tens of millions)

- $X$: the year

These data are interesting because there were outliers due to a mistake in the recording units for the years 1964 to 1969. Kloke and McKean (2012) analyzed these data using a rank regression, which we do as well.

Figure 25.20 is a scatterplot of the Belgium telephone data. The recorded number of phone calls between for the years 1964 to 1969 clearly illustrate an outlying behavior. Both the rank regression fit and ordinary least squares fit are overlaid on this scatterplot. Clearly the rank regression fit provides a more robust fit to these data as it is not being influenced by those outlying values. Just as with fitting linear regression models, we can also obtain summary results from the fitted model. For example, the standard errors and test statistics for the regression coefficients can be obtained:

```
##############################
Call:
rfit.default(formula = calls ~ year, data = telephone)

Coefficients:
             Estimate    Std. Error  t.value p.value
(Intercept) -283.967577  156.266852  -1.8172 0.08284 .
year           0.145683    0.079667   1.8287 0.08104 .
---
Signif. codes:  0 *** 0.001 ** 0.01 * 0.05 . 0.1   1
Overall Wald Test: 3.344 p-value: 0.05394
##############################
```

Based on this output, we would claim that the effect due to the year on the number of calls is marginally significant with a $p$-value of 0.08104.

## 25.15   Mixtures of Regressions

Suppose we have a large dataset consisting of the heights of males and females. The heights for the males will (on average) be higher than that of the females. A histogram of this data would clearly show two distinct bumps or modes. Knowing the gender labels of each subject would allow one to account for that subgroup in the analysis being used. However, what happens if the gender label of each subject was lost or unobserved? The setting where data appear to be from multiple subgroups, but there is no label providing such identification, is the focus of the area called **mixture modeling**. Texts devoted to the topic of mixture modeling include Titterington et al. (1985), Lindsay (1995), and McLachlan and Peel (2000).

There are many issues one should be cognizant of when building a mixture model. In particular, maximum likelihood estimation can be quite complex since the likelihood does not yield closed-form solutions and there are identifiability issues, however, the use of a Newton–Raphson or EM algorithm usually provides a good solution. Another option is to use a Bayesian approach with Markov Chain Monte Carlo (MCMC) methods, but this too has its own set of complexities. While we do not explore these issues, we illustrate how a mixture model can occur in the regression setting.

A **mixture-of-linear-regressions** model can be used when it appears that there is more than one regression line that could fit the data due to some underlying characteristic; i.e., a latent or unobserved variable. Suppose we have $n$ observations which belong to one of $k$ groups. If we knew to which group an observation belonged (i.e., its label), then we could write down explicitly the linear regression model given that observation $i$ belongs to group $j$:

$$y_i = \mathbf{x}_i^T \boldsymbol{\beta}_j + \epsilon_{i_j},$$

such that $\epsilon_{i_j}$ is normally distributed with mean 0 and variance $\sigma_j^2$. Notice how the regression coefficients and variance terms are different for each group. However, now assume that the labels are unobserved. In this case, we can only assign a probability that observation $i$ came from group $j$. Specifically, the density function for the mixture-of-linear-regressions model is:

$$f(y_i) = \sum_{j=1}^{k} \lambda_j (2\pi\sigma_j^2)^{-1/2} \exp\left\{ -\frac{1}{2\sigma_j^2}(y_i - \mathbf{x}_i^T\boldsymbol{\beta}_j)^2 \right\},$$

such that $\sum_{j=1}^{k} \lambda_j = 1$. Estimation is done by using the likelihood (or log-likelihood) function based on the above density. For maximum likelihood, one typically uses an EM algorithm (Dempster et al., 1977). Bayesian estimation can also be performed, which is the focus of Hurn et al. (2003). For more details on mixtures-of-linear-regressions models, including computational aspects, see DeVeaux (1989), Turner (2000), and Benaglia et al. (2009).

It should be noted that variants on mixtures-of-regressions models appear in many areas. For example, in economics it is called **switching regimes** (Quandt, 1958). In the social sciences it is called **latent class regressions** (Wedel and DeSarbo, 1994). As we noted earlier, the neural networking terminology calls this model (without the hierarchical structure) the **mixture-of-experts** problem (Jacobs et al., 1991).

**Example 25.15.1.** *(Ethanol Fuel Data)*
Brinkman (1981) presented data from an experiment in which ethanol was burned in a single cylinder automobile test engine. The dataset consists of $n = 88$ observations. The variables in this study are:

- $Y$: equivalence ratio for the engine at compression ratios from 7.5 to 18

- $X$: peak nitrogen oxide emission levels

Figure 25.21(a) gives a plot of these data.

A plot of the equivalence ratios versus the measure of nitrogen oxide is given in Figure 25.21(a). Suppose one wanted to predict the equivalence ratio from the amount of nitrogen oxide emissions. Clearly there are two groups of data where separate regressions appear appropriate – one with a positive trend and one with a negative trend. Thus, we analyze these data using a mixture of linear regressions, which was also the model used in Hurn et al. (2003).[8]

Figure 25.21(b) is again a plot of the data, but with estimates from an EM algorithm overlaid. EM algorithm estimates for this data are as follows:

---

[8]We note that these data have also been analyzed using nonlinear and nonparametric regression models where the roles of the variables have been flipped.

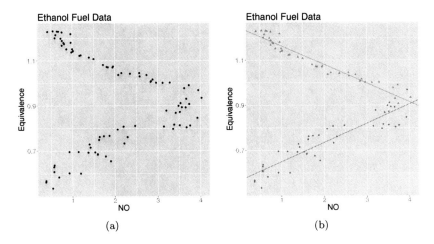

(a)                                                        (b)

**FIGURE 25.21**
(a) Plot of the ethanol fuel data with equivalence ratio as the response and the measure of nitrogen oxide emissions as the predictor. (b) Plot of the data with EM algorithm estimates from a 2-component mixture of regressions fit overlaid with different symbols indicating most probable component membership.

```
#############################
$lambda
[1] 0.4897245 0.5102755

$beta
            comp.1       comp.2
beta.0 0.5649857   1.24708156
beta.1 0.0850231  -0.08299983

$sigma
[1] 0.04331334 0.02414119

$loglik
[1] 122.0384
#############################
```

The regression coefficient estimates for each component are given above, which correspond to the different colored lines in Figure 25.21(b). Moreover, the mixing proportions indicate that the two component populations in the data occur with proportions of, approximately, 0.49 and 0.51, respectively. The most probable component to which each observation belongs is also indicated by the different symbols for the points in Figure 25.21(b).

## 25.16 Copula Regression

Suppose that the $p$-dimensional random vector $\mathbf{X} = (X_1, \ldots, X_p)^T$ has continuous marginals; i.e.,

$$U_i \equiv F_{X_j}(x) = P(X_j \le x)$$

for $j = 1, \ldots, p$. A **copula** is the joint cumulative distribution of $\mathbf{U} = (U_1, \ldots, U_p)^T$:

$$\mathcal{C}(u_1, \ldots, u_p) = P(U_1 \le u_1, \ldots, U_p \le u_p).$$

The copula links the marginal distributions together to form the joint distribution. Thus, the copula $\mathcal{C}$ gives us all of the information about the dependence structure between the components of $\mathbf{X}$. Moreover, the importance of copulas is emphasized by the following famous theorem:

**Theorem 2.** (Sklar's Theorem) Let $\mathcal{H}$ be a $p$-dimensional distribution function with marginal distribution functions $F_{X_j}$ for the random variables $X_1, \ldots, X_p$. Then there exists a copula $\mathcal{C}$ such that

$$\mathcal{H}(x_1, \ldots, x_p) = \mathcal{C}(F_{X_1}(x_1), \ldots, F_{X_p}(x_p)).$$

Conversely, given a copula $\mathcal{C}$ and marginals $F_{X_j}$, then $\mathcal{H}$ is a $p$-dimensional distribution function. Furthermore, if the $F_{X_j}$ are continuous, then $\mathcal{C}$ is unique.

Suppose now that we have $n$ dependent variables $Y_1, \ldots, Y_n$ that depend on $p$-dimensional and $q$-dimensional predictor vectors $\mathbf{X}_1, \ldots, \mathbf{X}_n$ and $\mathbf{Z}_1, \ldots, \mathbf{Z}_n$, respectively. Suppose further that the marginal cumulative distribution function of a single $Y_i$ is denoted by $F(\cdot|\mathbf{X}_i, \mathbf{Z}_i)$, which is parameterized in terms of a location parameter $\mu_i$ and a (possible) dispersion parameter $\psi_i$. In particular, let $h_1(\cdot)$ and $h_2(\cdot)$ be the link functions with regression coefficient vectors $\boldsymbol{\beta}$ and $\boldsymbol{\gamma}$ such that

$$h_1(\mu_i) = \mathbf{X}_i^T \boldsymbol{\beta} \quad \text{and} \quad h_2(\psi_i) = \mathbf{Z}_i^T \boldsymbol{\gamma}.$$

In practice, it is possible that the vector of predictors in the parameterization of the dispersion parameter has some or all of the variables in $\mathbf{Z}_i$ the same as those in $\mathbf{X}_i$.

Let $y_1, \ldots, y_n$, $\mathbf{x}_1, \ldots, \mathbf{x}_n$, and $\mathbf{z}_1, \ldots, \mathbf{z}_n$ be the realizations of the quantities we introduced above. In **Gaussian copula regression** (Masarotto and Varin, 2012), the dependence between the $n$ response variables is modeled with a Gaussian copula. This gives the joint cumulative distribution function

$$P(Y_1 \le y_1, \ldots, Y_n \le y_n) = \Phi_n(\epsilon_1, \ldots, \epsilon_n; \mathbf{R}),$$

where $\epsilon_i = \Phi^{-1}\{F(y_i|\mathbf{x}_i, \mathbf{z}_i)\}$ such that $\Phi(\cdot)$ is, again, the standard normal

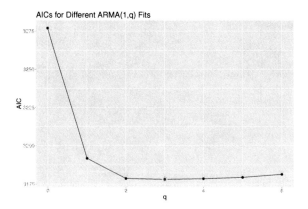

**FIGURE 25.22**
Plot of AIC values versus the moving avereage order $q$ for determining the best ARMA$(1, q)$ copula correlation to use. The value at $q = 3$ has the minimum AIC.

cumulative distribution function and $\Phi_n(\cdot; \mathbf{R})$ is the $n$-dimensional normal cumulative distribution function with correlation matrix $\mathbf{R}$. Choices of the correlation matrix include those discussed in the context of generalized estimating equations in Chapter 21. Estimation for Gaussian copula regression models can be performed using maximum likelihood estimation. We refer to Masarotto and Varin (2012) for details on estimation of Gaussian copula regression models and refer to the text by Nelsen (2006) for a general treatment of copulas.

**Example 25.16.1.** *(Cardiovascular Data, cont'd)*
We continue with the cardiovascular data that were analyzed in Example 13.5.2. We consider a Gaussian copula regression model with a marginal Gaussian distribution and an ARMA$(1, q)$ copula correlation. We use $p = 1$ based on the order of the lag found in the analysis of Example 13.5.2. We determined $q$ by fitting ARMA$(1, q)$ models for $q = 0, 1, \ldots, 6$ and selecting the model fit having the minimum AIC. These results are shown in Figure 25.22, which indicate a moving average order of 3 should be used.

Below are the results from the fitted copula regression model:

```
#############################
Coefficients marginal model:
            Estimate Std. Error z value Pr(>|z|)
(Intercept) 79.76918    3.58302  22.263  < 2e-16 ***
tempr        0.03295    0.05148   0.640    0.522
part         0.13885    0.02877   4.827 1.39e-06 ***
sigma        9.35690    0.94687   9.882  < 2e-16 ***
```

```
Coefficients Gaussian copula:
    Estimate Std. Error z value Pr(>|z|)
ar1  0.92166    0.02235  41.234  < 2e-16 ***
ma1 -0.49902    0.05205  -9.587  < 2e-16 ***
ma2  0.20140    0.04749   4.241 2.23e-05 ***
ma3 -0.07893    0.04923  -1.603    0.109
---
Signif. codes:  0 '***' 0.001 '**' 0.01 '*' 0.05 '.' 0.1 ' ' 1

log likelihood = 1581.1,  AIC = 3178.1
###############################
```

The top-half of the output gives the estimated regression coefficients and the bottom-half give the coefficients for the Gaussian copula with an ARMA$(1, 3)$ correlation structure. In the top-half, the average weekly temperature variable is not statistically significant. Rerunning the analysis with this variable omitted results in an AIC of 3176.5, only slightly lower than the AIC of the model with temperature included (3178.1). We decided to keep temperature in the model for the remaining analysis.

Figure 25.23 gives various diagnostic plots for this copula regression fit. Figure 25.23(a) is a plot of the time-ordered quantile residuals, which does not indicate any trends that may need to be explained further by a different correlation copula structure. Figure 25.23(b) is a normal Q-Q plot of the quantile residuals, which also looks good and does not indicate any significant departures from normality. Figure 25.23(c) is a plot of the quantile residuals versus the fitted values, which look very good and are not indicative of any sort of lack of fit. However, Figure 25.23(d) is a plot of the fitted values from the copula regression model versus the observed values of cardiovascular mortality. Clearly the points are dispersed about the $y = x$ line. This indicates that the predictive ability of this model could stand to be improved. As noted in Example 13.5.2, a number of other variables are available with this dataset. Exploring other copula regression models using these variables as predictors could help improve the predicted values plotted in Figure 25.23(d).

## 25.17 Tensor Regression

The era of big data has produced increasingly more problems involving multidimensional arrays. Some applied problems having data with multidimensional arrays include functional MRI (fMRI) data (time $\times$ $x$-axis $\times$ $y$-axis $\times$ $z$-axis), image sequences in computer vision (pixel $\times$ illumination $\times$ expression $\times$ viewpoints), and climate dynamics (temperature $\times$ latitude $\times$ longitude $\times$ time). If we have responses and/or predictors that are multidimensional arrays, the question is how to model such data.

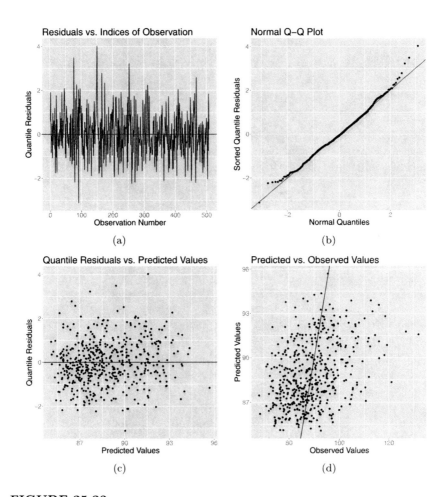

**FIGURE 25.23**
(a) Plot of the time-ordered quantile residuals from the copula regression fit.
(b) Normal Q-Q plot of the quantile residuals. (c) Plot of the quantile residuals
versus the copula regression fitted values. (d) Plot of the fitted values versus
the observed values for the cardiovascular data.

First, consider **tensors**, which are geometric objects that obey certain transformation rules (e.g., linear relations) between vectors, scalars, and other tensors. When given a coordinate basis, a tensor can be represented as a multidimensional array. Therefore, we can develop a notion of **tensor regression** when one has responses and/or predictors that are tensors. In order to mathematically develop some details of the tensor regression problem, we need to present some definitions. The definitions below follow portions of the material presented in Yu and Liu (2016).

Let $\mathcal{W}$ represent a tensor. Each dimension of a tensor is called a **mode**. An **$m$-mode unfolding** of the tensor $\mathcal{W}$ along mode $m$ transforms a tensor into a matrix, say, $\mathcal{W}_{(m)}$, by treating $m$ as the first mode and then subsequently concatenating each mode. This notion of **tensor matricization** is somewhat similar to when we vectorize a matrix. The **$m$-mode product** between tensor $\mathcal{W}$ and matrix $\mathbf{V}$ on mode $m$ is defined as $(\mathcal{W} \times_m \mathbf{V})_{(m)} = \mathbf{V}\mathcal{W}_{(m)}$. Finally, a **Tucker decomposition** decomposes a tensor into a set of matrices and one small core tensor; i.e., $\mathcal{W} = \mathcal{S} \times_1 \mathbf{V}_1 \times_2 \cdots \times_n \mathbf{V}_n$, where $\mathcal{S}$ is the core tensor.

Let $\mathcal{X}$, $\mathcal{Y}$, and $\mathcal{B}$ denote $D$-dimensional predictor, response, and coefficient tensors such that the $D$-dimensions are given by $p_1 \times \cdots \times p_D$. This implies that $\mathcal{B}$ has $\prod_{i=1}^{D} p_i$, which is ultrahigh dimensional and will far exceed the sample size $n$. The general tensor regression problem is to solve the following:

$$\hat{\mathcal{B}} = \underset{\mathcal{B}}{\mathrm{argmin}} \ \mathcal{L}(\mathcal{B}; \mathcal{X}, \mathcal{Y}), \tag{25.12}$$

such that $\mathrm{rank}(\mathcal{B}) \leq R$. The above is a loss function that we use to estimate $\mathcal{B}$ subject to the constraint that $\mathcal{B}$ has a Tucker rank decomposition of at most $R$; i.e., $\mathcal{B}$ has a low-dimensional factorization $\mathcal{B} = \mathcal{S} \times_1 \mathcal{V}_1 \times_2 \cdots \times_D \mathcal{V}_D$ with core tensor $\mathcal{S}$ and orthonormal projection matrices $\mathcal{V}_1, \ldots, \mathcal{V}_D$. The dimensionality of the core tensor $\mathcal{S}$ is at most $R$.

As noted in Yu and Liu (2016), there are various types of tensor regression models that have been discussed in the literature, all of which are special cases of the problem in (25.12). Perhaps the most intuitive model is the one due to Zhou et al. (2013). The model is

$$\mathcal{Y} = \mathrm{vec}(\mathcal{X})^{\mathrm{T}}\mathrm{vec}(\mathcal{B}) + \mathcal{E},$$

where $\mathcal{E}$ is a normal random error tensor. Estimation of the parameters in this model can be broken down into a sequence of low dimensional optimizations using the method of **alternating least squares** (de Leeuw et al., 1976). We refer to Zhou et al. (2013) for more details about this tensor regression model, including its use in the analysis of neuroimaging data.

**Example 25.17.1.** *(NASA Data)*
The National Aeronautics and Space Administration (NASA) provided data on geographic and atmospheric measures over Central America for the American Statistical Association's Data Expo at the 2006 Joint Statistical Meetings. The aim of the competition was to provide graphical summaries of important

features of these data. Murrell (2010) presents details of the data and the competition. The top three winning analyses were Cho and Chun (2010), Hobbs et al. (2010), and Eden et al. (2010), respectively. The variables in this dataset are:

- $Y$: level of ozone abundance (measured in Dobson units)

- $X_1$: the monthly mean percent of the sky covered by clouds at altitudes above 6.5 km ("high-clouds")

- $X_2$: the monthly mean percent of the sky covered by clouds at altitudes between 3.24 km and 6.5 km ("mid-clouds")

- $X_3$: the monthly mean percent of the sky covered by clouds at altitudes below 3.24 km ("low-clouds")

- $X_4$: monthly mean air pressure (in millibars)

- $X_5$: monthly mean surface temperature (in degrees Kelvin)

- $X_6$: monthly mean air temperature (in degrees Kelvin)

The data were measured on a $24 \times 24$ grid with a resolution of 2.5 degrees of latitude and longitude. As noted above, the measurements are monthly means (i.e., 12 measurements) over the years 1995 to 2000, inclusive. There are 110 missing values for the low-cloud measurement $X_3$. We did a local mean imputation where we took the mean of the five values immediately before and the five values immediately after the missing measurement.

As discussed in Murrell (2010), The competition involving these data focused on data visualization. We will use these data to provide an illustration of how a tensor regression could be developed. We emphasize that since this is an emerging research topic, the available computational tools and general statistical methodology for estimating tensor regression models are still in their infancy. Moreover, there are various formulations for tensor regression models as we briefly outlined earlier. Thus, the analysis we present here is more of a *proof of concept* regarding how to fit a tensor-type regression model.

For these data, we model ozone as a function of the remaining predictors. The tensors are constructed for each variable such that $\mathcal{X}$ and $\mathcal{Y}$ have dimensions $24 \times 24 \times 12 \times 6$, where $j = 1, \ldots, 6$ is the index of the predictor variables. Thus, the dimensionality of this problem is 41,472. Zhou et al. (2013) showed that a low rank tensor regression model provides a strong recovery of many low rank signals for their MRI data. In particular, they used a rank-3 decomposition. To simplify our problem, we use the core tensor after performing a Tucker decomposition of 3 modes for each dimension. We chose 3 simply because Zhou et al. (2013) used a rank-3 decomposition in their application; we provide no additional justification for the appropriateness of such an assumption for our data.

We then proceed to vectorize the corresponding tensors and use ordinary least squares to obtain coefficient estimates. The estimates are as follows:

```
##############################
, , 1

            I2
I1              [,1]    [,2]    [,3]
  (Intercept) -9.398  36.353  -6.633
  cloudhigh    0.254   1.597  -1.004
  cloudmid     0.155  -2.129  -1.339
  cloudlow    -0.364  -3.478   0.182
  pressure    -0.069   0.040  -2.989
  surftemp     1.778  -0.050  -2.327
  temperature -0.626  -2.201  -0.185

, , 2

            I2
I1              [,1]     [,2]     [,3]
  (Intercept) 44.438  -91.554  -16.932
  cloudhigh    0.512   -0.778   -0.259
  cloudmid     0.621   -1.041   -0.845
  cloudlow     0.600    0.564   -1.059
  pressure    -0.092    0.522    0.207
  surftemp     1.478   -2.779    1.822
  temperature  4.949   -4.834    2.832

, , 3

            I2
I1              [,1]     [,2]    [,3]
  (Intercept) -11.283  13.474  -6.988
  cloudhigh    -0.414   0.699  -0.082
  cloudmid      0.229  -3.441  -0.425
  cloudlow     -0.292   1.759   1.021
  pressure     -0.115   0.033   0.645
  surftemp     -2.608  -1.967  -0.279
  temperature  -0.180   2.574  -3.203

attr(,"class")
[1] "tensor"
##############################
```

There are three slices of coefficient estimates above, one pertaining to each of the 3 modes used in the Tucker decomposition. While it is challenging to provide interpretation of individual coefficients, we can make some general assessments. For example, the magnitude of each coefficient (in absolute value) tends to be the same for a given predictor across the three slices of the tensor fit. The only estimates that are fairly different across the slices are those for the intercept term. While no assessment about the statistical significance of these effects is presented, results from such tensor-type regression models could help build better models for prediction to indicate areas of more extreme ozone abundance.

# Part V

# Appendices

# A

## *Steps for Building a Regression Model*

This appendix outlines the basic steps for building a regression model. It is recommended that from time to time you revisit these steps as the corresponding techniques are introduced throughout the text. This outline provides a helpful framework for connecting the material with the broader scientific method.

1. After establishing a research hypothesis, proceed to design an appropriate experiment or experiments. Identify variables of interest, which variable(s) will be the response, and what levels of the independent variables you wish to cover in the study. If costs allow for it, then a pilot study may be helpful (or necessary).

2. Collect the data and make sure to "clean" it for any bugs; e.g., entry errors. If data from many variables are recorded, then variable selection and screening should be performed.

3. Consider the regression model to be used for studying the relationship and assess the adequacy of such a model. Initially, a linear regression model might be investigated. But as this text shows, there are numerous regression models and regression strategies for handling different data structures. How you assess the adequacy of the fitted model will depend on the type of regression model that is being used as well as the corresponding assumptions. For linear regression, the following need to be checked:

   (a) Check for normality of the residuals. This is often done through a variety of visual displays, but formal statistical testing can also be performed.

   (b) Check for constant variance of the residuals. Again, visual displays and formal testing can both be performed.

   (c) After time-ordering your data, assess the independence of the observations. Independence is best assessed by looking at a time-ordered plot of the residuals, but other time series techniques exist for assessing the assumption of independence.

   Regardless, checking the assumptions of your model as well as the model's overall adequacy is usually accomplished through residual diagnostic procedures.

4. Look for any outlying or influential data points that may be affecting the overall fit of your current model. Care should be taken with

how you handle these points as they could be legitimate in how they were measured. While the option does exist for excluding such problematic points, this should only be done after carefully considering if such points are recorded in error or are truly not representative of the data you collected. If any corrective actions are taken in this step, then return to Step 3.

5. Assess multicollinearity; i.e., linear relationships amongst your independent variables. Multicollinearity can provide incorrect estimates as well as other issues. If you proceed to omit variables or observations which may be contributing to multicollinearity, then return to Step 3.

6. Assess the predictability and overall goodness-of-fit of your estimated model. If indicated, modifications to the model may be in order, such as using a different functional form or down-weighting certain observations. If you must take such actions, then return to Step 3 afterwards.

# B

## Refresher on Matrices and Vector Spaces

Suppose we have the following matrix:

$$\mathbf{A}_{r \times c} = \begin{pmatrix} a_{1,1} & a_{1,2} & \cdots & a_{1,c} \\ a_{2,1} & a_{2,2} & \cdots & a_{2,c} \\ \vdots & \vdots & \ddots & \vdots \\ a_{r,1} & a_{r,2} & \cdots & a_{r,c} \end{pmatrix},$$

where the subscript $r \times c$ pertains to the dimensions of the matrix. This is often suppressed except when to clarify. Then, the following are some general definitions and properties pertaining to $\mathbf{A}$:

- The **order** of $\mathbf{A}$ is $r \times c$, which is the number of rows and the number of columns. These are also called the **dimensions** of the matrix.

- $\mathbf{A}$ is a **square matrix** if $r = c$.

- The **transpose** of $\mathbf{A}$ (written $\mathbf{A}^{\mathrm{T}}$), is obtained by writing all rows as columns in the order which they appear so that the columns become rows. For example, if $\mathbf{A} = \begin{pmatrix} 1 & 12 \\ 2 & 0 \\ 7 & 3 \end{pmatrix}$, then $\mathbf{A}^{\mathrm{T}} = \begin{pmatrix} 1 & 2 & 7 \\ 12 & 0 & 3 \end{pmatrix}$.

- When $\mathbf{A}$ is a square matrix, $\mathbf{A}$ is also a **symmetric matrix** if $\mathbf{A} = \mathbf{A}^{\mathrm{T}}$.

- When $\mathbf{A}$ is a square matrix, $\mathbf{A}$ is also a **diagonal matrix** if all off-diagonal elements are 0.

- The **trace** of $\mathbf{A}$ (written as $\mathrm{tr}(\mathbf{A})$) is the sum of all the diagonal elements when $\mathbf{A}$ is square.

- The **identity matrix** (written as $\mathbf{I}$), is a diagonal matrix with all diagonal entries equal to 1. For example, the $3 \times 3$ identity matrix is $\mathbf{I}_3 = \begin{pmatrix} 1 & 0 & 0 \\ 0 & 1 & 0 \\ 0 & 0 & 1 \end{pmatrix}$, where we have included the subscript 3 to indicate the dimension of the matrix.

- A **null matrix** (written as $\mathbf{0}$) is a matrix with all entries equal to 0.

- $\mathbf{A}$ is an **idempotent matrix** if $\mathbf{A}^2 = \mathbf{A}$.

575

- $\mathbf{A}$ is a **nilpotent matrix** if $\mathbf{A}^2 = \mathbf{0}$.

- $\mathbf{A}$ is a **unipotent matrix** if $\mathbf{A}^2 = \mathbf{I}$.

- When $\mathbf{A}$ is a square matrix, $\mathbf{A}$ is also an **orthogonal matrix** if $\mathbf{A}\mathbf{A}^{\mathrm{T}} = \mathbf{A}^{\mathrm{T}}\mathbf{A} = \mathbf{I}$.

- When $\mathbf{A}$ is a square matrix, the **inverse** of $\mathbf{A}$ is $\mathbf{A}^{-1}$ such that $\mathbf{A}^{-1}\mathbf{A} = \mathbf{A}\mathbf{A}^{-1} = \mathbf{I}$.

- The **Associative Law of Addition** holds: $\mathbf{A} + \mathbf{B} + \mathbf{C} = (\mathbf{A} + \mathbf{B}) + \mathbf{C} = \mathbf{A} + (\mathbf{B} + \mathbf{C})$, where $\mathbf{A}$, $\mathbf{B}$, and $\mathbf{C}$ are of the same order.

- The **Associative Law of Multiplication** holds: $\mathbf{A}\mathbf{B}\mathbf{C} = (\mathbf{A}\mathbf{B})\mathbf{C} = \mathbf{A}(\mathbf{B}\mathbf{C})$, where $\mathbf{A}$, $\mathbf{B}$, and $\mathbf{C}$ have compatible dimensions.

- The **Commutative Law of Addition** holds: $\mathbf{A} + \mathbf{B} = \mathbf{B} + \mathbf{A}$, where $\mathbf{A}$ and $\mathbf{B}$ have compatible dimensions.

- The **Commutative Law of Multiplication** does not hold: $\mathbf{A}\mathbf{B} \neq \mathbf{B}\mathbf{A}$.

- If $\mathbf{A}$ is $r \times c$ and $\mathbf{B}$ is $p \times q$, then the **Kronecker product** (denoted by $\otimes$) of $\mathbf{A}$ and $\mathbf{B}$ is an $rp \times cq$ block matrix, which is written as

$$
\mathbf{A} \otimes \mathbf{B} = \begin{bmatrix}
a_{11}b_{11} & a_{11}b_{12} & \cdots & a_{11}b_{1q} & \cdots & \cdots & a_{1c}b_{11} & a_{1c}b_{12} & \cdots & a_{1c}b_{1q} \\
a_{11}b_{21} & a_{11}b_{22} & \cdots & a_{11}b_{2q} & \cdots & \cdots & a_{1c}b_{21} & a_{1c}b_{22} & \cdots & a_{1c}b_{2q} \\
\vdots & \vdots & \ddots & \vdots & & & \vdots & \vdots & \ddots & \vdots \\
a_{11}b_{p1} & a_{11}b_{p2} & \cdots & a_{11}b_{pq} & \cdots & \cdots & a_{1c}b_{p1} & a_{1c}b_{p2} & \cdots & a_{1c}b_{pq} \\
\vdots & \vdots & & \vdots & \ddots & & \vdots & \vdots & & \vdots \\
\vdots & \vdots & & \vdots & & \ddots & \vdots & \vdots & & \vdots \\
a_{r1}b_{11} & a_{r1}b_{12} & \cdots & a_{r1}b_{1q} & \cdots & \cdots & a_{rc}b_{11} & a_{rc}b_{12} & \cdots & a_{rc}b_{1q} \\
a_{r1}b_{21} & a_{r1}b_{22} & \cdots & a_{r1}b_{2q} & \cdots & \cdots & a_{rc}b_{21} & a_{rc}b_{22} & \cdots & a_{rc}b_{2q} \\
\vdots & \vdots & \ddots & \vdots & & & \vdots & \vdots & \ddots & \vdots \\
a_{r1}b_{p1} & a_{r1}b_{p2} & \cdots & a_{r1}b_{pq} & \cdots & \cdots & a_{rc}b_{p1} & a_{rc}b_{p2} & \cdots & a_{rc}b_{pq}
\end{bmatrix}.
$$

- The **determinant** of a square matrix $\mathbf{A}$ (written $|\mathbf{A}|$), is a scalar. For an $n$-order matrix, the determinant is the sum of $n!$ signed products of the elements of $\mathbf{A}$. Some properties include:

    - $|\mathbf{A}^{\mathrm{T}}| = |\mathbf{A}|$.
    - If two rows of $\mathbf{A}$ are the same, then $|\mathbf{A}| = 0$.
    - $|\mathbf{A}\mathbf{B}| = |\mathbf{A}||\mathbf{B}|$.
    - $|\mathbf{A}\mathbf{B}| = |\mathbf{B}\mathbf{A}|$.
    - For $\mathbf{A}$ orthogonal, $|\mathbf{A}| = \pm 1$.
    - For $\mathbf{A}$ idempotent, $|\mathbf{A}| = 0$ or $1$.

- **A** is a **singular matrix** if $|\mathbf{A}| = 0$ and therefore has no inverse.

- The **eigenvalues** of a symmetric matrix $\mathbf{A}_{n \times n}$ are the roots of the characteristic polynomial, which is defined as $|\lambda \mathbf{I}_n - \mathbf{A}_{n \times n}| = \mathbf{0}$. The set of these eigenvalues, $\{\lambda_1, \ldots, \lambda_n\}$, is called the **spectrum** of the matrix $\mathbf{A}_{n \times n}$. The corresponding **eigenvectors** $\{\mathbf{p}_1, \ldots, \mathbf{p}_n\}$ are found by solving $(\mathbf{A}_{n \times n} \lambda_i \mathbf{I}_n) \mathbf{p}_i = \mathbf{0}$ for $i = 1, 2, \ldots, n$.

It is also important to establish some concepts regarding linear independence. Suppose $\mathbf{b}$ and $\mathbf{x}$ are vectors of appropriate dimensions. We are interested in solving a system of linear equations $\mathbf{A}\mathbf{x} = \mathbf{b}$.

- Suppose $\mathbf{a} = \mathbf{0}$ is the only vector for which $a_1 \mathbf{x}_1 + a_2 \mathbf{x}_2 + \ldots + a_n \mathbf{x}_n = \mathbf{0}$. Then, provided none of the $\mathbf{x}_i$ are null vectors, the $\mathbf{x}_i$ are said to be **linearly independent**.

- For a square matrix $\mathbf{A}$, the columns of $\mathbf{A}$ are linearly independent if and only if $\mathbf{A}\mathbf{x} = \mathbf{0}$ only for $\mathbf{x} = \mathbf{0}$ if and only if $\mathbf{A}$ is nonsingular.

- The **rank** of a matrix $\mathbf{A}$ (written rank($\mathbf{A}$)), is the number of linearly independent rows (columns) of a matrix. Specifically,

  - **A** is of **full rank** when it is a square matrix with all linearly independent rows (columns).

  - **A** is of **full row rank** when it has all linearly independent rows.

  - **A** is of **full column rank** when it has all linearly independent columns.

- Suppose that $\mathbf{A}$ is an $r \times c$ matrix and not necessarily of full rank. Then the $c \times r$ matrix $\mathbf{A}^-$ is the **generalized inverse** of $\mathbf{A}$ if the following four conditions are met:

  1. $\mathbf{A}\mathbf{A}^-\mathbf{A} = \mathbf{A}$
  2. $\mathbf{A}^-\mathbf{A}\mathbf{A}^- = \mathbf{A}^-$
  3. $(\mathbf{A}\mathbf{A}^-)^{\mathrm{T}} = \mathbf{A}\mathbf{A}^-$
  4. $(\mathbf{A}^-\mathbf{A})^{\mathrm{T}} = \mathbf{A}^-\mathbf{A}$

- A set $\mathcal{S} \subset \mathbb{R}^n$ is a **subspace** if it is closed under the operation of taking finite linear combinations; i.e., for all $c_i \in \mathbb{R}$, if $\mathbf{x}_i \in \mathcal{S}$, then $\sum_i c_i \mathbf{v}_i \in \mathcal{S}$. Note that the $n$-dimensional vector of zeroes, $\mathbf{0}$, is in every subspace. $\mathbb{R}^n$ is the $n$-dimensional space of real numbers.

- The set consisting only of the $\mathbf{0}$ vector is called the **trivial subspace**.

- Let $\mathbf{x}_i \in \mathcal{S}$. The $\mathbf{x}_i$ **span** $\mathcal{S}$ if every vector in $\mathcal{S}$ is a finite linear combination of the $\mathbf{x}_i$.

- The $\mathbf{x}_i$ are a **basis** for $\mathcal{S}$ if they are linearly independent and they span $\mathcal{S}$.

- The $\mathbf{x}_i$ form an **orthonormal basis** for $\mathcal{S}$ if they are a basis and, in addition, $\mathbf{x}_i^{\mathrm{T}}\mathbf{x}_i = 1$ and $\mathbf{x}_i^{\mathrm{T}}\mathbf{x}_j = 0$ when $i \neq j$. Every non-trivial subspace has an orthonormal basis and any two bases for a subspace have the same number of vectors called the **dimension** of $\mathcal{S}$ and is written as $\dim(\mathcal{S})$.

- The $n \times p$ matrix $\mathbf{X}$ is a **basis matrix** for $\mathcal{S}$ if the columns of $\mathbf{X}$ form a basis for $\mathcal{S}$. $\mathbf{X}$ is an **orthonormal basis matrix** for $\mathcal{S}$ if the columns of $\mathbf{X}$ form an orthonormal basis for $V$. Some properties of basis and orthonormal basis matrices include:

  - If $\mathbf{X}$ is a basis matrix, then its columns are linearly independent and, hence, $\mathbf{X}^{\mathrm{T}}\mathbf{X} > 0$ (positive definite) and is invertible.
  - If $\mathbf{X}$ is an orthonormal basis matrix, then $\mathbf{X}^{\mathrm{T}}\mathbf{X} = \mathbf{I}_p$.
  - If $\mathbf{X}$ is a basis matrix for $\mathcal{S}$, then $\mathbf{a} \in \mathcal{S}$ if and only if $\mathbf{a} = \mathbf{X}\mathbf{b}$ for some $\mathbf{b} \in \mathbb{R}^p$.

- Let $\mathbf{u}, \mathbf{v} \in \mathbb{R}^n$ and let $\mathcal{S}_U$ and $\mathcal{S}_V$ be the respective subspaces of $\mathbb{R}^n$ to which those vectors belong. $\mathbf{u}$ and $\mathbf{v}$ are said to be **orthogonal** (written $\mathbf{u} \perp \mathbf{v}$) if $\mathbf{u}^{\mathrm{T}}\mathbf{v} = 0$, $\mathbf{u}$ is orthogonal to $\mathcal{S}_V$ (written $\mathbf{u} \perp \mathcal{S}_V$) if $\mathbf{u}^{\mathrm{T}}\mathbf{v} = 0 \ \forall \ \mathbf{v} \in \mathcal{S}_V$, and $\mathcal{S}_U$ and $\mathcal{S}_V$ are orthogonal subspaces (written $\mathcal{S}_U \perp \mathcal{S}_V$) if $\mathbf{u}^{\mathrm{T}}\mathbf{v} = 0 \ \forall \ \mathbf{u} \in \mathcal{S}_U, \ \forall \ \mathbf{v} \in \mathcal{S}_V$.

- The **Euclidean norm** $\| \cdot \|$ means for an $n$-dimensional vector $\mathbf{x}$ and an $n \times p$ dimensional matrix $\mathbf{X}$, we have

$$\|\mathbf{x}\| = \sqrt{\sum_{i=1}^{n} x_i^2} \Rightarrow \|\mathbf{x}\|^2 = \sum_{i=1}^{n} x_i^2,$$

which is the inner product of $\mathbf{v}$ with itself, and

$$\|\mathbf{X}\|^2 = \mathbf{X}^{\mathrm{T}}\mathbf{X}.$$

- An $n \times n$ square matrix $\mathbf{X}$ is an **orthogonal matrix** if $\mathbf{X}^{\mathrm{T}}\mathbf{X} = \mathbf{X}\mathbf{X}^{\mathrm{T}} = \mathbf{I}_n$.

- An $r \times c$ square matrix $\mathbf{X}$ with $r > c$ is a **semi-orthogonal matrix** if its columns are orthogonal and have norm 1 so that $\mathbf{X}^{\mathrm{T}}\mathbf{X} = \mathbf{I}_c$.

- For any subspace $\mathcal{S} \in \mathbb{R}^n$ and for all $\mathbf{y} \in \mathbb{R}^n$, the **orthogonal projection** of $\mathbf{y}$ onto $\mathcal{S}$ is the vector $\mathbf{P}_{\mathcal{S}}\mathbf{y}$ such that $\mathbf{P}_{\mathcal{S}}\mathbf{y} \in \mathcal{S}$ and $(\mathbf{y} - \mathbf{P}_{\mathcal{S}}\mathbf{y}) \perp \mathcal{S}$. The matrix $\mathbf{P}_{\mathcal{S}}$ is called the **projection matrix**, which is a symmetric, idempotent matrix such that $\mathrm{tr}(\mathbf{P}_{\mathcal{S}}) = \mathrm{rank}(\mathbf{P}_{\mathcal{S}}) = \dim(\mathcal{S})$. Furthermore, if $\mathbf{X}$ is a basis matrix for $\mathcal{S}$, then

$$\mathbf{H} = \mathbf{X}(\mathbf{X}^{\mathrm{T}}\mathbf{X})^{-1}\mathbf{X}^{\mathrm{T}}$$

is a projection matrix.

- The **orthogonal complement** of $\mathcal{S} \in \mathbb{R}^n$, written $\mathcal{S}^{\perp}$, is the set of all vectors in $\mathbb{R}^n$ that are orthogonal to $\mathcal{S}$.

# C

## Some Notes on Probability and Statistics

**Statistics** are determined using data to characterize various aspects underlying the population of interest that is being studied. This appendix provides a brief overview of some relevant statistical terminology and probability basics. While not everything outlined here is emphasized throughout the text, it is good to know about this material as it is part of the foundation for the regression techniques discussed in this text.

A key assumption for many statistical procedures is that we have **statistical independence** (also called **stochastic independence**), which means that the value of one observation does not influence the value of any other. This is best accomplished through an appropriate sampling plan; e.g., simple random sampling.

Various assumptions can be made about the population of interest, which is assumed to be characterized by the unknown random variables $X_1, \ldots, X_n$. **Random variables** are unknown quantities, but they are used to couch the underlying process or characteristic of interest. For example, we might assume that the $X_i$ are *iid* $\mathcal{N}(\mu, \sigma^2)$, where $\mu$ and $\sigma^2$ are the unknown population mean and variance, respectively. The values $x_1, \ldots, x_n$ are called **realizations** of the random variables. Using the realizations (i.e., the observed data) we can proceed with estimation of population parameters. Specifically, an **estimator** is a statistics used to estimate the parameter and an **estimate** is the calculated (or realized) value using the data.

While the observations are assumed to be drawn from a certain distribution, the sample statistics also have a distribution, which we call a **sampling distribution**. The sampling distribution is characterized by a mean and standard deviation, which we call the **standard error**. The standard error of a statistic is usually obtained by taking the positive square root of the (estimated) variance of the statistic; e.g., see the standard error formulas for the sample regression coefficients in Chapter 3.

Suppose $X$ is a discrete random variable that can take on any value from the set of natural numbers; i.e., $0, 1, 2, \ldots$. A **probability mass function** (or **pmf**) $p(\cdot)$ measures the mass of probability and must satisfy:

1. $p(x; \boldsymbol{\theta}) \geq 0$ for all $x$,
2. $\sum_{x=0}^{\infty} p(x; \boldsymbol{\theta}) = 1$,

such that $\boldsymbol{\theta}$ is a vector of unknown parameters.

Suppose $X$ is a continuous random variable that can take on any value from the set of real numbers; i.e., $(\infty, +\infty)$. A **probability density function** (or **pdf**) $f(\cdot)$ measures the mass of probability and must satisfy:

1. $f(x; \boldsymbol{\theta}) \geq 0$ for all $x$,

2. $\int_{x=-\infty}^{+\infty} f(x; \boldsymbol{\theta}) dx = 1$,

such that $\boldsymbol{\theta}$ is, again, a vector of unknown parameters.

A **cumulative distribution function** (or **cdf**) $F(\cdot)$ measures the cumulative probability of a random variable occurring up to and including the value of $x$. It is mathematically defined as:

$$F(x) = \mathrm{P}\{X \leq x\} = \begin{cases} \sum_{t=0}^{x} p(t; \boldsymbol{\theta}), & \text{for } X \text{ discrete}; \\ \int_{t=-\infty}^{x} f(t; \boldsymbol{\theta}) dt, & \text{for } X \text{ continuous.} \end{cases}$$

The cdf also has the following four properties:

1. $F(x)$ is nondecreasing on the support of $X$.

2. $F(x)$ is right-continuous everywhere.

3. $\lim_{x \to -\infty} F(x) = 0$.

4. $\lim_{x \to +\infty} F(x) = 1$.

Suppose we have a sample of random variables $X_1, \ldots, X_n$ which are *iid* according to the distribution with density $f(\cdot; \theta)$, where $f(\cdot; \theta)$ is characterized by the parameter $\theta$. The **joint pdf** (or **joint pmf** for the discrete case) is given by:

$$f(\mathbf{x}; \theta) = \prod_{i=1}^{n} f(x_i; \theta).$$

When realizations of these random variables become available (i.e., the actual measurements taken) then the above expression is written as a function of $\theta$. In other words, we simply use the notation $\mathcal{L}(\theta; x_1, \ldots, x_n)$ instead of $f(\mathbf{x}; \theta)$. The quantity $\mathcal{L}(\theta; x_1, \ldots, x_n)$ is called the **likelihood function**. From here, we attempt to maximize the likelihood by finding

$$\hat{\theta} = \operatorname*{argmax}_{\theta} \mathcal{L}(\theta; x_1, \ldots, x_n).$$

Oftentimes, it easier to find $\hat{\theta}$ by maximizing the **loglikelihood function** $\ell(\theta; x_1, \ldots, x_n)$, which is simply the natural logarithm of the likelihood function.

Finally, the following is an important theorem which is the foundation for how we characterize the sample mean of many quantities of interest, provided certain assumptions are made:

**Theorem 3.** (Central Limit Theorem) If $X_1, \ldots, X_n$ are *iid* according to a distribution have mean $\mu < \infty$ and variance $\sigma^2 > 0$, then

$$\frac{\sqrt{n}(\bar{X} - \mu)}{\sigma} \xrightarrow{D} \mathcal{N}(0, 1).$$

In other words, the quantity on the left-hand side has approximately a standard normal distribution.

# Bibliography

K. P. Adragni and R. D. Cook. Sufficient Dimension Reduction and Prediction in Regression. *Philosophical Transactions of The Royal Society A - Mathematical Physical and Engineering Sciences*, 367(1906):4385–4405, 2009.

C. C. Aggarwal. *Outlier Detection*. Springer, New York, NY, 2013.

J. Aldrich. Correlations Genuine and Spurious in Pearson and Yule. *Statistical Science*, 10(4):364–376, 1995.

M. Aldrin, E. Bølviken, and T. Schweder. Projection Pursuit Regression for Moderate Non-Linearities. *Computational Statistics and Data Analysis*, 16 (4):379–403, 1993.

D. M. Allen. The Relationship Between Variable Selection and Data Augmentation and a Method for Prediction. *Technometrics*, 16(1):125–127, 1974.

T. Allison and D. V. Cicchetti. Sleep in Mammals: Ecological and Constitutional Correlates. *Science*, 194(4266):732–734, 1976.

N. S. Altman. An Introduction to Kernel and Nearest-Neighbor Nonparametric Regression. *The American Statistician*, 46(3):175–185, 1992.

T. Amemiya. Regression Analysis when the Dependent Variable Is Truncated Normal. *Econometrica*, 41(6):997–1016, 1973.

T. Amemiya. Tobit Models: A Survey. *Journal of Econometrics*, 24(1/2): 3–61, 1984.

T. Amemiya. *Advanced Econometrics*. Harvard University Press, Cambridge, MA, 1985.

T. W. Anderson and D. A. Darling. A Test of Goodness of Fit. *Journal of the American Statistical Association*, 49(268):765–769, 1954.

D. F. Andrew. A Robust Method for Multiple Linear Regression. *Technometrics*, 16(4):523–531, 1974.

D. W. K. Andrews. Heteroskedasticity and Autocorrelation Consistent Covariance Matrix Estimation. *Econometrica*, 59(3):817–858, 1991.

583

F. J. Anscombe. Contribution to Discussion on Professor Hotelling's Paper. *Journal of the Royal Statistical Society, Series B*, 15(2):193–232, 1953.

L. Anselin. Spatial Regression. In A. S. Fotheringham and P. A. Rogerson, editors, *The SAGE Handbook of Spatial Analysis*, pages 255–275. SAGE Publications, Los Angeles, CA, 2009.

M. Aoyagi and S. Watanabe. Stochastic Complexities of Reduced Rank Regression in Bayesian Estimation. *Neural Networks*, 18(7):924–933, 2005.

K. G. Ashton, R. L. Burke, and J. N. Layne. Geographic Variation in Body and Clutch Size of Gopher Tortoises. *Copeia*, 2007(2):355–363, 2007.

B. R. Avalos, J. L. Klein, N. Kapoor, P. J. Tutschka, J. P. Klein, and E. A. Copelan. Preparation for Marrow Transplantation in Hodgkin's and Non-Hodgkin's Lymphoma Using Bu/CY. *Bone Marrow Transplantation*, 12(2): 133–138, 1993.

G. J. Babu and E. D. Feigelson. Astrostatistics: Goodness-of-Fit and All That! In C. Gabriel, C. Arvisete, D. Ponz, and E. Solano, editors, *Astronomical Data Analysis Software and Systems XV ASP Conference Series*, volume 351, San Francisco, CA, 2006. Astronomical Society of the Pacific.

R. A. Bailey. *Design of Comparative Experiments*. Cambridge University Press, New York, NY, 2008.

R. A. Bailey and L. J. Simon. Two Studies in Automobile Insurance Ratemaking. *ASTIN Bulletin*, 1(4):192–217, 1960.

A. Bąk and T. Bartłomowicz. Conjoint Analysis Method and Its Implementation in conjoint R Package. In J. Pociecha and R. Decker, editors, *Data Analysis Methods and Its Applications*, pages 239–248. Wydawnictwo C. H. Beck, Warsaw, Poland, 2012.

S. Banerjee, B. P. Carlin, and A. E. Gelfand. *Hierarchical Modeling and Analysis for Spatial Data*. Chapman & Hall/CRC Monographs on Statistics and Applied Probability. Chapman & Hall, Boca Raton, FL, 2nd edition, 2014.

V. Barnett and T. Lewis. *Outliers in Statistical Data*. Wiley Series in Probability and Mathematical Statistics. Wiley, New York, NY, 3rd edition, 1994.

R. M. Baron and D. A. Kenny. The Moderator-Mediator Variable Distinction in Social Psychological Research - Conceptual, Strategic, and Statistical Considerations. *Journal of Personality and Social Psychology*, 51(6):1173–1182, 1986.

J. P. Barrett. The Coefficient of Determination - Some Limitations. *The American Statistician*, 28(1):19–20, 1974.

A. Basilevsky. Factor Analysis Regression. *The Canadian Journal of Statistics*, 9(1):1, 1981.

D. Bates, M. Mächler, B. M. Bolker, and S. C. Walker. Fitting Linear Mixed-Effects Models Using lme4. *Journal of Statistical Software*, 67(1):1–48, 2015. URL https://www.jstatsoft.org/article/view/v067i01.

D. M. Bates and D. G. Watts. *Nonlinear Regression Analysis and Its Applications*. Wiley Series in Probability and Mathematical Statistics. John Wiley & Sons, Inc., Hoboken, NJ, 1988.

G. E. Battese, R. M. Harter, and W. A. Fuller. An Error-Components Model for Prediction of County Crop Areas Using Survey and Satellite Data. *Journal of the American Statistical Association*, 83(401):28–36, 1988.

G. Belenky, N. J. Wesensten, D. R. Thorne, M. L. Thomas, H. C. Sing, D. P. Redmond, M. B. Russo, and T. J. Balkin. Patterns of Performance Degradation and Restoration During Sleep Restriction and Subsequent Recovery: A Sleep Dose-Response Study. *Journal of Sleep Research*, 12(1):1–12, 2003.

R. E. Bellman. *Adaptive Control Processes: A Guided Tour*. Princeton University Press, Princeton, NJ, 1961.

D. A. Belsley. *Conditioning Diagnostics: Collinearity and Weak Data in Regression*. John Wiley & Sons, Inc., New York, NY, 1991.

D. A. Belsley, E. Kuh, and R. E. Welsch. *Regression Diagnostics: Identifying Influential Data and Sources of Collinearity*. Wiley Series in Probability and Statistics. Wiley, Hoboken, NJ, 1980.

T. Benaglia, D. Chauveau, D. R. Hunter, and D. S. Young. mixtools: An R Package for Analyzing Finite Mixture Models. *Journal of Statistical Software*, 32(6):1–29, 2009. URL http://www.jstatsoft.org/v32/i06/.

S. M. Bendre. Masking and Swamping Effects on Tests for Multiple Outliers in Normal Sample. *Communications in Statistics - Theory and Methods*, 18 (2):697–710, 1989.

J. Berkson. Are There Two Regressions? *Journal of the American Statistical Association*, 45(250):164–180, 1950.

C. M. Bishop. *Neural Networks for Pattern Recognition*. Oxford University Press, Oxford, UK, 1995.

P. D. Bliese and R. R. Halverson. Individual and Nomothetic Models of Job Stress: An Examination of Work Hours, Cohesion, and Well-Being. *Journal of Applied Social Psychology*, 26(13):1171–1189, 1996.

A. J. Blustin, D. Band, S. Barthelmy, P. Boyd, M. Capalbi, S. T. Holland, F. E. Marshall, K. O. Mason, M. Perri, T. Poole, P. Roming, S. Rosen, P. Schady, M. Still, B. Zhang, L. Angelini, L. Barbier, A. Beardmore, A. Breeveld, D. N. Burrows, J. R. Cummings, J. Canizzo, S. Campana, M. M. Chester, G. Chincarini, L. R. Cominsky, A. Cucchiara, M. de Pasquale, E. E. Fenimore, N. Gehrels, P. Giommi, M. Goad, C. Gronwall, D. Grupe, J. E. Hill, D. Hinshaw, S. Hunsberger, K. C. Hurley, M. Ivanushkina, J. A. Kennea, H. A. Krimm, P. Kumar, W. Landsman, V. La Parola, C. B. Markwardt, K. McGowan, P. Mészáros, T. Mineo, A. Moretti, A. Morgan, J. Nousek, P. T. O'Brien, J. P. Osborne, K. Page, M. J. Page, D. M. Palmer, A. M. Parsons, J. Rhoads, P. Romano, T. Sakamoto, G. Sato, G. Tagliaferri, J. Tueller, A. A. Wells, and N. E. White. Swift Panchromatic Observations of the Bright Gamma-Ray Burst GRB 050525a. *The Astrophysical Journal*, 637 (2):901–913, 2006.

K. A. Bollen. *Structural Equations with Latent Variables*. Wiley, New York, NY, 1989.

G. E. P. Box and D. R. Cox. An Analysis of Transformations. *Journal of the Royal Statistical Society, Series B*, 26(2):211–252, 1964.

G. E. P. Box and D. A. Pierce. Distribution of Residual Autocorrelations in Autoregressive-Integrated Moving Average Time Series Models. *Journal of the American Statistical Association*, 65(332):1509–1526, 1978.

G. E. P. Box, J. S. Hunter, and W. G. Hunter. *Statistics for Experimenters: Design, Innovation, and Discovery*. John Wiley & Sons, Inc., Hoboken, NJ, 2$^{nd}$ edition, 2005.

G. E. P. Box, G. M. Jenkins, and G. C. Reinsel. *Time Series Analysis: Forecasting and Control*. John Wiley & Sons, Inc., Hoboken, NJ, 4$^{th}$ edition, 2008.

W. J. Braun and D. J. Murdoch. *A First Course in Statistical Programming with R*. Cambridge University Press, New York, NY, 2008.

L. Breiman. Bagging Predictors. *Machine Learning*, 24(2):123–140, 1996.

L. Breiman and J. H. Friedman. Estimating Optimal Transformations for Multiple Regression and Correlation. *Journal of the American Statistical Association*, 80(391):580–598, 1985.

L. Breiman, J. H. Friedman, R. A. Olshen, and C. J. Stone. *Classification and Regression Trees*. Chapman and Hall, New York, NY, 1984.

N. Breslow. Covariance Analysis of Censored Survival Data. *Biometrics*, 30 (1):89–99, 1974.

T. S. Breusch. Testing for Autocorrelation in Dynamic Linear Models. *Australian Economic Papers*, 17(31):334–355, 1978.

T. S. Breusch and A. R. Pagan. A Simple Test for Heteroscedasticity and Random Coefficient Variation. *Econometrica*, 47(5):1287–1294, 1979.

N. D. Brinkman. Ethanol Fuel – A Single-Cylinder Engine Study of Efficiency and Exhaust Emissions. *SAE Transactions*, 90(No 810345):1410–1424, 1981.

P. J. Brockwell and R. A. Davis. *Time Series: Theory and Methods*. Springer, New York, NY, 2$^{nd}$ edition, 1991.

C. C. Brown. On a Goodness-of-Fit Test for the Logistic Model Based on Score Statistics. *Communications in Statistics - Theory and Methods*, 11 (10):1087–1105, 1982.

M. B. Brown and A. B. Forsythe. Robust Tests for the Equality of Variances. *Journal of the American Statistical Association*, 69(346):364–367, 1974.

P. J. Brown. *Measurement, Regression, and Calibration*. Oxford Statistical Science Series. Clarendon Press, New York, NY, 1994.

R. L. Brown, J. Durbin, and J. M. Evans. Techniques for Testing the Constancy of Regression Relationships Over Time (with discussion). *Journal of the Royal Statistical Society, Series B*, 37(2):149–192, 1975.

M. Bruche and C. González-Aguado. Recovery Rates, Default Probabilities, and the Credit Cycle. *Journal of Banking and Finance*, 34(4):754–764, 2010.

S. F. Buck. A Method of Estimation of Missing Values in Multivariate Data Suitable for Use with an Electronic Computer. *Journal of the Royal Statistical Society, Series B*, 22(2):302–306, 1960.

E. Bura. Dimension Reduction via Parametric Inverse Regression. In Y. Dodge, editor, $L_1$ *Statistical Procedures and Related Methods*, pages 215–228. Institute of Mathematical Sciences, Hayward, CA, 1997.

E. Bura and R. D. Cook. Estimating the Structural Dimension of Regressions via Parametric Inverse Regression. *Journal of the Royal Statistical Society, Series B*, 63(2):393–410, 2001.

A. C. Cameron and P. K. Trivedi. *Regression Analysis of Count Data*. Cambridge University Press, New York, NY, 2$^{nd}$ edition, 2013.

B. P. Carlin and T. A. Louis. *Bayesian Methods for Data Analysis*. Texts in Statistical Science. Chapman and Hall/CRC Press, Boca Raton, FL, 3$^{rd}$ edition, 2008.

R. J. Carroll and D. Ruppert. *Transformation and Weighting in Regression*. Chapman & Hall/CRC Monographs on Statistics and Applied Probability. Chapman & Hall, New York, NY, 1988.

R. J. Carroll, J. Fan, I. Gijbels, and M. P. Wand. Generalized Partially Linear Single-Index Models. *Journal of the American Statistical Association*, 92 (438):477–489, 1997.

R. J. Carroll, D. Ruppert, L. A. Stefanski, and C. M. Crainiceau. *Measurement Error in Nonlinear Models: A Modern Perspective.* Chapman & Hall/CRC Monographs on Statistics and Applied Probability. Chapman & Hall, Boca Raton, FL, 2$^{nd}$ edition, 2006.

C. M. Carvalho. High-Dimensional Sparse Factor Modeling: Applications in Gene Expression Genomics. *Journal of the American Statistical Association*, 103(484):1438–1456, 2008.

E. Castillo, A. S. Hadi, N. Balakrishnan, and J. M. Sarabia. *Extreme Value and Related Models with Applications in Engineering and Science.* Wiley Series in Probability and Statistics. Wiley, Hoboken, NJ, 2005.

N. R. Chaganty. An Alternative Approach to the Analysis of Longitudinal Data Via Generalized Estimating Equations. *Journal of Statistical Planning and Inference*, 63(1):39–54, 1997.

W. Chang, J. Cheng, J.J. Allaire, Y. Xie, and J. McPherson. *shiny: Web Application Framework for R*, 2016. URL https://CRAN.R-project.org/package=shiny. R package version 0.13.2.

O. Chapelle, B. Schölkopf, and A. Zien, editors. *Semi-Supervised Learning.* Adaptive Computation and Machine Learning. MIT Press, Cambridge, MA, 2006.

S. Chatterjee and A. S. Hadi. *Regression Analysis by Example.* John Wiley & Sons, Inc., Hoboken, NJ, 5$^{th}$ edition, 2012.

C. W. S. Chen, J. S. K. Chan, R. Gerlach, and W. Y. L. Hsieh. A Comparison of Estimators for Regression Models with Change Points. *Statistics and Computing*, 21(3):395–414, 2011.

G. Chen, R. A. Lockhart, and M. A. Stephens. Box-Cox Transformations in Linear Models: Large Sample Theory and Tests of Normality (with discussion). *The Canadian Journal of Statistics*, 30(2):1–59, 2002.

N. Chernov. *Circular and Linear Regression: Fitting Circles and Lines by Least Squares.* Chapman & Hall/CRC, Boca Raton, FL, 2010.

D. Chirot and C. Ragin. The Market, Tradition and Peasant Rebellion: The Case of Romania in 1907. *American Sociological Review*, 40(4):428–444, 1975.

S.-H. Cho and H. Chun. Visualizing Abnormal Climate Changes in Central America from 1995 to 2000. *Computational Statistics*, 25(4):555–567, 2010.

G. A. Colditz, T. F. Brewer, C. S. Berkey, M. E. Wilson, E. Burdick, H. V. Fineberg, and F. Mosteller. Efficacy of BCG Vaccine in the Prevention of Tuberculosis: Meta-Analysis of the Published Literature. *Journal of the American Medical Association*, 271(9):698–702, 1994.

S. Coles. *An Introduction to Statistical Modeling of Extreme Values.* Springer, London, UK, 2001.

P. Comon. Independent Component Analysis - A New Concept? *Signal Processing*, 36(3):287–314, 1994.

R. D. Cook. Detection of Influential Observations in Linear Regression. *Technometrics*, 19(1):15–18, 1977.

R. D. Cook. *Regression Graphics: Ideas for Studying Regressions Through Graphics.* John Wiley & Sons, Inc., New York, NY, 1998a.

R. D. Cook. Principal Hessian Directions Revisited. *Journal of the American Statistical Association*, 93(441):84–94, 1998b.

R. D. Cook. Discussion of "An Adaptive Estimation of Dimension Reduction Space" by Y. Xia, H. Tong, W. K. Li, and L.-X. Zhu. *Journal of the Royal Statistical Society, Series B*, 64(3):397–398, 2002.

R. D. Cook and H. Lee. Dimension Reduction in Binary Response Regression. *Journal of the American Statistical Association*, 94(448):1187–1200, 1999.

R. D. Cook and B. Li. Dimension Reduction for the Conditional Mean in Regression. *The Annals of Statistics*, 30(2):455–474, 2002.

R. D. Cook and B. Li. Determining the Dimension of Iterative Hessian Transformation. *The Annals of Statistics*, 32(6):2501–2531, 2004.

R. D. Cook and L. Ni. Sufficient Dimension Reduction via Inverse Regression: A Minimum Discrepancy Approach. *Journal of the American Statistical Association*, 100(470):410–428, 2005.

R. D. Cook and S. Weisberg. *Residuals and Influence in Regression.* Monographs on Statistics and Applied Probability. Chapman & Hall/CRC, New York, NY, 1982.

R. D. Cook and S. Weisberg. Comment on "Sliced Inverse Regression for Dimension Reduction" by K.-C. Li. *Journal of the American Statistical Association*, 86(414):328–332, 1991.

R. D. Cook and S. Weisberg. *An Introduction to Regression Graphics.* Wiley Series in Probability and Mathematical Statistics. John Wiley & Sons, Inc., New York, NY, 1994.

R. D. Cook, B. Li, and F. Chiaromonte. Dimension Reduction in Regression without Matrix Inversion. *Biometrika*, 94(3):569–584, 2007.

R. D. Cook, B. Li, and F. Chiaromonte. Envelope Methods for Parsimonious and Efficient Multivariate Regression (with Discussion). *Statistica Sinica*, 20(3):927–960, 2010.

J. A. Cornell. *Experiments with Mixtures: Designs, Models, and the Analysis of Mixture Data*. John Wiley & Sons, Inc., New York, NY, 3rd edition, 2002.

D. R. Cox. Regression Models and Life-Tables. *Journal of the Royal Statistical Society, Series B*, 34(2):187–220, 1972.

N. A. C. Cressie. *Statistics for Spatial Data*. Wiley, New York, NY, 2nd edition, 1993.

N. A. C. Cressie and C. K. Wikle. *Statistics for Spatio-Temporal Data*. Wiley, New York, NY, 2011.

F. Cribari-Neto and A. Zeileis. Beta Regression in R. *Journal of Statistical Software*, 34(2):1–24, 2010. URL https://www.jstatsoft.org/article/view/v034i02.

E. L. Crow, F. A. Davis, and M. W. Maxfield. *Statistic Manual*. Dover, Mineola, NY, 1960.

M. Crowder. On the Use of a Working Correlation Matrix in Using Generalised Linear Models for Repeated Measures. *Biometrika*, 82(2):407–410, 1995.

W. Dai, Q. Yang, G.-R. Xue, and Y. Yu. Boosting for Transfer Learning. In *Proceedings of the 24th International Conference on Machine Learning*, pages 193–200, New York, NY, 2007. ACM.

A. C. Davison. *Statistical Models*. Cambridge Series in Statistical and Probabilistic Mathematics. Cambridge University Press, Cambridge, UK, 2003.

A. C. Davison and D. V. Hinkley. *Bootstrap Methods and their Applications*. Cambridge Series in Statistical and Probabilistic Mathematics. Cambridge University Press, Cambridge, UK, 1997.

C. de Boor. *A Practical Guide to Splines*. Applied Mathematical Sciences, Volume 27. Springer-Verlag, New York, NY, revised edition, 2001.

S. de Jong. SIMPLS: An Alternative Approach to Partial Least Squares Regression. *Chemometrics and Intelligent Laboratory Systems*, 18(3):251–263, 1993.

J. de Leeuw, F. W. Young, and Y. Tokane. Additive Structure in Qualitative Data: An Alternating Least Squares Method with Optimal Scaling Features. *Psychometrika*, 41(4):471–503, 1976.

J. de Leeuw, K. Hornik, and P. Mair. Isotone Optimization in R: Pool-Adjacent-Violators Algorithm (PAVA) and Active Set Methods. *Journal of Statistical Software*, 32(5):1–24, 2009. URL http://www.jstatsoft.org/v32/i05/.

M. H. DeGroot and M. J. Schervish. *Probability and Statistics*. Addison-Wesley, Boston, MA, 4$^{\text{th}}$ edition, 2012.

A. P. Dempster, N. M. Laird, and D. B. Rubin. Maximum Likelihood from Incomplete Data via the EM Algorithm. *Journal of the Royal Statistical Society, Series B*, 39(1):1–38, 1977.

R. D. DeVeaux. Mixtures of linear regressions. *Computational Statistics and Data Analysis*, 8(3):227–245, 1989.

W. E. Diewart and T. J. Wales. Quadratic Spline Models for Producer's Supply and Demand Functions. *International Economic Review*, 33(3):705–722, 1992.

W. J. Dixon. Ratios Involving Extreme Values. *The Annals of Mathematical Statistics*, 22(1):68–78, 1951.

N. R. Draper and H. Smith. *Applied Regression Analysis*. John Wiley & Sons, Inc., Hoboken, NJ, 3$^{\text{rd}}$ edition, 1998.

M. Drton and M. Plummer. A Bayesian Information Criterion for Singular Models. *Journal of the Royal Statistical Society, Series B*, 79(2):1–38, 2017.

N. Duan and K.-C. Li. Slicing Regression: A Link-Free Regression Method. *The Annals of Statistics*, 19(2):505–530, 1991.

P. K. Dunn and G. K. Smyth. Randomized Quantile Residuals. *Journal of Computational and Graphical Statistics*, 5(3):236–244, 1996.

W. D. Dupont and W. D. Plummer, Jr. Power and Sample Size Calculations for Studies Involving Linear Regression. *Controlled Clinical Trials*, 19(6):589–601, 1998.

J. Durbin and G. S. Watson. Testing for Serial Correlation in Least Squares Regression. I. *Biometrika*, 37(3/4):409–428, 1950.

J. Durbin and G. S. Watson. Testing for Serial Correlation in Least Squares Regression. II. *Biometrika*, 38(1/2):159–177, 1951.

J. Durbin and G. S. Watson. Testing for Serial Correlation in Least Squares Regression. III. *Biometrika*, 58(1):1–19, 1971.

D. D. Dyer and J. P. Keating. On the Determination of Critical Values for Bartlett's Test. *Journal of the American Statistical Association*, 75(370):313–319, 1980.

R. L. Dykstra. An Algorithm for Restricted Least Squares Regression. *Journal of the American Statistical Association*, 78(384):837–842, 1983.

S. K. Eden, A. Q. An, J. Horner, C. A. Jenkins, and T. A. Scott. A Two-Step Process for Graphically Summarizing Spatial Temporal Multivariate Data in Two Dimensions. *Computational Statistics*, 25(4):587–601, 2010.

S. Efromovich. Orthogonal Series Density Estimation. *Wiley Interdisciplinary Reviews: Computational Statistics*, 2(4):467–476, 2010.

B. Efron. The Efficiency of Cox's Likelihood Function for Censored Data. *Journal of the American Statistical Association*, 72(359):557–565, 1977.

B. Efron and R. J. Tibshirani. *An Introduction to the Bootstrap*. Chapman & Hall/CRC Monographs on Statistics and Applied Probability. Chapman & Hall, Boca Raton, FL, 1993.

B. Efron, T. Hastie, I. Johnstone, and R. Tibshirani. Least Angle Regression. *The Annals of Statistics*, 32(2):407–499, 2004.

F. Eicker. Asymptotic Normality and Consistency of the Least Squares Estimators for Families of Linear Regressions. *The Annals of Mathematical Statistics*, 34(2):447–456, 1963.

J. G. Eisenhauer. Regression Through the Origin. *Teaching Statistics*, 25(3): 76–80, 2003.

W. Enders. *Applied Econometric Time Series*. Wiley Series in Probability and Statistics. Wiley, New York, NY, 2nd edition, 2003.

R. F. Engle, C. W. J. Grainger, J. Rice, and A. Weiss. Semiparametric Estimates of the Relation Between Weather and Electricity Sales. *Journal of the American Statistical Association*, 81(394):310–320, 1986.

P. L. Espinheira, S. L. P. Ferrari, and F. Cribari-Neto. On Beta Regression Residuals. *Journal of Applied Statistics*, 35(4):407–419, 2008.

R. C. Fair. A Theory of Extramarital Affairs. *Journal of Political Economy*, 86(1):45–61, 1978.

F. Famoye. Restricted Generalized Poisson Regression Model. *Communications in Statistics - Theory and Methods*, 22(5):1335–1354, 1993.

J. Fan and I. Gijbels. *Local Polynomial Modelling and Its Applications*. Monographs on Statistics and Applied Probability. Chapman & Hall/CRC Press, London, UK, 1996.

J. Fan and W. Zhang. Statistical Methods with Varying Coefficient Models. *Statistics and Its Interface*, 1(1):179–195, 2008.

J. J. Faraway. *Linear Models with R*. Chapman & Hall/CRC Texts in Statistical Science. Taylor & Francis, Boca Raton, FL, 3rd edition, 2014.

J. J. Faraway. *Extending the Linear Model with R: Generalized Linear, Mixed Effects and Nonparametric Regression Models*. Chapman & Hall/CRC Texts in Statistical Science. Taylor & Francis, Boca Raton, FL, 2016.

J. J. Faraway and J. Sun. Simultaneous Confidence Bands for Linear Regression with Heteroscedastic Errors. *Journal of the American Statistical Association*, 90(431):1094–1098, 1995.

R. E. Fay and R. A. Herriot. Estimation of Income from Small Places: An Application of James-Stein Procedures to Census Data. *Journal of the American Statistical Association*, 74(366):269–277, 1979.

E. D. Feigelson and G. J. Babu. *Modern Statistical Methods for Astronomy with R Applications*. Cambridge University Press, New York, NY, 2012.

S. L. P. Ferrari and F. Cribari-Neto. Beta Regression for Modelling Rates and Proportions. *Journal of Applied Statistics*, 31(7):799–815, 2004.

R. A. Fisher. *The Design of Experiments*. Hafner, New York, NY, 1935.

G. M. Fitzmaurice, N. M. Laird, and J. H. Ware. *Applied Longitudinal Analysis*. Wiley Series in Probability and Statistics. Wiley, Hoboken, NJ, 2nd edition, 2011.

J. Fox and S. Weisberg. *An R Companion to Applied Regression*. SAGE Publications, Thousand Oaks, CA, 2nd edition, 2010.

N. Freemantle, J. Cleland, P. Young, J. Mason, and J. Harrison. $\beta$ Blockade After Myocardial Infarction: Systematic Review and Meta Regression Analysis. *British Medical Journal*, 318(7200):1730–1737, 1999.

Y. Freund and R. E. Schapire. A Decision-Theoretic Generalization of Online Learning and an Application to Boosting. *Journal of Computer and System Sciences*, 55(1):119–139, 1997.

J. H. Friedman. Multivariate Adaptive Regression Splines. *The Annals of Statistics*, 19(1):1–67, 1991.

J. H. Friedman and N. I. Fisher. Bump Hunting in High-Dimensional Data. *Statistics and Computing*, 9(2):123–143, 1999.

J. H. Friedman and W. Stuetzle. Projection Pursuit Regression. *Journal of the American Statistical Association*, 76(376):817–823, 1981.

A. Fujita, J. R. Sato, M. A. A. Demasi, M. C. Sogayar, C. E. Ferreira, and S. Miyano. Comparing Pearson, Spearman and Hoeffding's D Measure for Gene Expression Association Analysis. *Journal of Bioinformatics and Computational Biology*, 7(4):663–684, 2009.

W. A. Fuller. *Measurement Error Models*. Wiley, New York, NY, 1987.

F. Galton. Opening Address by Francis Galton, F.R.S. *Nature*, 32(830):507–510, 1885.

F. Galton. Regression Towards Mediocrity in Hereditary Stature. *The Journal of the Anthropological Institute of Great Britain and Ireland*, 15:246–263, 1886.

S. Gates. Review of Methodology of Quantitative Reviews Using Meta-Analysis in Ecology. *Journal of Animal Ecology*, 71(4):547–557, 2002.

A. Gelman and J. Hill. *Data Analysis Using Regression and Multilevel/Hierarchical Models*. Analytical Methods for Social Research. Cambridge University Press, New York, NY, 2006.

A. Gelman, J. B. Carlin, H. S. Stern, D. B. Dunson, A. Vehtari, and D. B. Rubin. *Bayesian Data Analysis*. Texts in Statistical Science. Chapman and Hall/CRC Press, Boca Raton, FL, 3$^{rd}$ edition, 2013.

M. Gilli and E. Këllezi. An Application of Extreme Value Theory for Measuring Financial Risk. *Computational Economics*, 27(2):207–228, 2006.

H. Glejser. A New Test for Heteroskedasticity. *Journal of the American Statistical Association*, 64(325):315–323, 1969.

L. G. Godfrey. Testing Against General Autoregressive and Moving Average Error Models when the Regressors Include Lagged Dependent Variables. *Ecnometrica*, 46(6):1293–1301, 1978.

S. M. Goldfeld and R. E. Quandt. Some Tests for Homoscedasticity. *Journal of the American Statistical Association*, 60(310):539–547, 1965.

H. Goldstein. *Multilevel Statistical Models*. Wiley Series in Probability and Statistics. Wiley, West Sussex, UK, 4$^{th}$ edition, 2011.

B. Gompertz. On the Nature of the Function Expressive of the Law of Human Mortality, and on a New Mode of Determining the Value of Life Contingencies. *Philosophical Transactions of the Royal Society of London*, 115:513–583, 1825.

F. A. Graybill and H. K. Iyer. *Regression Analysis: Concepts and Applications*. Duxbury Press, Belmont, CA, 1994.

P. J. Green and B. W. Silverman. *Nonparametric Regression and Generalized Linear Models: A Roughness Penalty Approach*. Monographs on Statistics and Applied Probability. Chapman & Hall/CRC Press, Boca Raton, FL, 1993.

P. J. Green and B. S. Yandell. Semi-Parametric Generalized Linear Models. In R. Gilchrest, B. Francis, and J. Whittaker, editors, *Proceedings of the GLIM 85 Conference*, pages 44–55, Berlin, Germany, 1985. Springer-Verlag.

W. H. Greene. *Econometric Analysis*. Macmillan Publishing Company, New York, NY, 2$^{nd}$ edition, 1993.

B. M. Greenwell and C. M. Schubert-Kabban. investr: An R Package for Inverse Estimation. *The R Journal*, 6(1):90–100, 2014.

F. E. Grubbs. Procedures for Detecting Outlying Observations in Samples. *Technometrics*, 11(1):1–21, 1969.

Y. Grunfeld. *The Determinants of Corporate Investment*. PhD thesis, University of Chicago, 1958. Unpublished Ph.D. Dissertation.

A. S. Hadi. A New Measure of Overall Potential Influence in Linear Regression. *Computational Statistics and Data Analysis*, 14(1):1–27, 1992.

J. D. Hamilton. *Time Series Analysis*. Princeton University Press, Princeton, NJ, 1994.

F. Hampel. On the Philosophical Foundations of Statistics: Bridges to Huber's Work, and Recent Results. In H. Rieder, editor, *Robust Statistics, Data Analysis, and Computer Intensive Methods: In Honor of Peter Huber's 60th Birthday*, pages 185–196. Springer, New York, NY, 1996.

J. W. Hardin and J. M. Hilbe. *Generalized Estimating Equations*. Chapman & Hall, Boca Raton, FL, 2$^{nd}$ edition, 2013.

W. K. Härdle and T. M. Stoker. Investigating Smooth Multiple Regression by the Method of Average Derivatives. *Journal of the American Statistical Association*, 84(408):986–995, 1989.

W. K. Härdle, H. Liang, and J. Gao. *Partially Linear Models*. Springer, Berlin, Germany, 2000.

W. K. Härdle, S. Huet, E. Mammen, and S. Sperlich. Bootstrap Inference in Semiparametric Generalized Additive Models. *Econometric Theory*, 20(2): 265–300, 2004a.

W. K. Härdle, M. Müller, S. Sperlich, and A. Werwatz. *Nonparametric and Semiparametric Models*. Springer, Berlin, Germany, 2004b.

D. Harrison and D. L. Rubinfeld. Hedonic Housing Prices and the Demand for Clean Air. *Journal of Environmental Economics and Management*, 5 (1):81–102, 1978.

A. C. Harvey and P. Collier. Testing for Functional Misspecification in Regression Analysis. *Journal of Econometrics*, 6(1):103–119, 1977.

T. Hastie and R. Tibshirani. *Generalized Additive Models*. Monographs on Statistics and Applied Probability. Chapman & Hall/CRC Press, Boca Raton, FL, 1990.

T. Hastie and R. Tibshirani. Varying-Coefficient Models. *Journal of the Royal Statistical Society, Series B*, 55(4):757–796, 1993.

T. Hastie, R. Tibshirani, and J. Friedman. *The Elements of Statistical Learning: Data Mining, Inference, and Prediction*. Springer, New York, NY, 2nd edition, 2009.

T. Hastie, R. Tibshirani, and M. Wainwright. *Statistical Learning with Sparsity: The Lasso and Generalizations*. Monographs on Statistics and Applied Probability. Chapman & Hall/CRC, New York, NY, 2015.

D. M. Hawkins and D. Olive. Applications and Algorithms for Least Trimmed Sum of Absolute Deviations Regression. *Computational Statistics and Data Analysis*, 32(2):119–134, 1999.

D. Hedeker and R. D. Gibbons. *Longitudinal Data Analysis*. Wiley Series in Probability and Statistics. Wiley, Hoboken, NJ, 2006.

H. Heinzl and M. Mittlböck. Pseudo R-squared Measures for Poisson Regression Models with Over- or Underdispersion. *Computational Statistics and Data Analysis*, 44(1–2):253–271, 2003.

L. J. Hendrix, M. W. Carter, and D. T. Scott. Covariance Analysis with Heterogeneity of Slopes in Fixed Models. *Biometrics*, 38(3):226–252, 1982.

J. A. Herriges and C. L. Kling. Nonlinear Income Effects in Random Utility Models. *The Review of Economics and Statistics*, 81(1):62–72, 1999.

T. P. Hettmansperger and J. W. McKean. *Robust Nonparametric Statistical Methods*. Chapman Hall, New York, NY, 2nd edition, 2011.

J. M. Hilbe. *Negative Binomial Regression*. Cambridge University Press, New York, NY, 2nd edition, 2011.

J. M. Hilbe. *Modeling Count Data*. Cambridge University Press, New York, NY, 2014.

C. Hildreth and J. Y. Lu. Demand Relations with Autocorrelated Disturbances. *Technical Bulletin 276*, Michigan State University Agricultural Experiment Station, 1960.

J. L. Hintze and R. D. Nelson. Violin Plots: A Box Plot-Density Trace Synergism. *The American Statistician*, 52(2):181–184, 1998.

T. K. Ho. Random Decision Forests. In *Proceedings of the 3rd International Conference on Document Analysis and Recognition*, pages 14–16, Montreal, Quebec, 1995. IEEE.

J. Hobbs, H. Wickham, H. Hofmann, and D. Cook. Glaciers Melt as Mountains Warm: A Graphical Case Study. *Computational Statistics*, 25(4):569–586, 2010.

W. Hoeffding. A Non-Parametric Test of Independence. *The Annals of Mathematical Statistics*, 19(4):546–557, 1948.

A. E. Hoerl and R. W. Kennard. Ridge Regression: Biased Estimation for Nonorthogonal Problems. *Technometrics*, 12(1):55–67, 1970.

A. E. Hoerl and R. W. Kennard. Ridge Regression Iterative Estimation of the Biasing Parameter. *Communications in Statistics - Theory and Methods*, 5 (1):77–88, 1975.

A. E. Hoerl, R. W. Kennard, and K. F. Baldwin. Ridge Regression: Some Simulations. *Communications in Statistics - Simulation and Computation*, 4(2):105–123, 1975.

C. C. Holt. Forecasting Seasonals and Trends by Exponentially Weighted Moving Averages. *International Journal of Forecasting*, 20(1):5–10, 2004.

J. L. Horowitz and W. K. Härdle. Testing a Parametric Model against a Semiparametric Alternative. *Econometric Theory*, 10(5):821–848, 1994.

D. G. Horvitz and D. J. Thompson. A Generalization of Sampling Without Replacement From a Finite Universe. *Journal of the American Statistical Association*, 47(260):663–685, 1952.

D. W. Hosmer, T. Hosmer, S. Le Cressie, and S. Lemeshow. A Comparison of Goodness-of-Fit Tests for the Logistic Regression Model. *Statistics in Medicine*, 16(9):965–980, 1997.

D. W. Hosmer, S. Lemeshow, and R. X. Sturdivant. *Applied Logistic Regression*. Wiley Series in Probability and Statistics. Wiley, Hoboken, NJ, 3rd edition, 2013.

H. Hotelling. New Light on the Correlation Coefficient and Its Transforms. *Journal of the Royal Statistical Society, Series B*, 15(2):193–232, 1953.

T. Hothorn and B. S. Everitt. *A Handbook of Statistical Analyses Using R.* CRC Press, Boca Raton, FL, 3rd edition, 2014.

J. C. Hsu. *Multiple Comparisons: Theory and Methods.* Chapman & Hall/CRC Press, Boca Raton, FL, 1996.

B. Hu, J. Shao, and M. Palta. Pseudo-$R^2$ in Logistic Regression Model. *Statistica Sinica*, 16(3):847–860, 2006.

M.-C. Hu, M. Pavlicova, and E. V. Nunes. Zero-Inflated and Hurdle Models of Count Data with Extra Zeros: Examples from an HIV-Risk Reduction Intervention Trial. *The American Journal of Drug and Alcohol Abuse*, 37 (5):367–375, 2011.

X. Huang and C. W. Oosterlee. Generalized Beta Regression Models for Random Loss Given Default. *The Journal of Credit Risk*, 7(4):45–70, 2011.

P. J. Huber. Robust Estimation of a Location Parameter. *The Annals of Mathematical Statistics*, 35(1):73–101, 1964.

P. J. Huber. The Behavior of Maximum Likelihood Estimation Under Nonstandard Conditions. In J. Neyman and L. M. LeCam, editors, *Proceedings of the Fifth Berkeley Symposium on Mathematical Statistics and Probability*, volume 1, pages 221–233. University of California Press, 1967.

P. J. Huber and E. M. Ronchetti. *Robust Statistics*. Wiley, Hoboken, NJ, 2nd edition, 2009.

M. Hurn, A. Justel, and C. P. Robert. Estimating Mixtures of Regressions. *Journal of Computational and Graphical Statistics*, 12(1):55–79, 2003.

A. Hyvärinen and E. Oja. Independent Component Analysis: Algorithms and Applications. *Neural Networks*, 13(4-5):411–430, 2000.

J. Ichimura. Semiparametric Least Squares (SLS) and Weighted SLS Estimation of Single-Index Models. *Journal of Econometrics*, 58(1-2):71–120, 1993.

B. Iglewicz and D. C. Hoaglin. *How to Detect and Handle Outliers*, volume 16. ASQC Quality Press, Milwaukee, WI, 1993.

A. J. Izenman. Reduced-Rank Regression for the Multivariate Linear Model. *Journal of Multivariate Analysis*, 5(2):248–264, 1975.

R. A. Jacobs, M. I. Jordan, S. J. Nowlan, and G. E. Hinton. Adaptive mixtures of local experts. *Neural Computation*, 3(1):79–87, 1991.

L. A. Jaeckel. Estimating Regression Coefficients by Minimizing the Dispersion of the Residuals. *The Annals of Mathematical Statistics*, 43(5): 1449–1458, 1972.

G. James, D. Witten, T. Hastie, and R. Tibshirani. *An Introduction to Statistical Learning: With Applications in R*. Springer, New York, NY, 2013.

S. Johansen. *Likelihood-Based Inference in Cointegrated Vector Auotoregressive Models*. Oxford University Press, Oxford, UK, 1995.

P. W. M. John. *Statistical Design and Analysis of Experiments*. MacMillan Company, New York, NY, 1971.

R. A. Johnson and D. W. Wichern. *Applied Multivariate Statistical Analysis.* Pearson, Upper Saddle River, NJ, 6$^{th}$ edition, 2007.

M. I. Jordan and L. Xu. Convergence Results for the EM Approach to Mixtures of Experts Architectures. *Neural Networks*, 8(9):1409–1431, 1995.

K. G. Jöreskog and M. van Thillo. LISREL: A General Computer Program for Estimating a Linear Structural Equation System Involving Multiple Indicators of Unmeasured Variables. Technical Report ETS-RB-72-56, Educational Testing Service, 1972.

J. Jurečková. Nonparametric Estimate of Regression Coefficients. *The Annals of Mathematical Statistics*, 42(4):1328–1338, 1971.

C. Jutten and J. Herault. Blind Separation of Sources, Part I: An Adaptive Algorithm Based on Neuromimetic Architecture. *Signal Processing*, 24(1): 1–10, 1991.

J. D. Kalbfleisch and R. L. Prentice. *The Statistical Analysis of Failure Time Data.* Wiley, New York, NY, 1980.

G. V. Kass. An Exploratory Technique for Investigating Large Quantities of Categorical Data. *Journal of the Royal Statistical Society, Series C*, 29(2): 119–127, 1980.

M. G. Kendall. A New Measure of Rank Correlation. *Biometrika*, 30(1/2): 81–93, 1938.

G. Kimeldorf and G. Wahba. Some Results on Tchebycheffian Spline Functions. *Journal of Mathematical Analysis and Applications*, 33(1):82–95, 1971.

L. Kish. *Survey Sampling.* Wiley, New York, NY, 1965.

C. Kleiber and A. Zeileis. *Applied Econometrics with R.* Use R! Springer, New York, NY, 2008.

J. P. Klein and M. L. Moeschberger. *Survival Analysis: Techniques for Censored and Truncated Data.* Statistics for Biology and Health. Springer-Verlag, New York, NY, 2$^{nd}$ edition, 2003.

R. B. Kline. *Principles and Practice of Structural Equation Modeling.* The Guilford Press, New York, NY, 4$^{th}$ edition, 2016.

J. D. Kloke and J. W. McKean. Rfit: Rank-based Estimation for Linear Models. *The R Journal*, 4(2):57–64, 2012.

J. Kmenta. *Elements of Econometrics.* Macmillan Publishing Company, New York, NY, 2$^{nd}$ edition, 1986.

R. Koenker. *Quantile Regression.* Econometric Society Monographs. Cambridge University Press, Cambridge, UK, 2005.

R. Koenker and G. Bassett, Jr. Robust Tests for Heteroscedasticity Based on Regression Quantiles. *Econometrica,* 50(1):43–61, 1982.

S. Konishi and G. Kitagawa. *Information Criteria and Statistical Modeling.* Springer Series in Statistics. Springer, New York, NY, 2008.

K. Krishnamoorthy and T. Mathew. *Statistical Tolerance Regions: Theory, Applications, and Computation.* Wiley, Hoboken, NJ, 2009.

K. Krishnamoorthy, P. M. Kulkarni, and T. Mathew. Multiple Use One-Sided Hypotheses Testing in Univariate Linear Calibration. *Journal of Statistical Planning and Inference,* 93(1–2):211–223, 2001.

R. O. Kuehl. *Design of Experiments: Statistical Principles of Research Design.* Duxbury Press, Pacific Grove, CA, 2nd edition, 2000.

M. H. Kutner, C. J. Nachtsheim, J. Neter, and W. Li. *Applied Linear Statistical Models.* McGraw-Hill/Irwin, New York, NY, 5th edition, 2005.

D. Lambert. Zero-Inflated Poisson Regression, with an Application to Defects in Manufacturing. *Technometrics,* 34(1):1–14, 1992.

K. Lange. *Optimization.* Springer Texts in Statistics. Springer, New York, NY, 2nd edition, 2013.

E. L. Lehmann. *Nonparametrics: Statistical Methods Based on Ranks.* Springer, New York, NY, revised edition, 2006.

I. E. Leppik, F. E. Dreifuss, T. Bowman, N. Santilli, M. Jacobs, C. Crosby, J. Cloyd, J. Stockman, N. Graves, T. Sutula, T. Welty, J. Vickery, R. Brundage, R. Gumnit, and A. Gutierres. A Double-Blind Crossover Evaluation of Progabide in Partial Seizures. *Neurology,* 35(4):285, 1985.

H. Levene. Robust Tests for Equality of Variances. In I. Olkin, S. G. Ghurye, W. Hoeffding, W. G. Madow, and H. B. Mann, editors, *Contributions to Probability and Statistics: Essays in Honor of Harold Hotelling,* pages 278–292. Stanford University Press, 1960.

B. Li and S. Wang. On Directional Regression for Dimension Reduction. *Journal of the American Statistical Association,* 102(479):997–1008, 2007.

B. Li, H. Zha, and F. Chiaromonte. Contour Regression: A General Approach to Dimension Reduction. *The Annals of Statistics,* 33(4):1580–1616, 2005.

J. Li and R. Valliant. Survey Weighted Hat Matrix and Leverages. *Survey Methodology,* 35(1):15–24, 2009.

K.-C. Li. Sliced Inverse Regression for Dimension Reduction (with Discussion). *Journal of the American Statistical Association*, 86(414):316–327, 1991.

K.-C. Li. On Principal Hessian Directions for Data Visualization and Dimension Reduction: Another Application of Stein's Lemma. *Journal of the American Statistical Association*, 87(420):1025–1039, 1992.

K.-Y. Liang and S. L. Zeger. Longitudinal Data Analysis Using Generalized Linear Models. *Biometrika*, 73(1):13–22, 1986.

D. Liao and R. Valliant. Variance Inflation Factors in the Analysis of Complex Survey Data. *Survey Methodology*, 38(1):53–62, 2012.

B. G. Lindsay. *Mixture Models: Theory, Geometry and Applications*, volume 5 of *NSF-CBMS Regional Conference Series in Probability and Statistics*. Institute of Mathematical Statistics and the American Statistical Association, 1995.

R. J. A. Little and D. B. Rubin. *Statistical Analysis with Missing Data*. Wiley Series in Probability and Mathematical Statistics. Wiley, Hoboken, NJ, 2nd edition, 2002.

W. Liu. *Simultaneous Inference in Regression*. Chapman & Hall/CRC Monographs on Statistics and Applied Probability. Taylor & Francis, Boca Raton, FL, 2011.

G. M. Ljung and G. E. P. Box. On a Measure of a Lack of Fit in Time Series Models. *Biometrika*, 65(2):297–303, 1978.

C. Loader. *Local Regression and Likelihood*. Statistics and Computing. Springer, New York, NY, 1999.

R. H. Lock. 1993 New Car Data. *Journal of Statistics Education*, 1(1), 1993. URL `http://www.amstat.org/publications/jse/v1n1/datasets.lock.html`.

W.-Y. Loh. Classification and Regression Trees. *Wiley Interdisciplinary Reviews: Data Mining and Knowledge Discovery*, 1(1):14–23, 2011.

S. L. Lohr. *Sampling: Design and Analysis*. Advanced Series. Brooks/Cole, Cengage Learning, Boston, MA, 2nd edition, 2010.

J. S. Long. The Origins of Sex Differences in Science. *Social Forces*, 68(3):1297–1316, 1990.

D. P. MacKinnon. *Introduction to Statistical Mediation Analysis*. Multivariate Applications Series. Taylor and Francis, New York, NY, 2008.

D. P. MacKinnon, J. L. Krull, and C. M. Lockwood. Equivalence of the Mediation, Confounding and Suppression Effect. *Prevention Science*, 1(4): 173–181, 2000.

D. P. MacKinnon, A. J. Fairchild, and M. S. Fritz. Mediation Analysis. *Annual Review of Psychology*, 58:593–614, 2007.

G. S. Maddala. *Limited-Dependent and Qualitative Variables in Econometrics*. Cambridge University Press, Cambridge, UK, 1983.

C. L. Mallow. Some Comments on $C_p$. *Technometrics*, 15(4):661–675, 1973.

I. Markovsky and S. Van Huffel. Overview of Total Least-Squares Methods. *Signal Processing*, 87(10):2283–2302, 2007.

G. Masarotto and C. Varin. Gaussian Copula Marginal Regression. *Electronic Journal of Statistics*, 6:1517–1549, 2012.

L. F. Masson, G. McNeill, J. O. Tomany, J. A. Simpson, H. S. Peace, L. Wei, D. A. Grubb, and C. Bolton-Smith. Statistical Approaches for Assessing the Relative Validity of a Food-Frequency Questionnaire: Use of Correlation Coefficients and the Kappa Statistic. *Public Health Nutrition*, 6(3):313–321, 2003.

N. Matloff. *The Art of R Programming*. No Starch Press, San Francisco, CA, 2011.

D. Mavridis and I. Moustaki. The Forward Search Algorithm for Detecting Aberrant Response Patterns in Factor Analysis for Binary Data. *Journal of Computational and Graphical Statistics*, 18(4):1016–1034, 2009.

P. McCullagh and J. A. Nelder. *Generalized Linear Models*. Chapman and Hall, Boca Raton, FL, 2nd edition, 1989.

M. B. McElroy. Goodness of Fit for Seemingly Unrelated Regressions: Glahn's $R^2_{y,x}$ and Hooper's $\bar{r}^2$. *Journal of Econometrics*, 6(3):381–387, 1977.

C. E. McHenry. Computation of a Best Subset in Multivariate Analysis. *Journal of the Royal Statistical Society, Series C*, 27(3):291–296, 1978.

G. J. McLachlan and D. Peel. *Finite Mixture Models*. Wiley, New York, 2000.

J. S. Menet, J. Rodriguez, K. C. Abruzzi, and M. Rosbash. Nascent-Seq Reveals Novel Features of Mouse Circadian Transcriptional Regulation. *eLife*, 1(e00011):1–25, 2012.

X.-L. Meng, R. Rosenthal, and D. B. Rubin. Comparing Correlated Correlation Coefficients. *Psychological Bulletin*, 111(1):172–175, 1992.

L. Michaelis and M. L. Menten. Die Kinetik der Invertinwirkung. *Biochemistry Zeitung*, 49:333–369, 1913.

A. J. Miller. Selection of Subsets of Regression Variables. *Journal of the Royal Statistical Society, Series A*, 147(3):389–425, 1984.

D. C. Montgomery. *Design and Analysis of Experiments.* John Wiley & Sons, Inc., Hoboken, NJ, 8th edition, 2013.

D. C. Montgomery, E. A. Peck, and G. G. Vining. *Introduction to Linear Regression Analysis.* Wiley Series in Probability and Statistics. Wiley, Hoboken, NJ, 5th edition, 2013.

D. C. Montgomery, C. L. Jennings, and M. Kulahci. *Introduction to Time Series and Forecasting.* Wiley Series in Probability and Statistics. Wiley, Hoboken, NJ, 2nd edition, 2015.

P. A. P. Moran. The Interpretation of Statistical Maps. *Journal of the Royal Statistical Society, Series B*, 10(2):243–251, 1948.

P. A. P. Moran. A Test for the Serial Independence of Residuals. *Biometrika*, 37(1/2):178–181, 1950.

P. Mukhopadhyay. *Small Area Estimation in Survey Sampling.* Narosa Publishing House, New Delhi, India, 1998.

J. Mullahy. Specification and Testing of Some Modified Count Data Models. *Journal of Econometrics*, 33(3):341–365, 1986.

D. Muller, C. M. Judd, and V. Y. Yzerbyt. When Moderation is Mediated and Mediation is Moderated. *Journal of Personality and Social Psychology*, 89(6):852–863, 2005.

P. Murrell. The 2006 Data Expo of the American Statistical Association. *Computational Statistics*, 25(4):551–554, 2010.

R. H. Myers, D. C. Montgomery, and C. M. Anderson-Cook. *Response Surface Methodology: Process and Product Optimization Using Designed Experiments.* Wiley Series in Probability and Statistics. John Wiley & Sons, Inc., Hoboken, NJ, 3rd edition, 2009.

E. A. Nadaraya. On Estimating Regression. *Theory of Probability and Its Applications*, 9(1):141–142, 1964.

M. S. Ndaro, X.-Y. Jin, T. Chen, and C.-W. Yu. Splitting of Islands-in-the-Sea Fibers (PA6/COPET) During Hydroentangling of Nonwovens. *Journal of Engineered Fibers and Fabrics*, 2(4):1–9, 2007.

R. B. Nelsen. *An Introduction to Copulas.* Springer, New York, NY, 2nd edition, 2006.

W. D. Nelson and G. J. Hahn. Linear Estimation of a Regression Relationship from Censored Data Part I – Simple Methods and Their Application. *Technometrics*, 14(2):247–269, 1972.

W. K. Newey and K. D. West. A Simple, Positive Semi-definite, Heteroskedasticity and Autocorrelation Consistent Covariance Matrix. *Econometrica*, 55 (3):703–708, 1987.

NIST. NIST/SEMATECH e-Handbook of Statistical Methods, 2011. URL `http://www.itl.nist.gov/div898/handbook/`. Accessed: 2017-01-25.

M. Oliveira, R. M. Crujeiras, and A. Rodríguez-Casal. NPCirc: An R Package for Nonparametric Circular Methods. *Journal of Statistical Software*, 61(9): 1–26, 2014. URL `http://www.jstatsoft.org/v61/i09/`.

R. A. Olshen, E. N. Biden, M. P. Wyatt, and D. H. Sutherland. Gait Analysis and the Bootstrap. *The Annals of Statistics*, 17(4):1419–1440, 1989.

G. Osius and D. Rojek. Normal Goodness-of-Fit Tests for Multinomial Models with Large Degrees of Freedom. *Journal of the American Statistical Association*, 87(420):1145–1152, 1992.

W. Pan. Akaike's Information Criterion in Generalized Estimating Equations. *Biometrics*, 57(1):120–125, 2001.

D. Pardoe and P. Stone. Boosting for Regression Transfer. In J. Fürnkranz and T. Joachims, editors, *Proceedings of the 27th International Conference on Machine Learning (ICML-10)*, pages 863–870, Haifa, Israel, 2010. Omnipress.

H. Passing and W. Bablok. A New Biometrical Procedure for Testing the Equality of Measurements from Two Different Analytical Methods. Application of Linear Regression Procedures for Method Comparison Studies in Clinical Chemistry, Part I. *Journal of Clinical Chemistry & Clinical Biochemistry*, 21(11):709–720, 1983.

H. Passing and W. Bablok. Comparison of Several Regression Procedures for Method Comparison Studies and Determination of Sample Sizes. Application of Linear Regression Procedures for Method Comparison Studies in Clinical Chemistry, Part II. *Journal of Clinical Chemistry & Clinical Biochemistry*, 22(6):431–445, 1984.

H. Passing, W. Bablok, R. Bender, and B. Schneider. A General Regression Procedure for Method Transformation. Application of Linear Regression Procedures for Method Comparison Studies in Clinical Chemistry, Part III. *Journal of Clinical Chemistry & Clinical Biochemistry*, 26(11):783–790, 1988.

J. C. Pinheiro and D. M. Bates. *Mixed-Effects Models in S and S-PLUS*. Springer-Verlag, New York, NY, 2000.

L. Powsner. The Effects of Temperature on the Durations of the Developmental Stages of *Drosophila Melanogaster*. *Physiological Zoology*, 8(4):474–520, 1935.

S. J. Prais and C. B. Winsten. Trend Estimators and Serial Correlation. *Cowles Commission Discussion Paper No. 383*, University of Chicago, 1954.

P. N. Price, A. V. Nero, and A. Gelman. Bayesian Prediction of Mean Indoor Radon Concentrations for Minnesota Counties. *Health Physics*, 71(6):922–936, 1996.

F. Pukelsheim. *Optimal Design of Experiments*. John Wiley & Sons, Inc., New York, NY, 1993.

R. E. Quandt. The Estimation of the Parameters of a Linear Regression System Obeying Two Separate Regimes. *Journal of the American Statistical Association*, 53(284):873–880, 1958.

M. H. Quenoille. Problems in Plane Sampling. *Journal of the American Statistical Association*, 20(3):355–375, 1949.

J. R. Quinlan. *C4.5: Programs for Machine Learning*. Morgan Kaufmann Publishers, Inc., San Francisco, CA, 1993a.

J. R. Quinlan. Combining Instance-Based and Model-Based Learning. In *Proceedings on the Tenth International Conference of Machine Learning*, pages 236–243, University of Massachusetts, Amherst, 1993b. Morgan Kaufmann.

R Core Team. *R: A Language and Environment for Statistical Computing*. R Foundation for Statistical Computing, Vienna, Austria, 2016. URL https://www.R-project.org/.

J. O. Ramsay and B. W. Silverman. *Applied Functional Data Analysis: Methods and Case Studies*. Springer, New York, NY, 2002.

J. O. Ramsay and B. W. Silverman. *Functional Data Analysis*. Springer, New York, NY, 2nd edition, 2005.

J. B. Ramsey. Tests for Specification Errors in Classical Linear Least Squares Regression Analysis. *Journal of the Royal Statistical Society, Series B*, 31 (2):350–371, 1969.

J. N. K. Rao. *Small Area Estimation*. Wiley, Hoboken, NJ, 2003.

V. R. Rao. *Applied Conjoint Analysis*. Springer, New York, NY, 2014.

S. W. Raudenbush and A. S. Bryk. *Hierarchical Linear Models: Applications and Data Analysis Methods*. Advanced Quantitative Techniques in the Social Sciences. SAGE Publications, Thousand Oaks, CA, 2nd edition, 2002.

C. Reimann, M. Äyräs, V. Chekushin, I. Bogatyrev, R. Boyd, P. de Caritat, R. Dutter, T. E. Finne, J. H. Halleraker, Ø. Jæger, G. Kashulina, O. Lehto, H. Niskavaara, V. Pavlov, M. L. Räisänen, T. Strand, and T. Volden. *Environmental Geochemical Atlas of the Central Barents Region*. Special Publication, Geological Survey of Norway, Trondheim, Norway, 1998.

C. Reimann, P. Filzmoser, and R. G. Garrett. Factor Analysis Applied to Regional Geochemical Data: Problems and Possibilities. *Applied Geochemistry*, 17(3):185–206, 2002.

W. A. Renner, Jr. The Relationship Between Selected Physical Performance Variables and Football Punting Ability. Master's thesis, Virginia Polytechnic Institute and State University, Blacksburg, VA, 1983.

A. Rettinger, M. Zinkevich, and M. Bowling. Boosting Expert Ensembles for Rapid Concept Recall. In *Proceedings of the 21st National Conference on Machine Learning (ICML-10)*, pages 464–469, Boston, MA, 2006. AAAI Press.

F. J. Richards. A Flexible Growth Function for Empirical Use. *Journal of Experimental Botany*, 10(2):290–300, 1959.

M. S. Ridout and P. Besbeas. An Empirical Model for Underdispersed Count Data. *Statistical Modelling*, 4(1):77–89, 2004.

B. D. Ripley and M. Thompson. Regression Techniques for the Detection of Analytical Bias. *Analyst*, 112(4):377–383, 1987.

J. M. Robins and S. Greenland. Identifiability and Exchangeability for Direct and Indirect Effects. *Epidemiology*, 3(2):143–155, 1992.

J. L. Rodgers and W. A. Nicewander. Thirteen Ways to Look at the Correlation Coefficient. *The American Statistician*, 42(1):59–66, 1988.

B. Rosner. Percentage Points for a Generalized ESD Many-Outlier Procedure. *Technometrics*, 25(2):165–172, 1983.

P. J. Rousseeuw and M. Hubert. Regression Depth. *Journal of the American Statistical Association*, 94(446):388–402, 1999.

P. J. Rousseeuw and A. M. Leroy. *Robust Regression and Outlier Detection*. Wiley Series in Probability and Statistics. Wiley-Interscience, Hoboken, NJ, 2003.

D. B. Rubin. Inference and Missing Data. *Biometrika*, 63(3):581–592, 1976.

M. V. Rudorfer. Cardiovascular Changes and Plasma Drug Levels After Amitriptyline Overdose. *Journal of Toxicology - Clinical Toxicology*, 19 (1):67–78, 1982.

D. Ruppert, M. P. Wand, and R. J. Carroll. *Semiparametric Regression*. Cambridge Series in Statistical and Probabilistic Mathematics. Cambridge University Press, Cambridge, England, 2003.

T. A. Ryan and B. L. Joiner. Normal Probability Plots and Tests for Normality. Technical report, Department of Statistics, The Pennsylvania State University, 1976.

T. A. Ryan, B. L. Joiner, and B. Ryan. *Minitab Student Handbook*. Duxbury Press, North Scituate, MA, 1976.

A. Santos, O. Anjos, M. E. Amaral, N. Gil, H. Pereira, and R. Simões. Influence on Pulping Yield and Pulp Properties of Wood Density of *Acacia melanoxylon*. *Journal of Wood Science*, 58(6):479–486, 2012.

H. Scheffé. *The Analysis of Variance*. Wiley, New York, NY, 1959.

H. Scheffé. A Statistical Theory of Calibration. *The Annals of Statistics*, 1 (1):1–37, 1973.

G. Schmidt, R. Mattern, and F. Schüler. Biomechanical Investigation to Determine Physical and Traumatological Differentiation Criteria for the Maximum Load Capacity of Head and Vertebral Column With and Without Protective Helmet Under Effects of Impact. Technical Report Final Report Phase III, Project 65, EEC Research Program on Biomechanics of Impacts, Institut für Rechtsmedizin, Universität Heidelberg, Germany, 1981.

D. P. Schneider, P. B. Hall, G. T. Richards, M. A. Strauss, D. E. Vanden Berk, S. F. Anderson, W. N. Brandt, X. Fan, S. Jester, J. Gray, J. E. Gunn, M. U. SubbaRao, A. R. Thakar, C. Stoughton, A. S. Szalay, B. Yanny, D. G. York, N. A. Bahcall, J. Barentine, M. R. Blanton, H. Brewington, J. Brinkmann, R. J. Brunner, F. J. Castander, I. Csabai, J. A. Frieman, M. Fukugita, M. Harvanek, D. W. Hogg, Ž. Ivezić, S. M. Kent, S. J. Kleinman, G. R. Knapp, R. G. Kron, J. Krzesiński, D. C. Long, R. H. Lupton, A. Nitta, J. R. Pier, D. H. Saxe, Y. Shen, S. A. Snedden, D. H. Weinberg, and J. Wu. The Sloan Digital Sky Survey Quasar Catalog. IV. Fifth Data Release. *The Astronomical Journal*, 134(1):102–117, 2007.

D. Schoenfeld. Partial Residuals for the Proportional Hazards Regression Model. *Biometrika*, 69(1):239–241, 1982.

R. E. Schumacker and R. G. Lomax. *A Beginner's Guide to Structural Equation Modeling*. Routledge, New York, NY, 2nd edition, 2016.

A. J. Scott and D. Holt. The Effect of Two-Stage Sampling on Ordinary Least Squares. *Journal of the American Statistical Association*, 77(380):848–854, 1982.

D. W. Scott. *Multivariate Density Estimation: Theory, Practice, and Visualization*. Wiley Series in Probability and Statistics. Wiley, Hoboken, NJ, 2nd edition, 2015.

G. A. F. Seber and C. J. Wild. *Nonlinear Regression*. Wiley Series in Probability and Mathematical Statistics. John Wiley & Sons, Inc., Hoboken, NJ, 2003.

K. F. Sellers and G. Shmueli. A Flexible Regression Model for Count Data. *The Annals of Applied Statistics*, 4(2):943–961, 2010.

P. K. Sen. Estimates of the Regression Coefficient Based on Kendall's Tau. *Journal of the American Statistical Association*, 63(324):1379–1389, 1968.

D. Sentürk and H.-G. Müller. Covariate-Adjusted Regression. *Biometrika*, 92 (1):75–89, 2005.

D. Sentürk and H.-G. Müller. Inference for Covariate Adjusted Regression Via Varying Coefficient Models. *The Annals of Statistics*, 34(2):654–679, 2006.

R. Serfling. Depth Functions in Nonparametric Multivariate Inference. In R. Y. Liu, R. Serfling, and D. L. Souvaine, editors, *DIMACS Series in Discrete Mathematics and Theoretical Computer Science - Data Depth: Robust Multivariate Analysis, Computational Geometry and Applications*, volume 72, pages 1–16. The American Mathematical Society, 2006.

R. Serfling and S. Wang. General Foundations for Studying Masking and Swamping Robustness of Outlier Identifiers. *Statistical Methodology*, 20: 79–90, 2014.

T. A. Severini and J. G. Staniswalis. Quasi-Likelihood Estimation in Semi-parametric Models. *Journal of the American Statistical Association*, 89 (426):501–511, 1994.

S. S. Shapiro and M. B. Wilk. An Analysis of Variance Test for Normality (Complete Samples). *Biometrika*, 52(3/4):591–611, 1965.

S. J. Sheather. *A Modern Approach to Regression with R*. Springer Texts in Statistics. Springer, New York, NY, 2009.

J. Shults and N. R. Chaganty. Analysis of Serially Correlated Data Using Quasi-Least Squares. *Biometrics*, 54(4):1622–1630, 1998.

J. Shults and J. M. Hilbe. *Quasi-Least Squares Regression*. Chapman & Hall/CRC, Boca Raton, FL, 2014.

R. H. Shumway and D. S. Stoffer. *Time Series Analysis and Its Applications: With R Examples*. Springer Texts in Statistics. Springer, New York, NY, 3rd edition, 2011.

R. H. Shumway, R. S. Azari, and Y. Pawitan. Modeling Mortality Fluctuations in Los Angeles as Functions of Pollution and Weather Effects. *Environmental Research*, 45(2):224–241, 1988.

M. W. Sigrist, editor. *Air Monitoring By Spectroscopic Techniques*. Chemical Analysis Series, vol. 197. John Wiley & Sons, New York, NY, 1994.

B. W. Silverman. Some Aspects of the Spline Smoothing Approach to Non-Parametric Regression Curve Fitting. *Journal of the Royal Statistical Society, Series B*, 47(1):1–52, 1985.

B. W. Silverman. *Density Estimation for Statistics and Data Analysis.* Monographs on Statistics and Applied Probability. Chapman & Hall/CRC Press, London, UK, 1998.

A. B. Simas, W. Barreto-Souza, and A. V. Rocha. Improved Estimators for a General Class of Beta Regression Models. *Computational Statistics and Data Analysis*, 54(2):348–366, 2010.

J. W. Smith, J. E. Everhart, W. C. Dickson, W. C. Knowler, and R. S. Johannes. Using the ADAP Learning Algorithm to Forecast the Onset of Diabetes Mellitus. In R. A. Greenes, editor, *Proceedings of the Symposium on Computer Applications in Medical Care*, pages 261–265, Washington, DC, 1988. IEEE Computer Society Press.

A. J. Smola and B. Schölkopf. A Tutorial on Support Vector Regression. *Statistics and Computing*, 14(3):199–222, 2004.

G. K. Smyth. Australasian Data and Story Library (OzDASL). http://www.statsci.org/data, 2011.

M. E. Sobel. Asymptotic Confidence Intervals for Indirect Effects in Structural Equation Models. *Sociological Methodology*, 13:290–312, 1982.

C. Spearman. The Proof and Measurement of Association Between Two Things. *The American Journal of Psychology*, 15(1):72–101, 1904.

P. E. Speckman. Kernel Smoothing in Partial Linear Models. *Journal of the Royal Statistical Society, Series B*, 50(3):413–436, 1988.

D. Staiger and J. H. Stock. Instrumental Variables Regression with Weak Instruments. *Econometrica*, 65(3):557–586, 1997.

T. A. Stamey, J. N. Kabalin, J. E. McNeal, I. M. Johnstone, F. Freiha, E. A. Redwine, and N. Yang. Prostate Specific Antigen in the Diagnosis and Treatment of Adenocarcinoma of the Prostate: II. Radical Prostatectomy Treated Patients. *Journal of Urology*, 141(5):1076–1083, 1989.

T. D. Stanley and S. B. Jarrell. Meta-Regression Analysis: A Quantitative Method of Literature Surveys. *Journal of Economic Surveys*, 3(2):161–170, 1989.

P. C. Stark, L. M. Ryan, J. L. McDonald, and H. A. Burge. Using Meteorologic Data to Predict Daily Ragweed Pollen Levels. *Aerobiologia*, 13(3):177–184, 1997.

C. M. Stein. Estimation of the Mean of a Multivariate Normal Distribution. *The Annals of Statistics*, 9(6):1135–1151, 1981.

J. H. Stock and M. W. Watson. *Introduction to Econometrics (Update).* Pearson, New York, NY, 3rd edition, 2015.

M. Stone and R. J. Brooks. Continuum Regression: Cross-Validated Sequentially Constructed Prediction Embracing Ordinary Least Squares, Partial Least Squares and Principal Components Regression. *Journal of the Royal Statistical Society, Series B*, 52(2):237–269, 1990.

T. A. Stukel. Generalized Logistic Models. *Journal of the American Statistical Association*, 83(402):426–431, 1988.

J. Sun and C. R. Loader. Simultaneous Confidence Bands for Linear Regression and Smoothing. *The Annals of Statistics*, 22(3):1328–1345, 1994.

M. Tableman and J. S. Kim. *Survival Analysis Using S: Analysis of Time-to-Event Data*. Chapman & Hall/CRC, Boca Raton, FL, 2004.

J. A. Tawn. Estimating Probabilities of Extreme Sea-levels. *Journal of the Royal Statistical Society, Series C*, 41(1):77–93, 1992.

P. F. Thall and S. C. Vail. Some Covariance Models for Longitudinal Count Data with Overdispersion. *Biometrics*, 46(3):657–671, 1990.

H. Theil. A Rank-Invariant Method of Linear and Polynomial Regression Analysis. *Proceedings of the Royal Netherlands Academy of Sciences*, 53 (I,II,III):386–392,521–525,1397–1412, 1950.

T. M. Therneau, P. M. Grambsch, and T. R. Fleming. Martingale-Based Residuals for Survival Models. *Biometrika*, 77(1):147–160, 1990.

D. L. Thistlewaite and D. T. Campbell. Regression-Discontinuity Analysis: An Alternative to the Ex Post Facto Experiment. *Journal of Educational Psychology*, 51(6):309–317, 1960.

F. T. Thompson and D. U. Levine. Examples of Easily Explainable Suppressor Variables in Multiple Regression Research. *Multiple Linear Regression Viewpoints*, 24(1):11–13, 1997.

S. G. Thompson and J. P. T. Higgins. How Should Meta-Regression Analyses Be Undertaken and Interpreted? *Statistics in Medicine*, 21(11):1559–1573, 2002.

S. K. Thompson. *Sampling*. Wiley Series in Probability and Statistics. Wiley, New York, NY, 3rd edition, 2012.

G. L. Tietjen and R. H. Moore. Some Grubbs-Type Statistics for the Detection of Several Outliers. *Technometrics*, 14(3):583–597, 1972.

D. M. Titterington, A. F. M. Smith, and U. E. Makov. *Statistical Analysis of Finite Mixture Distributions*. Wiley, New York, 1985.

J. Tobin. Estimation of Relationships for Limited Dependent Variables. *Econometrica*, 26(1):24–36, 1958.

M. A. Treloar. *Effects of Puromycin on Galactosyltransferase in Golgi Membranes.* M.Sc. Thesis, University of Toronto, 1974.

E. Tuğran, M. Kocak, H. Mirtagiogğlu, S. Yiğit, and M. Mendes. A Simulation Based Comparison of Correlation Coefficients with Regard to Type I Error Rate and Power. *Journal of Data Analysis and Information Processing,* 3 (3):87–101, 2015.

N. Turck, L. Vutskits, P. Sanchez-Pena, X. Robin, A. Hainard, M. Gex-Fabry, C. Fouda, H. Bassem, M. Mueller, F. Lisacek, L. Puybasset, and J.-C. Sanchez. A Multiparameter Panel Method for Outcome Prediction Following Aneurysmal Subarachnoid Hemorrhage. *Intensive Care Medicine,* 36(1):107–115, 2010.

T. R. Turner. Estimating the Propagation Rate of a Viral Infection of Potato Plants via Mixtures of Regressions. *Journal of the Royal Statistical Society, Series C,* 49(3):371–384, 2000.

J. M. Utts. The Rainbow Test for Lack of Fit in Regression. *Communications in Statistics - Theory and Methods,* 11(24):2801–2815, 1982.

V. N. Vapnik. *The Nature of Statistical Learning Theory.* Statistics for Engineering and Information Science. Springer, New York, NY, 2$^{nd}$ edition, 2000.

J. M. Ver Hoef and P. L. Boveng. Quasi-Poisson vs. Negative Binomial Regression: How Should We Model Overdispersed Count Data? *Ecology,* 88 (11):2766–2772, 2007.

W. Viechtbauer. Conducting Meta-Analyses in R with the metafor Package. *Journal of Statistical Software,* 36(3):1–48, 2010. URL https://www.jstatsoft.org/article/view/v036i03.

C. Viroli. On Matrix-Variate Regression Analysis. *Journal of Multivariate Analysis,* 111:296–309, 2012.

G. Wahba. *Spline Models for Observational Data,* volume 59 of *CBMS-NSF Regional Conference Series in Applied Mathematics.* Society for Industrial and Applied Mathematics, 1990.

W. A. Wallis. Tolerance Intervals for Linear Regression. In J. Neyman, editor, *Second Berkeley Symposium on Mathematical Statistics and Probability,* pages 43–51, Berkeley, CA, 1946. University of California Press.

L. Wasserman. *All of Nonparametric Statistics.* Springer Texts in Statistics. Springer, New York, NY, 2006.

R. L. Wasserstein and N. A. Lazar. The ASA's Statement on *p*-values: Context, Process, and Purpose. *The American Statistician,* 70(2):129–133, 2016.

G. S. Watson. Smooth Regression Analysis. *Sankhyā: The Indian Journal of Statistics, Series A*, 26(4):359–372, 1964.

R. W. M. Wedderburn. Quasi-Likelihood Functions, Generalized Linear Models, and the Gauss-Newton Method. *Biometrika*, 61(3):439–447, 1974.

M. Wedel and W. S. DeSarbo. A review of recent developments in latent class regression models. In R. Bagozzi, editor, *Advanced Methods of Marketing Research*, pages 352–388. Blackwell Publishing, London, UK, 1994.

W. W. S. Wei. *Time Series Analysis: Univariate and Multivariate Methods.* Pearson, Boston, MA, 2$^{nd}$ edition, 2005.

F. Westad. Independent Component Analysis and Regression Applied on Sensory Data. *Journal of Chemometrics*, 19(3):171–179, 2005.

H. White. A Heteroskedasticity-Consistent Covariance Matrix Estimator and a Direct Test for Heteroskedasticity. *Econometrica*, 48(4):817–838, 1980.

H. Wickham. *ggplot2: Elegant Graphics for Data Analysis.* Springer-Verlag New York, 2009. ISBN 978-0-387-98140-6. URL http://ggplot2.org.

H. Wickham. *Advanced R.* Chapman & Hall/CRC The R Series. Taylor & Francis, Boca Raton, FL, 2014.

P. R. Winters. Forecasting Sales by Exponentially Weighted Moving Averages. *Management Science*, 6(3):324–342, 1960.

I. H. Witten, E. Frank, and M. A. Hall. *Data Mining: Practical Machine Learning Tools and Techniques.* Morgan Kaufmann Publishers, Burlington, MA, 3$^{rd}$ edition, 2011.

S. N. Wood. Thin-Plate Regression Splines. *Journal of the Royal Statistical Society, Series B*, 65(1):95–14, 2003.

Y. Xia, H. Tong, W. K. Li, and L.-X. Zhu. An Adaptive Estimation of Dimension Reduction Space. *Journal of the Royal Statistical Society, Series B*, 64(3):363–410, 2002.

C. Yale and A. B. Forsythe. Winsorized Regression. *Technometrics*, 18(3):291–300, 1976.

F. Yao, H.-G. Müller, and J.-L. Wang. Functional Linear Regression Analysis for Longitudinal Data. *The Annals of Statistics*, 33(6):2873–2903, 2005.

K. K. W. Yau, A. H. Lee, and A. S. K. Ng. Finite Mixture Regression Model with Random Effects: Application to Neonatal Hospital Length of Stay. *Computational Statistics and Data Analysis*, 41(3/4):359–366, 2003.

D. S. Young. Regression Tolerance Intervals. *Communications in Statistics - Simulation and Computation*, 42(9):2040–2055, 2013.

D. S. Young. "Bond. James Bond". A Statistical Look at Cinema's Most Famous Spy. *CHANCE*, 27(2):21–27, 2014.

D. S. Young and D. R. Hunter. Mixtures of Regressions with Predictor-Dependent Mixing Proportions. *Computational Statistics and Data Analysis*, 54(10):2253–2266, 2010.

R. Yu and Y. Liu. Learning from Multiway Data: Simple and Efficient Tensor Regression. In *Proceedings on the 33rd International Conference of Machine Learning*, pages 373–381, Brookline, MA, 2016. Microtome Publishing.

A. Zeileis, C. Kleiber, and S. Jackman. Regression Models for Count Data in R. *Journal of Statistical Software*, 27(8):1–25, 2008. URL http://www.jstatsoft.org/v27/i08/.

A. Zellner. An Efficient Method of Estimating Seemingly Unrelated Regressions and Tests for Aggregation Bias. *Journal of the American Statistical Association*, 57(298):348–368, 1961.

H. Zhou, L. Li, and H. Zhu. Tensor Regression with Applications in Neuroimaging Data Analysis. *Journal of the American Statistical Association*, 108(502):540–552, 2013.

H. Zou and T. Hastie. Regularization and Variable Selection via the Elastic Net. *Journal of the Royal Statistical Society, Series B*, 67(2):301–320, 2005.

# Index

**A**

AAPE, *see* Average of the absolute percent errors (AAPE)

ACF, *see* Autocorrelation function (ACF)

ACF and PACF plots, 226

Achieved significance level (ASL), 204–205

Activation function, 492

AdaBoost, 490

Added variable plot, *see* Partial regression plots

Additive model (AM), 457

Adjusted variable plot, *see* Partial regression plots

Advanced methods
  ARIMA models, 230–233
  exponential smoothing, 233–234
  spectral analysis, 234–235

AIC, *see* Akaike's Information Criterion (AIC)

Air Passengers Data, 245–246

Aitken estimator, 229

Akaike's Information Criterion (AIC), 254

All possible regressions, *see* Best subset procedure

Alternating least squares, 567

Alternative hypothesis ($H_A$ or $H_1$), 25

AM, *see* Additive model (AM)

Amemiya's Prediction Criterion (APC), 255

AMISE, *see* Asymptotic MISE (AMISE)

Amitriptyline Data, 449–453

Amplitude, in spectral analysis, 235

Analysis of covariance, 265–266

Analysis of variance (ANOVA)
  constructing ANOVA table, 73–77
  examples, 78–81
  formal lack of fit, 77–78

ANCOVA, *see* Analysis of covariance

Anderson-Darling test, 60

Andrews's Sine, 199

Animal Weight Data, 516–517, 540–542, 545–546

Annual Maximum Sea Levels Data, 548

ANOVA; *see also* Analysis of variance
  $k$-way, 265
  models, 263, 265
  one-way, 265
  two-way, 265

ANOVA for multiple linear regression
  ANOVA table, 121–122
  extra sums of squares, 124–125
  general linear $F$-test, 122–123
  lack-of-fit testing in the multiple regression setting, 123–124
  partial measures and plots, 125–128
  Simulated Data, 131–136
  Thermal Energy Data, 129–131

Anscombe residuals, 388, 393, 418

Antitonic regression, 545

APC, *see* Amemiya's Prediction Criterion (APC)

Approximate inference procedures, 367–370

Approximate standard errors, 368

AR(1), *see* First-order autoregression AR(1)

AR(2), *see* Second-order autoregression AR(2)

ARIMA models, *see* Autoregressive integrated moving average (ARIMA models)

AR($k$), $k^{\text{th}}$–order autoregression AR($k$)

Arsenate Assay Data, 192–193, 212–216

AR(1) structure, 423

ASL, *see* Achieved significance level (ASL)

Associative Law of Addition, 576

Associative Law of Multiplication, 576

Assumptions
    diagnosing validity of, 51–53
    invalid, consequences of, 50–51
    linear regression, 49–50

Asymptotic MISE (AMISE), 312–313

Asymptotic theory, in nonlinear regression, 367–368

Attenuation bias, *see* Regression dilution attenuation

Attributes, 143, 478

Auditory Discrimination Data, 144–146

Autocorrelated, 219

Autocorrelation, testing and remedial measures of, 224–240

Autocorrelation function (ACF), 222

Autocorrelation parameter, 220

Autocovariance function, 222

Automobile features data, 304–306

Autoregressive integrated moving average (ARIMA models), 230–233

Autoregressive model for the errors, 220

Autoregressive structures, 219–221

Auxiliary variables, 527

Averaged shifted histogram estimator, 329

Average of the absolute percent errors (AAPE), 253–254

**B**

Backcasting method, 233

Backfitting algorithm, 457–458

Backward pass approach, 487

Backward search method, 163

Backward selection, in stepwise methods, 255–256

Bagging, 490–491

Bandwidth, 312

Bartlett's test, 5, 65

Baseline cumulative hazard function, 344

Baseline hazard, 339

Basic area level (type A) model, 529

Basic matrix, 321

Basic set, 164

Basic unit level (type B) model, 530

Basis matrix, 578

Bayesian inference, 538

Bayesian Information Criterion (BIC), 254

Bayesian linear regression, 539–540

Bayesian regression, 538–542

Bayes' Theorem, 538

Belgian Food Expenditure Data, 543–544

Belgium Telephone Data, 559–560

Berkson model, 180

Best linear unbiased estimator (BLUE), 90

Best regression line, for sample, 11

Best subset procedure, 252–254

Beta function, 341

Beta regression, 420–421

Beta regression model, 420–421

Biased regression, regression shrinkage and dimension reduction
    Automobile Features Data, 304–306

partial least squares, 294–296
principle components regression,
    292–294
Prostate Cancer Data, 301–304
reduction methods and
    sufficiency, 296–301
regression shrinkage and
    penalized regression,
    287–292
Biasing constant, 289
Bias-variance tradeoff, 288, 482
BIC, *see* Bayesian Information
    Criterion (BIC)
Big data, 477
Binary logistic regression, 375
Binomial regression, 378
Biochemistry Publications Data,
    405–409
Black Cherry Tree Data, 69–72
Blood Alcohol Concentration Data,
    188–190
BLUE, *see* Best linear unbiased
    estimator (BLUE)
Bone Marrow Transplant Data,
    355–356
Bonferroni Joint Confidence
    Intervals, 35–36, 95
Bonferroni Joint Confidence Intervals
    for $(\beta_0, \beta_1)$, 29
Bonferroni Joint Prediction
    Intervals, 37, 95
Bonferroni's inequality, 29
Boosted regression, 489–490
Boosting, *see* Boosted regression
Bootstrap aggregating, *see* Bagging
Bootstrapping method, 205
Bootstrap replications, 205
Bootstrap samples, 205
Boston Housing Data, 473–475,
    499–501, 552–554
Box-Cox transformations, 59
Box-Jenkins backshift operator, 231
Box-Jenkins models, 230
Box-Pierce test, 227
Breakdown values, 203

Breusch-Godfrey test, 225–226
Breusch-Pagan test, 65–66
Brown-Forsythe test, 64
Brown's test, 382
*B*-spline, 322
*B*-spline basis, 322
Building regression model, steps for,
    573–574
Bump hunting, 486

**C**
Calibration and regulation intervals,
    39–40
Calibration confidence interval,
    39–40
Calibration (or inverse regression),
    39
Canadian Auto Insurance Data,
    429–431
C and $\bar{\text{C}}$, 380
Canonical analysis, 272
Canonical link function, 415
Canonical parameter, 415
Cardiovascular Data, 238–240,
    564–565
CART, *see* Classification and
    regression trees (CART)
Categorical variable, 137
Cauchit link, 414
cdf, *see* Cumulative distribution
    function (cdf)
Ceiling function, 208
Censored regression models, 337–338
Censoring, 336
Census of Agriculture Data, 530–532
Centered **X** matrix, 114
Central composite design, 270
Central mean subspace (CMS), 297
Central subspace (CS), 297
Change points, 307
Cheese-Tasting Experiment Data,
    403–405
Child nodes, 484
Chi-square goodness-of-fit test, 61
Chi-square outlier test, 157

Circular-circular regression, 554
Circular-linear regression, 554
Circular random variable, 554
Circular regression, 554–557
Classical measurement error model, 179
Classification, 478–482
Classification and regression trees (CART), 486
Classification trees, 483
Classifier, 478
Clean instruments, 185
Clustering, 478
Cluster sampling, 527
CMS, *see* Central mean subspace (CMS)
Cochrane-Orcutt procedure, 227–228
Coded variable, 141–143
Coefficient estimation, 4
Coefficient interpretations, 138–140
Coefficient of determination, 12
Coefficient of multiple determination, 122
Coefficient of partial correlation, 126–127
Coefficient of partial determination, 125
Coefficient of variation (CV), 253
Cofounding variable, 520
Collinearity-influential observations, 114
Commutative Law of Multiplication, 576
Complementary log-log link, 414
Completely missing data, 533
Components, in mixture experiment, 273
Computer-Assisted Learning Data, 210–212
Computer Repair Data, 21–23, 44–45, 79
Concave regression, 545
Concomitant variable, *see* Covariate
Concordant pairs, 16

Conditional autoregressive processes, 552
Conditional mean function, 325
Condition number, 113
Confidence interval, 26–27, 35, 95
Confidence interval for the slope $(\beta_1)$, 28
Confidence level, 26
Confidence regions, 443
Conjoint analysis, 143–144
Consolidation and model selection methods
  best subset procedures, 252–254
  crossvalidation, 249–251
  prediction sum of squares, 251–252
  Punting Data, 256–260
  statistics from information criteria, 254–255
  stepwise procedures for identifying models, 255–256
Constant error variance, tests for, 63–66
Constant-variance-covariance matrix condition, 297
Constrained response function, 377
Contaminated instruments, 184
Contour directions, 300
Contour regression, 300
Control, regression effect and, 18
Control, regression methods and, 4
Convex minorant, 546
Convex regression, 545
Conway-Maxwell-Poisson regression model, 396
Cook's D (Cook's distance), 161
Copula, 563
Copula regression, 563–565
Correlated errors and autoregressive structures
  advanced methods, 230–236
  Air Passengers Data, 245–246
  Cardiovascular Data, 238–240
  Google Stock Data, 236–238
  Mouse Liver Data, 246–247

Natural Gas Prices Data, 240–245

properties of error terms, 221–224

testing and remedial measures for autocorrelation, 224–230

time series and autoregressive structures, 219–221

Correlation matrix of $\beta$ 92

Correlation model, inference on, 33–35

Correlation transformation, 112

Correlation *vs.* regression, 14–16

Covariate-adjusted regression, 511

Covariance ratio (CovRatio), 162

Covariate, 265

Covariate space, *see* X-space

CovRatio, *see* Covariance ratio (CovRatio)

Cox proportional hazards regression, 343–344

Cox-Snell residuals, 344

Cracker Promotion Data, 278–281

Credible interval, 540

Credit Loss Data, 431–434

Critical value/critical region approach, 26

Crossvalidation estimate of prediction error, 250–251

Crossvalidation method, 249–251

CS, *see* Central subspace (CS)

Cubic, 97

Cubic spline, 320

Commutative Law of Addition, 576

Cumulative conditional failure function, 337

Cumulative distribution function (cdf), 580

Curse of dimensionality, 296–297

CV, *see* Coefficient of variation (CV)

**D**

Data mining, 477

Boston Housing Data, 499–501

classification and support vector regression, 478–482

ensemble learning methods for regression, 488–492

neural networks, 492–493

prediction trees and related methods, 483–488

Prima Indian Diabetes Data, 493–499

Simulated Motorcycle Accident Data, 493

Data splitting, 249

Data subsetting, 124

Data transformations, 57–59

Davison-Hinkley modified residuals, 206–207

Decision rule, 26

Decision tree learning, 483

Decision trees, 483

Deductive imputation, 534–535

Degree or order of polynomial, 96

Delete-$d$ jackknife, 208–210

Delete-$d$ subsamples, 208

Deleted residuals, 155

Deming regression, 182–184

Density estimation, 312

Dependent variable, 4–5, 8

Derived predictor values, 492

Descriptions, regression methods and, 3–4

Design-based approaches, 527

Design effect, 527

Deterministic relationships, 3–4

Deviance, 387, 392

Deviance residuals, 345–346, 380, 388, 393, 417

Deviance statistic, 381

DFBETAS, *see* Difference in betas (DFBETAS)

DFDEV and DFCHI, 382

$df_E$, *see* Error degrees of freedom ($df_E$)

DFITs (difference in fits) residual, 160–161

Diagnostic procedures, 344–347
Diagonal matrix, 575
Diagonal values, 92
Difference in betas (DFBETAS), 162
Dimension reduction, 297, 445
Dimension reduction methods,
        296–301
Dimension reduction subspace
        (DRS), 297
Dimensions of the matrix, 575
Directional regression (DR), 301
Discordant pairs, 16
Disordinal interaction, 141
Dispersion parameter, 390
Distorting functions, 511–512
Dixon's Q test, 158–159
Double censoring, 336
Double exponential smoothing, 234
DR, *see* Directional regression
DRS, *see* Dimension reduction
        subspace (DRS)
Duality principle, 30
Dummy matrix, 143
Durable Goods Data, 356–359
Durbin-Watson test, 224–225
Dykstra's algorithm, 545

**E**
Earth, 487
EDR-directions, *see* Effective
        dimension reduction
        directions (EDR-directions)
EDR-space, *see* Effective dimension
        reduction space
        (EDR-space)
Effective degrees of freedom, 311–312
Effective dimension reduction
        directions
        (EDR-directions), 297
Effective dimension reduction space
        (EDR-space), 297
Effect modifiers, 461
Effects, *see* Independent variables
Effect size, 522
Efficiency, 203

Eicker-White estimator, 230
Eigenvalue methods, 112–113
Eigenvalues, 577
Elastic net regression, 291
EM algorithm, *see*
        Expectation-Maximization
        (EM) algorithm
Empirical directions, 300–301
Empirical estimator, 424
E(MSLOF), *see* Expected mean
        squares for lack-of-fit
        E(MSLOF)
Endogenous variable, 184
Ensemble methods for regression,
        488–489
Envelope regression model, 445
Envelopes, 445
Epilepsy Data, 434–438
Error degrees of freedom ($df_E$), 74
Error or deviation, 9
Errors-in-variables models, *see*
        Measurement error models
Error sum of squares (SSE), 74, 121
Error terms
        independence of, 53
        properties of, 221–224
ESD test, *see* Generalized extreme
        studentized deviate (ESD)
        test
Ethanol fuel data, 561–562
Euclidean norm, 578
EVA, *see* Extreme value analysis
        (EVA)
Exact tests, *see* Permutation tests
Exchangeable structure, 423
Exogenous variable, 184
ExpBoost, 490
Expectation-Maximization (EM)
        algorithm, 264
Expected mean squares E(MSR), 76
Expected mean squares for lack-of-fit
        E(MSLOF), 78
Expenditures data, 171–177
Expert networks, 488
Explicit models, 534

Exponential growth equation, for variables, 58
Exponential regression model, 362
Exponential smoothing, 233–234
Exposure, 386
Externally studentized residuals, *see* Studentized residuals
Extramarital Affairs Data, 350–352
Extrapolating, 4
Extra sums of squares, 124–125
Extreme value analysis (EVA), 546

**F**
Factor analysis regression, 294
Factor levels, 263
(Factor) loadings matrix, 292
Factor regression model, 441
False negative, *see* Type II error
False positive, *see* Type I error
Fay-Herriot model, 530
Feasible generalized least squares, 229–230
Feature measurements, 477
Feature selection, 478
Fiber strength data, 78–81
Finite population correction factor, 527
First differences procedure, 228–229
First-order autoregression AR(1), 221
(First-order) spatial autoregressive model, 551
Fisher's z-transformation, 34
Fixed-effect meta-regression, 523
Fixed effects, 263
Fixed point method, 289
Floor function, 207–208
Forecasting, 233
Formal lack of fit, 77–78
Forward pass approach, 487
Forward search method, 163–164
Forward selection, in stepwise methods, 255, 256
Forward stagewise regression, 292
Fourier analysis, 235

Frequency, in spectral analysis, 235
Frequentist's view, 538
Fruit Fly Data, 426–429
*F*-test, 63
Full model, 123
Functional data, 510
Functional linear regression analysis, 510–514
Functional measurement error model, 180

**G**
Gait data, 512–514
Galton, Sir Francis, 17
GAM, *see* (Generalized) additive models (GAM)
Gamma function, 341
Gamma-Ray Burst data, 327–328
Gamma regression, 418–419
Gamma regression model, 419
Gating networks, 488
Gaussian copula regression, 563
Gauss-Markov Theorem, 11, 90
Gauss-Newton algorithm, 366
GCV, *see* Generalized crossvalidation (GCV)
GEE, *see* Generalized estimating equations (GEE)
GEE models, *see* Generalized estimating equations (GEE) models
Generalization error, 482
(Generalized) additive models (GAM), 457–458
Generalized crossvalidation (GCV), 322
Generalized estimating equations (GEE), 422–426
Generalized estimating equations (GEE) models, 422–426
Generalized extreme studentized deviate (ESD) test, 159–160
Generalized extreme value distribution, 546–547

Generalized extreme value regression
  methods, 546–548
Generalized extreme value regression
  model, 547
Generalized least squares, 229
Generalized linear models
  Beta regression, 420–421
  Canadian Auto Insurance Data,
    429–431
  Credit Loss Data, 431–434
  Epilepsy Data, 434–438
  Fruit Fly Data, 426–429
  Gamma regression, 418–419
  Generalized estimating
    equations, 422–426
  Generalized linear models and
    link functions, 413–418
  Inverse Gaussian (Normal)
    regression, 419–420
Generalized logistic curve model, *see*
  Richards' curve model
(Generalized) partial linear model
  (GPLM), 458–459
(Generalized) partial linear partial
  additive models
  (GPLPAM), 460
Generalized partial linear
  single-index model, 459
Generalized Poisson regression
  model, 395–396
Generalized regression estimators,
  528–529
Generalized variance, 443
Generalized varying-coefficient
  model, 461
General linear $F$-statistic, 123
General linear $F$-test, 122–123
Geographically weighted regression
  (GWR), 550
Geometric distribution, 396
Geometric regression model, 396
Glejser's test, 164–165
Goldfeld-Quandt test, 165
Gompertz growth curve model, 363
Gompit link, 414

Goodman-Kruskal Gamma, 16
Goodness-of-Fit, 387, 392
Goodness-of-Fit measures, 52
Goodness-of-Fit tests, 381–382
Google Stock Data, 236–238
GPLM, *see* Generalized partial linear
  model (GPLM)
GPLPAM, *see* (Generalized) partial
  linear partial additive
  models (GPLPAM)
Gradient, 365
Gradient methods, 366
Graphical regression, 300
Grubbs' test, 156–157
GWR, *see* Geographically weighted
  regression (GWR)

**H**
HAC estimator, *see*
  Heteroskedasticity and
  autocorrelation (HAC)
  estimator
Hadi's influence measure, 161
Hampel's Method, 199
Harmonic regression, 235
Harvey-Collier test, 166–167
Hat matrix, 154
Hat values, 380, 388–389, 394
Hazard rate, 337
Hessian matrix, 366
Heteroskedasticity, 195
Heteroskedasticity and
  autocorrelation (HAC)
  estimator, 230
Hex maps, 554
Hierarchical mixture-of-experts
  (HME), 488
Hierarchical regression models, 503
Hierarchy principle, 99
Hildreth-Lu procedure, 228
Hinge function, *see* Truncated power
  function
Histogram of residuals, 52, 53
HME, *see* Hierarchical
  mixture-of-experts (HME)

Hoeffding's $D$ statistic, 15, 22
Hoerl-Kennard iterative method, 289–290
Holdout method of crossvalidation, 250
Holt's method, *see* Double exponential smoothing
Holt-Winters exponential smoothing, 234
Homogeneity of slopes, 265
Homoskedasticity, 49, 51, 195
Horvitz-Thompson estimator, 529
Hosmer-Lemeshow test, 382
Hospital Stays Data, 409–412
Hotelling-Lawley Trace, 442
Huber's loss, 482
Huber's Method, 200
Hurdle count regression models, 397
Hyperparameters, 538
Hypothesis testing, 25–30
Hypothesis test for the Intercept $(\beta_0)$, 28–29
Hypothesis test for the slope $(\beta_1)$, 28
Hypothesis Testing, 442–443

**I**

Idempotent matrix, 575
Identity matrix, 575
IHT, Iterative Hessian Transformation (IHT)
Implicit models, 534
Imputation, 534
Incomplete gamma function, 341
Independence, 49, 51
Independence of error terms, 53
Independence structure, 423
Independent component analysis, 294
Independent component regression, 294
Independent variables, 263
Independent variable(s), 4–5, 8
Index, 456
Index set, 220
Indicator variable, 137

Auditory Determination Data, 144–146
coded variable, 141–143
coefficient interpretations, 138–140
conjoint analysis, 143–144
interactions, 140–141
leave-one-out method, 137–138
Steam Output Data, 146–147
Tea Data, 147–149
Individual coefficients plot, *see* Partial regression plots
Individual predictor variables, testing contribution of, 94
Inference, 346–347, 389–390, 394–395
Inference for multiple $\beta_j$, 370
Inference for single $\beta_j$, 369
Inference on correlation model, 33–35
Influential values, outliers, and diagnostic tests, 153–154
comments on outliers and influential values, 167–168
diagnostic methods, 164–167
Expenditures Data, 171–177
masking, swamping, and search methods, 163–164
Punting Data, 168–171
residuals and measures of influence, 154–162
$\varepsilon$-Insensitive loss, 482
Instrumental variable, 184
Instrumental variables regression, 184–186
Interactions, 140–141
Intercept, 8
Internally studentized residuals, 54; *see also* Studentized residuals
Internal nodes, 484
Interval censoring, 336
Interpretation plot, 140–141
Intervals for mean response, 35–36
Intervals for new observation, 36–40
Intrinsically linear, 361–362

Invalid assumptions, consequences of, 50–51
Invariant subspace, 445
Inverse Gaussian (Normal) regression, 419–420
Inverse Gaussian regression model, 419–420
Inverse link, 414
Inverse Mills ratio, 338
Inverse regression, 39, 298
Inverse regression estimator, 298
Investment Data, 453–454
IRLS, *see* Iteratively reweighted least squares (IRLS)
Isotonic regression, 544
Isotropic, 323
Item nonresponse data, 533
Iterative Hessian Transformation (IHT), 299–300
Iteratively reweighted least squares (IRLS), 199

**J**
Jackknife, 208
Jacobian matrix, 365
James Bond Data, 373–374
Job Stress Data, 520–522
Joint pdf, 580
Joint pmf, 580

**K**
Kendall's $\tau_a$, 16
Kendall's $\tau_b$, 15–16
Kernel density estimator, 312
Kernel function, 312
Kernel regression, 312–315, 314
$K$-fold crossvalidation, 250
Knots, 307
Kola Project Data, 103–106
Kolmogorov-Smirnov test, 60
Kriging, 552
Kronecker delta, 326
Kronecker product, 576
$k^{\text{th}}$-order autoregression AR($k$), 221
Kurtosis, 61–62

$k$-way ANOVA, 265

**L**
Lack-of-fit sum of squares (SSLOF), 77
Lack-of-fit test, 52
Lack-of-fit testing, in the multiple regression setting, 123–124
Lag $k$ autocorrelation, 222
Laplace loss, 482
LARS, *see* Least angle regression (LARS)
LASSO, *see* Least absolute shrinkage and selection operator (LASSO)
Latent class regressions, 561
Latent variables, 186
Learner, 477
Least absolute deviation, 198
Least absolute shrinkage and selection operator (LASSO), 290–291
Least angle direction, 292
Least angle regression (LARS), 291–292
Least median of squares method, 202
Least quantile of squares method, 202
Least Squares, 441–442
Least trimmed sum of absolute deviations method, 203
Least trimmed sum of squares method, 202–203
Leave-one-out crossvalidation, 250
Leave-one-out method, 137–138
Left censoring, 336
Leptokurtic, 62
Levenberg-Marquardt method, 367
Levene test, 52, 63–64
Leverage, 154
LIDAR data, 331–334
Light Data, 371–373
Likelihood function, 580
Likelihood ratio test, 381
Linear, 9

Linear check function, 542–543
Linear-circular regression, 554
Linearity, 49, 50
Linearity condition, 297
Linear regression model, 3
    key theoretical assumptions,
        49–50
Linear smoothers, 311
Linear spline, 320
Linear Structural Relationships
    (LISREL), 186–187
Link functions, 383
LISREL, *see* Linear Structural
    Relationships (LISREL)
Ljung-Box Q test, 226–227
$L_1$-norm regression, *see* Least
    absolute deviation
$L_2$-norm regression, 198
Local least squares, 316
Local likelihood regression, 318
Locally estimated scatterplot
    smoother (LOESS),
    316–318
Local polynomial regression, 315–316
Local regression, 311
Local regression methods, 310–318
Local scoring algorithm, 458
LOESS, *see* Locally estimated
    scatterplot smoother
    (LOESS)
Logistic regression, 375–383
    binary logistic regression, 375,
        376–383
    nominal logistic regression, 375,
        383–384
    ordinal logistic regression,
        375–376, 384–385
Logit function, 377
Logit link, 413
Loglikelihood function, 580
Loglikelihood regression model, 318
Log link, 414
Log-log link, 414
Log-log regression equation, 58
Log odds, 377

Longitudinal data, 422
Loss functions, 479

**M**
Machine learning, 477
Mallow's $C_p$, 252–253
Mammal Sleep Data, 535–538
MAR, *see* Missing at random (MAR)
Margin, 478, 479
Marginal density function, 538
Marginal integration estimators, 458
Marginally significant, 26
Marquardt parameter, 36
MARS, *see* Multivariate adaptive
    regression splines (MARS)
Martingale residuals, 345
Masking, 163
Matrices, refresher on, 575–578
Matrix notation, 7
Matrix notation in regression, 87–91
Matrix theorem, 109
Matrix-variate regression, 439
MAVE, *see* Minimum average
    variance estimation
    (MAVE)
Maximum normed residual test, *see*
    Grubbs' test
MCAR, *see* Missing completely at
    random (MCAR)
McElroy's $R^2$, 448
McHenry's best subset algorithm,
    254
$m$-dependent non-stationary
    structure, 423
$m$-dependent stationary structure,
    423
Mean DRS, 297
Mean imputation, 534
Mean integrated squared error
    (MISE), 312
Mean of population, notation for, 7,
    8
Mean of sample, notation for, 7, 8
Mean parameter, 390

Mean squared error (MSE), 12, 73, 74, 121, 122

Mean square for regression (MSR), 75

Measurement equations, *see* Linear Structural Relationships (LISREL)

Measurement errors
Arsenate Assay Data, 192–193
Blood Alcohol Concentration Data, 188–190
characteristics of, 179–180
estimation in presence of, 180–182
instrumental variables regression, 184–186
models, 179–180
orthogonal and Deming regression, 182–184
structural equation modeling, 186–188
U.S. Economy Data, 190–192

Measurement models, dilution, 188

Measures of association, 15

Median absolute deviation, 317

Median smoothing regression, 325

Mediated moderation, 519

Mediation regression model, 517–522

Mediator variable, 517

M-envelope, 445

M-estimators, 198

Meta-analysis, 522

Meta-regression models, 522–526

Meta-regressions, 522

Michaelis-Menten models, 362–363

Minimum average variance estimation (MAVE), 299

MISE, *see* Mean integrated squared error (MISE)

Missing at random (MAR), 533

Missing completely at random (MCAR), 533

Missing-data indicator matrix M, 533

Mixed and regression models
ANVOVA, 265–268

Cracker Promotion Data, 278–281
mixed effects models, 263–264
mixture experiments, 273–275
Odor Data, 281–283
response surface regression, 268–273
Sleep Study Data, 276–278
Yarn Fiber Data, 283–286

Mixed effects models, 263–264

Mixed models, 264

Mixed regressive spatial autoregressive model, 551

Mixture experiments, 273–275

Mixture modeling, 560

Mixture-of-experts, 561

Mixture-of-linear-regressions, 560–561

Mixtures of regressions, 560–562

m-mode product, 567

m-mode unfolding, 567

Mode, 567

Model assessment, in crossvalidation method, 249

Model-based approaches, 526

Model-based estimator, 425

Model selection, in crossvalidation method, 249

Moderated mediation, 519

Moderator variables, 519

Modified Bessel function of the first kind of order, 556

Modified Levene tests, 64–65

Monotone regression, 544–546

Monotonic regression, 544

Moran's $I$ statistic, 552

Mother Haar wavelet, 327

Motorette Data, 352–355

Mouse Liver Data, 246–247

MSE, *see* Mean squared error (MSE)

MSR, *see* Mean square for regression (MSR)

$M^{\text{th}}$-order splines, 319–320

Multiclass logistic regression, 383

Multicollinearity, 109

detecting and remedying,
110–114
sources and effects of, 109–110
structural, 114–115
Thermal Energy Data, 115–119
Tortoise Eggs Data, 119–120
Multidimensional splines, 322–323
Multilevel regression models,
503–510
Multinomial logistic regression
model, 383
Multiple linear regression
about the model, 85
estimates of model parameters,
86
interaction terms, 87
Kola Project Data, 103–106
matrix notation in regression,
87–91
model notation, 85–86
polynomial regression, 96–99
predicted values and residuals,
86–87
statistical intervals, 95–96
testing contribution of
individual predictor
variables, 94
Thermal Energy Data, 99–103
Tortoise Eggs Data, 106–108
variance-covariance matrix and
correlation **matrix of** $\beta$
92–93
Multiple time series regression
model, 220
Multiplier, 27
Multivariate adaptive regression
splines (MARS), 487
Multivariate analysis of variance
(MANOVA), 444
Multivariate kernel density estimator
(MVKDE), 314–315
Multivariate multiple regression
Amitriptyline Data, 449–453
estimation and statistical
regions, 441–446

examples, 449–454
Investment Data, 453–454
model, 439–441
reduced rank regression, 446–447
seemingly unrelated regressions,
447–448
Multivariate multiple regression
model, 439–441
MVKDE, *see* Multivariate kernel
density estimator
(MVKDE)

**N**
Nadaraya-Watson estimator, 314
Nave estimator, *see* Model-based
estimator
NASA Data, 567–570
Natural cubic splines, 320
Natural Gas Prices Data, 240–246
Natural splines, 320
Nearest-neighbor regression, 324–325
Negative binomial distribution, 390
Negative binomial regression,
390–396
Neighborhood, 316
Nested effects regression models, 503
Neural networks, 492–493
Newey-West estimator, 230
Newton's method, 366
1993 Car Sale Data, 68–69
The 1907 Romanian Peasant Revolt
Data, 462–473
NIPALS, *see* Nonlinear iterative
partial least squares
(NIPALS)
NMAR, *see* Not missing at random
(NMAR)
Node impurity measure, 483
Nominal logistic regression, 375,
383–384
Non-basic set, 164
Nonconstant error variance, 376
Nonconstant variance, 52
Nonfit, 515

Nonlinear iterative partial least
    squares (NIPALS), 295–296
Nonlinear least squares, 364
    few algorithms, 365–366
Nonlinear regression
    approximate inference
        procedures, 367–370
    James Bond Data, 373–374
    Light Data, 371–373
    models, 361–364
    nonlinear least squares, 364–366
    nonlinear regression models,
        9–10, 361–362
    Puromycin Data, 370–371
    unspecified functional form,
        363–364
Nonnormal error terms, 376
Nonparametric measures, 15
Nonparametric models, 455
Nonparametric regression, 310–311
Nonparametric regression
    procedures, 324–327
Non-random censoring, 336
Normal equations, 10
Normality, 49, 50–51
Normalized residuals, *see*
    Standardized residuals
Normal probability plot (NPP) of
    residuals, 52, 53
Normal theory, 367–368
Normit link, 414
Notation
    for population, 7, 8
    regression, 8
    for sample, 7, 8
Not missing at random (NMAR), 533
NPP, *see* Normal probability plot
    (NPP)
Null hypothesis (H0), 25
Null matrix, 575

**O**

Observed error/residual, 11
Odds for an event, 377
Odds ratio, 379

Odor Data, 281–283
Off-diagonal value, 92
Offset, 386
Offset term, 386, 392
One-way ANOVA, 265
One-way ANOVA model, 266–268
Optimization Model, 270
Order of autoregression, 221
Order of differencing, 232
Order statistics, 202
Ordinal interaction, 141
Ordinal logistic regression, 375–376,
    384–385
Ordinary least squares, 10–11
Ordinary least squares estimates,
    10–11
Ordinary least squares line, 10–11
Orthogonal complement, 578
Orthogonal regression, 182–184
Orthogonal series regression, 325–327
Orthogonal matrix, 576, 578
Orthonormal basis, 578
Orthonormal basis matrix, 578
Osius and Rojek normal
    approximation test, 382
Outlier accommodation, 167
Outlier identification, 167
Outlier labeling, 167
Overall variation, measuring, 12–13
Overdispersion, 386–387
Overfit model, 487

**P**

PACF, *see* Partial autocorrelation
    function (PACF)
Pairwise scatterplots, 110–111
Parameter, 7
Parametric inverse regression (PIR),
    298
Parametric models, 455
Partial autocorrelation function
    (PACF), 223–224
Partial correlation, 126–127
Partial deviance, 346
Partial least squares, 294–296

Partial leverage, 128
Partial leverage plots, 128
Partial linear model (PLM), 459
Partial linear partial additive model (PLPAM), 460
Partially missing data, 533
Partial measures and plots, 125–128
Partial $R^2$, 125–126
Partial regression plots, 127–128
Partial residual plots, 128
Partial residuals, 128
Passing-Bablok estimator, 201
Passing-Bablok regression, 201–202
Pasting, 487
Patient rule induction method (PRIM), 486–487
pdf, *see* Probability density function (pdf)
Pearson chi-square statistic, 381
Pearson correlation coefficient ($\rho$), 14–15, 22, 23
Pearson residuals, 380, 388, 393, 417
Peeling, 487
Penalized regression, 287–292
Penalized splines, 321
Penalized sum of squares, 287–288
Periodograms, 235
Permutation tests, 204
Phase, in spectral analysis, 235
PHD, *see* Principal Hessian directions (PHD)
Piecewise, nonparametric, and local regression methods
  Gamma-Ray Burst Data, 327–328
  local regression methods, 310–318
  nonparametric regression procedures, 324–327
  piecewise linear regression, 307–310
  Quasar Data, 328–331
  splines, 318–324
Piecewise, nonparametric, and local

regression methods LIDAR data, 331–334
Piecewise constant spline, 319
Piecewise linear regression, 307–310
Pillai's Trace, 442
Pima Indian Diabetes Data, 493–499
PIR, *see* Parametric inverse regression (PIR)
$p^k$ factorial design, 268–269
Platykurtic, 62
PLM, *see* Partial linear model (PLM)
Plot of residuals *vs.* fitted values, 51
Plots
  difficulties seen in, 56–57
  ideal appearance of, 54–56
Plots of residuals *vs.* fitted values, 53–57
PLPAM, *see* Partial linear partial additive model (PLPAM)
pmf, *see* Probability mass function (pmf)
Pointwise regression model, 511
Pointwise tolerance intervals, 37–39
Poisson distribution, 385
Poisson-gamma mixture, 392
Poisson regression, 385–390
Polynomial regression, 96–99, 268–269
Polytomous, 383
Pool adjacent violators algorithm, 545
Population, 7
Population model for simple linear regression, 8–10
Portmanteau test, *see* Ljung-Box Q test
Posterior distribution, 538–539
Potential effect modifiers, 522
Potential function, 161
Potential-residual plot, 162
Power, 30–33
Power and Box-Cox transformations, 58–59
Power curve equation, 58
Power link, 414

Power of statistical test, 31
Power of test, 31
Power transformations, 55–59
PPR, *see* Projection pursuit
    regression (PPR)
Practical significance *vs.* statistical
    significance, 30
Prais-Winsten procedure, 228
Predicted value/predicted fit, 11
Predicted values *vs.* plot of residuals,
    86–87
Prediction, regression methods and,
    4
Prediction intervals, 36, 95
Prediction issues, 230
Prediction regions, 443
Prediction sum of squares (PRESS),
    251–252
Prediction trees, 483–486
Predictors, 8
PRESS, *see* Prediction sum of
    squares (PRESS)
Pricewise linear regression, 307–310
(Principal component) scores matrix,
    292–294
Principal Hessian directions (PHD),
    298–299
Principle components regression,
    292–294
Prior distribution, 538
Probability density function (pdf),
    580
Probability mass function (pmf), 579
Probit function, 382
Probit link, 414
Probit regression, 382–383
Product kernel, 315
Profile, 143
Projection matrix, 578
Projection pursuit, 462
Projection pursuit regression (PPR),
    461–462
Prostate Cancer Data, 301–304
Pruning, 484

Pseudo-maximum likelihood
    estimation, 456
Pseudo-$R^2$, 382, 387, 392–393
Pulp Property Data, 45–47, 216–217
Punting Data, 168–171, 256–260
Pure error sum of squares (SSPE), 77
Pure error test, *see* Lack-of-fit test
Puromycin Data, 370–371
$p$-value, 25–26

**Q**

QIC, *see* Quasi-likelihood under the
    independence model
    criterion (QIC)
$q^{\text{th}}$-degree polynomial regression,
    115
Quadratic, 96
Quadratic loss, 482
Quadratic spline, 320
Quantile regression, 542
Quasar Data, 328–331
Quasi-least squares regression, 425
Quasi-likelihood function, 396
Quasi-likelihood under the
    independence model
    criterion (QIC), 425
Quasi-Poisson regression, 396

**R**

$\rho$, *see* Pearson correlation coefficient
    $(\rho)$
$R^2$, 12–13
$R^2$-adjusted, 252
Radial bias functions, 323
Radon Data, 506
Rainbow test, 166
Ramsey's regression equation
    specification error test
    (RESET), 166
Ramsey's reset, 165–166
Random coefficients regression
    models, 503
Random effects, 264
Random-effects meta-regression,
    523–524

Random forests, 491–492

Randomization, regression effect and, 18

Randomization tests, *see* Permutation tests

Randomized quantile residuals, 418

Random term, 413

Random variables, 7, 579

Rank regression, 557–560

Rate ratios, 389, 394

Ratio estimation and linear regression, 527–528

Ratio estimator, 528

Raw residuals, 379, 388, 393, 417

Realizations, 7

Recursive partitioning, 483

Recursive residuals, 156

Reduced bank regression, 446–447

Reduced model, 123

Reducing subspace of M, 445

Reference group, 383

Reflected pair, 487

Regression analysis, defined, 3

Regression assumptions
  Black Cherry Tree Data, 69–71
  consequence of invalid assumptions, 50–51
  data transformations, 57–59
  diagnosing validity of assumptions, 51–53
  linear, key assumption about, 49–50
  1993 Car Sale Data, 68–69
  plots of residuals *vs.* fitted values, 53–57
  skewness and kurtosis, 61–62
  Steam Output Data, 66–68
  tests for constant error variance, 63–66
  tests for normality, 61–66

Regression coefficient, 30

Regression continuity design, 310

Regression degrees of freedom (dfR), 5

Regression depth, 515–517

Regression dilution attenuation, 188

Regression effect/regression toward the mean, 16–18

Regression equation, 8

Regression estimators, 528

Regression fallacy, 17–18

Regression imputation, 534

Regression methods, uses of, 3–5

Regression methods for analyzing survey data, 526–532

Regression models, 8–9, 263

Regression models, basics
  Computer Repair Data, 21–23
  measuring overall variation from sample line, 12–13
  ordinary least squares, 10–11
  population model for simple linear regression, 8–10
  regression effect, 16–18
  regression notation, 8
  regression through the origin, 13–14
  regression *vs.* correlation, 14–16
  Steam Output Data, 20–21
  Toy Data, 18–20

Regression models with censored data
  Bone Marrow Transplant Data, 355
  censored regression model, 337–338
  Cox proportional hazards regression, 343–344
  diagnostic procedures, 344–347
  Durable Goods Data, 356–359
  Extramarital Affairs Data, 350–352
  Motorette Data, 352–355
  survival and reliability analysis, 335–337
  survival (reliability)regression, 339–343
  truncated regression models, 347–350

Regression models with discrete
    responses
    Biochemistry Publications Data,
        405–409
    Cheese-Tasting Experiment
        Data, 403–405
    Hospital Stays Data, 409–412
    logistic regression, 375–385
    negative binomial regression,
        390–396
    Poisson regression, 385–390
    specialized models involving
        zero counts, 396–398
    Sportsfishing Survey Data,
        402–403
    Subarachnoid Hemorrhage Data,
        398–402
Regression notation, 8
Regression shrinkage, 287–292
Regression splines, 320–321
Regression through the origin
    (RTO), 13–14
Regression trees, 483
Regression *vs.* correlation, 14–16
Regressogram, 324
Regularization procedure, 321
Regularization term, 321
Regulation interval, 40
Relationships, deterministic, 3–4
Reliability analysis, 335
Reliability function, 337
Reliability ratio, 182
Remedial measures for
        autocorrelation, testing
        and, 224–230
REML, *see* Restricted maximum
        likelihood (REML)
Replicates, 77
Reproducing kernel Hilbert space,
        324
Resampling techniques for **B**,
        203–210
RESET, *see* Ramsey's regression
        equation specification error
        test (RESET)

Residual-based (r-based) Hessian
        matrix, 299
Residual function, 161
Residuals and measures of influence,
        154–162
Residual term, 86
Resistant regression methods,
        202–203
Resistant statistics, 202
Response, 8; *see also* Dependent
        variable
Response-based (y-based) Hessian
        matrix, 299
Response surface model (RSM),
        268
Restricted least squares regression,
        545
Restricted maximum likelihood
        (REML), 265
Richards' curve model, 363
Ridge analysis, 273
Ridge regression, 288–290
Ridge regression estimates, 289
Ridge trace, 290
Right censoring, 336
RMSE, *see* Root mean square root
        (RMSE)
Robust estimator, *see* Empirical
        estimator
Robust LOESS, 317
Robust regression methods, 197–201
Root mean square root (RMSE), 12
Root node, 483
Rose diagram, 556
Roy's Greatest Root, 442
RSM, *see* Response surface model
        (RSM)
R Student residuals, *see* Studentized
        deleted residuals
$r^{\text{th}}$ central moment, 62
$r^{\text{th}}$ (sample) moment, 61
RTO, *see* Regression through the
        origin (RTO)
Ryan-Joiner test, 61

**S**

Sample, 7
Sample design, 527
Sample standard deviation, 12
Sample statistic, 26
Sample variance, 12
Sampling distribution, 579
Sandwich estimator, *see* Empirical
    estimator
Saturated model, 345
SAVE, *see* Sliced average variance
    estimation (SAVE)
SBC, *see* Schwartz's Bayesian
    Criterion (SBC)
sBIC, *see* Singular BIC (sBIC)
Scheff joint prediction, 37
Scheff joint prediction intervals, 96
Schoenfeld residuals, 345
Schwartz's Bayesian Criterion
    (SBC), 254
Score function, 558
Score residuals, 345
Screening Response Model, 270
Search methods, *see* Backward
    search method; Forward
    search method
Second-order autoregression AR(2),
    221
Seemingly unrelated regressions
    (SUR), 447–448
SEM, *see* Structural equation
    modeling (SEM)
Semi-orthogonal matrix, 578
Semiparametric least squares, 456
Semiparametric models, 455
Semiparametric regression
    Boston Housing Data, 473–475
    (Generalized) additive models,
        457–458
    (Generalized) partial linear
        models, 458–459
    (Generalized) partial linear
        partial additive models, 460
    1907 Romanian Peasant Revolt
        Data, 462–473

projection pursuit regression,
    461–462
single-index models, 456–457
varying coefficient models,
    460–461
Semiparametric regression models,
    455
Semistudentized residuals, *see*
    Standardized residuals
Semi-supervised learning, 477
Sequential replacement regression, in
    stepwise methods, 256
Serially correlated, 219
Shrinkage estimate, 287
Significance level ($\alpha$-level), 25
SIM, *see* Single-index models (SIM)
Simple linear regression
    equation, 9
    population model for, 8–10
Simple linear regression model, for
    individuals in larger
    population, 9
Simple meta-regression, 523
Simple random sampling, 527
Simplex centroid design, 273–275
Simplex lattice design, 273
Simulated Motorcycle Accident
    Data, 493
Simultaneous confidence intervals,
    443
Simultaneous prediction intervals,
    443
Single exponential smoothing,
    233–234
Single-hidden-layer feedforward
    neural network, 492
Single-index models (SIM), 456–457
Single-layer perception, 492
Singular BIC (sBIC), 447
Singular value decomposition (SVD)
    method, 292
Simulated Data, 131–136
SIR, *see* Sliced inverse regression
    (SIR)

Silverman's rule of thumb,
313–314Kernel regression
Skewness, 61–62
Slack variables, 481
Sleep Study Data, 276–278
Sliced average variance estimation
(SAVE), 298
Sliced inverse regression (SIR), 298
Slope, 8
Small area models, 529–532
Small areas, 529
Smoothing matrix, 311
Smoothing parameters, 233
Smoothing splines, 321
Sobel's test, 519
Softmax function, 492
Softmax regression, 383
Somer's D, 16
Span, 316
spatial autoregressive coefficient, 551
Spatial dependency, 550
Spatial error, 551
Spatial heterogeneity, 550
Spatial lag, 550
Spatial moving average processes,
552
Spatial regression, 548–554
Spearman's rank correlation
coefficient, 15
Speckman estimator, 459
Spectral analysis, 234–235
Spectral density, 235
Splines, 318–324
Sportsfishing Survey Data, 402–403
Square matrix, 575
SSCP, *see* Sum of squares and
cross-products (SSCPs)
SSCPE, *see* Sum of squares and
cross-products for the errors
(SSCPE)
SSCPR, *see* Sum of squares and
cross-products due to
regression (SSCPR)
SSCPTO, *see* Sum of squares and

cross-products for total
(SSCPTO)
SSLOF, *see* Lack-of-fit sum of
squares (SSLOF)
SSPE, *see* Pure error sum of squares
(SSPE)
SSTO, 74, 77
Standard error, 27, 579
Standard Errors of $\hat{\beta}_0$ and $\hat{\beta}_1$, 27
Standardized residuals, 155
Standardized $\mathbf{X}$ matrix, 111–112
Standardized $Y$ vector, 111–112
Standard Tobit model, 338
Standardized ridge regression
estimates, 289
Stationary stochastic process, 231
Statistic, 7
Statistical depth functions, 515
Statistical independence, 579
Statistical inference, 25
Computer Repair Data, 44–45
hypothesis testing and
confidence intervals, 25–30
inference on correlation model,
33–35
intervals for mean response,
35–36
intervals for new observation,
36–40
power, 30–33
Pulp Property Data, 45–47
Steam Output Data, 41–45
Statistical intervals, 95–96
Statistical learning, 477
Statistically significant, 26
Statistical significance, practical
significance *vs.*, 30
Statistics, 579
Statistics from information criteria,
254–255
Steam Output Data, 20–21, 41–44,
66–68, 78–79, 146–147
Steepest Ascent Model, 270
Stepwise procedures for identifying
models, 255–256

Stepwise regression, in stepwise methods, 256
Stochastic independence, 579
Stochastic process, 231
Stochastic regression imputation, 534
Stratified sampling, 527
Structural dimension, 297
Structural equation modeling (SEM), 186–188
Structural measurement error model, 180
Structural multicollinearity, 114–115
Studentized deleted residuals, 155–156
Studentized deviance residuals, 380
Studentized Pearson residuals, 380
Studentized residuals, 380, 389, 394
Stukel's score test, 382
Subarachnoid Hemorrhage Data, 398–403
Sufficient dimension reduction methods, 297
Sum of squared observed errors, 11
Sum of squares and cross-products due to regression (SSCPR), 444
Sum of squares and cross-products for total (SSCPTO), 444
Sum of squares and cross-products (SSCPs), 444
Sum of squares and cross-products for the errors (SSCPE), 444
Sum of squares due to regression (SSR), 75, 121
Supervised learning, 477
Support Vector Machines (SVMs), 478–482
Support vector regressions (SVRs), 480
Support vectors, 478
Suppressor variable, 520
SUR, *see* Seemingly unrelated regressions (SUR)
Survey weighted least squares, 527

Survey weighted least squares estimator, 527
Survival analysis, 335
Survival function, 337
Survival (Reliability) regression, 339–343
SVD method, *see* Singular value decomposition (SVD) method
SVMs, *see* Support Vector Machines (SVMs)
SVRs, *see* Support vector regressions (SVRs)
Swamping, 163
Switching regimes, 561
Symmetric matrix, 575
Systematic sampling, 527
Systematic term, 413

**T**
$\tau^{\text{th}}$ quantile, 542
$\tau^{\text{th}}$ regression quantile, 543
Tea Data, 147–149
Tensor matricization, 567
Tensor product basis, 323
Tensor regression, 565–570, 567
Tensors, 567
Terminal nodes, 484
Test error, 482
Test sample, in crossvalidation method, 250
Tests for constant error variance, 63–66
Tests for normality, 61–66
Test statistic, 25
Theil-Sen estimator, 201
Theil-Sen regression, 201
Thermal Energy Data, 99–103, 115–119, 129–131
Thin-plate spline, 323
Thresholding, 327
Tietjen-Moore test, 158
Tikhonov regularization, 291
Time series, overview of, 219–221
Tobit models, 338

Toeplitz matrix, 423
Tolerance , VIF and, 111
Tolerance intervals, 96
Tortoise Eggs Data, 106–108, 119–120
Total degrees of freedom (dfT), 74
Total least squares, 182
Total sum of squares (SSTO), 12, 74, 77, 121
Toy Data, 18–20
TrAdaBoost, 490
Training data, 477
Training error, 482
Training sample, in crossvalidation method, 249
Trajectory data, 422
Treatment contrast, 138
Treatments, *see* Independent variables
Tree-based methods, 483
Trial-and-error approach, 57–59
Trimmed mean, 64
Trimming, 202
Trivial subspace, 577
Truncated power basis, 319
Truncated power function, 308
Truncated regression models, 347–350
Tube formula, 312
Tuberculosis Vaccine Literature Data, 524–526
Tucker decomposition, 567
Tukey's Biweight, 200
SLS, *see* Two-stage least squares (2SLS)
Two-stage least squares (2SLS), 185–186
Two-way ANOVA, 265
Type I error, 31–32
Type II error, 31–32
Type II or failure-truncated, 336
Type II regression problem, 179
Type I or time-truncated, 336
Type I regression problem, 179

**U**
Underdispersion, 395–396
Unit nonresponse data, 533
Unspecified functional form, in nonlinear regression, 363–364
Unstandardized difference in fits residual, 160–161
Unstructured correlation matrix, 423–424
Unsupervised learning, 477
U.S. Economy Data, 190–192
Utility, 143

**V**
Validation sample, in crossvalidation method, 249
Variable selection or screening, regression methods and, 4–5
Variance components regression models, 503
Variance-covariance matrix and correlation matrix of $\beta$ 92–93
Variance-decomposition proportion, 113–114
Variance inflation factor (VIF), 111–112
Varying-coefficient models, 460–461
Varying dispersion beta regression model, 421
Varying-intercept, varying-slope model, 505–506
Varying-intercept model, 504–505
Varying-intercept model with no predictors, 504
Varying-slope model, 505
Vector, 478
Vector spaces, refresher on, 575–578
VIF, *see* Variance inflation factor (VIF)

**W**
Wald test, 379
Wavelet basis, 326

Wavelet regression, 326–327
(Weighted) average derivatives
    estimator, 456
Weighted least squares, 195–197
Weighted least squares and robust
    regression methods
    Arsenate Assay data, 212–216
    Computer-Assisted Learning
        Data, 210–212
    Pulp Property Data, 216–217
    resistant regression methods,
        202–203
    robust least squares, 195–197
    robust regression methods,
        197–201
    Theil-Sen and Passing-Bablok
        regression, 201–202
Weighting function, 199
White's test, 105
Wilks' Lambda, 442
Wind Direction Data, 556–557

Winsorized regression, 200–201
Winsorizing residuals, 200
Working correlation structure, 422
Working-Hotelling confidence bands,
    36, 95
Working model, 527
Working responses, 415–417

**X**
X-space, 154

**Y**
Yarn Fiber Data, 283–286
Yule-Walker equations, 223

**Z**
Zero-counts models, 396–398
Zero-inflated count regression
    models, 397–398
Zero-truncated count regression
    models, 397